Lianas
of the
Guianas

Lianas of the Guianas
A guide to woody climbers in the tropical forests of Guyana, Suriname and French Guiana

LM Publishers
Parallelweg 37
1131 DM Volendam
The Netherlands

info@lmpublishers.nl
www.lmpublishers.nl

© 2017 – LM Publishers

Authors
Bruce Hoffman, Ph.D.
Sofie Ruysschaert

Contributions by
Mark J. Plotkin, Ph.D.
Frits van Troon
Joeri Zwerts

Editing
David Stone

Cover and graphic design

 / NANCY VALIES

Icon art
Kristian Johnson Michiels

Production
High Trade BV

ISBN 978 94 6022 224 5

Disclaimer
The information provided within this book is for general informational purposes only. While we try to keep the information up-to-date and correct, there are no representations or warranties, express or implied, about the completeness, accuracy, reliability, suitability or availability with respect to the information, products, services, or related graphics contained in this book for any purpose.

A Word from the Sponsor
In a country like Suriname with a pristine tropical rainforest, it is important to be aware of the richness of the forest. Lianas are very well known, but it was surprising to read about the number and types of liana species in our country. It is therefore with much pleasure that we have agreed to sponsor the printing of this publication. It is our hope that this reference work shall be well received by the general public and especially our youth. We hope they will gain much knowledge on the subject of lianas. We thank and congratulate everyone who has given their utmost to put this reference work together.

The Suriname Conservation Foundation
Henna J. Uiterloo LLB
Executive Director

With the generous support of
Amazon Conservation Team
Alberta Mennega Stichting (Netherlands)
Stichting het van Eeden Fonds (Netherlands)
Hugo de Vries Fonds (Netherlands)
Naturalis Biodiversity Center (Netherlands)

Lianas
of the
Guianas

A guide to woody climbers in the tropical forests of Guyana, Suriname and French Guiana

Bruce Hoffman, Ph.D. and Sofie Ruysschaert
with contributions by Mark J. Plotkin, Ph.D.,
Frits van Troon, and Joeri Zwerts

FOREWORD

Schlegelia violacea (Schlegeliaceae), with flowering branch extending down from host tree.

FOREWORD

According to the eminent Neotropical botanist Thomas Croat, the presence of lianas is the single most important physiognomic feature that distinguishes tropical from temperate forests. This distinction holds true in the collective mind of the general public as well: ever since Johnny Weissmuller's Tarzan first swung through the jungle, we all associate woody vines with tropical forests.

Because they often flower and fruit in the canopy far above the forest floor, lianas are notoriously difficult to collect and study. Famed 20th-century ethnobotanist Richard Schultes searched seven years before he was able to find the fabled ayahuasca vine in flower, and he then had to fell seven trees to access the blossoms, so interwoven was the liana in the crowns of these forest giants. Challenges like these long hindered the study of these tropical vines by non-indigenous scholars, hence the likely immediate worth of any published research that attempts to digest their enormous diversity for both academic and lay readers.

Recent years have seen a plethora of field guides devoted to explaining and identifying tropical organisms: palms, birds, reptiles, freshwater fishes and others. With such manuals in hand, both the amateur and the professional biologist can walk through these forests and identify flora and fauna as never before. However, prior to the herculean effort leading to this publication, a regional guide to these challenging and mysterious life forms has not been available.

The race is on to better understand tropical rainforests and their component species in the face of relentless deforestation and extractive pressures. It is my great hope that this pioneering research by my longtime associate Dr. Bruce Hoffman and his local colleagues—fellow scientists and indigenous mentors—will make a significant contribution to our better valuing, utilizing and protecting these magnificent species and the rainforests in which they thrive.

Dr. Mark J. Plotkin

ACKNOWLEDGEMENTS

Aristolochia stahelii (Aristolochiaceae), in flower at Brownsberg Nature Park in Suriname

ACKNOWLEDGEMENTS

The kind support of many people, institutions and foundations has made it possible for this book, like a liana, to germinate and grow upwards through the understory to see the light of day. The guide began as a "Lianas of Suriname" project. When the project was expanded to a regional focus, the morphological diversity and number of taxa greatly increased. The original inspiration for a "liana guide" came from Frits van Troon, a Saramacca Maroon parabotanist and treespotter with a remarkable, hybrid scientific-local knowledge and a special place in his heart for lianas. He remains one of the most knowledgeable people on forest trees and lianas in Suriname. We owe much gratitude to him for his valuable contribution during the initial fieldwork and continued motivation to pass his knowledge on to others. Biologist Marc van Roosmalen – best known as author of the classic *Fruits of the Guianan Flora*" – also contributed to the initial Suriname fieldwork.

Knowledgeable collaborators of the Trio village of Kwamalasamutu in Suriname graciously supported the project as parabotanists and ethnobotanical informants. These include Amashina, Wuta, Korotai, Okoi, Kamainja, and Natara. On the upper Suriname River (Gran Lio), Saramaccan colleagues also participated in fieldwork, including Sopokuja (Awaradam), Zevee (Stonhuku village), & Johan (Godowata village).

Many thanks are due to Minu Parahoe, Mark Plotkin, David Stone (editor) and many others at the Amazon Conservation Team who have supported this project through thick and thin. Nancy Valies, the graphic designer and a native daughter of Suriname, has done a remarkable job - integrating elements of art and science in a beautiful way. We express great appreciation to Ron Smit and Peter Sanches at LM Publishers and the Suriname Conservation Fund for their generous support. Research, data processing and travel in the Netherlands was funded by grants from the Albert Mennega Stichting, van Eden Fonds and the Hugo de Vries Fonds. Exhaustive, high quality technical support in Suriname was cheerfully provided by Chantal van den Berg, Sarah Crabbe, Sara Fuste, and Sara Svensson.

Several institutions and their current or former staff made their expertise and networks available to facilitate the development of this guide. William Hawthorne at Kew Royal Botanic Gardens graciously allowed for the use and adaptation of entries from his 'plant characteristics' glossary. At the Naturalis Biodiversity Center (NBC - National Herbarium of the Netherlands), Tinde van Andel and Niels Raes have been invaluable supporters. Niels and Renske Ek provided early inspiration and liana data from their 'Climbers of Guyana' website. Sylvia de Moto Oliveira, the Flora of the Guianas journal editor, assisted with contacting experts and providing literature. Various (former) Dutch taxonomists have kindly assisted with this guide, either directly or through their publications, including Cees Berg, Anne Görts-van Rijn, Eric Gouda, Marion Janson-Jacobs, Jan Lindeman, Paul & Hiltje Maas, Bep Mennega, Andre van Proosdijk, A. C. de Roon, and Lubbert Westra.

The (former) staff of the Suriname National Herbarium have helped with processing and curating liana plant specimens, including Usha Raghoenandan, Marga Werkhoven, Dorothy Traag, Gisla Jairam-Doerga, Joelaika Behari-Ramdas, Angela Grant, and Maureen Playfair. (Former) staff of the Smithsonian Institution (U.S. National Herbarium, Biodiversity of the Guianas program) have assisted greatly with data, publications, processing and identification of herbarium specimens, including Pedro Acevedo, Sara Alexander, Debby Bell, John Boggan, Bob DeFilipps, Christian Feuillet, Vicki Funk, Lynn Gillespie, Tom Hollowell, Carol Kelloff, Shirley Maina, Harold Robinson, Rusty Russell, Larry Skog, and Dieter Wasshausen. Several other (taxonomic) experts at herbaria and universities worldwide have contributed to documentation of the Guianas climber diversity in this guide, including: Lucile Allorge, Bill & Christiane Anderson, Daniel Austin, Gerardo Aymard, Rupert Barneby, Paul Berry, Richard Cowan, Georges Cremers, Tom Croat, Stephen Dressler, Jean-Jacques de Granville, Enrique Forero, Pierre Grenand, Andrew Henderson, Paul Hiepko, Eduardo Lleras, Lucia Lohmann, Julio A. Lombardi, Scott Mori, Gilberto Morillo, Michael Nee, Sir Ghillean Prance, Marie F. Prévost, Odile Poncy, Clive A. Stace, and Dennis W. Stevenson.

In addition to images provided by the authors and contributors, the visuals and taxonomic coverage of the guide were greatly improved by several major contributors of photographs, including Olivier Gaubert (Flore de Guyane website, French Guiana) [http://floredeguyane.piwigo.com], the Flora da Reserva Ducke project (Mike Hopkins, Manaus, Brazil), and Pieter Teunissen (plant ecologist and botanist, Suriname). Additional, no less critical photographs provided in smaller numbers by various photographers are listed in the appendices. The artistic content of the book was greatly enriched by plant illustrations from Bobbi Angel, Omar Kasijo, the University of Michigan herbarium (Malpighiaceae) and artists of the NBC, including Hendrik Rypkema and Wim Hekking. The guide icons, created specifically for this project, are the work of the fine artist Kris Johnson Michiels [www.inkimage.net].

Finally, last but certainly not least: many thanks to all of our current and former colleagues not mentioned here, and friends and family for their continued support to create an enjoyable (working) environment that allowed this guide to be completed.

Thank you; gran-tangi; merci beaucoup.

Frits van Troon (Saramacca Maroon parabotanist) at Tonka Island, Suriname, showing a 'forwrufutu titei' liana (Dolichandra unguis-cati, Bignoniaceae).

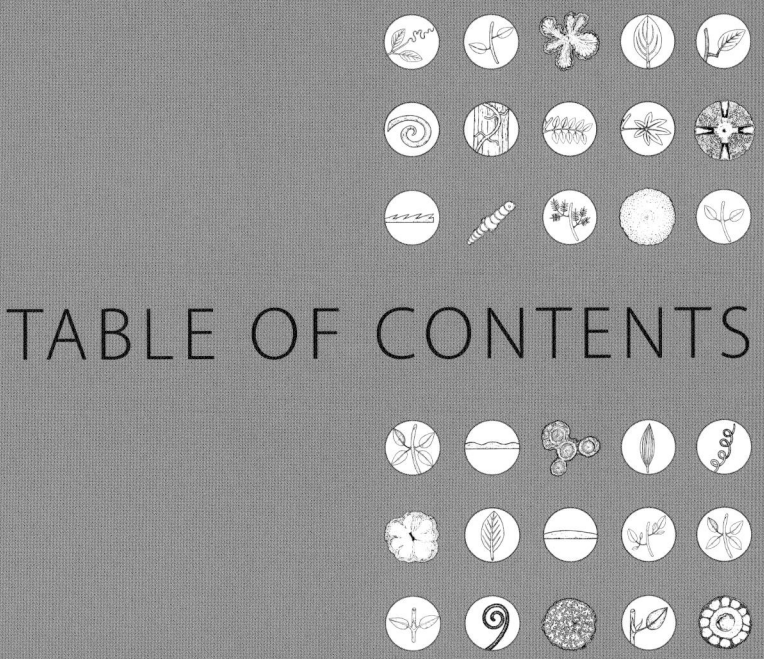

TABLE OF CONTENTS

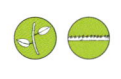
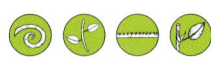

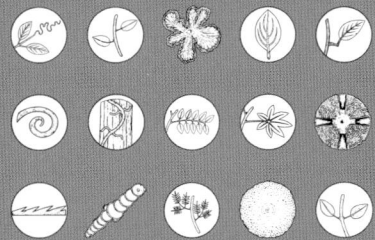

LIANAS OF THE GUIANAS

Guide

BRUCE HOFFMAN AND JOERI ZWERTS

INTRODUCTION

Overview of the Guide

Lianas (woody vines) and other high-climbing plants, experienced in tropical forests mostly as mysterious bare woody stems disappearing into the canopy, are notoriously difficult to observe, document, and identify. The aim of this guide is to provide readers with a greater understanding and (hopefully) appreciation of liana diversity, ecology, and related human significance in the Guianas, and thereby increase knowledge and appreciation of the region's tropical forests as a whole. The intended audience is broad, including students, scientists, resource managers, forest inventory teams, international and national institutions, tourists, forest-dependent communities and the general public. The information in this guide was compiled from a wide variety of sources, including empirical research of the authors and colleagues, herbarium specimens, flora and field guides, scientific websites and databases, forestry specialists, ethnobotanical references, indigenous and tribal peoples, illustrators and photographers.

The front of the guide consists of three brief chapters: a general introduction to climbing plant diversity and the Guianas, an illustrated botanical glossary, and a `how-to-use' chapter with tables of the icons used in family chapters and other identification aids. A vegetative key to liana families and genera is in development but could not be included in this volume. The core of the book includes forty-four family chapters in alphabetical order, with 175 genera and 384 species in total. Detailed descriptions are provided for all documented families and genera with woody climbers and approximately 38% of the 1,023 species. Scientific names are based upon the modern `APG III' classification (Bremer et al. 2009) and the Taxonomic Names Resolution Service v. 4.0. (TNRS 2016). In the appendices, readers will find a comprehensive checklist of herbaceous and woody climbers in the Guianas (excluding ferns), text-based botanical glossary, common names index, bibliography, and author image credits.

We hope that this guide will serve as a useful tool for pragmatic purposes such as forest inventories, sustainable logging, ecotourism, non-timber forest product development and scientific research. We also hope that it will be an enjoyable book for those simply wanting to discover more about these fascinating, iconic life forms of tropical forests.

The Significance of Lianas

Climbers are species that require the support of another plant or surface for growth during at least one part of their life cycle. The climbing habit occurs throughout the world with an increasing abundance towards the equator and within lowland tropical forests (Gentry 1991). Woody vines, commonly known as lianas, are iconic symbols of tropical forest ecosystems. Using active climbing mechanisms such as twining shoots, tendrils or adhesive roots and passive mechanisms such as spines, rough surfaces, and angled branches, lianas take advantage of the biologically-expensive, vertical structure of trees to gain relatively inexpensive access to the light-rich canopy. The resources that are thus conserved can be invested elsewhere to gain a competitive advantage in growth and reproduction. Once in the canopy, lianas often extend horizontally and may produce more leaf area than the crowns of the trees they occupy (Putz 1983).

The success of the climbing strategy is revealed by the widespread occurrence of climbing species in unrelated plant groups. Aside from a rich representation among angiosperms (flowering plants), climbers can also be found amongst neotropical palms (Desmoncus), ferns, orchids, and gymnosperms (Gnetum). In the Neotropics, three distantly-related angiosperm plant families typically dominate in species diversity, including the Bignoniaceae, Sapindaceae and Leguminosae (Fabaceae). The evolution of the climbing habit in forests may have resulted from the diverse surfaces and spaces presented by multi-story forest canopies, interactions with diverse pollinators and seed dispersers, or the obvious adaptive benefit of escaping the shaded forest floor. On average, woody climbers compose around 25% of the woody stems and species in rainforests, forming a key element in forest diversity, structure and composition (Schnitzer, et al. 2012, Gentry 1991).

Lianas play an important role in forest ecology, gap dynamics and forestry operations. Dormant seeds, suppressed seedlings and vegetative sprouts from fallen lianas, in combination with high growth rates, result in the rapid proliferation of lianas in forest gaps. For trees, lianas can impede growth rates and regeneration, and increase mortality. To minimize damage to commercial species, lianas are often cut from trees by foresters before felling. Due to their narrow stems and low wood density, lianas sequester relatively little carbon compared to the trees whose growth and turnover rate they influence negatively. Therefore the net carbon sequestration of a forest becomes lower with increased liana occupation (Laurance 1997). Lianas appear to be out-competing trees and becoming more abundant in the face of global warming and increasing seasonality in some tropical forests (Phillips, et al. 2002).

Throughout the tropical world, forests and their lianas provide a bounty of ecosystem and economic services in the form of food, raw materials, water, oxygen, medicine, biodiversity, tourism and carbon sequestration, to name but a few. In the canopy, lianas grow laterally and establish important canopy pathways for animals to move through the forest. The negative impact of lianas upon tree growth is counterbalanced by the horizontal lattice that helps keep trees from falling over during storms. Lianas are also an important part of the food web providing leaves, flowers, fruits, nectar, water as well as nesting material and habitat. To maintain these services, it is important to understand the system from which they are derived, a system in which climbing plants play an unmistakable role in a complex ecological web of direct and indirect interactions.

The Guianas Region

This guide focuses upon woody climbers native to three territories in northeastern South America: French Guiana, Guyana, and Suriname. Guyana (formerly British Guiana) and Suriname (formerly Dutch Guiana) are independent countries, while French Guiana remains an overseas department of France. These three culturally diverse and unique territories are known collectively as "the Guianas", although the underlying Precambrian rock of the Guiana Shield formation extends over a much greater area, including Brazil, Venezuela and eastern Colombia. The greater region is associated with ancient, nutrient-poor soils, translucent 'black water' creeks, savannas and relatively low diversity forests compared to western Amazonia. The western Guiana Shield, including parts of Guyana, has sandstone tepui plateaus to 3000 meters, spectacular waterfalls, and a highly endemic flora (Steyermark et al. 1995). The eastern Guiana Shield (French Guiana, Suriname) is less precipitous, with broad expanses of lowland tropical forest interrupted by scattered mountain ranges, granitic domes and a single tepui reaching 1300 meters. From European colonial times to the present, population growth and development in the Guianas region has remained concentrated in accessible areas along the Atlantic coast and major rivers. Although unsustainable extractive activities such as gold-mining are encroaching, the interior forests of the Guianas have been largely spared by difficulty of access, and the region composes one of the world's largest remaining contiguous expanses of tropical forest. Human presence in the inland forests consists largely of proud indigenous and tribal (Maroon) forest communities, with largely subsistence-based livelihoods.

As revealed in the species accounts of this guide, lianas and other climbers are ubiquitous across the Guianas in a wide variety of habitats. Sun-loving, slender lianas thrive in open areas and forest edges, while shade-tolerant, robust lianas are abundant in old growth forests. This guide is likely to be useful over a much greater area than the Guianas, as many of the described taxa occur in adjoining areas and in the Neotropics at large.

Map of the Guianas region, including Guyana, Suriname, and French Guiana

Climbers of the Guianas: Diversity, Growth Forms, Unique Characters

Climbing plants in the tropics and the Guianas exhibit great variety, including high taxonomic diversity (many families and species), diverse growth forms, and diverse morphological adaptations for a climbing or crawling way of life. Even species of the same family or genus may exhibit very different climbing solutions, and some species opportunistically switch from self-supporting to a climbing form when the opportunity arises. We estimate 59 plant families and 253 genera of the Guianas to include at least one climbing species. Figure 2 shows the species richness and distribution of climber growth forms (both woody and herbaceous) within 21 plant families. Some families represent only one climbing growth form (Bignoniaceae, Clusiaceae), while others have multiple climbing growth forms among their species (Apocynaceae, Convolvulaceae). In this comparison, the non-climbing species were excluded (e.g., Leguminosae trees). The Apocynaceae, Bignoniaceae and Leguminosae families have the greatest number of climbing species, with 123, 103 and 96 species respectively. However, the dissimilar, herbaceous monocot family Araceae has the fourth largest number of climbing species at 77. In the following few pages we provide an overview of common climber growth forms.

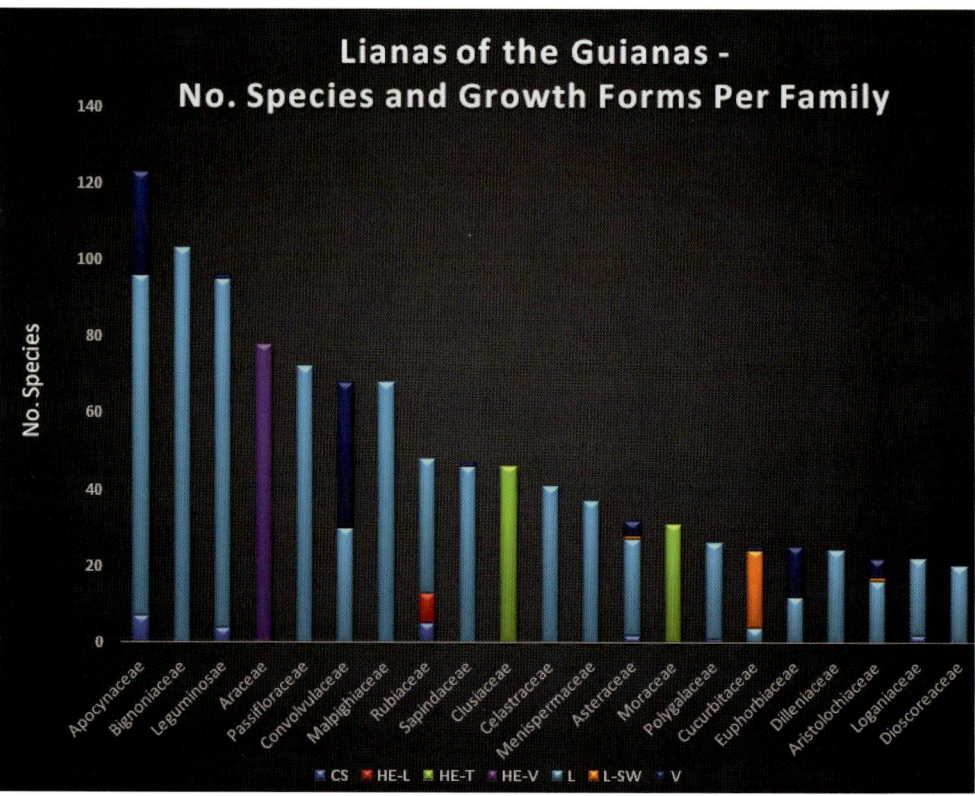

Figure 2. The number of species and growth form distributions among the 21 most climber-rich plant families in the Guianas Climber Checklist. Growth form definitions: CS, climbing shrub; HE-L, liana-like hemi-epiphyte; HE-T, tree-like hemi-epiphyte; HE-V, vine-like (herbaceous) hemi-epiphyte; L, liana woody, slender to robust; L-SW, sub-woody lianas; V, vines (herbaceous).

Lianas (L) are climbers that are rooted in the soil, have a main stem, require support at some point in their life cycle and produce at least some tissues with wood. Stem diameters range from less than 0.5 cm diameter to greater than 30 cm diameter. Lianas of stem diameter > 0.5 cm are the main focus of this book as they are the most widely-dispersed, locally abundant and ecologically significant forest climbers. Climbers with fine "woody" stems less than 0.5 cm diameter can be considered perennial vines or vine-like lianas (L-V).

The stem tissues of many forest climbers exists somewhere along a continuum between 'woody' and 'herbaceous'. **Sub-woody lianas (L-SW),** including many Cucurbitaceae and Vitaceae species, often have robust, green stems that appear to produce a soft wood. Desmoncus is a liana genus in the palm family that, like all monocots, does not have "true wood". Yet Desmoncus stems are tough, and they are included as lianas in this guide.

1. Pyrostegia venusta (Bignoniaceae), with orange tubular flowers, lobes recurved at maturity [BH]

2. A robust liana, Croton pullei [BH]

3. Petrea bracteata [BH]

4. Gurania reticulata [PT]

5. Subwoody liana stem with tendril, Cayaponia sp. (Cucurbitaceae) [PT]

Climbing shrubs (CS) are woody, terrestrial plants that lack a single main stem and use spreading branches to lean and climb, often rather clumsily, over other plants. They can be aided in their attachment by hooks and spines. Some Gouania (Rhamnaceae), Piper (Piperaceae) and Randia (Rubiaceae) species fit this description. Some canopy-occupying shrubs and cactus species occupy a growth form gray area somewhere between epiphytes, hemi-epiphytes and lianas.

Vines (V) are similar in many respects to lianas, but lack wood and generally remain lower in the canopy or in open areas. The flowers of vines are often large and showy. Two vine-rich plant families in the Guianas are the Apocynaceae and the Convolvulaceae. Conspicuously non-woody, slender vines are not included in this guide. Definitions can be tricky because the woody stems of lianas and basally woody climbers commonly produce herbaceous shoots (e.g., Mendoncia species).

6.   Lomariopsis japurensis (Lomariopsidaceae), a common climbing fern in the Guianas [BH]

7.   Hylocereus scandens, (Cactaceae) epiphytic or climbing cactus [PT]

8.   Piper hispidum, (Piperaceae) herb or shrub, sometimes weakly climbing [PT]

9.   Randia armata (Rubiaceae), climbing shrub with spines [PT]

10. Dichorisandra hexandra (Commelinaceae), a weak, herbaceous climber [BH]

11. Ipomoeae batatoides (Convolvulaceae), a vine or very slender liana [PT]

12. Dioscorea amazonum (Dioscoreaceae), perennial vine with large tubers [BH]

'Hemi-epiphyte' is a term for plants that live perched upon trees like (epiphytic) orchids, yet also have an aerial root connection to the ground, unlike most orchids. Three types of hemi-epiphytes are recognized in this guide.

Tree-like "primary" hemi-ephiphytes (HE-T), including Clusia (Clusiaceae), Coussapoa (Urticaceae), Ficus (Moraceae) and Oreopanax (Araliaceae) species, establish as robustly woody epiphytes in the canopy and then extend roots to the forest floor. Strangler figs, comprising many species, are well known for engulfing and killing host trees in the tropics.

Vine-like "secondary" hemi-epiphytes (HE-V), including herbaceous Araceae and Cyclanthaceae species, and the Vanilla orchid genus, begin life on the forest floor and climb up tree trunks with clasping roots to reach higher light conditions. Heteropsis, commonly known as cipó, kamina, or nibbi, is a common green-stemmed hemi-epiphyte on tree trunks that produces extremely strong, pencil-thick, woody roots.

Liana-like hemi-epiphytes (HE-L) can be similar to either HE-V or HE-T species. They are robustly woody to sub-woody and may begin life on the ground or in the canopy. Representatives include species of Adelobotrys (Melastomataceae), Schlegelia (Schlegeliaceae), Schradera (Rubiaceae), Ericaceae (Cavendishia, Satyria), Marcgraviaceae (Marcgravia, Marcgraviastrum, Souroubea) and Solanaceae (Markea, Solandra). All HE types are included in the Guianas climber checklist, but only HE-L taxa are covered in the family chapters.

True epiphytes are plants that never root in the soil. They complete an entire life cycle anchored to a host plant, upon which they germinate as seeds or propagules and become rooted. The majority of epiphytes in the Guianas belong to the Orchidaceae, Bromeliaceae and fern families. Parasitic plants extract water and nutrients from a 'host' plant upon which they completely depend for their survival. Many species in the Loranthaceae family (fowru doti) are parasites that occupy tree crowns.

13. Clusia grandiflora (Clusiaceae), a tree-like hemi-epiphyte with woody aerial roots [BH]

14. Clusia palmicida, rooted at host tree base [BH]

15. Heteropsis flexuosa (Araceae), a vine-like hemi-epiphyte with woody aerial roots (upper left in photo) [BH]

16. Philodendron solimoesense (Araceae), a vine-like hemi-epiphyte with aerial roots [BH]

17. Vanilla planifolia (Orchidaceae), a vine-like hemi-epiphyte in the orchid family [BH]

18. Guzmania lingulata (Bromeliaceae), a true epiphyte - perches upon host trees or other surfaces without extending aerial roots to the soil [PT]

	Growth form (habit or habitus)	Code	No. species (1,307 total)
Table 1. Number of species by growth form in the Guianas climber checklist (Appendix 2).			
Non-woody	Vines and creepers	V	90 (7%)
Sub-woody	Lianas, soft or woody only at base	L-SW	55 (4%)
	Vine-like hemi-epiphytes	HE-V	103 (8%)
Woody	Lianas, slender but woody, < 0.5 cm diam.	L-V	174 (13%)
	Lianas, small to robust, > 0.5 cm diam.	L	727 (56%)
	Liana-like hemi-epiphytes	HE-L	44 (3%)
	Tree-like hemi-epiphytes	HE-T	88 (7%)
	Clambering or climbing shrub	CS	26 (2%)

In Table 1, we provide species counts for six climber growth forms in the Guianas climber checklist, regardless of family. Small to robust lianas are the dominant growth form (56%).

THE ETHNOBOTANY
of Lianas

Korotai, a Trio shaman in Kwamalasamutu village, Suriname, drinking medicinal water from the cut stem of a liana - Doliocarpus dentatus (Dilleniaceae), also known as kapadula (GU), sakëtaitu (Tr), and watra tité (Sr).

THE ETHNOBOTANY OF LIANAS

Western botanists have often deemed lianas too difficult to access, leaving them in the forest rather than collecting them for the herbaria. Indigenous peoples in these same forests, however, can often distinguish lianas by the appearance of the stem or the fragrance of the bark. Their sophistication at identifying these species is matched only by their astonishing ability to make full use of them. In tropical South America, indigenous peoples have employed the components of lianas as food, arrow and fish poisons, cordage, dyes, stimulants, hallucinogens, euphoriants, and sources of potable water.

Documented western encounters with indigenous botanical knowledge of the Guianas trace back at least to the early 18th century, notably in the writings of Edward Bancroft, a sailor, ethnobotanist, physician, business tycoon, American diplomat and double agent for the British Crown. Born poor in western Massachusetts in 1735, Bancroft went to sea as a teenager in search of fame, fortune and adventure. History then records him practicing medicine (seemingly without having attended medical school) in what was then British Guiana. Based on his experience and observations, in 1769 he published a book entitled "Essay on the Natural History of Guiana" that contained numerous accounts of local plants, animals and peoples, including a thirty-page discussion of curare. Bancroft noted that each Amerindian tribe had its own recipe for arrow poison, and he recorded the curare of the Akawaio tribe of northwestern Guiana—perhaps the first detailed account of the deadly poison three decades before Alexander von Humboldt famously observed the process in neighboring Venezuela.

More prominently, in 1804, the naturalist Charles Waterton set sail for British Guiana to oversee his family's plantations near Georgetown. After eight years near the coast, he departed for the interior, in search of what was said to be the deadliest curare of the Guianas, that made by the Macushi tribe. Waterton's expedition—which he conducted barefoot, and in the rainy season—was successful. Unlike the poison collected by Bancroft, which was made from Strychnos guianensis, the seemingly more toxic Macushi curare was prepared from the liana now known as Strychnos toxifera, and confirmed by the far more meticulous Schomburgk brothers who traveled in Waterton's footsteps several decades later. Waterton brought samples back to England, and facilitated a number of experiments that eventually led to the use of Amazonian curares as anesthetics in abdominal surgery.

Ethnobotanical studies will undoubtedly reveal further intriguing applications for rainforest lianas, if they can be conducted before the tide of western culture sweeping through Amazonia obliterates such indigenous knowledge. By way of example, approximately thirty percent of the Trio indigenous ethnopharmacopeia in Suriname is derived from liana species. To the few living scientists who have directly immersed themselves in learning at the feet of indigenous botanical scholars of the Guianas— Dr. Bruce Hoffman and I both count ourselves among their scant ranks—the value of perpetuating the regional shamans' knowledge and their habitats could not be more unambiguous.

Dr. Mark J. Plotkin

ILLUSTRATED GLOSSARY

Unique liana structures

This chapter provides labeled illustrations for structures unique to climbers and for flowering plants in general.

Climbing mechanisms

All climbers can be categorized by one or more climbing mechanisms. This guide groups mechanisms within five major categories (See following illustrations and icon tables for climbing mechanisms in the next chapter):

a) Tendril climbers – Tendrils are long, slender organs developed from modified leaves, stems or branches to clasp surrounding structures, providing an elastic and robust form of attachment. The location, shape (3D spiral, 2D coil) and number of tendrils are important in plant identification.

b) Twining climbers - Twining plants revolve their shoots in a spiral motion as a means of exploration, attaching to objects and climbing. Some plants have tendril-like shoots or leaves that grasp objects instead of spiraling around them (e.g. Salacia, Securidaca).

c) Hook/Spine/Hair climbers – Plants use this mechanism to hold onto the host tree with downward pointing adhesive hairs, rough surfaces, hooks, or spines. The spines are often paired and represent modified stipules.

d) Clambering branch and petiole climbers – This method involves sharply angled branches or elbow-like or twisted petioles. Both hook and branch climbing are passive methods, and plants are more likely to lose their grip compared to twiners, tendril climbers, and root climbers.

e) Root climbers – These plants cling to a support with aerial roots that emerge from the stem. They grow into irregular surfaces or attach themselves with sticky glandular secretions. This mechanism requires close contact with the host tree's main trunk, and climbers typically occupy only lower levels of the canopy. Most root climbers are vine-like or liana-like secondary hemi-epiphytes.

Stem anatomy

See woody cross-section pattern icons here and icon tables in the next chapter.

Flexibility in lianas is due partly to their slender stems, but also results from their shapes and internal anatomy. Liana stems often have fewer fibers and an abundance of large-diameter, water-conducting xylem vessels (wood). This allows them to pipe more water and thus produce more leaf area than might be expected for their stem size. The xylem and phloem (sugar transport) tissues of liana stems are often uniquely arranged (compared to trees), a condition sometimes termed 'anomalous growth' or a 'cambial variant' (Carlquist 1991, Pace et al. 2009). With the vascular tissues spread out within the stem, tree fall damage is less likely to be fatal to the liana, because even if a stem is split in half, there should still be some functioning cables to deliver sugars, nutrients and water. This unique anatomy often results in unique patterns visible to the naked eye which can be useful in identification. However, we request that users of this guide not destroy lianas just to observe the stem cross-section. A lengthwise slash or cutting of a small side branch is usually sufficient to reveal the internal anatomical pattern.

Woody Stem cross-section patterns icons

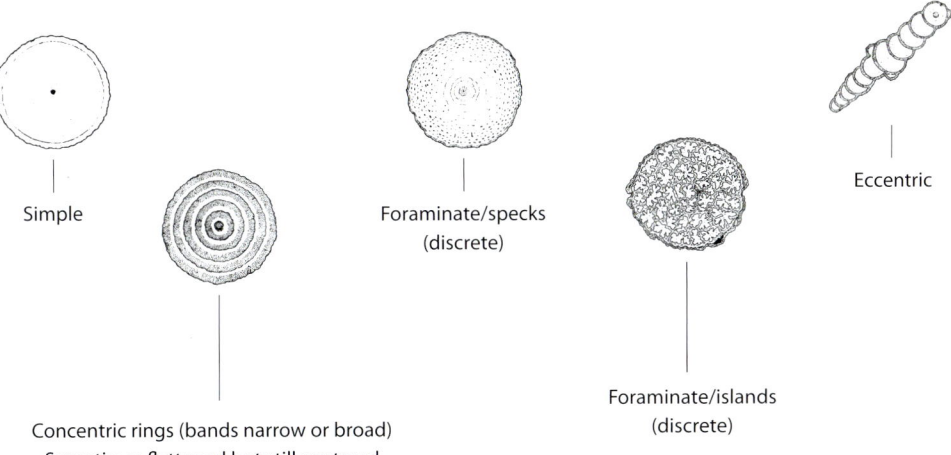

Simple

Concentric rings (bands narrow or broad)
Sometimes flattened but still centered

Foraminate/specks
(discrete)

Foraminate/islands
(discrete)

Eccentric

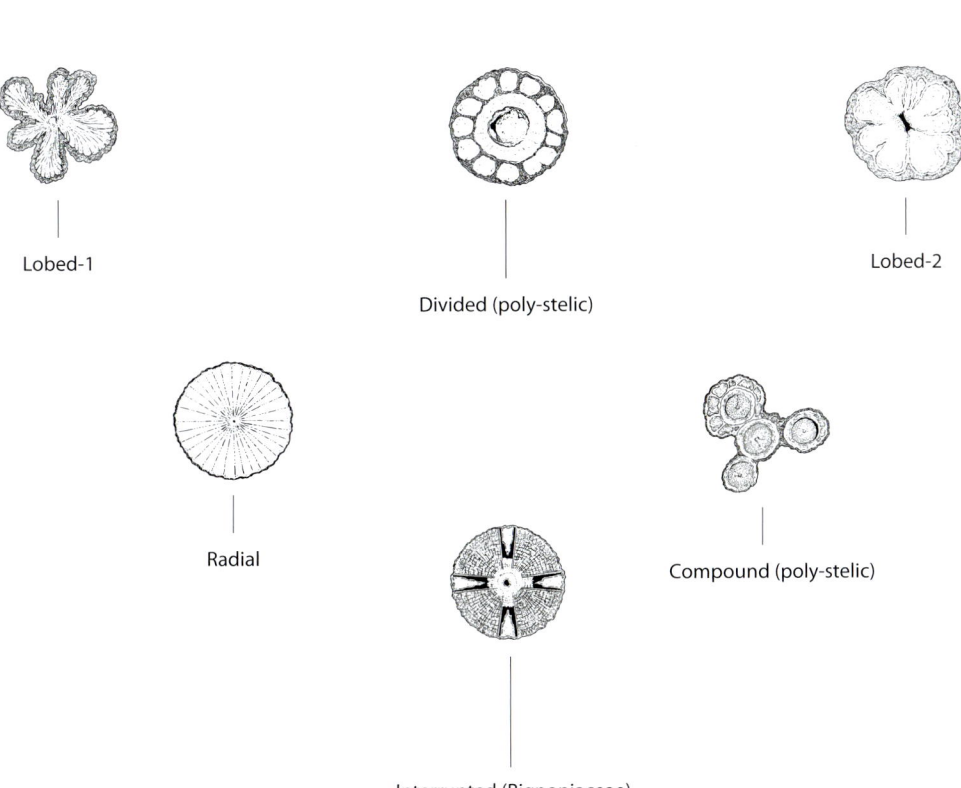

Lobed-1

Divided (poly-stelic)

Lobed-2

Radial

Interrupted (Bignoniaceae)

Compound (poly-stelic)

1. CLIMBING MECHANISM - HOOK CLIMBERS

Rubiaceae - Uncaria

Paired tendril-hooks or spines from modified
axillary stipules (coiled 2D or straight)

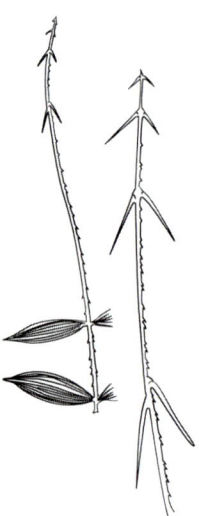

Loganiaceae - Strychnos

Woody tendril-hook from modified axillary
stipules paired or solitary (coiled 2D)

Arecaceae - Desmoncus

Barbed, reflexed hooks from modified leaflets at tail-like apex
of palm leaf

1. CLIMBING MECHANISM - TENDRILS

Passifloraceae - Passiflora

Tendril at petiole base (spiral)

Curcurbitaceae

Tendril at 90 degrees to
petiole base (spiral)

Vitaceae - Cissus

Tendril opposite petiole base (spiral)

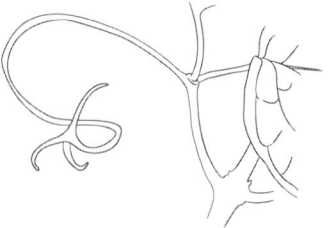

Tendril a modified leaflet on a compound leaf
(simple or 2-3 branched, rarely many branched)

Sapindaceae

Long, forked tendril from a modified
flowering branchlet (coiled 2D or spiral)

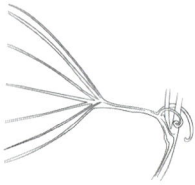

Smilacaceae - Smilax

Paired tendril from opposite
sides of petiolar sheath (spiral)

Illustrations redrawn with permission from Bobbi Angell illustrations

1. CLIMBING MECHANISM - TENDRILS

Leguminosae - Bauhinia
Woody tendril-hook from a modified branchlet (coiled 2D)

2. LEAVES - SCHEMATIC

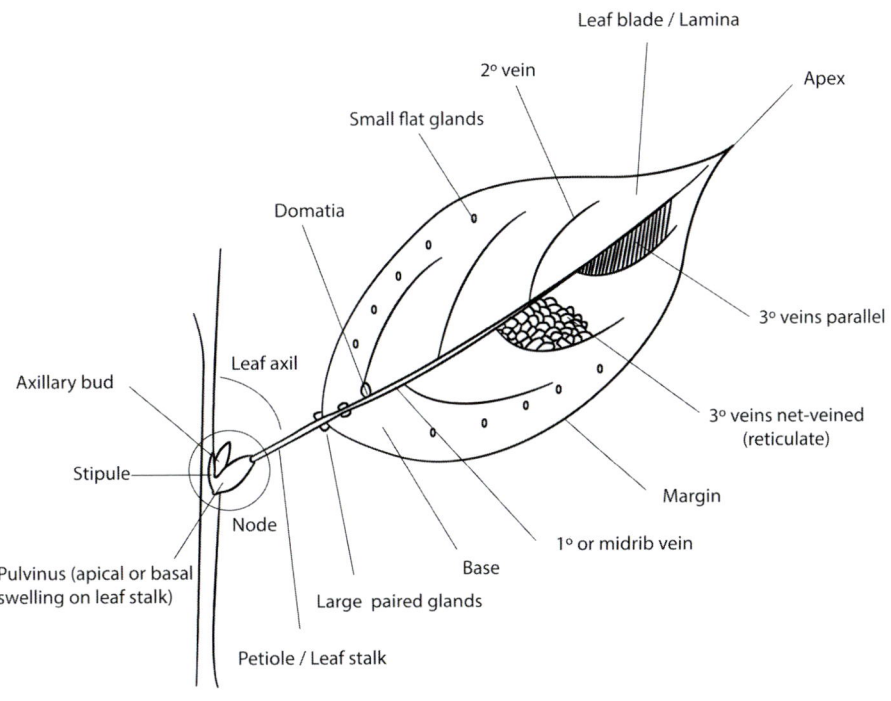

2. LEAVES - SHAPE

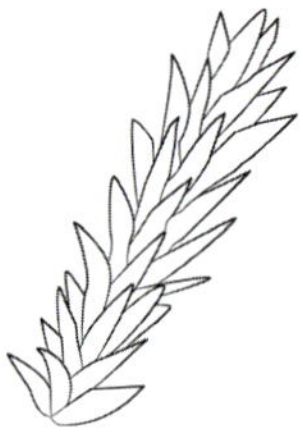
Awl shaped

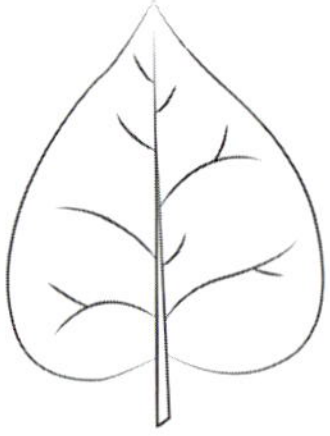
Cordate

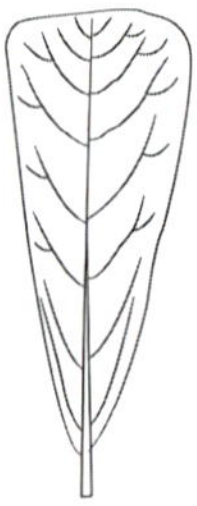
Cuneate

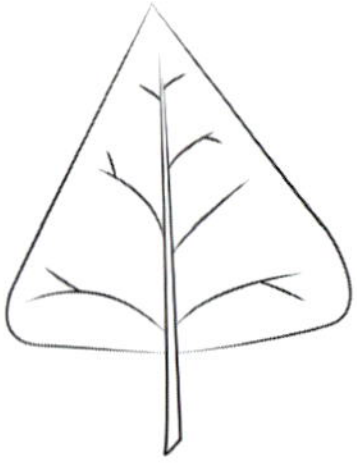
Deltoid

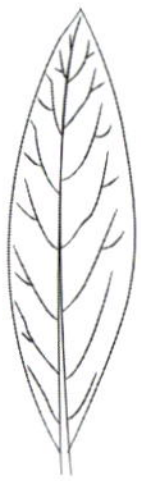
Elliptic

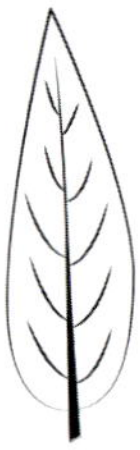
Lanceolate

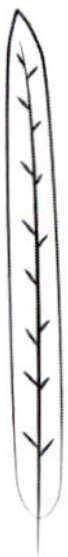
Linear

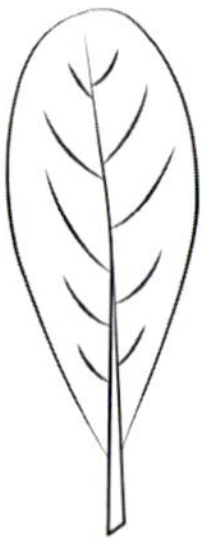
Oblanceolate

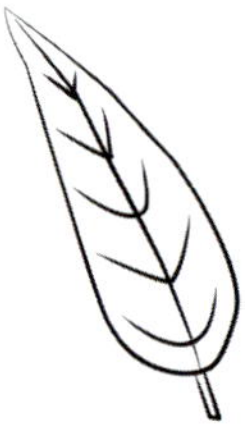
Oblong

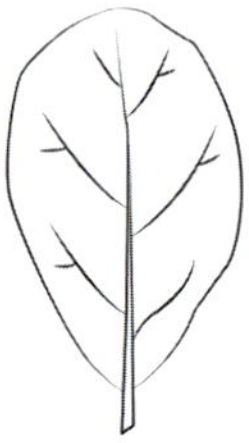
Obovate

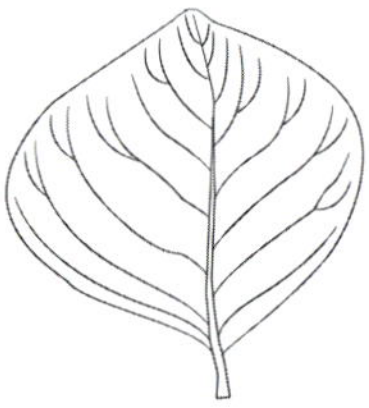
Orbiculate

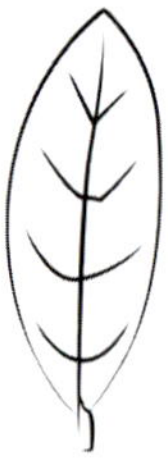
Oval

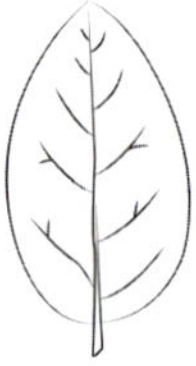
Ovate

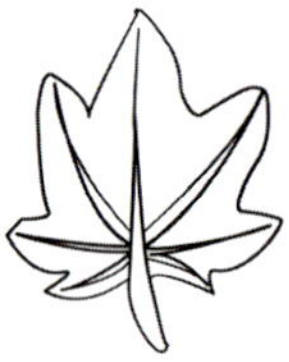
Palmate Lobed

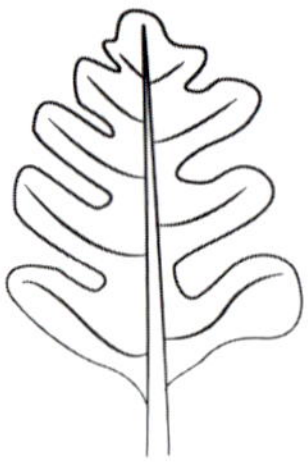
Pinnate Lobed

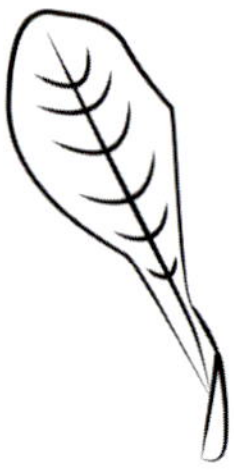
Spatulate

2. LEAVES - APEX

Acuminate

Acute

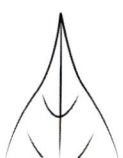

Attenuate

Cuspidate

Emarginate

Long tapering

Mucronate

Notched

Obtuse

Rounded

Spine tipped

Truncate

2. LEAVES - BASE

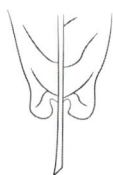

Auriculate base

Cordate base

Cuneate base

Oblique base

Round base

Truncate base

2. LEAVES - SECONDARY VEIN PATTERNS

Curving upwards with
marginal loops

Straight to margin

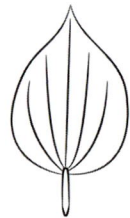

Palmate from base

Monocot type parallel

3. STIPULES

Intrapetiolar

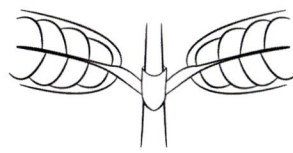

Interpetiolar leaves opposite

Ochrea

4. HAIRS

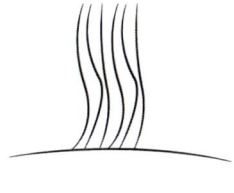

Simple

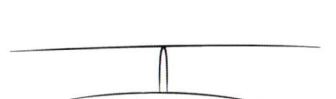

T-shaped: 2-armed, horizontal, see-saw like

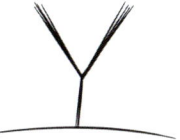

Y-shaped: 2-armed, raised

Dendritic: multi-branched, tree-like

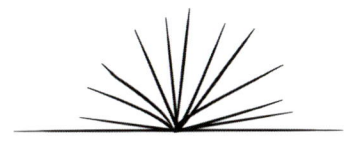

Stellate: star-shaped

5. INFLORESCENE - TYPES

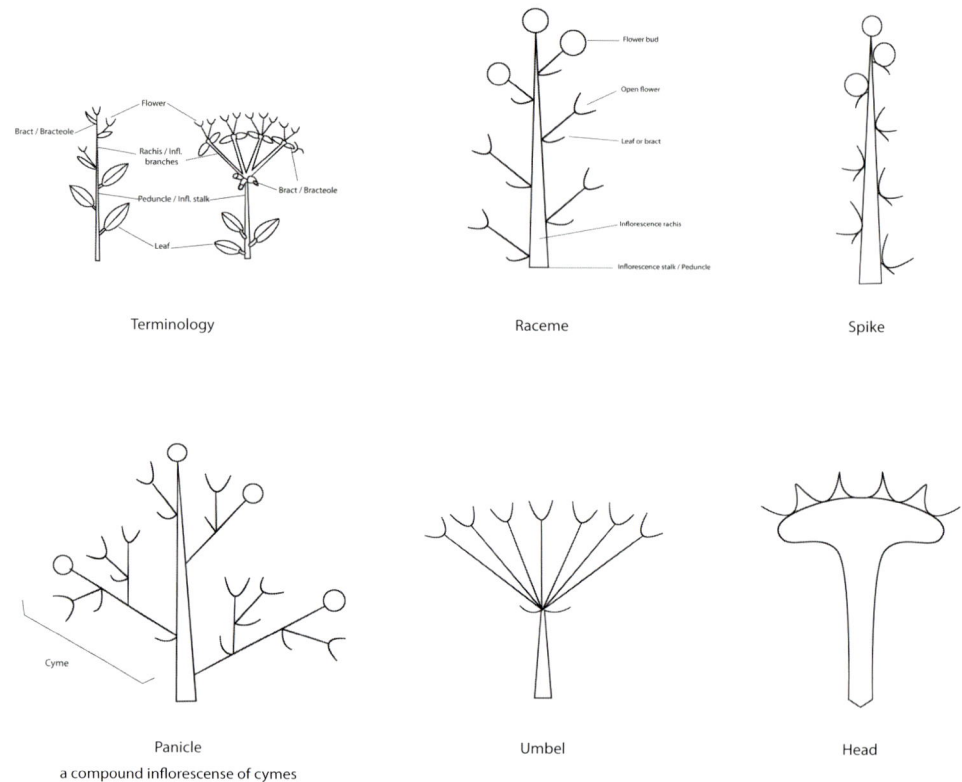

Terminology

Raceme

Spike

Cyme

Panicle
a compound inflorescense of cymes

Umbel

Head

6. FLOWER - FRUIT

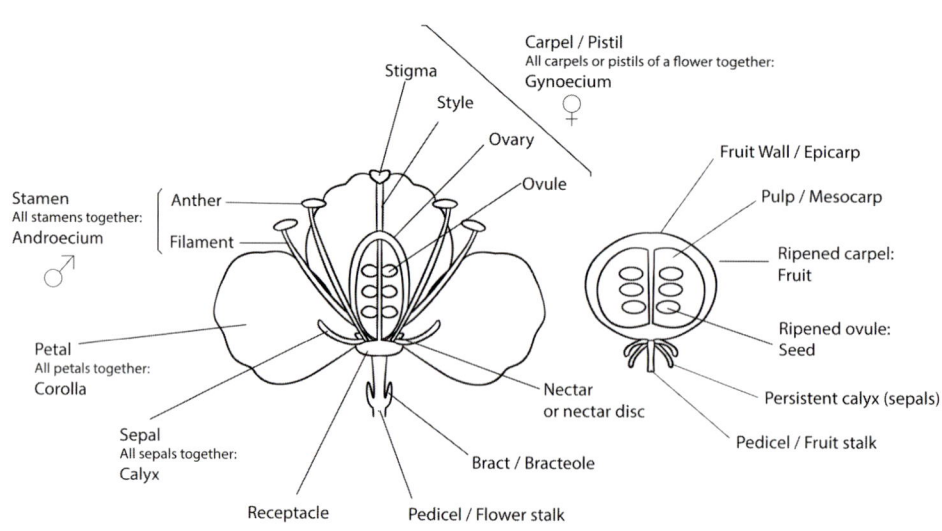

7. FLOWER - SYMMETRY

Radial, 4-parted (actinomorphic)

Radial, 3-parted (actinomorphic)

Bilateral (zygomorphic)

Irregular

7. FLOWER - FORM

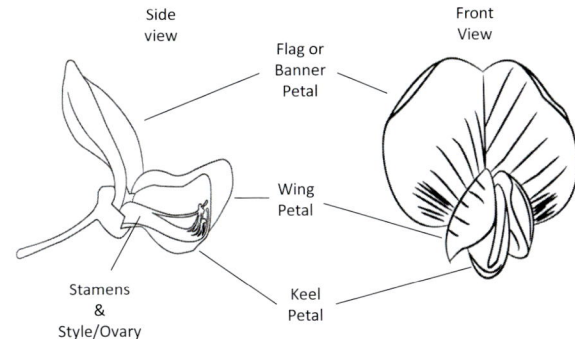

Side view

Flag or Banner Petal

Front View

Wing Petal

Stamens & Style/Ovary

Keel Petal

Leguminosae (Papilionoideae sub-family)

Pea-like flower

Tubular, urceolate, urn-shaped

Petals free, rotate (radial symmetry)

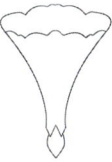

Tubular, funnel-shaped

Tubular, salverform, trumpet-shaped

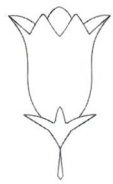

Tubular, campanulate, bell-shaped

Tubular, 2-lipped, bi-labiate

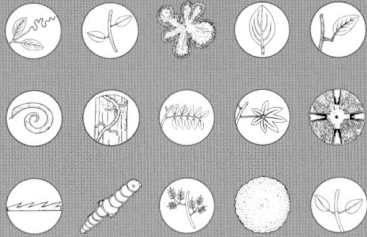

IDENTIFICATION
How to Use this Book

IDENTIFICATION

The information in this guide can be accessed and applied in many ways. The book is designed to be visually browsed to directly match specimens with images, or for the reader to use icons and other guide tools to narrow the field of potential correct family or genus identifications. There is no one-size-fits-all formula for climbing plant identification, because the category encompasses so much variation across different plant families and ecosystems. This book follows the practical family and genera identification methods of Gentry (1993) and others, by prioritizing growth form, leaf arrangement, and other vegetative characters. For many specimens, the combination of 2-3 vegetative characters allows for identification to family or genus. Flowers and fruits are less of a priority because they are rarely present and accessible in tropical plants. Fertile characters are still described because they are often highly useful when present. Many low-stature climbers flower and fruit throughout the year, making fertile characters more useful for identification.

On the front page of every family chapter, the **character icons** are presented in eight vertical positions indicating one or more alternative (horizontal) character states (Figure 3). From top to bottom, the multi-character positions include: climbing mechanism, leaf arrangement, leaf margin type, leaf vein pattern, stem cross-section pattern, stipule pattern, sap color and consistency, and seed dispersal mechanisms. Icon definitions and linked plant families are presented in Tables 2-7. Priority guiding icons have a light green color. Character icons are also provided within family chapters at the genus level when appropriate.

Taxonomic descriptions. Each family chapter begins with a simple summary paragraph for the family or genus level (Figure 4). The presentation of text fields varies greatly between family chapters due to natural variation in taxonomic structure. For example, the Cannabaceae family has only one Guianan climbing genus and species, Celtis iguanae. The family and genus receive relatively brief coverage, and most information is collapsed into the species description. By contract, the Menispermaceae family chapter (Figure 5) has multiple genera with one-to-many species. In this case, the family has a "full treatment" with dedicated fields for all characters including ecology, distribution, use, and notes. This approach was chosen primarily to save space.

Descriptive data includes climbing mechanism, number of genera/species, common names, vegetation zones, local and global distribution, uses and notes. The information fields for detailed botanical descriptions include **General** (widespread characters, such as hairs or sap), **Stems, Leaves, Inflorescences,** and **Fruits. Bold text** indicates especially important terms for recognizing a taxon and comparing it with other groups. Simplified botanical terms are often followed by a more technical term in brackets. Novice users can thereby learn more advanced terminology if desired, and professionals have some more familiar terminology available. For some characters, such as leaf shape, tip and base, botanical terminology is critical, and readers will have to consult the glossary.

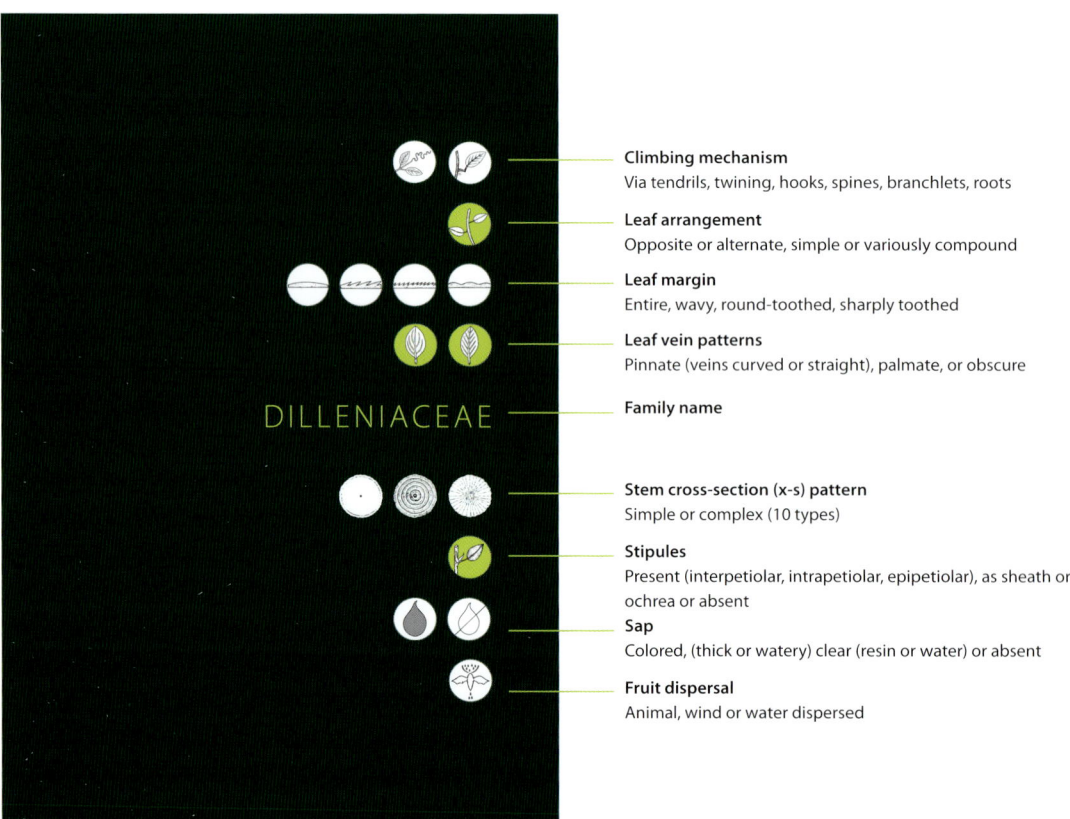

Figure 3. Front page of a family treatment with eight vertical position indicating alternative character states.

CANNABACEAE

Rarely lianas or scandent shrubs; commonly erect shrubs and trees. Widespread in tropical and temperate zones. Neotropics: 4 genera/20 species; Guianas: one genus with a climbing species. Most neotropical, woody Cannabaceae were formerly placed in the Ulmaceae (Elm family).

1. *Celtis* L.

Lianas or climbing shrubs; mostly erect shrubs or treelets. Climbing via twining stems, angled branchlets, and spines. Stipules are present. Leaves are simple, alternate, palmate-3-veined. Inflorescences with tiny flowers clustered in leaf axils; stamens 5; ovary superior. Fruits are yellow-orange drupes. Neotropics: 15 species; Guianas: 2 species, one climbing.

1.1. *Celtis iguanaea* (Jacq.) Sarg.

Common name: busi-lemki-maka (Sr)

Shrubby liana or climbing shrub; armed. **General:** surfaces rough to touch [*asperous*], hairless [*glabrous*] or lightly hairy, cystoliths (calcium carbonate) present in leaves, notable scent or sap about. **Stems: branchlets alternate, ± zigzag on main stem; stipules inconspicuous in leaf axils** (0.2-0.5 cm long) **or modified as 1-2 spines** (recurved on climbing branchlets, straight on normal branchlets), spines thick at base, to 1.5 cm long; older stems perfectly round, 3-8 cm diameter, with paired spines; bark light-colored, ± smooth, bearing lenticels;

stem x-s simple. **Leaves: 2-ranked** [*distichous*]; ovate to elliptic, 3-14 x 2.6-6.5 cm, **base unequal** [*asymmetrical*], rounded, truncate, or subcordate; apex acuminate to acute; margins entire, roughly toothed, or with a few teeth at tip; papery and rough; **veins strongly palmate at base, pinnate towards apex,** 2° veins looping at margins, raised below, tiny pockets [*domatia*] present in 2° vein axils; petiole standard, without swelling [*pulvinus*] at apex, to 0.8 cm long. **Inflorescences:** axillary, many-flowered (male only) or few-flowered (female + bisexual), bracts large, to 4 cm long; **flowers radial, tiny** (0.1 cm diameter), **green;** (4-)5-parted, margins frilly, corolla (petals) absent; stamens 5, free; ovary bottle-shaped, 2-chambered;

MENISPERMACEAE

Woody lianas and vines; rarely shrubs or treelets; plants male or female. Climbing via twining terminal shoots, tendril-like lateral shoots, and wiry, flexed petioles. Stipules are absent. Leaves are alternate, simple, entire, mostly palmate-veined. Inflorescences many branched or long and narrow, with small 3(4)-parted, radial flowers. Ovary is superior. Fruits are ovoid drupes.

General: Vegetation mostly **shiny green and smooth (hairless),** hairs simple when present; plant parts **bitter tasting** due to alkaloids, 'poison' scent noted on occasion; very rarely with colored sap (white-yellow) or onion scent.

Stems: Branchlets green; nodes smooth, without interpetiolar ridge or line, stipules and tendrils absent. Older stems robust, round or strongly flattened (e.g. Bauhinia, Machaerium); bark hard and fibrous with furrows, or soft and corky, often light brownish-yellow; **stem x-s pattern concentric, mostly off-center** (alternating color bands) or with **spoke-like radial rays** (Odontocarya); wood usually yellow.

Leaves: Shapes highly variable, margins entire (rarely wavy or few-toothed); leathery or subleathery, less commonly thin; **leaves 3-7 palmate-veined, palmate/pinnate-veined** (palmate at base, with alternate or opposite 2° lateral veins above base), **or fully pinnate-veined** (Telitoxicum), 3° and finer venation ladder-like [*scalariform*] or net-veined; **petiole (leaf stalk)** short to long, rigid, **apex commonly flexed or twisted**

and with a cylindrical swelling [*1st pulvinus*], base simple or with a half-globe-like swelling attached to stem [*2nd pulvinus*], **rarely twisted or flexed only at base** (Borismene, Disciphania, Odontocarya).

Inflorescences: Axillary, sub-axillary, or borne on old wood [*cauliflorous*]. Inflorescences unisexual, male flowers often in shortly-branched panicles, female flowers often in spikes or long racemes, sometimes with leafy bracts; **flowers tiny or small, radial, usually 3-parted** (4-parted in Cissampelos), up to 0.6 cm long, greenish-white, yellow, pink, orange, or brown; sepals 6-12, petals 6-12(0), free or fused into small inner ring; female flowers with 3-12 stamens, free or fused; female flowers with superior ovary carpels 3-6.

Fruits: Ovoid, single-seeded drupes [*drupe-like monocarps*], a very few or dozens per inflorescence, 1-3 per flower (6-15 in Sciadotenia); bright red, orange, yellow, tan, or purplish-black at maturity, often fuzzy, with a leathery skin [*exocarp*], fibrous or fleshy inner layer [*mesocarp*] and a bony, woody or papery inner wall [*endocarp*].

Figure 4. Example of a family chapter (Cannabaceae) with only one genus and one species. Most of the information is within the species account.

Figure 5. Example of a family chapter with many genera and many species. The family level receives a full account.

Common Names: Taxon descriptions often include common (local) names. The codes used in this book to indicate common name languages and cultures include the following:

Ak (Akawaio); Ar (Arawak); Are (Arekuna); Au (Aucaans or Ndjuka); Bo (Boni or Aluku); BP (Brazilian Portuguese); Ca (Carib); Du (Dutch); En (English); FG (French Guiana Creole); Fr (French); GU (Guyana Creole); Ja (Javanese); Kw (Kwinti); Mac (Macushi); Mat (Matawaai); Pal (Palikur); Par (Paramaccan); Pat (Patamona); Sa (Saramaccan); SD (Surinamese Dutch); Sp (Spanish); Sr (Sranantongo, Dutch Creole); Tr (Trio); War (Warau); Way (Wayana); Wpi (Wayapi); and Ww (Wai-wai).

Many Saramacca and Trio plant names were obtained directly from local residents in Suriname. Additional names were obtained largely from literature sources cited in the bibliography. Many language codes are strongly linked to a given territory in the Guianas, including French Guiana (FG, Fr, Pal, Wpi), Guyana (Ak, Are, En, GU, Mac, Pat, War, Ww) and Suriname (Au, Bo, Du, Ja, Kw, Mat, Par, Sa, SD, Sr, Tr, Way). Less territory-specific language codes in the Guianas include Ar, BP, Ca, and Sp.

Ecology: In this section, the primary information is on seed dispersal or pollination.

Distribution: For most species, information is provided on relative abundance, vegetation types, and distributions within the Guianas and the Neotropics. Vegetation type names (classification) are not based upon a single system, but were generalized for the three Guianas (e.g, non-flooded old-growth forest (terra firme), seasonally-flooded old-growth, (marsh forest) secondary forest, creek forest, riparian forest and swamp). Forest types are 'evergreen' unless mentioned otherwise. In a few cases, vegetation types in common usage only in Suriname are used (e.g., Lindemann & Molenaar 1955). The term 'savanna forest' indicates forests or woodlands of low stature on poor soils (e.g., deep white sand) with low water retention. The woody plants of savanna forest are often small in diameter, possess leathery leaves, and are composed of many shrubby species also found on savanna grasslands. The term 'mountain savanna' refers to forests with dense, small diameter, pole-like trees (often Myrtaceae) found on thin, stony soils (Pieter Teunissen, pers. comm.).

Uses: Ethnobotanists have documented many uses for the bark, wood or leaves of liana species in the Guianas (Fanshawe 1949, Plotkin 1986, Grenand et al., 2004, Hoffman 2009, van Andel 2000, van Andel & Ruysschaert 2011). Common traditional uses include medicines, arrow poisons, spiritual cleansing and protection, rope substitutes and male aphrodisiacs. The brief use citations given in this guide are based largely upon original research and compiled data 'Medicinal Plants of the Guianas' website (DeFillips & Maina 2001).

Field assessment of liana families and genera

A first step in identifying woody climbers in the forest is to learn recognition characters for distinctive growth forms and plant families. This is possible for many groups even without visible leaves or fertile parts. **Tree-like hemi-epiphytes** are conspicuous due to their large aerial roots produced from canopy perches. The roots may drop freely as straight thick cylinders (Clusia) or grow downwards like muscled arms hugging the host tree trunk (Coussapoa, Ficus, Oreopanax). The robust leaves of these plants are often conspicuous in the forest canopy or are present on the ground. Secondary forest trees such as Cecropia and Pourouma may have a similar look at the base with stilt roots.

Vine-like monocot hemi-epiphytes, including Araceae, Cyclanthaceae, and Orchidaceae, possess green, soft stems, hang onto tree trunks with clasping aerial roots, and drop slender feeding roots to the forest floor. Soft-stemmed, liana-like hemi-epiphytes include many taxa that are relatively easy to recognize, including climbing cacti (Cactaceae, look for spines), Gesneriaceae (Drymonia), Marcgraviaceae (Marcgravia, Souroubea), Rubiaceae (Hillia) and Solanaceae (Markea, Solandra). Slightly more difficult growth forms to distinguish are the **shrubby, woodier liana-like hemi-epiphytes,** including the Ericaceae (mostly higher elevation), Margraviaceae (Marcgraviastrum), Melastomataceae (Adelobotrys, Topobea), Rubiaceae (Schradera), and Schegeliaceae (Schlegelia). Other easily-recognized climbers include climbing green-stemmed pipers (Piperaceae), a gymnosperm liana genus (Gnetum, look for swollen nodes, jointed stems), climbing palms (Desmoncus, unique whip-like leaf extension with hooks), and spiny climbing shrubs (Seguieria, Randia, Celtis).

Once the more distinctive climbing growth forms are eliminated, what remains is still a very large group of **woody and subwoody lianas.** In many cases, the leaves and fertile parts of lianas are hidden, and only stems are visible. To ensure that useful characters are not missed, visually trace the route of leafless stems. Liana species often produce low leafy shoots where stems touch the ground, often in multiple spots in a given area. Also check the ground for fallen leaves, flowers and fruits. Unless canopy access is an option, ground-level stems should be observed for: **structure** (e.g., round, flattened, deeply channeled or lobed, with regular rings or knobs, twisted or straight); **bark** (e.g., texture, color, ridges or lenticels); and **internal stem characters** (slash to reveal sub-bark, wood, color, exudate/sap, scent). Observations of the cross-section anatomy can be useful in challenging cases. Slashed stems often take time to reveal their characters, and colors and scents may change rapidly with exposure to air.

DISTINCTIVE STEMS

Flattened: ribbon-like, wavy to ± round – Leguminosae (Bauhinia, Machaerium); flattened with off-center concentric circles (Menispermaceae)

Cabled: Multiple thick cables that sometimes separate – Malpighiaceae; shape of double-barrel shotgun in cross-section, rank "pig" or onion scent – Polygonaceae (Coccoloba); 3-6-9 cabled in cross-section – Sapindaceae; triangular in cross-section, texture soft – Passifloraceae

Subwoody stem: green, watery, with length-wise ridges and conspicuous rays within: Cucurbitaceae

Extreme-lobed in cross-section: Dichapetalaceae (Dichapetalum), Malpighiaceae

With corky, length-wise ridges (1-2 cm tall), medicinal scent in slash – Aristolochiaceae (Aristolochia)

With corky cones, white latex in slash, reddish bark with lenticels, two tiny triangular glands at leaf base – Apocynaceae (Forsteronia)

Corky, smooth, soft with complex cross-section of many islands – Acanthaceae (Mendoncia)

Mature stems remotely 4-6-sided, x-s with cross-like pattern, branchlets sharply 4-sided – Bignoniaceae

Bark softly woody, light-colored, round knobs at nodes – Bignoniaceae

Bark flaky, red to silver, x-s concentric rings with spoke-like rays – Dilleniaceae

Bark slightly flaky, red, with lengthwise ridge or lines – Leguminosae (Mimosa)

Bark dark brown to silvery, peeling in small vertical platelets, inner bark bright yellow – Leguminosae (Senna)

Bark smooth, skin-like, with orange sub-bark – Celastraceae (Salacia, Tontalea)

Stem round, 2-3 cm diameter, white-grey, bark smooth, twisting – Verbenaceae (Petrea)

Swollen, jointed nodes, exuding clear gum, sometimes small amounts of white sap – Gnetaceae (Gnetum)

ARMED STEMS (ALL LEAF TYPES):

Leguminosae - Acacia tenuifolia, Machaerium, Mimosa, Piptadenia
Rubiaceae – Uncaria, Randia
Ulmaceae - Celtis iguanea
Arecaceae - Desmoncus
Polygalaceae - Securidaca spinifex
Phytolaccaceae - Seguieria
Smilacaceae – Smilax

The opportunities for field identification of woody lianas greatly increase when leafy shoots or mature leafy branchlets are available (see the leaf-oriented icon tables). The first step is to determine one of four leaf arrangements, including: opposite-simple (OS), opposite-compound (OC), alternate-simple (AS), or alternate-compound (AC).

All native OC lianas in the Guianas are members of the Bignoniaceae family.

AC lianas are primarily Leguminosae, but also include Connaraceae (odd-pinnate leaves), Cucurbitaceae (coiled tendrils 90° to leaf axil), Sapindaceae (forked tendrils on branchlets & inflorescence rachis) and Vitaceae (coiled tendrils opposite leaf axil) (also a few herbaceous vines).

IDENTIFICATION

OS lianas include many of the most common and species-rich liana families besides the Bignoniaceae, including Celastraceae, Combretaceae, Malpighiaceae, and Rubiaceae. Critical characters for this group include sap (white latex – Apocynaceae; red or yellow latex – Celastraceae; resin – Combretaceae, Gnetaceae), venation (3-veined – Loganiaceae, Verbenaceae, Asteraceae), toothed (Celastraceae, Gesneriaceae, Asteraceae), hairs (silky and t-shaped – Malpighiaceae; white, dense and simple – Trigoniaceae), glands (petiole and leaf blade glands, Malpighiaceae); rough surfaces (Celastraceae-Prionostemma, Asteraceae, Lamiaceae-Aegiphila, Verbenaceae-Petrea); spines (Combretaceae, Rubiaceae – Uncaria, Randia, Chomelia, Nyctaginaceae – Pisonia); nodes swollen (Acanthaceae-Mendoncia, Gnetaceae-Gnetum); and stipules (4 linear stipules – Trigoniaceae, diverse often inconspicuous stipules – Malpighiaceae, stipular spines – Rubiaceae). In the OC group, only the Celastraceae has tendril-like, clasping, green leafy shoots.

The remaining families are AS leaf types, forming a very diverse, unique and sometimes challenging group for identification. Critical characters for the AS group include tendrils (see icon table); toothed leaves (see icon table); 3-veined leaves (icon table); "primitive" (medicinal or spicy) scent (Aristolochiaceae, Hernandiaceae, Annonaceae); glands on the petiole or leaf base (in taxa without tendrils: Euphorbiaceae (Omphalea and Plukentia), Polygalaceae-Moutabea; leaf venation palmate (see icon table); sap or latex (see icon table); spines (Smilacaceae, Ulmaceae, Phytolaccaceae-Seguieria, Solanaceae, Polygalaceae-Moutabea, Combretaceae, Malvaceae-Byttneria).

ICON TABLE 1 - LIANA CLIMBING MECHANISMS

Tendril-climber; tendril coiled in one plane, like a butterfly tongue or a watchspring. LEGUMINOSAE (BAUHINIA), LOGANIACEAE (STRYCHNOS), RHAMNACEAE (CELTIS), SAPINDACEAE

Tendril-climber; tendril spiral (common). CUCURBITACEAE, PASSIFLORACEAE, SAPINDACEAE, SMILACACEAE, VITACEAE

Tendril-climber; tendril replaces leaflet in compound, 3-foliolate leaf, sometimes branched. BIGNONIACEAE

Twiner; terminal shoot spirals: APOCYNACEAE, CELASTRACEAE, MALPIGHIACEAE, MENISPERMACEAE (35 families). Also shoot/infl. twisting and tendril-like: CELASTRACEAE, CONVOLVULACEAE (MARIPA), EUPHORBIACEAE (OMPHALEA), LEGUMINOSAE (MACHAERIUM), POLYGALACEAE (SECURIDACA)

Branch or petiole climber; passive jutting structures; also sometimes twisted petioles. (26 families) most opposite-leaved families; ASTERACEAE, BORAGINACEAE, CANNABACEAE, CONVOLVULACEAE (MARIPA), EUPHORBIACEAE (CROTON, MABEA), GNETACEAE, LEGUMINOSAE (ENTADA), MENISPERMACEAE, OLACACEAE, RHAMNACEAE, SOLANACEAE, VITACEAE

IDENTIFICATION

Hook-climber; passive use of hooks, spines, or other projections (uncommon).
ARECACEAE, CANNABACEAE, LEGUMINOSAE (MACHAERIUM, MIMOSA), LOGANIACEAE, MALVACEAE, NYCTAGINACEAE, PHYTOLACCACEAE (SEGUIERIA), RUBIACEAE (GUETTARDA, RANDIA, UNCARIA), SMILACACEAE

Root-climber; climbing via aerial roots from the stem.
BIGNONIACEAE (DOLICHANDRA, MANSOA), GESNERIACEAE, MARGRAVIACEAE, MELAS-TOMATACEAE (TOPOBEA), RUBIACEAE (SHRADERA), SCHLEGELIACEAE, SOLANACEAE (MARKEA, SOLANDRA) [*other tree-like and vine-like hemi-ephiphytes are in Guiana climber checklist*]

ICON TABLE 2 - LEAF TYPE

Alternate-simple (AS). One leaf per stem node, bud present in leaf axil. (25 families)
ANNONACEAE, ARISTOLOCHIACEAE, BORAGINACEAE, CANNABACEAE, CONVOLVULACEAE, CUCURBITACEAE, DILLENIACEAE, EUPHORBIACEAE, MENISPERMACAE, PASSIFLORACEAE, POLYGALACEAE, POLYGONACEAE, SMILACACEAE, SOLANACEAE, VITACEAE

Alternate-simple, palmately-lobed, secondary veins radiating like spokes from center with distinct lobes. If the lobes become stalked leaflets, then the leaf is compound (see below).
ARISTOLOCHIACEAE, CUCURBITACEAE, MENISPERMACEAE, PASSIFLORACEAE

Alternate-compound (AC), 3-foliolate leaf (a leaf with three leaflets).
CUCURBITACEAE, LEGUMINOSAE (CLITORIA, DIOCLEA, MUCUNA), SAPINDACEAE, VITACEAE

Alternate-compound, once-pinnate with 2 terminal leaflets (even-pinnate).
ARECACEAE (DESMONCUS), LEGUMINOSAE (ABRUS, BAUHINIA, SENNA)

Alternate-compound, once-pinnate with 1 terminal leaflet (odd-pinnate).
CONNARACEAE, LEGUMINOSAE (DALBERGIA, DUGUELIA, LONCHOCARPUS), SAPINDACEAE

Alternate-compound, bi-pinnate (twice compound).
LEGUMINOSAE (ACACIA, ENTADA, MIMOSA, PIPTADENIA)

IDENTIFICATION

Alternate-compound, palmate. Leaflets radiating like spokes from central point.
PASSIFLORACEAE

Opposite-simple (OS). Two leaves per stem node. 16 families:
ACANTHACEAE, APOCYNACEAE, ASTERACEAE, CELASTRACEAE, COMBRETACEAE, GESNE-
RIACEAE, GNETACEAE, LAMIACEAE, LOGANIACEAE, MALPIGHIACEAE, MELASTOMATACEAE,
NYCTAGINACEAE, RUBIACEAE, SCHLEGELIACEAE, TRIGONIACEAE, VERBENACEAE

Opposite-whorled-simple. 3 or more leaves per stem node.
APOCYNACEAE (ALLAMANDA), ASTERACEAE, RUBIACEAE

Opposite-compound (OC), 3-foliolate leaf.
BIGNONIACEAE

Opposite-compound, 3-foliolate or (usually) 2-foliolate leaf and tendril.
BIGNONIACEAE

Opposite-compound, 1-3-times. Palmately-compound (3-leaflet subunits) or pinnately compound (5-9-leaflet subunits).
BIIGNONIACEAE (ADENOCALYMMA, CUSPIDARIA PLEONOTOMA)

ICON TABLE 3 - **LEAF MARGINS**

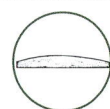

Entire. Leaf margin smooth, no teeth. Very common.

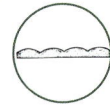

Lobed, rounded teeth [*cuneate*]
CELASTRACEAE, CUCURBITACEAE, MARCGRAVIACEAE, MALPIGHIACEAE (STIGMAPHYL-
LON), MALVACEAE, PASSIFLORACEAE, PHYTOLACCACEAE , VIOLACEAE, VITACEAE

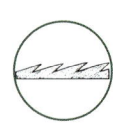

Toothed; teeth large, few to many.
ASTERACEAE, BIGNONIACEAE (RARE-STIZOPHYLLUM), CANNABACEAE, CELASTRACEAE,
DILLENIACEAE, GESNERIACEAE, MENISPERMACEAE (RARE), PASSIFLORACEAE, RHAMNACEAE
(GOUANIA), SAPINDACEAE, SOLANACEAE (RARE), VITACEAE

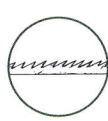

Toothed; teeth many, fine.
ASTERACEAE, BORAGINACEAE (VARRONIA), CANNABACEAE, CUCURBITACEAE, DILLENI-ACEAE, EUPHORBIACEAE, GESNERIACEAE, ICACINACEAE, MALVACEAE, MELOSTOMATACE-AE, PASSIFLORACEAE, SAPINDACEAE, VITACEAE

Wavy [*undulate*]; slightly to strongly wavy margin. Many species, most notably:
ASTERACEAE, CELASTRACEAE, CUCURBITACEAE, GESNERIACEAE, MALPIGHIACEAE, SMILACACEAE, MENISPERMACEAE, PASSIFLORACEAE, SOLANACEAE

ICON TABLE 4 - LEAF VEIN PATTERNS

Pinnate secondary (2°) veins curving upwards. Very common.

Pinnate 2° veins straight, parallel.
APOCYNACEAE, CONVOLVULACEAE, DILLENIACEAE, TRIGONIACEAE, URTICACEAE (COUSSAPOA), SAPINDACEAE

Obscure. 2° or 3° veins either invisible or very fine. Common in monocots and dicot hemi-epiphytes.
ARACEAE, CLUSIACEAE, MARCGRAVIACEAE, MELASTOMATACEAE, RUBIACEAE (HILLIA), GESNERIACEAE, NYCTAGINACEAE, SCHLEGELIACEAE, ORCHIDACEAE (VANILLA)

Palmate, 2° veins radiate from a point above the leaf base.
ASTERACEAE, RHAMNACEAE (AMPELOZIZYPHUS), LOGANIACEAE, MELASTOMATACEAE, MENISPERMACEAE

Palmate, 2° veins radiate from base of the leaf. Common in climbers.
ARISTOLOCHIACEAE, BIGNONIACEAE (LUNDIA), CANNABACEAE, CUCURBITACEAE, EUPHORBIACEAE, HERNANDIACEAE, LEGUMINOSAE, LOGANIACEAE, MALVACEAE, MELASTOMATACEAE, MENISPERMACEAE, OLACACEAE, PASSIFLORACEAE, RHAMNACEAE, SMILACACEAE, VITACEAE

Parallel. Common in monocot species.
ARECACEAE (DESMONCUS), CYCLANTHACEAE, ORCHIDACEAE

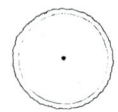 Simple – Standard dicot or monocot stem without any unique pattern.

 Concentric, bull's-eye pattern due to circular bands of xylem and phloem. Bands wide: APOCYNACEAE (ODONTADENIA), CELASTRACEAE, CONVOLVULACEAE, DILLENIACEAE, GNETACEAE, LEGUMINOSAE (MACHAERIUM). Bands narrow: ICACINACEAE, MENISPERMACEAE, PHYTOLACCACEAE (SEGUIERIA), POLYGALACEAE (MOUTABEA)

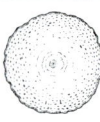 Foraminate-1 (black specks within, tiny 2° phloem islands). LOGANIACEAE (STRYCHNOS)

 Foraminate-2 (larger islands within). ACANTHACEAE (MENDONCIA), CONVOLVULACEAE (COLYCOBOLUS), ICACINACEAE

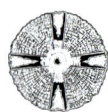 Interrupted with a diagnostic cross-like pattern, typically with 4 symmetrical arms [2° *phloem wedges*] at 90° points (i.e., 0°, 90°, 180°, 270°), also with arms in multiples of 4 (8-64), 6, 12, and irregular/many. BIGNONIACEAE

 Lobed-1 (naked lobes). DICHAPETALACEAE, MALPIGHIACEAE, PASSIFLORACEAE

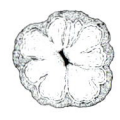 Lobed-2 (lobes enclosed within one stem). MALPIGHIACEAE, MALVACEAE , PASSIFLORACEAE

 Flattened/Eccentric (stem flattened, often off-center). LEGUMINOSAE (BAUHINIA, MACHAERIUM), ICACINACEAE, MENISPERMACEAE, POLYGONACEAE (COCCOLOBA)

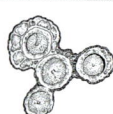 Compound (core + separate cables). MALPIGHIACEAE, SAPINDACEAE

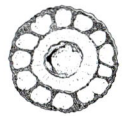 Divided (core + separate cables enclosed within one stem).
MALPIGHIACEAE

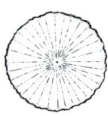 Radial (spokes radiating from center).
ANNONACEAE, ARISTOLOCHIACEAE, CELASTRACEAE, CUCURBITACEAE, DICHAPETALACEAE, MENISPERMACEAE (CISSAMPELOS), VERBENACEAE

ICON TABLE 6 - STIPULES

 Intrapetiolar – stipules at the base of petiole, often short-lived or inconspicuous:
CANNABACEAE, DICHAPETALACEAE, DILLENIACEAE, EUPHORBIACEAE, LEGUMINOSAE, MALPIGHIACEAE, MALVACEAE, PASSIFLORACEAE, RHAMNACEAE, SAPINDACEAE, TRIGO-NIACEAE, VIOLACEAE, VITACEAE. Stipule-like structures: APOCYNACEAE, ARISTOLOCHI-ACEAE, ASTERACEAE, BIGNONIACEAE

 Interpetiolar - stipules on stem between petioles, leaves mostly opposite.
LOGANIACEAE (RARE), MALPIGHIACEAE, RUBIACEAE

Stipules absent (no icon). Common. Check for scars left behind by fallen stipules.

 Ochrea (stipules present as circular sheath around bud and at nodes).
POLYGONACEAE (COCCOLOBA), SMILACACEAE (SMILAX) - Stipules fused as open sheath that bears 2 tendrils.

ICON TABLE 7 - SAP

 Colored sap of any thick or watery texture in slash or leaves. White: APOCYNACEAE, CLUSIA, CONVOLVULACEAE, EUPHORBIACEAE, FICUS, OLACACEAE, SAPINDACEAE.
Non-white: CELASTRACEAE, CLUSIA, CONNARACEAE, EUPHORBIACEAE, GESNERIACEAE, GNETACEAE, LEGUMINOSAE, RHAMNACEAE

 Clear water, sap or resin present.
COMBRETACEAE, CUCURBITACEAE, DILLENIACEAE, EUPHORBIACEAE (CROTON), GNETA-CEAE, ICACINACEAE, MALVACEAE, SAPINDACEAE

ICON TABLE 7 - SAP (CONTINUED)

 Sap absent (most). Without conspicuously colored sap, abundant water, or resin/gum. Very common.

ICON TABLE 8 - FRUIT DISPERSAL

 Animal dispersed

 Wind dispersed

 Water dispersed

An additional, rare dispersal mechanism is the explosive or ballistic type, with fruits that 'self-disperse' their seeds.
An example of this is the Mabea genus in the Euphorbiaceae.

PLANT FAMILIES WITH CONSPICUOUS GLANDS

Flat glands on leaf blade – Apocynaceae; Bignoniaceae; Convolvaceae (Ipomoea, Maripa); Cucurbitaceae (Cayaponia); Lamiaceae; Malpighiaceae; Passifloraceae, Polygalaceae (Moutabea).

1-2 raised glands at leaf base, rachis or petiole – Apocynaceae (Forsteronia); Euphorbiaceae (Croton pullei, Omphalea, Plukentia); Leguminosae (Acacia, Senna); Malpighiaceae; Passifloraceae; Rhamnaceae (Gouania).

Glands on the inflorescence or flowers – Bignoniaceae (glands often on calyx and corolla, volcano-crater-shaped-glands in Adenocalymma, peltate glands, saucer-shaped glands; Euphorbiace (Mabea - on inflorescence stalk, Omphalea - on inflorescence bracts); Malpighiaceae (oil glands on calyx); Malpighiaceae (Stigmaphyllon - on inflorescence bracts); Passifloraceae (along margin of inflorescence bracts).

Glands on the leaves (look at young leaves) – Mimosoids (Acacia tenuifolia, stalked glands), Caesalpins (Senna, stalked or cone-glands), Bignoniaceae (gland fields, lamina glands, pseudostipules), Euphorbiaceae (Croton pullei, Dalechampia, Mabea pearls), Malpighiaceae (Heteropterys, Hiraea, Mascagnia, Stigmaphyllon Tetrapterys, marginal glands), Passifloraceae, Cucurbitaceae (Cayaponia, row of glands on petiole base and lamina base), Verbenaceae (Aegiphila, disc-shaped glands), Moraceae (Ficus); Apocynaceae (Mandevilla rugellosa, adaxial surface); Convolvaceae (Ipomoea, Maripa glabra, sunken oil glands above); Polygalaceae (Moutabea, scattered below).

ACANTHACEAE

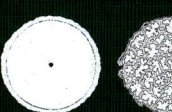

ACANTHACEAE

includes Mendonciaceae

Some subwoody lianas and vines; mostly erect herbs, shrubs, and treelets. Tropics and subtropics worldwide, ca. 250 genera/2500 species; Guianas - ca. 19 genera/62 species. Climbing taxa of the Guianas include Mendoncia, a native subwoody liana genus, Ruellia, a weakly climbing shrub, and Thunbergia, a popular non-native ornamental vine. Only Mendoncia is treated here; sometimes placed in a separate family, the Mendonciaceae.

1. Mendoncia Vell. ex Vand.

Woody to sub-woody lianas and vines with counter-clockwise twining stems; often hairy and slender. Leaves are opposite, simple and never toothed. Stipules absent. Inflorescences are axillary or terminal, with showy tubular flowers enclosed by leafy bracts; with four stamens and a superior ovary. Fruits are single-seeded drupes (explosive capsules in most Acanthaceae).

General: hairless to densely hairy; **stellate hairs sometimes present; vegetation rough** [*asperous*].

Stems: Branchlets oval or 4-sided, slender, usually twisted; **nodes with a horizontal line or swelling.** Older stems round and furrowed, twisted, woody mostly near base; bark soft, corky or smooth; **stem x-s pattern often "foraminate" (with many 'islands' within), diagnostic to genus.**

Leaves: +/- rounded to narrow; sub-leathery to papery; venation pinnate; all veins strongly raised below; petiole often twisted or curved.

Inflorescences: axillary, **flowers 1-5, infl. stalks long and hanging; enveloped by 2 large, persistent leafy bracts/bracteoles;** calyx inconspicuous; **corolla bright red and cylindrical or white and widening at mouth,** 5-lobed or 2-lipped; stamens 4, 2-ranked; ovary 2-chambered but only 1 seed developing, stigma two-lobed.

Fruits: Fleshy drupes, ovoid to ellipsoid, often compressed, maturing purplish-black; lightly to densely hairy; seeds 1.

Ecology: Forest gaps, trails, open areas; seed dispersal via animal (ingested by birds, monkeys).

Distribution: Neotropics, African tropics, and Madagascar; 60 species known; Guianas: 6 species known, 3 described here.

1.1. Mendoncia aspera Ruiz & Pavon

Subwoody liana or vine, brown-hairy [*sericeous*], **asperous. Stem:** branchlets round; nodes swollen; older stem round, twisted, furrowed, to 5 cm diameter, bark yellow-orange, thin. **Leaves:** ovate to round, 8-10 x 5-6 cm, base merging onto petiole [*decurrent*]; **veins visible below;** petiole to 3 cm long. **Inflorescences:** flower stalks to 4 cm long; bracteoles narrow-ovate, to 2.5 x 1.2 cm, ribbed, hairy outside; **corolla deep red, cylindrical, 4-5 cm long. Fruits:** ovoid, 1.2-1.5 cm diameter.

Uncommon; old-growth forest, disturbed areas. Guianas; widespread in the Amazon Basin.

1.2. Mendoncia bivalvis (L.f.) Merr.

Subwoody liana or vine; yellow-hairy [*hirsute, pilose*], **non-asperous. Stems:** branchlets round, with stiff, straight hairs; nodes swollen. **Leaves:** elliptic to ovate, 6-15 x 6-8 cm, base cuneate, apex often fine-pointed; papery, **veins invisible below;** petiole to 2 cm long. **Inflorescences:** flower stalks to 2.5 cm long; bracteoles broad, oblong-elliptic to ovate, 2-2.5 cm long, papery, veined; corolla **creamy-white, spreading, ca. 2.5 cm long. Fruits:** ovoid, 1.5-1.8 cm diameter.

Uncommon; old-growth forest, creek forest, disturbed areas. Guianas; widespread in the Amazon basin and E Andean slopes.

1.3. Mendoncia hoffmannseggiana Nees

Common name: makui pana (Tr) [= *Golden-handed tamarin ear (Saguinus midas)*]

Subwoody liana or vine; golden-hairy [*sericeous*], **asperous. Stems:** branchlets remotely 4-sided; nodes flattened. **Leaves:** narrowly to broadly elliptic, 5-12 x 4-8 cm, base cuneate, apex abruptly acuminate, with long fine point; sub-leathery to papery, lightly hairy above, golden-hairy below; veins visible below; petiole 1-2 cm long. **Inflorescences:** flower stalks to 6 cm long, densely golden-hairy [*pilose*]; bracteoles narrowly oblong, 3.5-4.5 x 0.8-0.9 cm, apex+/- curved; **corolla deep red to purple, cylindrical, 4-5 cm long. Fruits:** obovate, 1.5-2 cm diameter.

Uncommon; old-growth forest, submontane forest to 700 meters. Fr. Guiana, Suriname.

- Mendoncia hoffmannseggiana, flowering branches along trail in southern Suriname [BH]
- Mendoncia bivalvis with veined bracts/bracteoles, long hairs and white tubular flower [BH]
- Mendoncia aspera, older woody stem cross-section of unique "foraminate" type [BH]
- Mendoncia aspera, older woody stem with twisted furrows [BH]
- Mendoncia hoffmannseggiana plant with mature, single-seeded drupes [BH]

HERBARIUM BOSBEHEER SURINAME   Thunbergiac.
Mendoncia vs.aspera R.et P.
No: LH-183

4

5

ANNONACEAE

ANNONACEAE

Rarely lianas and climbing shrubs; mostly small- to medium-sized trees. Climbing via twining stems or clasping branchlets. Stipules are absent. Leaves alternate, simple, two-ranked [*distichous*], with entire margins and pinnate veins. Flowers with 3 sepals, 6 fleshy petals, and many stamens and ovaries [*carpels*]. Fruiting with many stalked fruitlets per flower or a single fused fruit.

General: Hairless to densely hairy, hairs simple, flattened (Guatteria) or erect (Annona); shoot, leaf and flower glands absent; sap clear when present.

Stems: Bark and twigs with spicy scent when cut; branchlets often 'zig-zag'; nodes simple, without ridge. **Older stem bark tough and fibrous, peeling in strips,** blackish-green with lenticel-lined furrows, **slash margin often forming a black ring;** wood hard, white to bright yellow; **stem x-s pattern often radial, with 4-8 rays.**

Leaves: Variable, elliptic to elliptic-oblong; leathery to papery; in lianas, the 1° and 2° veins are sunken above with **2° veins widely-spaced, looping at margins.**

Inflorescences: Terminal, axillary, opposite to leaf, or borne upon stem; **flower stalks unjointed (Annona) or jointed below middle (Guatteria);** flower stalks jointed (Annona) or jointed below middle (Guatteria) sepals 3; petals 6 in 2 whorls; with many stamens and ovaries [*carpels*] in tight spirals.

Fruits: Two distinct types: i) clusters of stalked fruitlets [*monocarps*] per flower, each containing 1-7 seeds (Guatteria); ii) a single multi-seeded fruit per flower [*syncarpous*] (Annona). Seeds with 'ruminate' endosperm - the seed coat grows inwards and a wavy pattern is visible in cross-section.

Ecology: Beetle-pollinated; seed dispersal via animal (monkeys, birds, rodents, fish).

Distribution: Flooded and non-flooded old-growth forest, especially in understory. Pantropical: 112 genera/ca. 2440 species; Guianas: 18-19 genera/126 species, lianas: 2 genera/2 species.

1. Annona L.

Rarely lianas and clasping treelets; mostly small- to medium-sized trees. Neotropics, rarely African tropics; Guianas: 16 species, one liana species known. Annona hypoglauca and A. sericea are small trees that occasionally climb or clamber.

1.1 Annona haematantha Miquel

Common name: ariminaimë (Tr); karampai (Ar)

Robust liana or climbing shrub; hairs erect and brown. **Stems:** branchlets and young shoots hairy; older stems round, 5-15 cm. diam., **slash with clear sap and bright yellow wood. Leaves:** elliptic to elliptic-obovate, 11-17 x 5-7 cm, base cuneate to rounded, apex short-acuminate; thin to papery, smooth-surfaced; **3° veins parallel and slanted to midvein [*oblique*]**; petiole long (5-10 cm). **Inflorescences:** 1-flowered, (sub) opposite to leaves; flower stalks unjointed; petals narrowly triangular, 2-2.5 cm long, pink to cream-colored with red tips, outer surface with rusty-brown hairs. **Fruits:** round to ovoid, 1.7-2.3 cm diam., maturing yellow; many seeded. **Use:** Wood and bark used as remedy against skin ulcers and coughs.

Rare; non-flooded and flooded old-growth forest. Guianas; also NE Brazilian Amazon.

2. Guatteria Ruiz & Pavon

Rarely lianas and climbing treelets; mostly small- to medium-sized trees. Neotropical; Guianas: ± 30 species, one climbing species known.

2.1. Guatteria scandens Ducke

Common name: apuku-titei (Sr); bosolijf (SD); cipó-uíra (Bra); ërëpurë (Tr); kasselerodang, kirikahu kirikawa (Ar); kiintongo (Au); krintongo (Sr); kofiballi (Ar); malakopesi (Bo); molokju (Way); murewa (Ca); ndulu-ndulu (Al/Bo); wanëkë (Tr); wime etni kamwi (Palik).

High-climbing liana, initially growing as a small tree; hairs flattened and brown. **Stems:** branchlets round, green, smooth and lightly hairy; older stems round, 3-5 cm diameter; **slash with black outline, inner bark with fiber network, wood whitish-yellow with dark yellow rays in x-s. Leaves:** elliptic to elliptic-oblong, 10-25 x 4-11 cm, base acute, apex acuminate, tip 0.5-2 cm long; leathery, **rough-surfaced above,** hairless, light green below; **3° veins net-veined [*reticulate*];** petiole 0.6-1.3 cm. **Inflorescences:** 1-few-flowered, borne on main stem or leafless branches; **flower stalks jointed below middle;** sepals 0.4-0.6 cm long; petals 6, fleshy, ovate, 1-4 x 0.7-1.6 cm, greenish-yellow, with dense brown hairs outside. **Fruits:** with 25-40 fruitlets per flower; fruitlets ellipsoid, 1.3-2.5 x 0.7-1.3 cm, stalks 0.5-1.2 cm long, ripening blackish-purple with dark red stalks; skin thin and tough; seeds 1(2) per fruitlet. **Use:** medicine/ritual - herbal baths to commune with forest spirits (Au); back pain, colds, heart problems (Tr).

Common; old-growth forest, white-sand forest, riparian forest. Guianas; also NE Amazonian Brazil.

ANNONACEAE

1. Annona haematantha climbing, with syncarpous fruit (Hoffman coll. no. 6370) [BH]
2. Guatteria scandens with many stalked fruitlets per flower [BH]
3. G. scandens, slash with inner bark fiber network and yellow wood [BH]
4. G. scandens, close-up of stalked fruitlets on stem [PT]

APOCYNACEAE

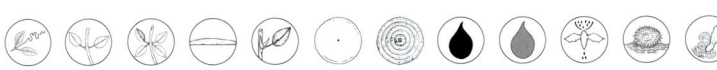

APOCYNACEAE

A diverse family including robust canopy lianas, sub-woody or herbaceous vines, and clambering shrubs; also erect herbs, shrubs, and canopy trees. Climbing via twining shoots, rarely with spurs at nodes or inflorescence branchlet hooks. Leaves are opposite or whorled, simple, and entire. Stipules are absent but replaced by stipule-like glands. Flowers are tiny to large, 5-parted, radial, fused with lobes often overlapping. Fruits are paired (single) dry follicles or fleshy globes. Following the APG classification, two formerly distinct families, the liana-rich Apocynaceae (s.s.) and vine-rich Asclepia-daceae (s.s.), are pooled here.

General: All cut parts with abundant, paint-like milky sap [*latex*], rarely with green, clear, watery-white or watery-orange sap. Surfaces hairless to densely hairy, rarely star-shaped (Forsteronia).

Stems: Branchlets round, flat or 4-sided in cross-section, often bearing lenticels; nodes with a conspicuous horizontal ridge or line and (often) **tiny stipule- or finger-like glands in leaf axils** [*colleters*]. Older stems 0.5 cm to > 20 cm diameter, +/- round; bark smooth to extremely knobby-corky; lenticels abundant or few; inner bark sometimes bright green; **stem x-s pattern simple or concentric** (Forsteronia, Odontadenia).

Leaves: Variable shape and size, often large; 1° and 2° veins raised below, 2° veins widely-spaced, either straight and parallel or curved with marginal loops, **parallel inter-secondary veins often present,** 3° veins usually net-veined [*re-ticulate*]; sometimes present: i) tiny, needle-like gland pairs at base of leaf above; ii) tiny hairy or hairless pockets [*domatia*] in 2° leaf axils below; iii) black dots on 3° veins.

Inflorescences: Axillary or terminal, single-to-many flowered; bracts present; flowers tiny to large (0.25-6 cm long), sepals or petals sometimes bearing linear colleter glands; calyx 5, free or 5-lobed; **corolla radial, tubular to bell-shaped,** often narrow and round in lower half, wider and spreading in upper half, lobes 5–either **overlapping like a pinwheel** and twisted in bud or straight; stamens 5, inserted on corolla, alternating with lobes, **anthers often closely encircling or pressed against the stigma** (Asclepiad species with complex, fused flower parts); 'corona' present or absent; nectar disc or lobes present; ovary superior, 2-parted [*2-carpellate*], with a complex style head and 2-lobed stigma.

Fruits: Two distinct fruit types: i) Paired or single, **dehiscent follicles,** +/- woody, narrow and pencil-like to fat, follicle pairs spread wide apart or parallel, straight or curved, sometimes joined at tip; seeds large or tiny, often with conspicuous hairtuft (coma), arils, or wings; ii) globose or ovoid, **indehiscent berries or drupes** with seeds immersed in pulp (Pacouria).

Ecology: Seed dispersal via wind or water.

Distribution: Global, mostly tropical: 415 genera/> 4500 species; Guianas: 11 genera/56 species of woody climbers (in the Asclepiadoideae subfamily, lianas: 2 genera/6 species; herbaceous vines: 12 genera/60 species).

Use: Due to showy flowers, the family includes some globally important ornamentals such as Allamanda and Plumeria. The thick white sap of many species contains poisonous, caustic, or medicinal substances. The latex is used in the Guianas as a sticky glue-trap to catch birds for the wildlife trade. The common name merki-titei or 'milk-vine' is applied to many Apocynaceae genera and is useful only for family identification.

Notes: The abundant, paint-like milky sap of Apocynaceae lianas and vines allows for rapid identification of the family. Clusiaceae is the only other family in the region with woody climbers that possess opposite, simple leaves and (sometimes) whitish latex. Clusiaceae is distinguished vegetatively from Apocynaceae by i) latex mostly orange or yellow; ii) tree-like hemi-epiphyte habit; iii) unmistakable leaves – leathery, obovate or obcordate, with entire margins and finely-parallel inter-secondary veins.

1. Odontadenia nitida habit, with large leaves and pinkish-orange flowers [BH]
2. Odontadenia sp. seed with hairtuft [STRI]
3. Cut stem of Apocynaceae species exuding white latex [BH]
4. Stipule-like colleter glands in Apocynaceae [FRD]
5. Forsteronia acouci leaf pockets [*domatia*] on leaf underside in vein axils [BH]
6. Odontadenia macrantha, paired fruits (follicles) [BH]
7. Odontadenia nitida, follicle with seeds [BH]
8. Forsteronia acouci, young woody stem [BH]
9. Allamanda cathartica, flower close-up [PT]
10. A. cathartica, cross-section of fruit with many flattened seeds [CB]

1. Allamanda L.

Shrubby lianas, sometime erect shrubs. Climbing via twining shoots.

General: All cut parts with abundant milky sap; bald or densely hairy.

Stems: Branchlets round or 4-sided; nodes with interpetiolar ridge or line, **stipule-like colleter glands usually present.** Older stems less than 3 cm diameter, outer bark smooth, brown, with lenticels; stem x-s pattern simple.

Leaves: Whorled, 3-4 per node, opposite to alternate towards apex; elliptic to oblanceolate, sub-leathery, 2° veins looping at margin, paired glands and domatia absent.

Inflorescences: Terminal or axillary, few-to-many flowered; **flowers large, showy, mostly yellow; 3-6.5 cm long,** radial [*actinomorphic*] or bilateral [*zygomorphic*], linear colleter glands usually absent; sepals free; corolla tube narrow below, funnel-shaped above, lobes twisted in bud, pointing counter-clockwise; stamens attached to lower tube, covered by hairs in throat, anthers unattached to style; ovary 1-chambered (syncarpous), style head 2-lobed.

Fruits: +/- globose capsules, covered with large prickles or smooth (only Allamanda setulosa), splitting along 2-valves; seeds numerous, flattened, concentric-winged.

Ecology: Dispersal via water or wind.

Distribution: NE South America: 14 species; Guianas: 4 species.

1.1. Allamanda blanchetii A.DC.

Liana; hairless to lightly hairy. **Stems:** branchlets round, with green, smooth bark. **Leaves:** 3 per node (rarely 4 or 5); obovate to oblong-lanceolate, 4.5-11 x 1.8-5.5 cm, base attenuate, apex acuminate; subleathery, shiny above, dull and ± hairy below; petiole 0.2-0.3 cm long. **Inflores-**cences: terminal, without scent; bracts 0.1-0.2 cm long; sepals 0.15-0.2 cm long, hairy outside; **corolla radial, yellow with violet lobes,** 2 cm wide, tube 5-6.5 cm long, lobes 3-4 cm long. **Fruits: capsule prickly,** ca. 5 x 3.5 cm, green turning brown, prickles to 1.3 cm long. **Uses:** Ornamental.

Uncommon. Cultivated as ornamental, possibly naturalized in open areas near coast. French Guiana only; native to Brazil.

1.2. Allamanda cathartica L.

Common Name: baruda balli (Ar); busi-kasterol (Ma); ingi brumtjie (Sa); kiraporan (Ca); pugassi (Dj); watramama-kasabatiki (Sr); wilkensbita (Sr)

Liana or scandent shrub; mostly hairless. **Stems:** branchlets round, with green to greenish-red, smooth bark. **Leaves:** 4 per node, rarely 3 or 5; obovate to oblanceolate, 8-16 x 2.5-6 cm, base obtuse to acute and extending onto petiole, apex long-acuminate; shiny above, dull below; petiole 0.3-0.7 cm long. **Inflorescences:** terminal; sepals ca. 1 cm long, hairy only outside; corolla radial, **yellow with orange stripes in throat,** 8-12 cm wide, tube 4.5-6.5 cm long, lobes 2-3 cm long. **Fruits: capsule prickly,** 5 x 3.5 cm, bright green, prickles to 1.5 cm long. **Use:** Widespread ornamental; medicine – milky sap as purgative and anti-bacterial. Sap causes rashes, itch, and blisters.

Common; along rivers; open areas with sunlight. Guianas; NE South America. Naturalized throughout the tropics.

11

1.3. Allamanda schottii Pohl

Common name: bush allamanda (GU).

Liana or shrub; hairless to lightly hairy. **Stems:** branchlets round, with green, smooth bark. **Leaves:** generally 4 per node, rarely 3-5; obovate to oblong-lanceolate, 6.5-13.5 x 1.6-3.5 cm, base acute and attenuate, apex shortly acuminate; blade thin and soft, shiny above, dull below; petiole 0.3-0.6 cm long. **Inflorescences:** axillary and/or terminal; sepals hairy, to 1.2 cm long, **with 2-8 colleter glands within;** corolla bilateral, **yellow with orange stripes in throat,** hairy inside, 1.5 cm diameter, tube 3.5-4.7 cm long, lobes < 1.5 cm long. **Fruits: capsule prickly,** 2.5-3 x 2-3 cm, prickles to 1.5 cm long. **Uses:** Ornamental.

Uncommon. Cultivated as ornamental, possibly naturalized in open areas near coast. French Guiana only; native to Brazil.

1.4. Allamanda setulosa Miq.

Liana or scandent shrub; hairy. **Stems: branchlets 4-sided, stipule-like colleter glands absent (present in other species). Leaves:** elliptic to oblanceolate, 8-9 x 2.5-3.5 cm, base obtuse to acute, apex acuminate to caudate, margins curled; **densely hairy below;** petiole 0.3-0.4 cm long. **Inflorescences:** sepals hairy, to 1.25 cm long, colleter glands absent; corolla solid yellow, hairy, tube 4.5-5.5 cm long, lobes 2-3 cm long.

Fruits: capsule without prickles, smooth and hairy, 7 x 5.5 cm, green.

Rare; savanna and riparian forest. Guyana (Rupununi savanna), Suriname (Suriname River).

2. Condylocarpon Desf.

Slender woody lianas. Climbing via twining shoots.

General: All cut parts with abundant milky sap; plants bald or densely hairy.

Stems: Branchlets round or flattened, slender, bark usually with lenticels, nodes with interpetiolar ridge or line, **stipule-like colleter glands absent.** Older stems round, reaching 2-3 cm diameter, outer bark smooth with scattered lenticels; stem x-s pattern simple.

Leaves: Elliptic-ovate, < 12 cm long; thin and soft to sub-leathery; 2° veins straight, marginal collecting vein present, parallel intersecondary veins conspicuous in one species; leaf-base glands and leaf pockets [*domatia*] absent.

Inflorescences: Terminal or axillary, flowers small and clustered on long flower stalks [*thyrses*]; linear colleter glands absent or inconspicuous; calyx 5-lobed; **corolla trumpet-shaped, small, < 0.4 cm long, white, yellow, or orange,** lobes pointing counter-clockwise; stamens attached mid-way on tube, anthers unattached to style; style head 2-lobed.

Fruits: Each fruit of 1-2, long and narrow 'follicles', +/- woody, hairless or hairs very long and bristly [*hirsute*], paired fruits unattached at tip; each follicle flat but with 1-5 swollen seed segments; seeds wingless, spindle-shaped.

Ecology: Seed dispersal via water.

Distribution: Lowland swamps or riparian forest, also non-flooded areas, one species adapted to coastal salt water. Neotropics, esp. N. Amazon and Guianan Shield, also Trinidad and Nicaragua: 7 species; Guianas: 3 species.

2.1. Condylocarpon amazonicum (Markgr.) Ducke

Large woody liana; conspicuously hairy. Stems: branchlets hairs long, golden- to reddish-brown. **Leaves:** elliptic to elliptic-oblong, 7-12 x 2.5-5 cm, base rounded, apex acuminate; sub-leathery, with dense, felt-like golden-brown to olive-green hairs below; petiole 0.5-1.0 cm. **Inflorescences:** with short grey to red-brown hairs; **corolla 0.35-0.4 cm long, cream-yellow,** scent said to be unattractive. **Fruits:** follicles 2, slender, 10-15 cm long, covered with long, straight, red-brown bottle-brush-like hairs; **with single seed in lower half (2 x 0.5 cm).**

Uncommon; old-growth lowland, riparian, and 2° forests, often on sandy soil. Guianas; also Venezuela, Amazonian Brazil, and Bolivia.

2.2. Condylocarpon intermedium Müll. Arg.

Common name: monkey bora (Gu).

Woody liana; mostly hairless. Stems: branchlets round to flattened, slender, bark brownish-grey, with lenticels. **Leaves:** ovate- to oblong-elliptic, 6.5-10 x 2.8-4.5 cm, base obtuse to rounded, apex acute to obtuse; thin to sub-leathery; petiole 0.5-1 cm long. **Inflorescences:** lightly hairy; corolla 0.35-0.4 cm long, cream-yellow. **Fruits:** follicles 2, ca. 10 cm long, green to yellowish-

APOCYNACEAE

brown, hairless; **with up to 5 single-seeded segments,** each 1.0-1.3 x 0.7-0.9 cm.

Uncommon; old-growth lowland and riparian forests, coastal wetlands, mangrove swamps. Guianas; also adjoining Brazil and Venezuela.

2.3. Condylocarpon myrtifolium (Miq.) Müll.Arg.

Woody liana; mostly hairless. Stems: branchlets round to flattened, reddish-brown, with lenticels; older stems round, twisted, with lengthwise furrows, +/- 2 cm diameter, outer bark silvery-grey, lenticels abundant. **Leaves:** oblong-elliptic, 6-10 x 2.0-3.5 cm, base obtuse to rounded, leaf tip long, to 1 cm; thin to sub-leathery, glossy; intersecondary veins conspicuous; petiole 0.2-0.4 cm long. **Inflorescences: corolla 0.30-0.35 cm long, white with yellow tube. Fruits: follicle 1,** ovoid, 4.0-5.0 x 1.5-2.0 cm, greenish-brown and hairless; number of seed segments unknown.

Uncommon; non-flooded and seasonally-flooded old-growth forest. Guyana, Suriname; also Amazonian Venezuela.

12

3. Forsteronia G.F.W.Mey.

Canopy lianas or scandent shrubs. Climbing via twining shoots; branching alternate or opposite.

General: All cut parts with abundant milky sap, stem slash rarely with clear **honey-like sap;** plants dimorphic, either bald or densely hairy, hairs simple or rarely stellate.

Stems: Branchlets round or flattened, lenticels present; **nodes with horizontal ridge and stipule-like colleter glands.** Older stems round, flattened or twisted, medium to large diameter; bark thick, sometimes with **conspicuous corky cones; stem x-s pattern simple or concentric.**

Leaves: Shape variable – elliptic to obovate; margins curled; thin to leathery; 2° veins curving upwards and looping at margin, 3° veins ± parallel, 90° to midrib; **paired leaf-base glands and domatia usually present.**

Inflorescences: Terminal and axillary, with **small white or yellowish flower clusters** on long flower stalks; flowers < 0.5 cm long; sepals each with one linear colleter gland attached within; corolla **lobes straight, not overlapping as in most of the family,** ring (annulus) present at throat, tube hairless to densely hairy; anthers tightly enclosing but not attached to the style/stigma.

Fruits: Each fruit of 2 long and slender, cylindrical follicles, slightly to widely separated, not attached at tip; seeds subround or flattened, each with a hairtuft.

Ecology: Seeds dispersed via wind.

Distribution: Wide habitat range including old-growth and secondary forest. Neotropical genus: 47 species; Guianas: 10 woody climbers, 4 described here.

3.1. Forsteronia acouci (Aubl.) A.DC.

Common Name: ëmapo petuku (Tr)

Shrubby liana; mostly hairless. **Stems:** branchlets round, bark reddish-brown, bearing lenticels; older stems round or flattened, bark with abundant lenticels and **large corky cones; slash often with clear, honey-colored sap;** stem x-s pattern **concentric. Leaves:** elliptic to narrow, 8-15 x 3-7 cm, base rounded (not sub-heart-shaped), apex abruptly acuminate; leathery to sub-leathery, dark green above, dull red-brown below; 3° veins ± parallel and at right angle to midrib; petiole 0.3-0.7 cm long; **leaf-base gland-pairs, hairless domatia present. Inflorescences:** many-flowered panicles, +/- 10 cm long, minutely hairy; corolla tiny, tube 0.1-0.25 cm long. **Fruits:** follicles widely separated, each 40-55 x 0.4-0.5 cm, connected at tip; seeds ovoid or flattened, 1.5 cm long, hairtuft +/- 4 cm long. **Notes:** This species is highly variable across the neotropics. F. acouci from the Guianas have thinner leaves and smaller, hairier flowers than Amazonian F. acouci.

Uncommon; riparian forest, occasionally in secondary growth and forest edges. Guianas; also Amazon basin, C America, S Mexico.

3.2 Forsteronia gracilis (Benth.) Müll.Arg.

Common Name: makwariballi (Ar)

Shrubby liana; mostly hairless. **Stems:** branchlets round, slender, bark brown with white lenticels; older stems round, **corky cones absent, stem x-s type simple. Leaves:** oblong-elliptic to ovate, 5-11 x 3-5 cm, **base +/- heart-shaped,** apex caudate-acuminate; thin; petiole slender, 0.4-0.7 cm long; **leaf-base gland-pairs, tufted or hairless domatia present. Inflorescences:** loose, delicate panicles, 10-15 cm long; corolla tiny, tube 0.05-0.1 cm long. **Fruits:** follicles moderately diverging, to 25 x 0.4 cm each, flattened between seeds, black, hairless, lengthwise lines present, base abruptly tapering, apex long-beaked; seeds 7 or 8, acuminate, to 2.5 cm long, hairtuft brown, silky, +/- 4 cm long.

Uncommon; non-flooded old-growth forest; Guyana, Suriname (Brownsberg; Tafelberg); also N. Brazil and Venezuela.

3.3. Forsteronia guyanensis Müll.Arg.

Common Name: ëmapo petuku (Tr); makwari-balli (Ar)

High climbing liana; non-fertile parts mostly hairless. **Stems:** branchlets round, with bark rings present above each node; older stems round, **corky cones absent,** inner bark pink, **x-s pattern simple. Leaves:** obovate-elliptic, 3-9 x 1.5-3.5 cm, base cuneate, apex abruptly acuminate to obtuse; leathery or sub-leathery, hairless; petiole short, 0.2-0.5 cm long; **leaf-base gland-pairs and hairless domatia present. Inflorescences:** loose, many-flowered panicles, 7-10 cm long, red-haired, surfaces often with tiny nipple-like bumps [*papillate*]; corolla tiny, round, tube 0.15-0.2 cm long. **Fruits:** follicles 2, slender and twisted, to 60 cm long, reddish-brown, thickened at 2 cm intervals; seeds numerous, round, acuminate, 1.5 cm long, hairtuft 2-3 cm long, brown-silky. **Notes:** The young, bright red roots of this species were observed growing in abundance across the forest floor in northwestern Guyana (van Andel 2001, Kariako Village).

Common; non-flooded old-growth forest, ridge forest; Guianas; also N Brazil, Peru.

3.4. Forsteronia obtusiloba Müll.Arg.

Shrubby liana; stellate hairs present in fruits only. **Stems:** branchlets stout, lenticel-covered, hairless; older stems round, outer bark with abundant lenticels, **corky cones absent, x-s pat-**

tern simple. **Leaves:** ovate to oblong-elliptic, 4.5-8 x 2-4 cm, base broadly obtuse or rounded (not sub-heart-shaped); apex acuminate; **thin to leathery, hairless;** petiole 0.4-0.6 cm long; **leaf-base gland-pairs inconspicuous, domatia absent. Inflorescences:** loose, many-flowered panicles, 7-10 cm long, with scent of jasmine; surfaces often with tiny nipple-like bumps (papillate); corolla tiny, round, tube less than 0.1 cm long. **Fruits:** follicles moderately diverging, each 40-55 x 0.4-0.5 cm, connected at tip, with white and brown stellate hairs; seeds ovoid or compressed, +/- 1.5 cm long, hairtuft +/- 4 cm long.

Uncommon; non-flooded old-growth forest, montane forest, forest edges; Guyana, Suriname; also adjoining Venezuela.

11. Allamand cathartica drawing [NBC]
12. Condylocarpon amazonicum with cluster of hairy, single-seeded follicles [BH]
13. Forsteronia acouci in flower [BB]
14. Forsteronia acouci, older stem with conical projections [PT]
15. Forsteronia sp. inflorescence [FRD]
16. Forsteronia acouci, branchlet with large lenticels, nodal ridge and triangular glands at petiole apex [BH]
17. Forsteronia sp, woody stem with white milk [BH]

4. Macropharynx Rusby

Shrubby lianas with large opposite leaves. Climbing via twining shoots. Neotropics: 3 species; Guianas: 1 species.

4.1. Macropharynx spectabilis (Stadelm.) Woodson

Shrubby liana. Cut parts with **watery-white sap that changes color (oxidizes) to orange;** plants hairy, hairs of two types mixed together - long, straight, white and short, matted, reddish-brown. **Stems:** Branchlets round or flattened; **nodes with interpetiolar ridge or line and stipule-like colleter glands** in leaf axils. Older stems round, robustly woody, **stem x-s type simple. Leaves:** Elliptic to +/- round, large, 10-30 x 5-14 cm, base rounded to heart-shaped, apex acuminate, margins often curled; shiny green and sparsely hairy above, **reddish-brown and rough-sandpapery below;** 2° veins very widely-spaced, curving and looping at margin, lined with hairs towards tip, 3° veins inconspicuous; petiole 3-5 cm long; leaf-base gland-pairs and domatia absent. **Inflorescences:** 2 flowering stalks per node, many- or few-flowered; flowers to 1.5 cm long; bracts linear and conspicuous, stipule-like colleter glands rare; calyx to 1 cm long, sepals linear; **corolla trumpet-shaped, 0.7-1.0 cm long, greenish-white,** lobes overlapping clockwise, hairless outside; stamens short, emerging at base of upper corolla tube, attached to stigma; ovary 2-locular, style thread-like, **stigma spindle-shaped. Fruits:** 2 stout, cylindrical follicles, 20-30 x 0.4-0.9 cm, with tiny reddish-brown hairs, tips free; seeds numerous, hair-tufted. **Ecology:** Seed dispersal via wind.

Common; non-flooded old-growth forest, secondary forest, and open areas. Guianas; widespread in the neotropics.

5. Mandevilla Lindl.

Slender woody and subwoody lianas, climbing shrubs, and herbaceous vines; also erect shrubs. Climbing via twining shoots and/or paired spurs (modified stipules) at nodes.

Common Name: merki-titei, wilkens-titei (Sr)

General: All cut parts with abundant milky sap; plants bald or densely hairy.

Stems: Branchlets round to oval, slender and winding; nodes **with interpetiolar line or ridge and stipule-like colleter glands.** Older stems round to flattened, 0.5-2 cm diameter, sometimes winged, often woody only at base; bark thin and often smooth, lenticels present; **stem x-s pattern simple.**

Leaves: Shape and size variable - often heart-shaped; 2° veins widely spaced, 3° veins net-veined [*reticulate*]; **leaf-base glands 2-4, scattered along midvein above,** domatia absent; petiole canal-shaped.

Inflorescences: Axillary or terminal, few- to many-flowered racemes; flowers variable in size; calyx 5-lobed, **bearing > 5 linear colleter glands on inner surface;** corolla funnel- or trumpet-shaped, white, pink, or yellow, lobes 5, overlapping clockwise; nectar disc present; stamens connected to style head; ovary 2-locular, style 5-sided.

Fruits: 2 cylindrical follicles, > 10 cm long and < 0.3-0.5 cm wide, often joined at tips; seeds numerous, round, hair-tufted.

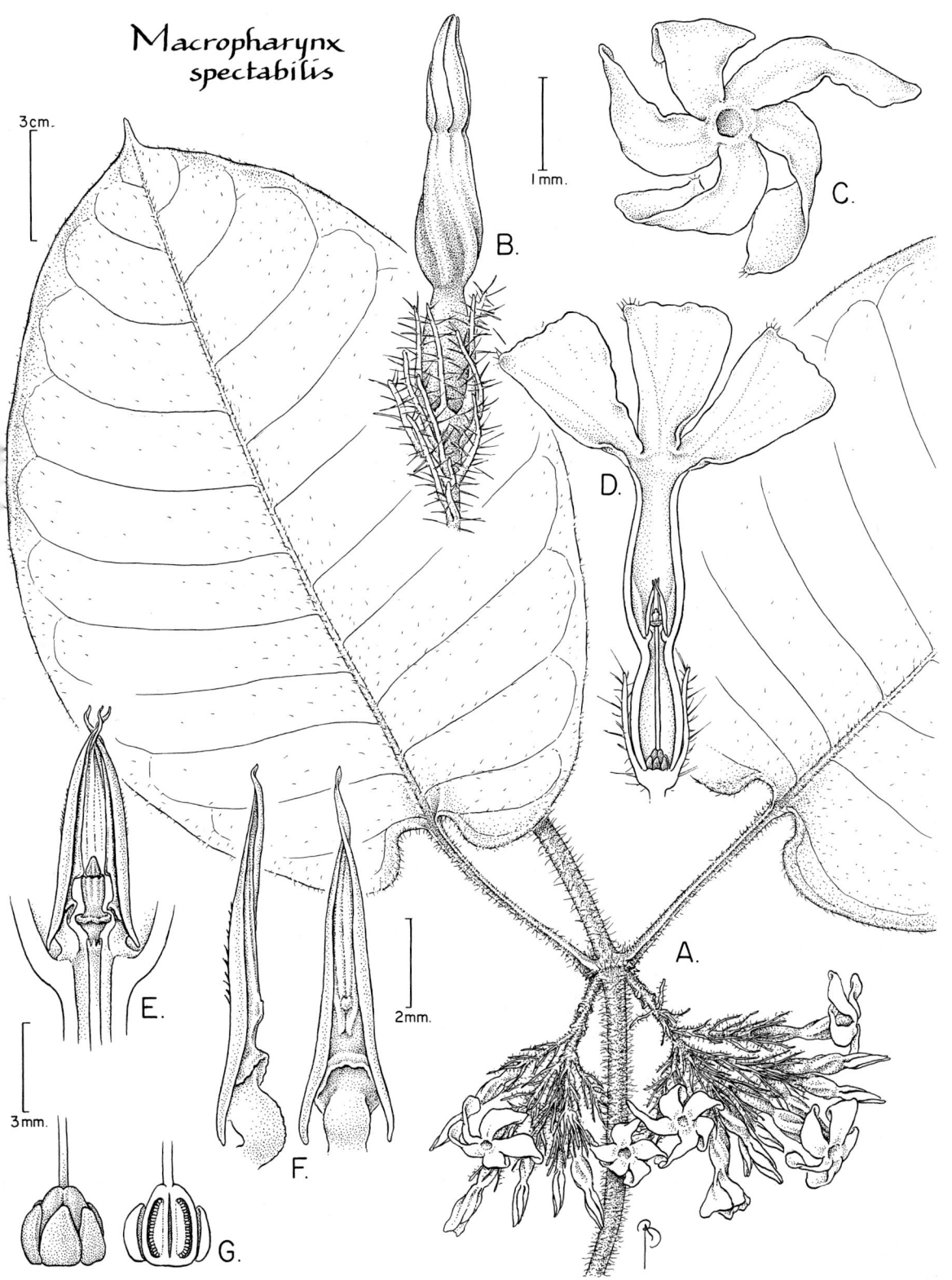

18. *Macropharynx spectabilis* drawing. a) habit; b) flower twisted in bud with spiny calyx; c) corolla with clockwise-turning lobes; d) flower cross-section, showing two chambers within, the stamens borne at the base of the upper chamber and forming cone around the stigma; e) closeup of 'gynoecium' and stamens; f) triangular anthers close-up; g) close-up of nectaries and immature seeds (ovules) within the ovary. [BA]

19

20

21

22

23

Ecology: Habitats variable, common in disturbed or open areas. Seed dispersal via wind.

Distribution: Widespread in the neotropics and subtropics, S America to Mexico, +/- 120 species; Guianas: 16 species; 4 slender woody lianas described here.

5.1. Mandevilla hirsuta (Rich.) K.Schum.

Subwoody liana or herbaceous vine; mostly densely hairy [*hirsute*]. **Stems:** round, bark brown; without paired spurs at nodes. **Leaves:** narrowly ovate, 8-10 x 3-4 cm, base cordate to auriculate, apex acuminate; thin and soft, minutely hairy above, minutely felt-like to densely hairy below; petiole 1-1.5 cm long. **Inflorescences:** axillary racemes to 10 cm long, densely hairy, with 6-15 flowers; bracts to 1 cm long; calyx lobes narrowly elliptic, hairy; **corolla funnel-shaped, 4-7 x 0.3-1.5 cm,** green at base, mostly yellow to cream above with red-orange to purple throat, hairless inside. **Fruits:** 2 follicles, straight, narrow, 12-13 x 0.3-0.5 cm, brown, with long, stiff hairs, curving at apex; seeds 0.8 x 0.15 cm, hairtuft 2 cm long.

Very common; open areas, riparian forest, savanna edges, and palm marshes. Guianas; also neotropics from Bolivia to Mexico.

5.2. Mandevilla rugellosa (Rich.) L.Allorge

Small liana; surfaces rough and lightly hairy. **Stems:** round, bark brown; without paired spurs at nodes. **Leaves:** elliptic, base ± cordate, apex acuminate; thin and soft, lightly hairy below. **Inflorescences:** axillary racemes to 30 cm long; flower stalks [*pedicels*] absent; bracts to 0.2 cm long, lightly hairy; calyx lobes to 0.2 cm long; **corolla trumpet-shaped, tube ca. 3 cm long, lobes 6-7 cm across,** mostly yellow to cream above with red lines in throat. **Fruits:** follicles straight, narrow, 12-20 x 0.3 cm, hairless; seeds 0.8 x 0.15 cm, hairtuft 2 cm long.

Common; open areas. Guianas; also Venezuela and Brazil.

5.3. Mandevilla surinamensis (Pulle) Woodson

Common Name: kuresah (Wa), tawimë (Tr)

Woody liana or shrub, hairless or lightly hairy. **Stems:** oval to flattened, lined or grooved lengthwise, bark reddish-black; **nodes with 4 recurved spurs, each to 0.5 cm long. Leaves:** elliptic, 4-8 x 2.5-4 cm, base obtuse to obscurely lobed, apex abruptly acuminate; leathery to subleathery, margin curled; petiole 0.5-1.0 cm. **Inflorescences:** axillary racemes to 7 cm long, with 3-7-flowers; calyx green, 0.4-0.5 cm long, **with 4 stipule-like colleter glands per sepal,** lobes toothed and persistent in fruit; **corolla funnel-shaped, 3.5-4 cm long,** tube green within, pink outside. **Fruits:** follicles curved, 16 x 0.3-0.4 cm, hairless, spindle-shaped; seeds 0.7-0.8 x 0.1 cm, hairtuft 2-2.5 cm long.

Rare; highland savanna, granite flats. Guianas; also Venezuela and Brazil.

5.4. Mandevilla symphitocarpa (G.Mey.) Woodson

Liana or subwoody vine; lightly to densely hairy. **Stems:** round, without paired spurs at nodes. **Leaves:** elliptic, 15-18 x 5-7 cm, base heart-shaped to auriculate, apex acuminate; thin and soft; petiole 1 cm long. **Inflorescences:** axillary racemes to 10 cm long, densely hairy, with 6-15 flowers; bracts to 1 cm long; calyx lobes narrow, hairy; **corolla funnel-shaped, 7-8 cm long,** green at base, mostly yellow to cream. **Fruits:** 2 follicles, straight, narrow, 30-35 x 0.3 cm; seeds 1 cm long, hairtuft 2.5-3 cm long.

Uncommon; open areas, forest edges. Guianas; also Venezuela, Trinidad, Brazil.

6. Marsdenia R.Br.

Small lianas to sub-woody and herbaceous vines; occasional erect shrubs. Climbing via twining shoots.

Common Name: merkitite (Sr)

General: All cut parts with abundant milky sap; plants either +/- hairless or densely hairy.

Shoots: Branchlets round, soft, lenticels absent or few; **nodes w/ interpetiolar line or ridge and stipule-like colleter glands.** Older stems round, small to medium diameter; **outer bark often softly woody or corky; sub-bark layer green, stem x-s pattern simple.**

Leaves: elliptic to ovate, wider than most Suriname asclepiads, occasionally cordate; soft and thin to leathery or fleshy; 2° veins widely-spaced and looping, 3° veins net-like; **petioles conspicuously long; two or more leaf-base gland-pairs present,** domatia absent.

Inflorescences: axillary, of diverse forms – umbel-like, raceme-like, dichasia-like, heads, few- to many-flowered; **sepals with 2-4 linear colleter glands within;** corolla rotate to tubular or trumpet-shaped, small (< 1 cm long in Suriname lianas), often hairy within and twisted in bud, style with cone-shaped head [Asclepiad terminology: *corona lobes* straight and fleshy; *pollinia* erect].

Fruits: **single elliptic-ovoid follicle, large, 10-22 x 5 cm,** smooth, with or without hairs; seeds numerous, often irregularly angled, margins winged, entire.

Ecology: Wind- or water-dispersed.

Distribution: Pantropical, +/- 120 species in the neotropics; Guianas: 5 species, 3 woodier climbers described here.

Notes: One of the few well-defined neotropical asclepiad genera (Morillo, 1978).

6.1. Marsdenia altissima (Jacq.) Dugand

Liana or vine; hairy. **Stems:** branchlets with velvety hairs; older stems round, soft, observed to 4 cm diameter, bark strongly wrinkled, ribbed, with green sub-bark layer. **Leaves:** widely ovate, ovate-deltoid or orbicular, largest leaves 10 cm x 6 cm, base cordate, apex abruptly acuminate or obtuse; soft and thin, **+/- hairy (tomentose) on both sides;** 2° veins widely-spaced, looping well before margin, 3° veins conspicuous and net-veined; petiole 1.5-3 cm long, +/- hairy; **leaf-base glands 7-12. Inflorescences:** densely (10-30) flowered, umbel-like cymes on 2-8 cm long infl. stalks, flowers small; calyx hairy; **corolla bowl-shaped, deep purple or reddish-brown,** tube constricted at throat, 0.45-0.5 cm long, hairy outside (tomentose), hairless inside. **Fruits:** single ovoid-round follicle, 16-22 x 3.5-6 cm, short and densely hairy (pilose), becoming hairless; pericarp hard; seeds oval, 1.6 cm long, hairtuft 3-4 cm long.

Common; old-growth and seasonal non-flooded forest, secondary forest. Guianas; widespread in tropical and sub-tropical South America.

6.2. Marsdenia macrophylla (Humb. & Bonpl. ex Schult.) E.Fourn.

Liana or vine. Stems: branchlets not densely hairy; older stems round, medium diameter, soft, +/- hairless, occasionally corky, bark with lengthwise lines or grooves and lenticels. **Leaves:** widely ovate, elliptic, or suborbicular, **largest leaves 18 x 13 cm (larger than M. altissima), base rounded,** apex acuminate; fleshy to subleathery, young leaves and veins moderately hairy; 2° veins widely spaced, almost parallel from midvein, prominent below, 3° inconspicuous above; petiole 0.5-2 cm long; leaf-base glands 6-12. **Inflorescences:** densely (10-30) flowered, umbel-like cymes on 4-12 cm long infl. stalks, flowers small; calyx

hairless; **corolla campanulate, whitish outside, dark red (purple) within,** 0.3-0.4 cm long, hairless throughout. **Fruits:** spindle-shaped follicles, 10-23 x 3-5 cm, hairless; pericarp thin and hard; seeds ovate, 1-1.3 cm long, hairtuft 4-7 cm long.

Very common; old-growth lowland forest, riverine forest, woodlands, bush islands, secondary forests. Guianas; adjoining Brazil and Venezuela, also from N. Argentina to Mexico.

6.3. Marsdenia rubrofusca E.Fourn

Liana or vine. Stems: older stems round, small to medium diameter; branchlets not densely hairy. **Leaves:** ovate-elliptic to lanceolate-elliptic, largest leaves 14 x 7.5 cm, base rounded or obtuse, apex apiculate; **soft and thin, hairless;** 2°

veins widely spaced, prominent below, initially straight and parallel to one another, 3° often perpendicular to 2°; petiole hairless, 0.5-4 cm long; leaf-base glands 4-5. **Inflorescences:** densely (4-30) flowered, umbel-like cymes on hairless, 1-5 cm long infl. stalks, flowers small; calyx hairless; **corolla blackish-purple, round, petals oblong** and eventually reflexed, tube 0.5-0.8 cm long, hairless outside and yellow-haired inside. **Fruits:** spindle-shaped follicles, usually swollen on one side, 12-16 x 5-6 cm, apiculate, thin, hairless; seeds irregularly ovate, very large, 3-5 cm long x 2.2-2.5 cm wide, hairtuft absent.

Common in riparian vegetation, growing in submerged areas and developing multiple adventitious roots. Guianas; also Venezuela and Amazonian Brazil, Colombia, and Peru.

7. Matelea Aubl.

Occasional lianas, mostly herbaceous or sub-woody vines; also erect or prostrate herbs. Climbing via twining shoots.

General: All cut parts with abundant milky sap; when present, hairs of two types mixed together – long, straight and short, matted.

Stems: Branchlets round, lenticels few, lightly to densely hairy, **nodes with interpetiolar line or ridge and without stipule-like colleter glands.** Older stems round, small- to medium-diameter, lenticels few; **bark in true liana species sometimes corky and fissured, stem x-s pattern simple.**

Leaves: Variable, shape narrow to oval; soft and thin, usually hairless; 2° veins widely spaced, curving upwards and looping, 3° veins net-like; leaf-base glands 2-5; domatia absent.

Inflorescences: Interpetiolar, 1 cyme per node; flowers 1-3 cm diameter; **calyx small, with 1-2 (-4) linear colleter glands per sepal;** corolla long-tubular, short-tubular with spreading lobes [*rotate*], or +/- bell-shaped [*campanulate*], yellowish-green, whitish-green, or yellow, hairless, petals thin, lobes net-veined; stigma head 5-sided. Asclepiadoideae terminology: *corona lobes* 'annular'; *pollinia* horizontally-oriented.

Fruits: Single spindle-shaped follicle, 5-8.5 x 1.5-2 cm, surface either smooth or bumpy [*tuberculate*], ribbed or sometimes winged; seeds numerous, hairtuft present or absent.

Ecology: Seeds wind- or water-dispersed.

Distribution: Neotropics: ± 300 species; Guianas: 25 species, 2 of the woodier species introduced here.

7.1. Matelea denticulata (Vahl) Fontella & E.A.Schwarz

Common Name: koni-koni boontje (SR)

Liana or vine. Stems: young branchlets densely hairy [*villose*]. **Leaves:** ovate, largest leaves 9 x 6 cm, base heart-shaped, apex acute; petiole +/- 1.5 cm long, moderately-hairy [*pilose*]. **Inflorescences:** 3 cm long with 2-5 cm long infl. stalk, flowers 2-5; corolla rotate, greenish-yellow, +/- 3 cm in diameter. **Fruits:** follicle evenly spindle-shaped, +/- 5 cm x 1.5-2 cm, green, hairless, smooth or tuberculate.

Common; old-growth lowland forest, secondary forest. Guianas; also adjacent Brazil and Venezuela to Mexico.

7.2. Matelea palustris Aubl.

Shrub, occasionally clambering. **Stems:** young branchlets +/- hairy [*puberulent*]. **Leaves:** oblong-lanceolate, 6-17.5 x 1.5-5.5 cm, base obtuse to cuneate, apex long-acuminate; petiole 1-2.3 cm long, ribbed, slightly hairy. **Inflorescences:** 4 cm long with 1 cm long infl. stalk, flowers 2-15; corolla rotate, greenish-white, +/- 1.2 cm in diameter. **Fruits:** follicle unevenly spindle-shaped, 5.5-8.5 x 1.5-2 cm, green, hairless, smooth or tuberculate.

Common; old-growth lowland forest, secondary forest. Guianas only.

7.3. Matelea surinamensis Morillo

Liana or vine. Stems: young branchlets densely hairy [*villose*]. **Leaves:** ovate, 5-10.5 x 2.5-5.5 cm, base cordate, apex acute or acuminate; petiole 2-3 cm long, densely hairy. **Inflorescences:** to 2 cm long with 0.5 cm infl. stalk, flowers 1-5; corolla +/- bell-shaped, yellow, +/- 2.5 cm diameter. **Fruits:** Fruit unknown.

Rare; old-growth lowland forest. Suriname only (Suriname River).

8. Mesechites Müll.Arg.

Slender lianas or vines. Climbing via twining shoots. Widespread in the New World tropics: 14 species total, Guianas: 1 species.

8.1. Mesechites trifidus (Jacq.) Müll.Arg.

Slender, woody liana; mostly hairless, all cut parts with abundant milky sap. **Stems:** branchlets slender, round, winding, green to reddish-brown; nodes with interpetiolar line or ridge and **many stipule-like colleter glands** in leaf axils; older stems slender, < 1 cm diameter, sub-woody, stem x-s pattern simple. **Leaves:** broadly oblong-lanceolate, 5-12 x 2-8 cm, base obtuse, apex acuminate; soft and thin; 2° veins prominent below, curving upwards, 3° veins net-like; petiole thick, +/- 1 cm long; **leaf-base glands 2-5;** domatia absent. **Inflorescences:** Axillary clusters (cymes); **flowers narrow and pointed,** 3-3.5 cm long; calyx with numerous linear colleter glands at base, lobes of different lengths, ciliate; **corolla tube with lower portion light pink to purplish,** +/- 2 cm long, and **upper portion green,** +/- 1.2 cm long; **lobes light green to yellow-green,** delicate, recurved, pointing clockwise, ca. 0.7 cm long; stamens narrow, hairy; style slender ca. 1.5 cm long. **Fruits:** follicles 2, thin, ca. 30 x 0.5 cm, connected at tips when young; seeds ca. 0.7 cm long; with +/- 2 cm long brownish hair-tuft. **Ecology:** Seed dispersal via wind.

Common in open areas, riparian zones, secondary forest. Guianas; also Brazil to Honduras.

29

30

APOCYNACEAE

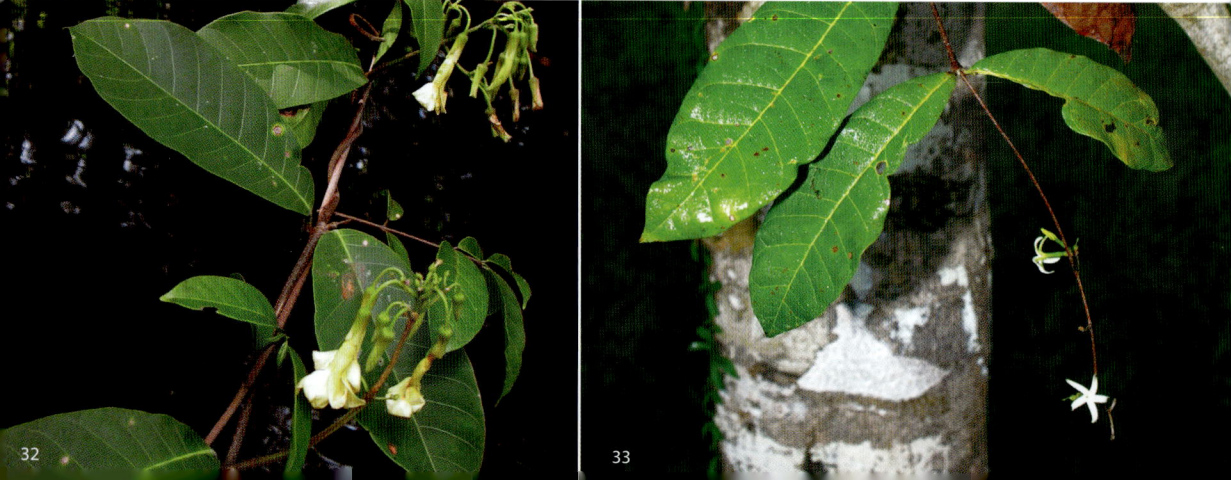

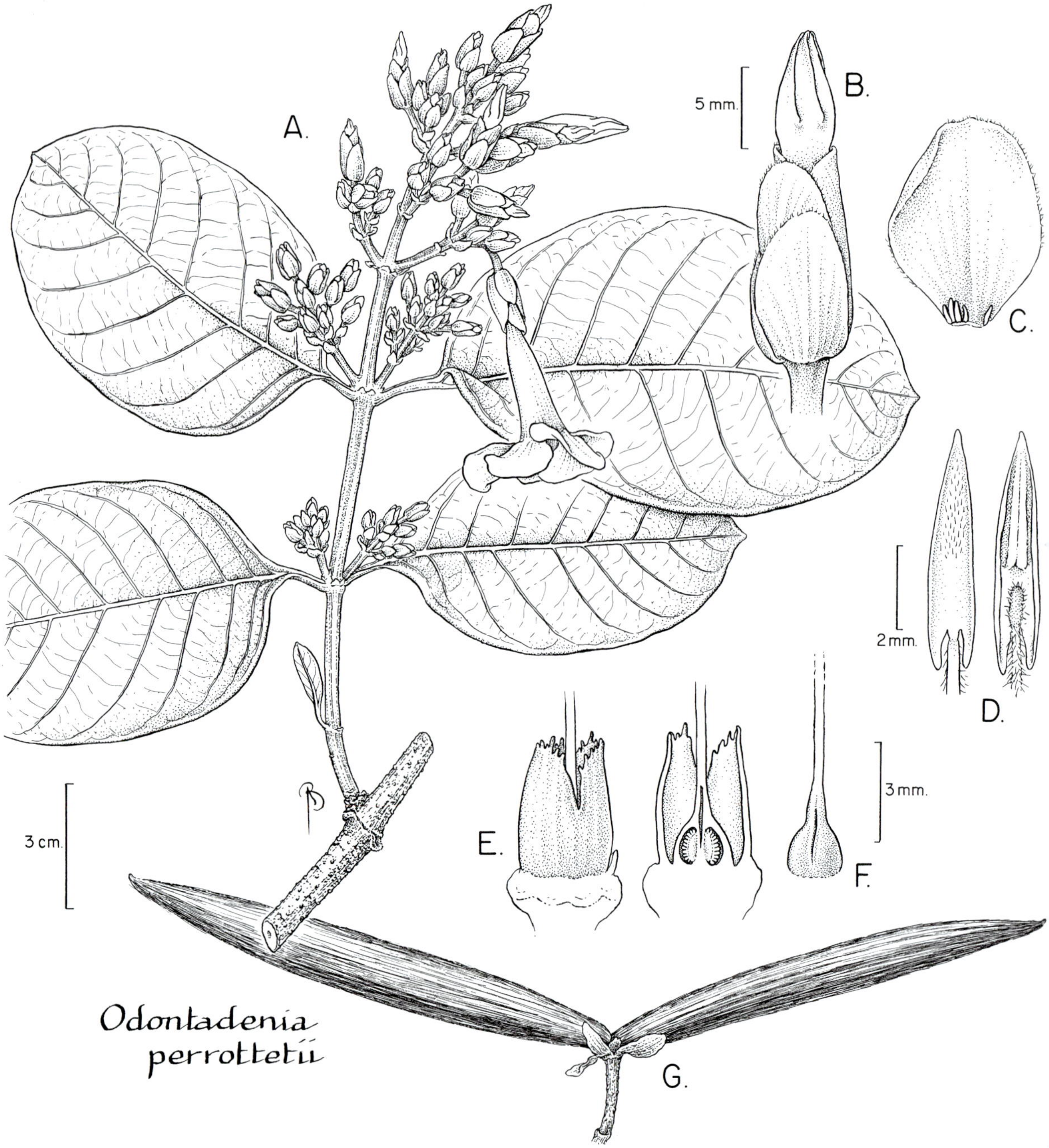
A.
B.
5 mm.
C.
D.
2 mm.
E.
F.
3 mm.
3 cm.
G.
Odontadenia perrottetii

31

32

33

9. Odontadenia Benth.

Robust canopy lianas, rarely herbs. Climbing via twining shoots, without tendril-like inflorescences or inflorescence 'hooks'.

General: All cut parts with abundant milky sap; **mostly hairless,** fertile parts occasionally hairy.

Stems: Branchlets round to oval, sometimes with a lengthwise groove; lenticels abundant; **nodes with interpetiolar ridge or line, stipule-like colleter glands present or (usually) absent.** Older stems round to flattened, smooth, bumpy [*tuberculate*] and/or with dense lenticels, nodes usually visibly swollen, **stem x-s pattern always concentric.**

Leaves: Robust, ovate to obovate, highly variable in size; leathery to thin and soft; **base usually cuneate, obtuse, rounded, or truncate, not heart-shaped** [*cordate*]; 2° venation well-spaced and mostly ± straight, **3° venation typically parallel and slanted towards midvein;** petiole ridged or canal-shaped; **leaf-base glands and domatia absent,** occasionally with black dots on 3° veins.

Inflorescences: Terminal or axillary, cymose, total inflorescence length less than that of nearest lower leaf; bracts and bracteoles scale-like, small; **sepals large,** usually unequal in length and overlapping [*imbricate*], **each with 1-5 tiny linear 'colleter' glands within; corolla funnel-shaped to trumpet-shaped,** cream, yellow, orange, reddish-yellow, or pink, **lobes overlapping clockwise,** tube mouth without a ring; stamens 5, shorter than tube, inserted on the corolla at variable heights, anthers attached to style head; nectar disc 5-lobed; ovary w/ 2 free carpels, style head shaped like a top [*turbinate*] and bi-lobed.

Fruits: Follicles 1 or 2, slender to thick and ovoid, hairless to densely hairy; seeds numerous, long and cylindrical, with a silky hairtuft.

Ecology: Some species hummingbird-pollinated. Seed dispersal via wind or water. Commonly producing showy floral displays along rivers.

Distribution: Neotropics and subtropics, from Brazil to Guatemala and the West Indies: 20 species; Guianas: 8 species; 6 described here.

9.1. Odontadenia geminata (Roem. & Schult.) Müll.Arg.

Shrubby climber or large liana. Stems: branchlets round, bark smooth, **colleter glands present. Leaves: elliptic-lanceolate,** 14-15 x 4-7 cm, base obtuse to truncate, apex acute to shortly acuminate; greyish-velvety below; petiole 0.5-0.7 cm long. **Inflorescences:** axillary, 2-8 flowered; infl. stalks 6-8 cm long; flower stalks 1.5-2.2 cm long; **sepals equal,** ovate, 0.4-0.6 cm long; **corolla trumpet-shaped,** yellow to pale cream; stamens inserted midway on tube. **Fruits: follicle 1,** cylindrical, 8-10 cm long, straight or curved; **smooth, hairless;** seed 2 cm long with 3.5 cm long hairtuft.

Uncommon; savanna, old-growth non-flooded forest, riverbanks, forest edges. Guianas; Amazonia.

9.2. Odontadenia macrantha (Roem. & Schult.) Markgr.

Common Name: ëpukuimë (Tr); kakhero (Ar); kiraporan (Ca); komo (Ca); taratité (Sr); topukeng (Ca)

Shrubby climber or liana. Stems: colleter glands absent; older stems round to oval, at least 2.5 cm diameter, with visible rings or nodes at ± 24 cm intervals; bark surface grey, bumpy, soft and fibrous; inner bark white. **Leaves: elliptic to broadly elliptic** (obovate), 12-22 x 4-10 cm, base acute to obtuse, apex acuminate; petiole 1.1-3 cm long. **Inflorescences:** axillary, few-to-many flowered; infl. stalk to 15 cm long; flower

APOCYNACEAE

stalks 2-2.5 cm long; **sepals equal to subequal,** ovate, 0.3-0.9 cm long, hairless; corolla funnel- to trumpet-shaped, orangish-yellow to reddish- yellow, 2.5-3.5 cm long, 1.8-2 cm diameter, lobes 2-3.5 cm long; stamens inserted at base. **Fruits: follicles 2 (1);** each narrow-ellipsoid, 15- 30 x 4-5 cm, slightly curved, apex acuminate; **smooth, hairless, finely-ribbed;** seed 4 cm long, hairtuft 4-6 cm long. Seeds water-dispersed.

Common; old-growth riparian and non-flooded forest. Guianas; also widespread in neotropics. Saramaccan Maroons use the latex as a glue to catch birds. The latex is cooked until it becomes black and very sticky (taratité = tar-vine).

9.3. Odontadenia nitida (Vahl) Müll.Arg.

Common Name: topukeng (Ca)

Shrubby climber or large liana. Stems: branch- lets round with smooth bark, colleter glands ab- sent. **Leaves: narrowly elliptic to obovate,** 6-17 x 2.2-8 cm, base rounded to obscurely cordate, apex obtuse to acuminate; subleathery, greyish- velvet below; petiole 0.8-1.5 cm long. **Inflores- cences:** axillary, few- to many-flowered; infl. stalk 3-8 cm long; flower stalk 0.7-2.6 cm long; **sepals ± equal,** ovate, 0.4-0.5 cm long, hairless, colleter glands present; **corolla funnel-shaped,** tube pink at base, spirally twisted, upper part cream- yellow and striped reddish-orange in throat, 2.3- 3.5 cm long, lobes ca. 1 cm long, pale yellow, stamens inserted midway in tube. **Fruits: folli- cle 1,** 19-22 x 0.8 cm, slightly curved; **smooth, hairless;** seed 2 cm long, hairtuft 3 cm long.

Common; old-growth riparian and non-flooded forest, forest edges, clay or sand savanna. Guianas; also W. Amazon, SE Brazil, Venezuela, Trinidad.

9.4. Odontadenia perrottetii (A.DC.) Woodson

Shrubby climber or large liana. Stems: colleter glands absent; older stem bark reddish-brown,

lenticels abundant. **Leaves: elliptic to obovate,** 5-16 x 1.6-6 cm, base rounded to broadly cune- ate; apex abruptly acuminate; thick and leathery; **3° veins with black dots;** petiole 0.5-1.6 cm long. **Inflorescences:** terminal and many-flow- ered or axillary and few-flowered; infl stalk 0.2- 2.2 cm long; flower stalks slender, 2-3 cm long; **sepals unequal,** 0.6-1.5 cm long (as long as lower corolla tube), ovate, hairless, linear colleter glands present within; **corolla funnel-shaped,** yellow, 5 cm long, lower portion narrow and short, lobes 1-1.5 cm; stamens inserted on upper corolla tube. **Fruits: follicles 2(1),** parallel; each 11-13 x 1.2-1.5 cm; **densely hairy;** seed 1.3 cm long, hairtuft 3 cm long.

Common; old-growth riparian and non-flooded forest, forest edge, clay or sand savanna. Guianas; also NE Brazil & Bolivia.

9.5. Odontadenia puncticulosa (A.Rich.) Pulle

Shrubby climber or large liana; similar to O. geminata. **Stems:** branchlets round, woody, with large lenticels, colleter glands absent; **older stem bark often extremely corky or bumpy. Leaves: elliptic to ovate,** 6.5-19 x 3-9 cm, base broadly acute to rounded, apex broadly acute; subleath- ery to leathery, **with black dots below on 3° veins;** petiole 0.5-2.1 cm long. **Inflorescences:** terminal or axillary, few- to many-flowered, lax, ± hairy [*puberulent*]; infl. stalk 1-7 cm long; flower stalk 0.7-3 cm long; **sepals unequal, ovate,** 0.3-0.6 cm long; **corolla funnel-shaped,** yellow or cream-yellow, 3-4 cm long, lower por- tion narrow and short, lobes 2-2.5 cm long; sta- mens inserted on upper corolla tube. **Fruits: fol- licles 2,** spreading; each cylindrical, 22 x 1.5 cm long, ribbed, apex slanted; **finely and densely hairy** [*puberulent/velutinous*]; seed 1.5 cm long, hairtuft 3-5 cm long.

Common; savanna, savanna woodland, and sec- ondary forest. Guianas; also Peru, Colombia, Brazil to Panama.

9.6. Odontadenia verrucosa (Roem. & Schult.) K.Schum. ex Markgr.

Common Name: ëepukuimë (Tr)

Shrubby climber or large liana. Stems: branchlets oval to round; colleter glands absent. **Leaves: elliptic to obovate,** 9-20 x 3-8 cm, base obtuse to attenuate, apex acuminate; papery to subleathery, occasionally with black dots on 3° veins below; petiole 0.7-2.4(3.1) cm long. **Inflorescences:** terminal to axillary, few- to many-flowered; infl stalk 1-6 cm long; flower stalks 0.5-2 cm long; **sepals unequal, ovate,** 0.6-0.7 cm long (in Suriname); **corolla funnel-shaped,** usually yellow or cream-yellow, hairless on outer surface (in Suriname); stamens inserted in lower part of corolla tube. **Fruits: follicles 2,** spreading, 19-30 cm long; **finely and densely hairy** [*puberulent*]; seed 2 cm long, hairtuft 5 cm long.

Common; savanna, savanna woodland, and secondary forest. Guianas; also ranging from Brazil to C America.

10. Pacouria Aubl. [Syn: Landolphia]

Woody canopy lianas. Climbing via twining shoots and unique tendril-like inflorescence branch 'hooks'. South American tropics: 2 species; Guianas: 1 species.

10.1. Pacouria guianensis Aubl.

Woody canopy liana. All cut parts with abundant milky sap; mostly hairless. **Stems:** branchlets round, hairs soft, velvety when young, bearing lenticels; nodes with interpetiolar ridge or line and **numerous stipule-like colleter glands;** older stems oval, furrowed, outer bark corky, lenticels dense; stem x-s pattern simple. **Leaves:** round to elliptic, 17-22 x 8-9 cm, base cuneate to heart-shaped, apex acuminate; leathery, minutely hairy below, **leaf-base glands and domatia absent;** 2° veins straight, in 9-14 pairs; 3° veins net-veined; petiole robust, 0.5-1 cm long, twisted. **Inflorescences:** terminal or axillary panicles with tendril-like branchlets, to 60 cm long with 7 cm long infl. stalk, 10-20-flowered; **flowers very small, 1-several;** calyx persistent, hairy, colleter glands absent; **corolla trumpet-shaped, greenish-white to yellowish-orange, tube 1-1.5 cm long, lobes straight, linear-oblong, ca. 1 cm long,** recurved with age, twisted in bud; stamens attached below middle of the corolla tube, anthers free of stigma head; ovary with thickened style head. **Fruits: large, globose 'berry',** 8-15 **cm diameter, blue-green with white-waxy layer,** maturing yellow-orange, smooth; pulp yellowish, sweet-tasting; seeds +/- 12, ellipsoid, 2.5 x 1.5-1.8 cm. **Ecology:** Seed dispersal via animal (through gut of **monkeys**).

Uncommon; lowland old-growth forest, especially in riparian zones. Guianas; also Brazil (Pará).

11. Prestonia R.Br.

Slender lianas or herbaceous climbers with woody bases. Climbing via twining shoots and opposite branchlets.

General: Cut parts with **milky white sap or clear sap;** hairless to densely hairy, hairs sometimes stinging.

Stems: Branchlets round, with lines or ridges running lengthwise; nodes with interpetiolar line or ridge and **numerous stipule-like colleter glands in leaf axils.** Older stems slender, to 1 cm diameter, round; outer bark sometimes corky and bearing lenticels; **stem x-s pattern simple.**

Leaves: Often large, oblong-elliptic to broadly ovate; thin to leathery, hairless to densely hairy, **leaf-base glands and domatia absent;** 3° veins net-like or parallel; petiole usually grooved.

Inflorescences: ± axillary, flowers in corymbs, racemes, or umbels, infl. rachis shorter than nearest lower leaf; calyx often as large as corolla; **corolla trumpet-shaped, medium-small (tube < 2 cm long),** light yellow or yellowish-green, with a **conspicuous ring in the throat,** lobes overlapping clockwise, corona often present; **calyx and corolla bearing linear, basal colleter glands within;** stamens either shorter or longer than tube, often attached to style head; ovary 2-chambered, stigma spindle-shaped or +/- head-like, sometimes longer than corolla tube.

Fruits: Two cylindrical follicles, often attached at tip; seeds numerous, oblong, hair-tufted.

Ecology: Seeds dispersed via wind.

Distribution: Widespread in the neotropics, including Mexico and the West Indies: 60 species, Guianas: 9 species; 4 of the woodier species treated here, all of them locally uncommon.

11.1. Prestonia annularis (L.f.) G.Don.

Small liana or vine; mostly hairless, sap clear.

Leaves: broadly ovate to oblong-elliptic, 10-12(15) x 6-8 cm, base obtuse to rounded, apex acute; leathery, hairless-shiny above; petiole 0.7-2 cm long. **Inflorescences:** racemes of 15-20 flowers, inflorescence stalks 2-4 cm long; calyx greenish-purple, hairless, persistent; corolla trumpet-shaped, **tube 1.4-1.5 x 0.3 cm,** yellow-brown, throat ring white with 5 yellow spots, corona bright yellow, corolla lobes pink to pink-yellow, 1 cm long, **Fruits:** 2 follicles, curved, 35-40 x 0.5-0.7 cm, diverging, tips occasionally connected; seeds +/- 1.8 x 0.2 cm, apex long-beaked, hairtuft 2-4 cm long, brownish-white.

Uncommon; open, low secondary forest. Guianas; also W Amazon to Andes, Colombia, N Brazil, Trinidad.

11.2. Prestonia megagros (Vell.) Woodson

Small liana or vine; densely reddish-brown-hairy; sap white. Stems: branchlets hairy [*hirtellous*]. **Leaves:** broadly oval to obovate, 11-20 x 5-12 cm; leathery, bullate; hairy above and below [*puberulous*]; petiole 1-1.4 cm long. **Inflorescences:** clusters of 20-40 flowers (corymbs), densely hairy [*hirtellous*]; calyx lobes persistent, 1-1.6 cm long, purplish-green; corolla trumpet-shaped, purplish-yellow, hairless to lightly hairy, **tube 1.3-1.4 cm long,** throat ring dark yellow, lobes 1-1.2 cm long, recurved. **Fruits:** unknown.

Uncommon; riparian forest. Guianas; also Brazil and Venezuela.

11.3. Prestonia quinquangularis (Jacq.) Spreng

Small liana or vine; lightly hairy [*puberulous*], **sap white. Leaves:** +/- elliptic, 6-16 x 2-8 cm, base obtuse, apex acuminate; thin; 2° veins

often purplish-red; petiole ± 2 cm long. **Inflorescences:** racemes of 10-20 flowers, inflorescence stalks 7-10 cm long; calyx lobes persistent, triangular, green; corolla trumpet-shaped, brown or greenish-yellow, **tube 1.5-2 cm long,** throat ring white, lobes yellow. **Fruits:** 2 follicles, straight, long, slender, 35-42 cm x 0.3-0.4 cm, tips connected; seeds 1 x 0.1 cm, hairtuft white, +/- 4.5 cm long.

Uncommon; riparian and semi-deciduous forest. Guyana, Suriname; widespread in South America.

11.4. Prestonia surinamensis Müll.Arg.

Common Name: sekemaran (Ca)

Small liana or vine, robust for genus; **densely golden-brown hairy** [*villous, hispid*]; **sap clear. Stems: branchlets with stinging hairs;** older stems to 1 cm diameter, with vertical white lenticels; slash green; living bark 0.2 cm thick. **Leaves:** broadly ovate to ovate-elliptic, 10-24 x 6-17 cm, base obtuse to rounded, apex acuminate; thin to sub-leathery, young leaves hairy above and below; petiole 0.7-3.5 cm long. **Inflorescences:** clusters of 10-40 flowers (corymbs), infl. stalks 2-4 cm long; calyx lobes persistent, hairy on both sides; corolla trumpet-shaped, yellowish-white, hairy outside, **tube 1.5-1.8 cm long,** annulus white; stigma extending out of tube. **Fruits:** follicles 2, stout and woody, 9-11 x 1.5-2.2 cm, hairy [*hirsute*], with 2 lateral ribs, apex uneven, abruptly acute, not connected at tips; seeds ovate, to 1.5 x 0.3 cm, hairtuft to 3.5 cm long.

Uncommon; riparian forest. Guianas; also Brazil.

12. Rhabdadenia Müll.Arg.

Slender lianas or herbaceous climbers with woody bases; erect shrubs. Climbing via twining shoots.

General: All cut parts with abundant milky sap; bald to lightly hairy.

Stems: Branchlets round to strongly flattened and grooved; nodes with interpetiolar ridge and colleter glands in leaf axils. Older stems round, up to 3 cm diameter; bark black or red-brown, often with conspicuous lenticels that peel in ribbons; wood light; stem x-s pattern simple.

Leaves: Variable, broadly oval to linear, thin and soft to leathery, drying bluish-green, leaf-base glands and domatia absent; veins prominent on both surfaces, 2° veins +/- straight, remote and thin, 3° veins finely net-veined below; petiole 0-1 cm.

Inflorescences: Axillary or terminal cymes, 1-per-node, 1-to-few-flowered, flower stalks 2-8 cm long; flowers to 5 cm long, linear colleter glands absent; calyx lobes persistent, conspicuous; **corolla funnel- or bell-shaped,** white, yellow, or pink, **tube very narrow at base,** mostly hairless; stamens inserted in lower tube, anthers attached to the stigma; ovary 2-locular, style thread-like, stigma spindle-shaped, surrounded by 5 triangular scales.

Fruits: Two straight, thin, cylindrical follicles, almost parallel; seeds numerous, pointed at both ends, with a white-silky hairtuft on a long apical beak.

Ecology: Seed dispersal via wind.

Distribution: Widespread in neotropics and subtropics, from Argentina to Mexico and Caribbean. 4 species total, 3 known in Suriname.

12.1. Rhabdadenia biflora (Jacq.) Müll.Arg.

Small lianas or vines, mostly hairless. **Stems:** branchlets round, winding, less than 0.4 cm diameter; **older stems to 3 cm diameter,** outer bark blackish, lenticels detachable in ribbons. **Leaves:** broadly oval (obovate) to lanceolate, 5-8 x 2-3.5 cm, **base obtuse,** apex rounded, apiculate; leathery, shiny above, dull below; petiole 1-1.5 cm. **Inflorescences:** corolla funnel-shaped, very narrow at base, white, white with yellow throat or yellow, tube 4 cm long. **Fruits:** follicles 2, 10-13 x 0.3-0.4 cm, hairless, dark brown; seeds large, 3-4 x 0.15 cm, hairtuft 4-5 cm long.

Uncommon; ridges, swamps, and riverbanks in coastal areas. Guianas; N South America to West Indies.

12.2. Rhabdadenia madida (Vell.) Miers

[Rhabdadenia macrostoma (Benth.) Müll.Arg, Rhabdadenia pohlii Mull.Arg]

Small lianas or vines, finely hairy [*pilose*]. **Leaves:** ovate-elliptic, 4-10 x 2-5 cm, **base with ear-like lobes** [*auriculate*]; apex mucronulate; thin and soft, shiny above, dull and hairy below; petiole 1 cm long. **Inflorescences:** corolla bell-shaped, white, pink or violet, tube 4-5 cm long. **Fruits:** follicles 2, 10-12 x 0.3-0.4 cm, hairless, reddish-green; seeds large, 2.3 x 0.13 cm, hairtuft 3 cm long.

Uncommon; ridges and swamps in coastal area, savannas. Guianas; also Venezuela, N Brazil, and Peru.

APOCYNACEAE

13. Secondatia A.DC.

Lianas or climbing shrubs. Climbing via twining shoots. Widespread in the New World tropics and subtropics. 7 species total, with 1 species known in the Guianas.

13.1. Secondatia densiflora A.DC.

Liana; vegetatively similar to Forsteronia. All cut parts with abundant milky sap; surfaces mostly hairless. **Stems:** Branchlets round, outer bark dark and smooth with white-tan, oval lenticels; **nodes with interpetiolar line/ridge and colleter glands;** older stems round, bark dark brown w/ large, raised lenticels; stem x-s type simple. **Leaves:** Broadly ovate-elliptic, larger leaves 7-9 x 4-6 cm, base rounded, apex with acuminate tip, 0.5-0.8 cm long; thin to papery; leaf-base glands and domatia absent; 2° veins slightly raised above, prominent below, curving upwards, 3° veins parallel, 90° to the midvein, visible above and below; petiole 1 cm long; **Inflorescences:** Terminal or axillary, ± 3 cm wide (dichasia); **flowers very small, < 1 cm long,** few to many; sepals broadly ovate, hairy, each with 1-2 linear colleter glands within; corolla trumpet-shaped, tube green, hairy, lobes white, overlapping clockwise; ovary surrounded by high, 5-lobed nectar disk; stigma head narrow, spindle-shaped. **Fruits: Two spindle-shaped follicles, ca. 14 x 3 cm,** hairless, green to reddish-brown; seeds spindle-shaped, 1.8-2 x 0.4-0.5 cm long, hairtuft 4.5-5 cm. **Ecology:** Seed dispersal via wind.

Common in riparian areas, secondary forest, and savanna-forest border. Guianas; also Brazil, Venezuela to Paraguay.

ARISTOLOCHIACEAE

ARISTOLOCHIACEAE

Mostly lianas and vines; rarely herbs or shrubs. Worldwide, 6-7 genera/500-600 species, mostly tropical with a few temperate species. Neotropics: 3 genera/250 species; only one genus known in the Guianas.

1. Aristolochia L.

Woody lianas and vines. Climbing via twining stems. Leaves are alternate, simple, entire, palmately-veined at least at base. Stipules are absent in the woody species of the Guianas. Flowers are very distinctive, often S-shaped like an old-fashioned smoking pipe; the ovary is inferior, 6-parted, with fused stamens and stigmas. Fruits are dry 6-sided capsules with small winged seeds.

General: Stems and leaves smooth, green, aromatic, with wintergreen or spicy [*Ranalean*] scent; flowers with rotten meat scent [*foetid*]; cut parts without colored sap or water.

Stems: Branchlets round to oval; nodes often with a thickened ridge; leafy 'pseudostipules' present in some vines. Older stems round to oval, 1-2 cm diameter (rarely 6-7 cm), **bark often thick-corky and fissured; stem x-s pattern 'radial', with a distorted pith and broad rays.**

Roots: Rhizomes and tubers strongly aromatic.

Leaves: Heart-shaped [*cordate*] to elliptic, occasionally 3-lobed, base cordate to truncate, margin always entire; **3-5-7-palmate-veined with** a very fine marginal vein pair on outermost edge, all veins prominent below, 3° veins ladder-like at base and along midvein, net-like [*reticulate*] elsewhere; gland dots present; **petioles long and twisted at base or wrapping onto stem.**

Inflorescences: Axillary or growing directly on stem; **flowers diagnostic,** 3 fused sepals forming an S-shaped to slightly curved tube with a flared opening [*limb*] above and a swollen chamber [*utricle*] below, often with raw-meat colors (reddish-brown with cream or yellow spots); petals absent or inconspicuous; bisexual - stamens and stigmas fused together [*gynostemium*]; stamens 6; **ovary inferior,** twisted, with 6 carpels and stigma lobes.

Fruits: **Elongate 6-sided capsule at maturity,** dry, woody to papery, with slits for seed dispersal, **resembling a "hanging basket"** (Gentry 1993); seeds winged or with an aril.

Ecology: Aristolochia species are rarely found with flowers or fruits and are poorly known. Fly pollination: flies are attracted to Aristolochia flowers by the scent of rotting flesh and trapped by hairs within the flower tube. The flies are forced to move down into the chamber at the base of the tube [*utricle*] where they are held prisoner and kept alive with nectar. If all goes well, the flies deliver pollen to the plant's receptive stigma. After pollination, the plant produces new pollen to dust the trapped flies, the hairs wither and the flies are released to cross-pollinate another individual of the species. Aristolochia seeds are most commonly wind-dispersed.

Distribution: Pan-tropical and subtropical; ± 550 species worldwide, ± 20 species in the Guianas. Nonflooded and flooded old growth forests, secondary forests, open disturbed areas.

Use: Aristolochia species are important in traditional medicine throughout the Guianas and the tropics. The stems and/or leaves have been used for fevers and malaria, diarrhea, snake bite, respiratory infections, female health (contraception, abortion, sterility), and male virility. Aristolochia species may contain aristolochic acids which have been shown to be carcinogens and mutagens (Arit et al. 2002).

Notes: Compare to other climbers with simple, alternate, palmately 3-veined leaves: Ampelozizyphus (Rhamnaceae), Dioscorea (Dioscoreaceae), Menispermaceae, Piper (Piperaceae), and Sparattanthelium (Hernandiaceae).

1.1. Aristolochia bukuti O.Poncy

Woody liana. Stems: branchlets hairless; older stems 1.5-2 cm diameter, bark corky. **Leaves:** widely triangular, 11-16 x 15-20 cm, base slightly cordate, apex acute to acuminate; hairless

and shiny above, grey-hairy below; lateral veins marginal at base; petiole thick, 5-7 cm long and twining. **Inflorescences:** cauliflorous, a dense cluster of short racemes, flowers 5 per raceme; perianth hairless, whitish with purple-brown veins; utricle 2-3 x 1.5 cm, tube 2 x 0.5 cm long, limb (flared mouth) sub-orbicular, 5-6 x 4-5 cm, apex acute, inside brown with golden speckles. **Fruit:** 6 cm long, outer wall thin, papery.

Rare; non-flooded old-growth forest. French Guiana only.

1.2. Aristolochia consimilis Masters

Woody liana. Stems: branchlets finely hairy; older stems oval, 1.5 cm diameter, bark corky. **Leaves:** **elliptic,** 7-11 x 4-7 cm, base deeply cordate, apex acute to acuminate; papery, hairy above and below, hairs lost with age; petiole 2 cm long and twining. **Inflorescences:** axillary, racemes hanging, each 2-4 cm long; perianth hairy, dark purple with green veins; utricle 0.7 cm long, tube 1-1.5 cm long, limb (flared mouth) with triangular lobe, +/- 1.5 x 1.5 cm, apex acute, yellow inside. **Fruit:** 5-6 cm long, hairless, each segment [*carpel*] with a lengthwise ridge, inner walls thin, comb-like; seeds heart-shaped, 0.5 x 0.5 cm, each with 2 papery wings. **Notes:** Similar to A. paramaribensis Duchartre, a vine common in secondary vegetation of the coastal Guianas.

Rare; old-growth forest. Fr. Guiana & Guyana (NW region).

1.3. Aristolochia cremersii

Woody liana. Stems: branchlets hairless; older stems round to oval, **6-7 cm in diameter (large),** bark corky. **Leaves:** widely triangular, 9-14 x 7-14 cm, base truncate to slightly cordate, apex acute to acuminate; hairless and shiny above, grey-hairy below; lateral veins marginal at base; petiole to 6 cm long and twining. **Inflorescences:** axillary or cauliflorous, racemes to 10 cm

long, flowers 10-15 per raceme; perianth hairless, yellow with brown veins; utricle globose, 0.3-0.5 cm wide, tube 2-2.5 cm long, limb with upper lobe brown, +/- linear, 3-3.5 x 0.2-0.4 cm. **Fruit:** 3-5 cm long; seeds ovoid, 0.5 x 0.3 cm.

Rare; vegetation zone unknown. Central French Guiana only.

1.4. Aristolochia daemoninoxia Masters

Common name: boyari (Ar); devildoer (Cr-GU), pauisima (Ak)

Woody liana. Stems: branchlets hairless; older stems round to oval, 2 cm diameter, bark corky. **Leaves:** triangular, 8-15 x 8-18 cm, base truncate to slightly cordate, apex acute to obtuse; hairless and shiny above, grey-hairy below; lateral veins marginal at base; petiole to 7 cm long and twining. **Inflorescences:** cauliflorous, racemes very short, flowers few per raceme; perianth hairy [*tomentose*], brownish-green with dark veins; utricle 2 x 1.3 cm, tube funnel-shaped, short, limb with upper lobe wider than long, 3 x 5-6 cm, apex truncate. **Fruit:** 7-10 cm long, surface rough, hairless, external wall thick, woody; seeds ovoid, 0.5 x 0.3 cm.

Rare; vegetation zone unknown. Guyana only.

1.5. Aristolochia iquitensis O.C.Schmidt

Woody or subwoody liana. Stems: branchlets hairless; older stems to 1 cm diameter, **bark thin, not corky. Leaves: elongate-elliptic,** 8-17 (-20) x 5-7 cm, base deeply cordate to auriculate, apex acuminate, margins strongly curved to parallel; papery, hairless above, lightly hairy below; lateral veins marginal only at the base; petiole 3-6 cm long and twining. **Inflorescences:** cauliflorous, 1-2 racemes, each 2-6 cm long with 10 or more flowers; perianth hairless, purplish-brown outside; tube funnel-shaped, short, limb with 2 lateral lobes, +/- triangular downward, yellow with brown markings. **Fruits:** 6-8 cm long, hairless; seeds ovoid, 0.5 x 0.3 cm.

Uncommon; non-flooded old-growth and secondary forest. Guianas; W and N Amazonia.

1.6. Aristolochia stahelii O.C.Schmidt

Common name: bukuti (Pal-FG); liane améré (Cr-FG); loango-titei (Sr)

Woody or subwoody liana. Stems: branchlets hairless; older stems 1-3 cm diameter, bark thick and corky, deeply-fissured. **Leaves:** broadly ovate, 7-13(-18) x 10-15(-22) cm, base cordate, apex obtuse to acuminate; papery, hairless above, **grey-hairy below;** lateral veins marginal at the base; petiole 5-7 cm long, not twining. **Inflorescences:** cauliflorous, racemes very short, to 1 cm long, each with 4-5 flowers; perianth hairless, with whitish pink and reddish veins outside; tube reduced and hidden by the limb (flared opening to flower). **Fruits:** 8-10 cm long, woody; seeds triangular, 0.6 x 0.3 cm. **Use:** medicinal - antidiabetic, emetic, fever/malaria.

Uncommon; non-flooded old-growth and secondary forest. Guianas; NE Amazonian Brazil.

1. Aristolochia stahelii, flowering on main stem [SR]
2. Aristolochia sp. flower growing with woody stem, stem cross-section, dissected flower, and mature and immature fruit capsules [BH]
3. Aristolochia sp., unique "Dutchman's pipe" flower form [BH]
4. Aristolochia sp. flower, dissected fly-trap chamber (utricle) at base of flower and fused male/female structure [*gynostemium*] [BH]
5. Aristolochia sp. with 6-sided immature capsule [BH]
6. Aristolochia sp. with dehisced capsule [BH]
7. Aristolochia stahelii flowers, growing on woody stem on forest floor [SR]
8. Aristolochia stahelii [SR]
9. Aristolochia bukuti, bark with corky ridges [PT]
10. Aristolochia sp. leaf, short-flexed petiole (Costa Rica) [RA]
11. Aristolochia sp. stem with corky outer bark and 'radial' cross-section with broad rays [BH]
12. Aristolochia sp. flower with neon blue hairs within tube [BH]
13. Aristolochia sp. leaf, 3-palmate-veined at base [BH]

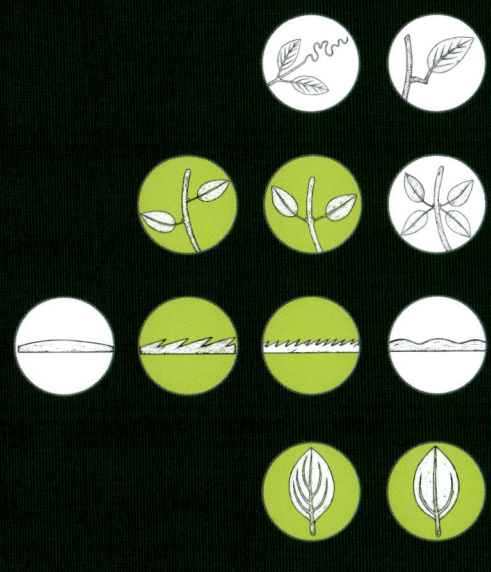

ASTERACEAE

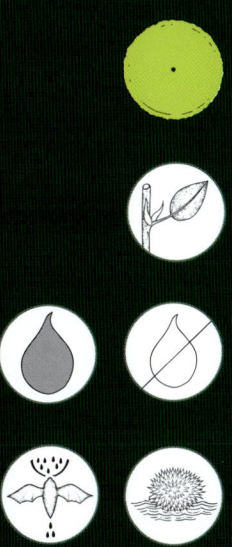

ASTERACEAE

A large, diverse family with a few woody lianas or climbing shrubs, and many sub-woody to herbaceous vines; mostly erect herbs, shrubs, and small trees. Climbing via twining shoots, twisted tendril-like leaf stalks, and recurved terminal branchlets. Stipules are absent. Leaves are opposite or rarely alternate (Piptocarpha), simple, and with entire, wavy or toothed margins. The basic inflorescence unit is a head of small flowers often appearing as a single flower. Fruits are mostly dry, single-seeded achenes and sometimes fleshy drupes.

General: Vegetation often **densely hairy, scaly or rough;** hairs simple, glandular and/or star-shaped [*stellate*]; often **strongly aromatic or resinous**.

Stems: Branchlets round, oval, or 4-6-sided, ribbed; **nodes commonly swollen or ringed, stipule-like 'appendages' often present** (true stipules absent); **pith often hollow or spongy-white.** Older stems mostly sub-woody (Stifftia robustly woody), lenticels present or absent; bark slash with a black layer along margin; stem x-s pattern simple.

Leaves: Shape and size variable; **margins toothed, wavy or entire.** Opposite-leaved species: **3-5-palmately veined from above the leaf base, pinnate towards apex.** Alternate-leaved species: pinnately-veined throughout (Piptocarpha), rarely palmate from above the leaf base (Stifftia).

Inflorescences: Flower heads [*capitula*] of few-to-many small flowers [*florets*]. Flowers are of two types: entirely tubular 'disc' flowers and 'ray' flowers with a ray-like horizontal extension of the corolla. Flower heads consist entirely of vertical 'disc' flowers [*discoid*] or of both types [*radiate*]. Asteraceae flower heads usually have rows of scale-bracts [*phyllaries*] at the base. Corolla 3-5-lobed; stamens forming tube around style, anthers with or without a stiff needle-like 'tail'; ovary inferior, containing a single ovule.

Fruits: Heads bearing many single-seeded, elongate, hair-tufted dry fruits or, rarely, bearing numerous ovoid, smooth drupes (Tilesia).

Ecology: Occurring across a wide range of habitats, often weedy; common in secondary forest, disturbed and rocky areas; rare in old-growth forest; high species endemism in the tepui highlands of Guyana and Venezuela. Bee- and bird-pollinated; seeds dispersed via wind (hair-tufts) or animal (bird — Tilesia).

Distribution: Cosmopolitan, ± 1600 genera/25,000 species; Guianas: 7-8 genera with woody and subwoody climbers; 6 genera and 14 species are described here.

Use: See Mikania (medicinal).

Notes: Infertile specimens are generally easy to identify to family. Brief comparisons for opposite-leaved Asteracs: Gesneriaceae (Codonanthe) — pinnate veins, fleshy; Loganiaceae (Strychnos) — palmate veins, climbing hooks; Melastomataceae — palmate veins, ladder-like cross-veins; Rubiaceae — pinnate veins, interpetiolar stipules. Compare alternate-leaved Asteraceae climbers with Solanaceae climbers — nodes smooth, petiole not twining, unique "rank" scent rather than aromatic scent or scent absent.

1. Calea L.

Woody climbers, scramblers and vines; mostly herbs, shrubs, or small trees. Widespread in the American tropics: 120 species; Guianas: 1 woody climber.

1.1. Calea solidaginea Kunth

Many-branched climbing shrub; hairy [*pilose*]. **Stems:** branchlets oval and lightly hairy; nodes un-swollen and without stipule-like 'appendages'. **Leaves:** opposite, **lanceolate to ovate-lanceolate,** 4-10 x 2-5 cm, base obtuse to heart-shaped, apex acuminate, margins recurved, remotely toothed to wavy; thin to papery, rough-surfaced and hairy above, glandular and lightly hairy below; **veins 3-palmate from near base;** petiole 0.5-1 cm long. **Inflorescences:** terminal and axillary, heads 'discoid', each 0.5-1.5 cm long, with 12-20 yellow disc flowers and 4-5 rows of lanceolate bract-scales; corolla tubular, 0.5 cm long, 5-lobed; anthers yellow. **Fruits:** achenes 0.2-0.3 cm long, hairtuft yellow-brown, ± 0.4 cm long.

Rare; savannas, trails, open areas, rock outcrops. Guianas; widespread in S American tropics.

2. Mikania Willd.

Mostly slender, sub-woody climbing shrubs or vines; some erect herbs or shrubs; often aromatic. Climbing via twining or scrambling shoots, twisted tendril-like petioles, and inflorescence branchlets.

Common name: brokobaka (Sr); kunaniimë (Tr)

General: Hairless or with simple hairs; cut parts aromatic, resinous-glandular.

Stems: **Branchlets round, oval, or 4-6-sided, ribbed, central chamber hollow or spongy-white; nodes swollen or ringed, stipule-like 'appendages' often present.** Older stems sub-woody, ribbed, bark slash with black outline; stem x-s pattern simple.

Leaves: **Always opposite;** elliptic to heart-shaped, **margins sometimes toothed;** papery to

2

ASTERACEAE

leathery, often green and rough above, hairless or densely felty-hairy below; **main veins 3-5-palmate from well above the leaf base or pinnate throughout.**

Inflorescences: Terminal or axillary, many branched and showy, often with flat-topped clusters of 'discoid' flower heads, **mostly white,** also green or pinkish-purple; nodes with leafy bracts; **each flower head with 4 tubular 'disc' flowers and 4 subtending scale-bracts;** style branches long and flower heads appearing 'fuzzy' or short and flower heads 'clean'.

Fruits: Achenes oblong, 5-sided, brown to black, smooth; hair-tuft with many soft yellow-brown hairs, ± equal to length of corolla.

Ecology: Seeds dispersed via wind (hair-tufts).

Distribution: Cosmopolitan; Mikania is the world's most diverse genus of Angiosperm climbers, with more than 400 species. Guianas: 20 species known (9 herbaceous vines, 11 subwoody lianas); 7 woodier species described here. Mikania micrantha is a widespread, weedy vine that exhibits the most important recognition characters of the genus.

Uses: Mikania species are popular in traditional medicine. They are commonly used against eye infections, rheumatism, skin complaints and snakebites in the Guianas. They are also part of herbal baths used in the Maroon Winti religion in Suriname (van Andel & Ruysschaert 2011).

2.1. Mikania banisteriae DC.

Liana or vine; hairs golden- to reddish-brown. **Leaves:** oval to ± heart-shaped, 8-20 x 5-10 cm, **base rounded, margins entire or with few, small teeth;** subleathery, densely soft-hairy below; **main veins palmate or pinnate;** petiole 1.5 cm long. **Inflorescences:** axillary and terminal, heads in widely-branching, climbing panicles, stalks with conspicuous leafy bracts; flowers white; style branches visible. **Fruits:** achenes cy-

lindrical or 5-sided, 0.3 cm long, with tiny hairs; hair-tuft reddish.

Very common and one of the most variable Mikania species; montane forest, river and creek forest, secondary forest, and disturbed areas. Guianas; widespread in tropical America.

2.2. Mikania gleasonii B.L.Rob.

High-climbing liana or vine; lightly hairy to hairless. **Leaves:** oval to ovate, 6-25 x 3-11 cm, **base rounded or acute, margins entire;** leathery, often gland-dotted; **main veins palmate;** petiole 1-5 cm long. **Inflorescences:** axillary and terminal, flower clusters on widely-spaced branchlets [*paniculate corymbs*], **ultimate units of 3 stalked heads;** flowers white; style branches visible. **Fruits:** achenes 0.4-0.6 cm long, hairless; hair-tuft reddish.

Common; swampy areas, along rivers, and in secondary forest. Guianas; widespread in tropical and subtropical America.

2.3. Mikania guaco Humb. & Bonpl.

Liana, climbing shrub, or vine; lightly hairy to hairless. **Leaves:** broadly ovate, 6-13 x 3.5-10 cm, base narrowing and extending onto petiole [*decurrent*], **margins entire, wavy or few-toothed;** papery, warty above, short-hairy below; **main veins palmate;** petiole 1-4 cm long. **Inflorescences:** axillary and terminal, with dense flower clusters on widely-spaced branchlets [*paniculate corymbs*], **ultimate units of 3 stalkless heads;** flowers white; style branches visible. **Fruits:** achenes 0.2-0.25 cm long, minutely hairy; hair-tuft reddish.

Common; swampy areas, along rivers, and in secondary forest. Guianas; widespread in tropical and subtropical America to the West Indies.

2.4. Mikania houstoniana (L.) B.L.Rob.

[Syn: Mikania hookeriana DC.]

Woody liana or vine; lightly hairy. **Leaves:** broadly ovate, 6-8 x 3.5-5 cm, **base cuneate to obtuse, margins entire;** fleshy to papery, with fine, soft hairs below [*puberulent*]; main veins palmate; petiole 2-3 cm long. **Inflorescences:** axillary and terminal, heads in long, narrow, dense panicles, 15-30 cm long; flowers white; style branches not visible. **Fruits:** achenes ribbed, 0.2-0.3 cm long, hairless; hair-tuft reddish, 0.5 cm long.

Common; non-flooded old-growth forest (lowland to 300 m). Guianas; widespread across tropical and subtropical America to the West Indies.

2.5. Mikania parviflora (Aubl.) H.Karst.

Slender liana or vine; also shrub or tree; lightly to densely hairy. **Leaves:** broadly lanceolate to ovate, 7-20 x 3-11 cm, **base rounded, margins entire;** leathery to papery, hairless or with fine, soft hairs below; **main veins palmate or pinnate;** petiole 0.5-2.5 cm long. **Inflorescences:** axillary and terminal, with dense flower clusters on widely-spaced branchlets [*paniculate corymbs*], **ultimate units of 3 stalkless heads;** flowers white; style branches visible. **Fruits:** achenes 5-angled, 0.3 cm long, hairless; hair-tuft reddish, 0.4 cm long.

Common, highly variable; non-flooded old-growth forest, secondary forest. Guianas; N Amazon and Bolivia.

2.6. Mikania psilostachya DC.

Common name: kaboeiakoro mibikoro (Ar); mananari (Ca)

Woody liana or vine; hairs brown, conspicuous. **Leaves:** elliptic to ovate, to 13 x 5 cm, base rounded, apex acute, **margins entire;** subleathery, **glandular and hairy below; main veins palmate or pinnate;** petiole 0.5-2 cm long. **Inflorescences:** axillary and terminal, long, dense racemes with leafy bracts, axes densely hairy, heads stalked; flowers white; style branches short. **Fruits:** achenes 0.3 cm long; hair-tuft reddish.

Common; non-flooded old-growth forest, savanna borders, riverine forest, and secondary forest. Guianas; N Amazon and Bolivia.

2.7. Mikania trinitaria DC.

Common name: kakutoru (Ar); tonolo epurlele (Ca)

Liana; mostly hairless. **Leaves:** elliptic to ovate, 7-11 x 3-4 cm, **base often asymmetrical,** apex acute, **margins entire; main veins pinnate;** petiole 1-2 cm long. **Inflorescences:** axillary and terminal, with dense flower clusters on narrowly-spaced branchlets [*paniculate corymbs*], **ultimate units of 3 stalkless heads;** flowers white, often glandular; style branches visible. **Fruits:** achenes 0.45-0.5 cm long, lightly hairy; hair-tuft whitish or reddish.

Uncommon; non-flooded old-growth forest, montane forest, savanna. Guianas; also N South America to Trinidad.

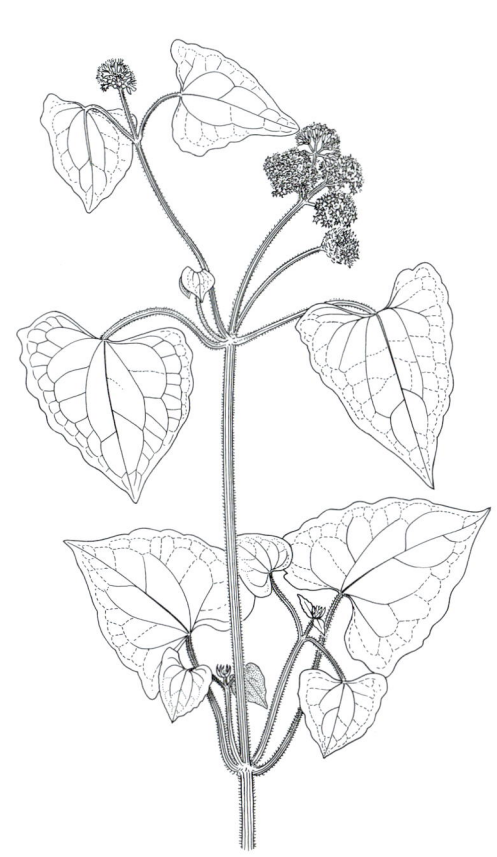

3. Pentacalia Cass.

Small woody lianas, climbing shrubs, epiphytes and vines. Neotropics: 220 species; Guianas: 1 species of woody climber.

3.1. Pentacalia freemanii (Britton & Greenm.) Cuatrec.

Woody vine or scrambler. Stems: branchlets oval, with lengthwise lines or ridges; nodes not swollen. **Leaves:** alternate, elliptic, 2-12 x 0.8-5 cm, **margin entire; sub-fleshy; venation pinnate,** sunken above, raised below; petiole 0.5-1.5 cm long. **Inflorescences:** axillary, **heads verti-** cal, with 15-22 yellow disc flowers and a single row of bract-scales; corolla 0.8-1 cm long, 5-lobed, hairless; anthers yellow. **Fruits:** achenes 5-angled, 0.15-0.3 cm long, hairless, hair-tuft white to brown-yellow, 0.6-1 cm long.

Rare; rocky ground, high elevations, 1100-2300 m. Guyana (Mt. Roraima); Venezuela & Trinidad.

4. Piptocarpha R.Br.

Woody lianas or climbing shrubs; also shrubs and small trees.

General: Surfaces often with dense greyish-white hairs [*tomentose*] **or scales, star-shaped hairs** [*stellate*] **usually present; not aromatic.**

Stems: Round, oval or angled, often ribbed, bark slash with black outline. Nodes often with stipule-like 'appendages'.

Leaves: **Alternate (rarely opposite),** broadly ovate to oblong, base often rounded, **margins entire or toothed towards apex;** ± leathery, green and rough above, **densely hairy or scaly below** (check for stellate hairs); **all veins pinnate,** with 2° veins arching upwards, sunken above and raised below; petiole often conspicuously twisted.

Inflorescences: **Flowering branches hanging downwards, side branches often at 90° to main flowering stem;** axillary, with compound clusters of flower heads, mostly white (rarely purple or yellow); **heads with 1-9 (22) tubular 'disc' flowers and many rows of scale-bracts below;** corolla usually hairless, with 5 linear lobes; anthers yellow, with long bristly tails; style branches long-haired.

Fruits: Achenes oblong, 3-10-sided, brown, smooth and often gland-dotted; hair-tuft with many soft yellow-brown hairs and a few outer bristles, +/- equal to corolla length.

Distribution: Tropical America: 46 species; Guianas: 5 species, 3 described here.

4.1. Piptocarpha opaca (Benth.) Baker

Woody vine or climbing shrub. Leaves: ovate-oblong, 8-14 x 4-6 cm, base round, apex acuminate; sub-leathery; **petiole 1-1.5 cm long,** hairy. **Inflorescences:** axillary, with **2-9 heads per inflorescence and 7-9 white disc flowers per head;** corolla ± 0.7 cm long. **Fruits:** achenes hairless, ± 0.3 cm long, hair-tuft white, to 0.6 cm long.

Rare; Guianas; also the greater Guiana Shield region.

4.2. Piptocarpha polycephala Baker

Woody liana or climbing shrub; very similar to P. opaca (only differences noted here). **Leaves: oblong-lanceolate,** 10-16 x 4-6.5 cm; **petiole 1-2 cm long. Inflorescences:** axillary, with **20-50 heads per inflorescence and 5-8 white disc flowers per head;** corolla ± 0.7 cm long.

Rare; Guianas; also the greater Guiana Shield region.

4.3. Piptocarpha triflora (Aubl.) Benn. ex Baker

High-climbing woody vine or climbing shrub. Leaves: elliptic, oblong or lanceolate, 8-14 x 3-5 cm, base round, apex acuminate; sub-leathery; **petiole 0.5-1.5 cm long,** hairy. **Inflorescences:** axillary and terminal, **heads in dense panicles with 1-3 white disc flowers per head;** corolla ± 0.4 cm long. **Fruits:** achenes gland-dotted, 0.4-0.5 cm long, hair-tuft white to yellow-brown, two-ranked, to 0.6 cm long.

Common; forms thickets along rivers and in non-flooded evergreen forest. Guianas; also N South America, Venezuela, Brazil and Colombia.

5. Stifftia J.C.Mikan.

Woody lianas. French Guiana and Brazil; 6 species; Guianas: 1 species.

5.1. Stifftia cayennensis H.Rob. & B.Kahn

Large woody liana. Stems: branchlets round, ribbed, with appressed hairs; older woody stems 5-6 cm diameter. **Leaves:** alternate, elliptic-oblong, 7-16 cm x 2.8-3.5 cm, base cuneate, apex acuminate, **margins entire to wavy;** sub-leathery; **veins palmate from above base (appearing as a 'leaf within a leaf');** petiole 1.0-1.2 cm long, sparsely hairy. **Inflorescences:** terminal; with oval leafy bracts to 2-cm long; **flower heads with 3 disc flowers** and several rows of overlapping bract-scales; corolla tubular, ± 2 cm long, 5-lobed, hairless. **Fruits:** achenes ± 0.9 cm long, slightly hairy; **hair-tuft large, to 2 cm long, reddish, multi-ranked.**

French Guiana only; first species of the genus known to occur outside of Brazil.

6. Tilesia L.

Lianas, scandent shrubs, vines, and herbs. Neotropics: 3 species; Guianas: 1 species.

6.1. Tilesia baccata (L.) Pruski
[Syn: Wulffia baccata (L.) Kuntze]

Common name: agbo-ndeku (Sa); kamararai (Ca); sukrutanta (Sr); warife (Ar).

Subwoody liana, clambering shrub or vine, reaching 1-3 m into canopy. **Stems: branchlets 4-sided,** green, with bristly-hairs, purple-spots, and soft white pith; older stems angular, brown, to 2-3 cm diameter. **Leaves:** elliptic to ovate, 10-15 x 4-10 cm, base sharply cuneate, apex acute or acuminate; **margin entire or ± toothed;** leathery to sub-leathery, **rough-surfaced above;** smooth, not densely hairy below; **3-5-palmate- or pinnate-veined;** petiole 2-3 cm long. **Inflorescences:** terminal and axillary, heads on stalks to 15 cm, with 2-layers of ovate bract-scales and **both ray and disc flowers;** ray flowers yellow-orange, 8-15, sterile; disc flowers tubular with black anthers; style branches hairy. **Fruits:** berry-like with oblong or pear-shaped, fleshy achenes, each 0.3-0.5 cm long. **Ecology:** Seed dispersal via animal (bird). **Use:** Leaves used for fever, hypertension, diabetes, general health, vaginal steam baths, herbal baths for cultural purposes. Also used as a weak fish poison.

Common; riparian zones, trails, naturalized in disturbed areas, urban gardens. Guianas; also widespread in tropical S America.

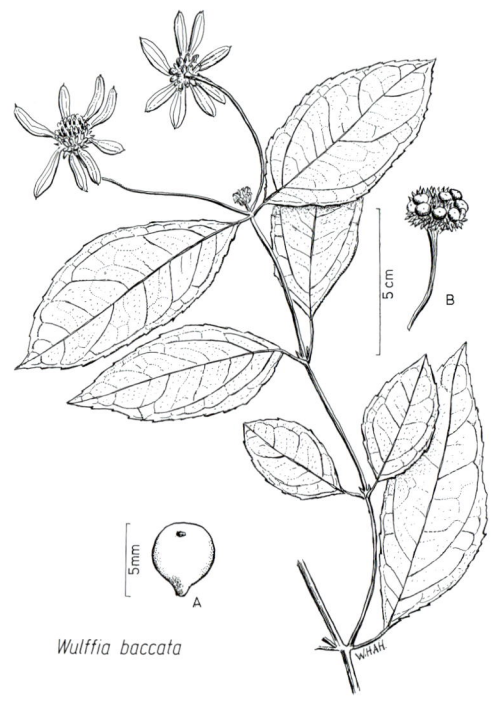

Wulffia baccata

1. Tilesia (Wulffia) baccata, fruiting branch [BH]

2. Mikania sp. node [BH]

3. Mikania sp. leaves with obscure teeth on margins [BH]

4. Mikania hookeriana, mature fruits, achenes with hair tufts [BH]

5. Mikania micrantha, achenes with hair tufts [BH]

6. Mikania congesta drawing, plant habit [NBC]

7. Piptocarpha triflora, densely hairy leaf undersides

8. T. baccata drawing, plant habit showing berry-like fruiting heads and containing ovoid fruits (achenes) [NBC]

9. T. baccata, close-up of inflorescence with disc and ray flowers [PT]

10. T. baccata, berry-like fruiting head with fleshy achenes [PT]

11. T. baccata, shrubby climbing habit [PT]

12. Mikania psilostachya slash [FRD]

13. Mikania guaco, node with swollen ridge or ring [BH]

ARECACEAE

ARECACEAE

Rarely lianas; mostly trees and treelets with leaves clasping the stem and forming a crown; often armed. Stipules absent. Leaves alternate and pinnately- or palmately-compound, small to giant, with a conspicuous mid-rib and few-to-many leaflets [*pinnae*]; leaflet veins are parallel. Inflorescences axillary, with tiny flowers on slender stems, held within 1-2 woody, protective spathes [*peduncular bracts*] that split open at maturity; flowers are 3-parted, bisexual or unisexual. Fruits berries or drupes with a fleshy to thin 'skin' [*exocarp, mesocarp*] and a hard seed coat layer [*endocarp*].

Pantropical and subtropical, extending into temperate zones. Americas: 67 genera/550 species; Guianas: 19 genera/70 species, with one climbing genus.

1. Desmoncus Mart.

Sub-woody climbing palms with long, slender, sprawling stems; Guianan species often multi-stemmed. Climbing via spines and a unique whip-like extension [*cirrus*] of the leaf midrib, with 1-2 robust hooks at nodes.

Common name: bambakka (Sa); bambam-akka (Sr); hold-me-back (Cr); jacitara (Bra); jamaraimë (Tr); kamuali (Ar); kamwari/karwari (Ca, Cr); voladora (Ven); weheyu (Ar).

General: Spines on leaves (incl. leaf sheath) and on the protective bract enclosing flowers/inflorescence. Spines of three primary types - black and straight (0.5-6 cm long), black and arching (0.5-6 cm long), or brown, short and recurved (< 2 cm long).

Stems: Round, +/- 2 cm diam., smooth, green and unarmed; completely enclosed by leaf base 'sheaths' that leave a circular scar visible on older stems; stem x-s pattern simple [*monocot stem*].

Leaves: Well-spaced along stem, pinnately-compound, with 6-23 leaflet pairs; leaflets subopposite or alternate, distinct or 2-3 clustered; texture thin and papery; a tubular sheath at the base of each leaf extends upwards to the next stem node [*ochrea*]. The 'whip-with-hooks' cirrus described above is not present on all leaves.

Inflorescences: Axillary, branchlets few to more than 25; with a single spathe [*peduncular bract*] on the inflorescence stalk, +/- woody, usually spiny outside, upper open part to 50 cm long; flowers unisexual, 3-parted, stamens 6; stigmas 3.

Fruits: Ovoid to elongate-ovoid drupes, up to 2 cm diam., skin ripening yellow to bright red; flesh thin, yellow; seeds 1, outer wall [endocarp] with 3 pores above the middle.

Ecology: Beetle-pollinated. Dispersed via animal (mostly birds).

Distribution: Coastal swamps, along rivers, on white or brown sand, disturbed and old-growth forest understory. Widespread in the Neotropics, including Trinidad and Tobago; ca. 7 species, 3 in the Guianas.

Use: Desmoncus stem fibers are used in basketry, fish traps, and crafts throughout the neotropics (Henderson et al. 1995). In the Guianas, Desmoncus is a subsistence fiber with little market value. Fibers lack the strength and durability of other forest craft fibers (Clusia, Heteropsis, Thoracocarpus). See medicinal uses for D. polyacanthos below.

Notes: Easily recognized due to the whip-like 'cirrus' and the typical palm family leaflets with parallel veins and diverse spines. Other monocot climbers with parallel veins (Araceae, Cyclanthaceae) climb mostly with aerial roots and lack the spines and woody bracts of Desmoncus. Characters of the spines and fruits are useful in distinguishing between species.

1.1. Desmoncus orthacanthos Mart.

Common name: alakule, arakure (Ca); dikke bambamaka (Sr).

Sprawling, robust climber reaching the canopy; bearing long and short spines, all spines straight (0.5-6 cm). **Stems:** 2-20 m long, 1.5-4 cm diam., unarmed. **Leaves:** 10-50 per stem; midrib 1-2 m long; midrib and tubular sheath with at least some black spines > 2 cm long; **leaflets 15-23 paired,** often clustered, blades linear to elliptic, long-acuminate, 20-30 x 4-5 cm; whip 50-80 cm long, with 5-8 pairs of strong hooks, each 1-8 cm long. **Inflorescences:** infl. branchlets > 25; **open part of spathe 40-50 cm x 8 cm, armed outside with 0.5-1.5 cm long straight spines.** **Fruits: elongate-ovoid,** 1.5-2 x 1-1.5 cm.

Common. Especially abundant along coastal rivers and swamps. Guianas; widespread in the neotropics from Bolivia to S Mexico.

1.2. Desmoncus phoenicocarpus Barb. Rodr.

Very slender climbing palm in the understory; bearing short spines, recurved or straight. **Stems:** 2-3 m long, to 1 cm diam. **Leaves:** numerous; midrib to 70 cm long, leaflets 4-8 paired, unclustered, blades elliptic, to 15 x 4.5 cm. **Inflorescence:** with 20 or fewer branchlets per infl.; **open part of spathe to 10 cm long, armed outside with straight, short, dark spines with white swollen bases.** **Fruits: ovoid,** 0.8-1 x 0.8-1 cm.

Uncommon. Lowland old-growth forest margins or disturbed forest, often on white sand soils. Guianas; also NE Amazonian Brazil and Venezuela.

1.3. Desmoncus polyacanthos Mart.

Common name: asitaremu (Ca); dunne bamba maka (Sr); kamowa (Pat); kamwari (small leaf) (Cr); piri (Tr, Wa).

Slender climbing palm in the understory; bearing both short-recurved and short-straight spines. Stems: 2-15 m long, 0.5-2 cm diam., unarmed. **Leaves:** 15-26 per stem, midrib 40-90 cm long and bearing short-recurved spines; tubular sheath with straight black spines; **leaflets 8-12 paired,** unclustered, blades lanceolate to elliptic, acuminate, 15-20 x 3-5 cm, unarmed or with short-recurved spines below; whip 35-50 cm long, with 4-6 pairs of strong hooks, each 1-5 cm long. **Inflorescences:** with 20 or fewer branchlets per infl.; **open part of spathe 23-30 x 3-4 cm, with many or few short-recurved, brown spines. Fruits:** ovoid, 1-1.5 x 0.8-1 cm. **Use:** Herbal bath with water from boiled roots to treat fevers, chicken pox, measles and itchy skin

in Guyana (Pat). The leaves are burned and ashes applied externally to treat fever and stomach pain in Suriname (Tr).

Widespread. Lowland old-growth forest, disturbed forest, brown or white sand, along rivers - but absent from coastal swamps. Wessels-Boer (1965) observed this species to be larger and spinier under conditions of increasing light and soil moisture. Guianas; widespread in tropical South America.

---

1. Desmoncus orthacanthos, robust stem, leaves and inflorescence [PT]
2. D. polyacanthos, silhouetted leaves [PT]
3. D. polyacanthos fruit cluster (close-up) [BH]
4. D. orthacanthos, close-up of paired hooks (modified leaflets) on 'cirrus' [PT]
5. Desmoncus polyacanthos [NBC]

ARECACEAE

BIGNONIACEAE

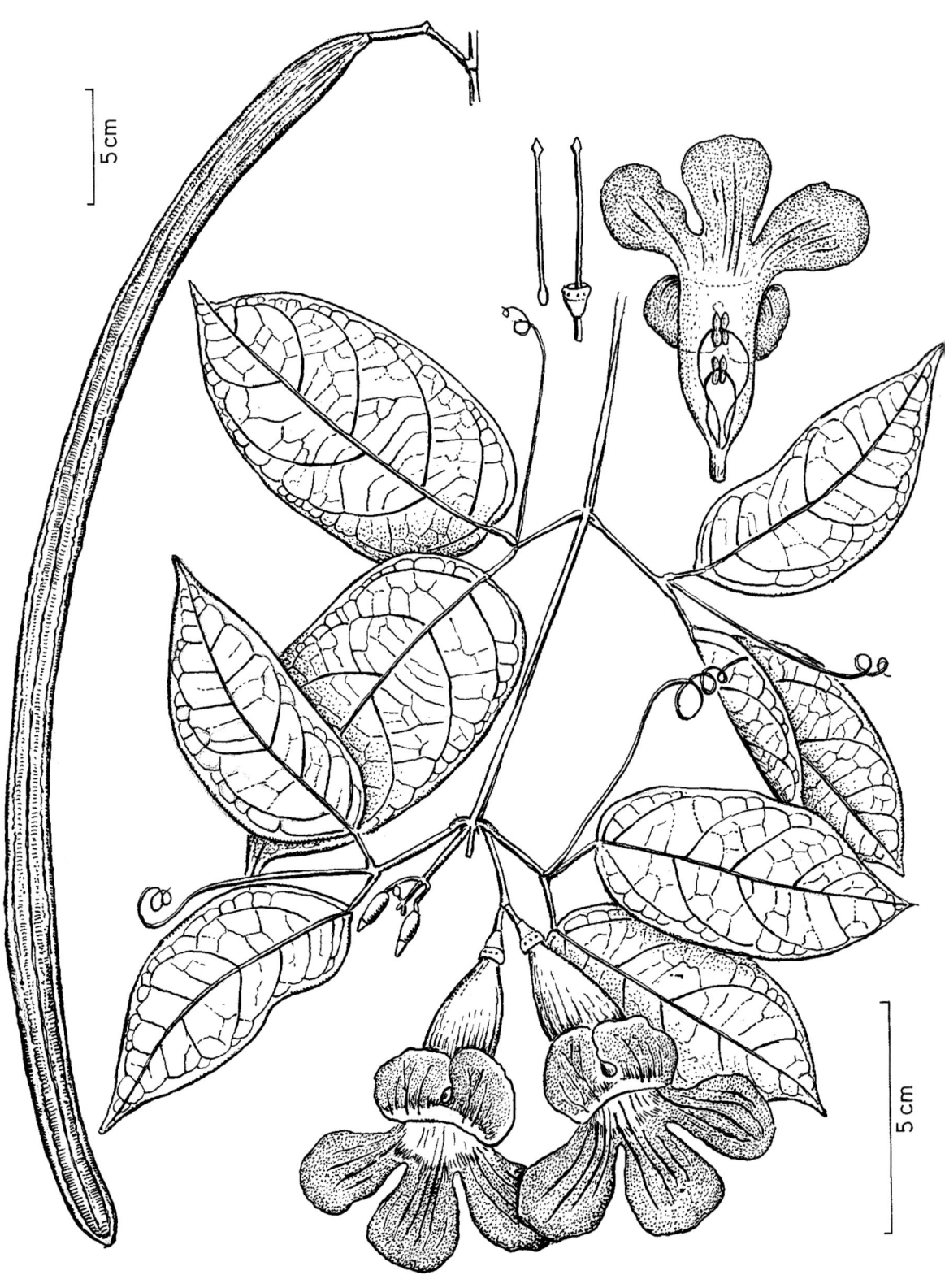

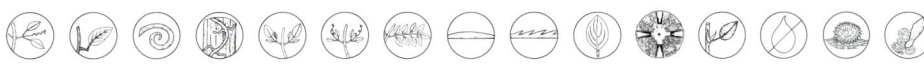

BIGNONIACEAE

A diverse family primarily of woody lianas and trees, with some shrubs, vines and herbs. All of the lianas are members of a single tribe within the family - the Bignonieae. Distinctive juvenile climbing shoots with small, flattened leaves and aerial roots occur in a few liana genera (Amphilophium, Dolichandra, Mansoa). Climbing is primarily via forked or unforked (simple) tendrils, which often replace a terminal leaflet. Leaves are opposite and compound, mostly with 2-3 leaflets, sometimes 2-3-times divided. Nodes are often with distinctive stipule-like 'prophylls' at leaf axils (true stipules absent). Flowers are showy and mostly large, with a conspicuous calyx, wide-mouthed to tubular and enclosing the corolla base, corolla tubular, funnel-shaped or bell-shaped, 5-lobed, with 4 fertile stamens, superior ovary and a 1-2 lobed stigma. Fruits are dry, 2(4)-valved capsules, long and narrow to short and squat, with winged or corky seeds.

General: Usually sparsely hairy, sometimes densely hairy or with scaly [*lepidote*] surface; hairs simple, forked [*dendroid*] or star-shaped [*stellate*]. **Glands of diverse shapes common:** including i) dome- or shield-shaped [*peltate*], ii) saucer-shaped like an open sauce pan with thickened edges [*patelliform*], or iii) volcano-crater-shaped (in Lohmann 2006 'volcano-glands'), with sloping edges, low rim, open pit within (Adenocalymma only). Lenticels few to dense, often raised. **Highly distinctive scents present in 5-6 species (almond, clove, garlic, onion, sweet).** Distinctive sap mostly absent (rarely sticky or soapy), living bark slash sometimes changing color [*oxidizing*] to dark yellow or orange.

Stems: **Branchlets in cross-section round, oval or 4(6)-sided;** bark peeling or falling away in some species; important node characters include: **i) stipule like axillary bud outgrowths [*prophylls*]** - shapes including leafy, triangular, blunt-pointed, fine-pointed [subulate], vertical-stacked, or miniature bromeliad-like - present or absent. **ii) 'interpetiolar' glands [*IP glands*]** - saucer-shaped glands clustered on stem in-between petioles - present or absent; and **iii) 'interpetiolar' ridge [*IP ridge*]** - a partial, long ridge or full ring around node - present or absent. Older stems round to remotely 4-sided, many species with conspicuous knob-like swellings at nodes and soft, fibrous bark; **stem x-s pattern 'interrupted' - with a diagnostic cross-like pattern,** typically with 4 symmetrical arms [*phloem wedges*] at 90° points (i.e., 0°, 90°, 180°, 270°), also with arms in multiples of 4 (8-64), 6, 12, and irregular/many. Branchlets and stems usu-

Bignoniaceae leaf types (all opposite, compound)

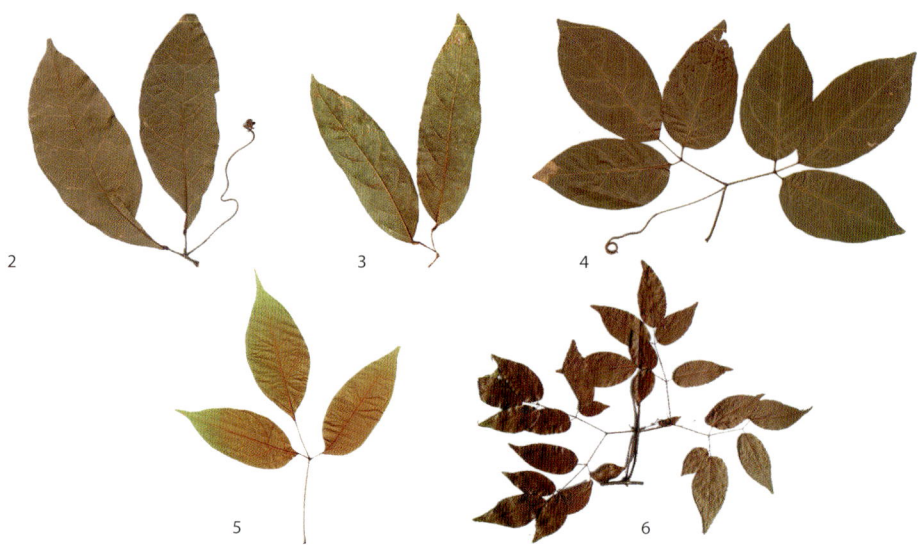

Bignoniaceae branchlet node characters

1. Illustration of Bignonia aequinoctialis, with flowers and fruits [NBC]
2. Leaves 2-foliolate with leaf tendril [FRD]
3. Leaves 2-foliolate without leaf tendril [FRD]
4. Leaves palmately-compound (1x divided) with leaf tendril [FRD]
5. Leaves 3-foliolate [FRD]
6. Leaves pinnately-compound (1x divided) [FRD]
7. Node with conspicuous interpetiolar (IP) glands, prophylls inconspicuous, Fridericia chica [BH]

8. Node with long, leafy prophylls and swollen petiole bases (Adenocalymma subincanum) [FRD]
9. Young shoot, tendrils 3-forked, IP ridge present, prophylls leafy [BH]
10. Node with conspicuous IP ridge and leafy, oval prophylls [BH]
11. Node with large, leafy prophylls and IP ridge [BH]
12. Node with less conspicuous glands in-between petioles [*IP glands*] [FRD]
13. Node with leafy prophylls, square stem and leaf tendril [BH]

Bignoniaceae glands and tendrils

Bignoniaceae woody stems

14. Leaf surface glands, Amphilophium elongatum [FRD]

15. Robust shoot with 3-forked tendrils [BH]

16. Young shoot, clasping and climbing with leaf tendrils [BH]

17. Leaf tendril clasping treelet [BH]

18. Robust woody stem growing in primary forest.

19. Stem x-s arms many, irregular [BH]

20. Typical for Bignoniaceae, robust stems with swollen nodes [BH]

21. Stem slash with swollen nodes, light colored bark, sub-bark green layer, and white wood, Mansoa alliacea [BH]

22. Woody stem, light-colored bark, Adenocalymma subincanum [FRD]

23. Stem x-s arms 4, fresh [BH]

24. Stem x-s arms multiples of 4, dry [NBC]

25. Stem x-s arms 4, fresh [FRD]

ally with a solid center or pith, but conspicuously hollow in Stizophyllum (some Anemopaegma).

Leaves: **Always compound and opposite,** 1-3-palmately-compound (leaves with 3-leaflets or 2-leaflets with or without a terminal tendril, rarely further divided to produce 9 or 27 leaflets with or without a terminal tendril), or 1-3-pinnately-compound (leaflets >3 per leaf and pinnately arranged); **tendrils unforked (simple), 2-3-forked near apex, or forking with multiple branches (multi-fid),** rarely bearing adhesive discs (Amphilophium, Mansoa) or claws (Dolichandra); **petiole mostly round or 4-sided** (rare petiole characters: swollen-jointed (Adenocalymma); saucer-shaped glands clustered at apex; modified as a robust tendril with no leaflets); leaflets of variable shapes, **margins entire** (young leaves rarely toothed); leathery to papery, surface colors above and below mostly the same [*concolorous*], rarely contrasting [*discolorous*]; glands dense, scattered or absent, sometimes clustered in vein axils, transparent gland dots [*pellucid-punctate glands*] rarely present (Stizophyllum); domatia in vein axils present or absent; veins pinnate, 1° and 2° veins mostly sunken above, 2° veins widely-spaced and looping at margins, higher order veins parallel to 2° veins (ladder-like), perpendicular to 2° veins (90° angle), or net-veined [*reticulate*].

Inflorescences: Axillary or terminal, racemes or panicle/thyrses, often presenting **showy, many-flowered displays;** flowers usually large (> 5 cm long); bracts and sub-bracts [*bracteoles*] common; **calyx conspicuous, tubular, cup-shaped or bell-shaped, often bearing glands,** margin entire, flat [*truncate*], wavy or frilly, 2-lipped/split, 5-lobed, finely toothed [*denticulate*] or double-layered; **corolla tubular/funnel-form,** tube straight or bending, often flattened in mid-section, **5-lobed, lobes evenly radial or ± 2-lipped (2 lobes above, 3 lobes below),** colors diverse (white, yellow, pink, magenta, light purple), throat often contrasting in color, colored nectar guides (markings) present or absent; nectar disc present (absent); **stamens 4** (2+2), with two pairs of different lengths [*didynamous*], a fifth poorly developed [*staminode*], anther sacs typically straight, rarely U-shaped; **ovary superior,** unstalked (stalked), surface smooth, bumpy [*papillate*] or scaly [*lepidote*], hairy or hairless, 2-chambered [2-*locular*], ovules (immature seeds) in 1-2-many series per chamber, stigma 1-2-lobed, circular, elliptic or rhomboid (slanted rectangular).

Fruits: **Woody or leathery capsules, dry, linear and ± flat to elliptic/ovoid and thicker; 2-valved and splitting open along both margins,** split is typically parallel (length-wise) to the inner wall [*septum*], rarely perpendicular (horizontal) or both (Dolichandra steyermarkii), each valve (side) with a midrib, marginal ribs or wings often present; surface smooth, bumpy [*tuberculate*], or spiny [*spiny*]. **Seeds typically flat, with a small, brown, dry seed body and very broad wings,** wings tissue-like and translucent (light shines through) to woodier and opaque (light does not shine through).

Ecology: Bignoniaceae lianas are a conspicuous and ecologically important component of forests in the Guianas and the Neotropics at large. The family includes more woody climber species than any other plant family in the neotropics. Many species produce abundant flowers and fruits throughout the year, especially notable along rivers. Pollination via insects (bees, moths, butterflies) and hummingbirds; seed dispersal via wind with the large wings, but a few species produce corky seeds with reduced wings likely specialized for water dispersal. The unique organization of stem tissues (cross-like pattern in x-s) in the family is likely an adaptation promoting lengthwise splitting rather than uncontrolled stem breakage due to tree falls.

Distribution: Widespread in the American tropics and subtropics. In recent decades, the Bignonieae tribe included 47 genera/360 species in the Neotropics and 28 genera/± 100 species in the Guianas. For this volume, we follow Lohmann (2006) and Lohmann & Taylor (2014) with 21 genera in the Neotropics, 18 genera/102 species in the Guianas.

Use: Traditional uses have been documented for most Bignon liana genera, in the Guianas and throughout the Neotropics (van Andel & Ruysschaert 2011, Gentry 1992, Grenand 1987, Hoffman 2009, Phillips 1991, Plotkin 1986). The family is especially known as a source of medicines, aphrodisiacs, poisons, strong scents, stimulants and hallucinogens, reflecting the presence of bioactive compounds. Additional uses include ritual and magic, ornamentals, dyes, condiments, lashing fibers for local construction, basketry, tourist goods and other crafts. In the Guianas, some of the more useful species include Bignonia nocturna (=Tanaecium nocturnum), Dolichandra species (= Macfadyena), Fridericia chica (= Arrabidaea chica), Mansoa species, Martinella obovata, Pyrostegia venusta and Tanaecium bilabiatum (Arrabidaea bilabiata).

Notes: As the only liana group with opposite, compound leaves, the Bignoniaceae is one of the easiest liana family to identify. Genera with many leaflets (>2-3) include certain Adenocalymma, Cuspidaria, and Pleonotoma. If leaves are unavailable, the unique stem cross-section pattern is very useful. Non-Bignon lianas with showy, tubular flowers (e.g., Acanthaceae, Apocynaceae, Gesneriaceae) have simple leaves and lack the unique x-s pattern. A former Bignoniaceae genus, Schlegelia, with simple, opposite leaves and berry-like fruits, is now placed in its own family, the Schlegeliaceae (in this volume). In the Guianas, traditional peoples generally recognize Bignoniaceae lianas as a distinct group, but distinguish, through names, many fewer Bignon taxa than professional taxonomists. Unsurprisingly, common species with conspicuous scents (e.g., Mansoa - garlic vines) or highly distinctive structures (e.g., Dolichandra, cat's-claw lianas) receive more specific local names.

Lohmann (2006) analyzed morphological and molecular characters to radically revise relationships and genus definitions in the tribe. She found that many flower and fruit characters used by past taxonomists were too similar across the group for effective comparisons. For example, in the Guianas only Amphilophium, Bignonia and Tynanthus have truly unique corollas and only Dolichandra has a unique fruit (4-parted). Other characters, such as branchlet shape, tendril type, prophyll type, stem cross-section patterns and gland presence/type are more useful and are included below. Many descriptions in this chapter are adapted from Lohmann & Taylor (2014). In Table 1, we list the currently accepted genera in the Guianas and the various genera that species were transferred from (or kept in). Some of the "old" genus names (concepts) are now invalid but many remain valid.

Some morphological characters are extremely common in Bignoniaceae climbers (see list below). To avoid repetition, we do not mention these very common character states in the text – they can be assumed for most species. On the other hand, we do mention the less common characters states when present (shown in parentheses below).

1. Plant without strong scent (rarely with strong scent).
2. Pith solid (rarely hollow).
3. Epidermis/bark not peeling (rarely peeling or shredding).
4. Tendrils without discs or claws (rarely with discs or claws)
5. Leaflet surface without translucent dots (rarely with translucent dots)
6. Petiole not modified as a robust tendril (rarely modified)
7. Petioles and leaflet stalks not jointed or swollen (rarely jointed and/or swollen)
8. Petioles without glands at apex (rarely with glands)
9. Nectar disc present (rarely absent)
10. Anther sacs straight (rarely u-shaped)
11. Ovary unstalked (rarely stalked)
12. Fruit capsule splitting into 2 valves (rarely 4 valves)

Table 1. Bignoniaea genera currently accepted and pre-2006 source genera in the Guianas.

Current Genera (Lohmann 2006, 2014)	Source Genera
1. Adenocalymma (16 sp.)	Adenocalymma, Memora
2. Amphilophium (11 sp.)	Amphilophium, Distictis, Distictella, and Pithecoctenium
3. Anemopaegma (13 sp.)	Anemopaegma
4. Bignonia (9 sp.)	Cydista, Clytostoma, Mussatia, Potamoganos, Roentgenia, Tanaecium
5. Callichlamys (1 sp.)	Callichlamys
6. Cuspidaria (2 sp.)	Arrabidaea, Cuspidaria
7. Dolichandra (3 sp.)	Macfadyena, Parabignonia
8. Fredericia (21 sp.)	Arrabidaea
9. Lundia (3 sp.)	Lundia
10. Mansoa (4 sp.)	Mansoa
11. Martinella (1 sp.)	Martinella
12. Pachyptera (1 sp.)	Mansoa
13. Pleonotoma (5 sp.)	Pleonotoma
14. Pyrostegia (1 sp.)	Pyrostegia
15. Stizophyllum (2 sp.)	Stizophyllum
16. Tanaecium (5 sp.)	Arrabidaea, Ceratophytum, Paragonia, Tanaecium,
17. Tynanthus (3 sp.)	Tynanthus
18. Xylophragma (1 sp.)	Xylophragma

1. Adenocalymma Mart. ex Meisn.

Woody lianas or shrubs. Climbing via leaf tendrils.

General: Hairs simple when present; 'volcano-crater' glands unique in genus; inner bark slash changing color (oxidizing) in some species.

Stems: Branchlets round (remotely 4-sided), lenticels abundant; nodes with diverse prophylls (narrow-pointed, awl-shaped [*subulate*] to leafy, often bearing glands); IP glands typically absent, IP ridge typically present. Older stems round to remotely 4-sided, often with conspicuous knobby swellings at nodes; bark white or yellow, soft; **stem x-s arms 4.**

Leaves: **2-3-foliolate or further divided** [*2-3-palmately-compound or 1-4-pinnately-compound*], **tendrils simple** (3-forked); **petiole round (x-s), conspicuously jointed and swollen.** Leaflets mostly elliptic-ovate to ovate (rhomboid, tiny, legume-like); leathery to papery, sometimes with a white, thickened margin upon drying; **leaflet stalks [*petiolules*] often swollen and jointed, sometimes dark red.**

Inflorescences: Terminal or axillary racemes; bracts and bracteoles common; **calyx** cup-shaped, two-lipped or entire, **often bearing volcano-crater glands,** often minutely toothed; **corolla funnel-shaped** with a constricted base, to 10 cm long, 5-lobed (not 2-lipped), **solid yellow to yellowish-orange (red, white),** lightly hairy outside, nectar guides absent, marginal glands present; stamens held within corolla tube; ovary surface smooth and scaly [*ovules in 1-series per chamber*].

Fruits: **Capsules variable, linear and very long** (to 60 cm) **to ovoid-elliptic, with a minor to prominent midrib,** flattened and constricted between seeds or inflated, wingless, valves convex, woody to leathery, hairless to hairy [*villose*], often **bearing volcano-crater glands,** drying greyish-tan; seeds ± winged to corky and wingless, seed body smooth and hairless.

Ecology: Pollination via bees and hummingbirds. Seed dispersal via wind or water. Commonly in liana tangles on riverbanks with showy, bright yellow flowers.

Distribution: Widespread in the Neotropics and subtropics, from Paraguay to Mexico and the Lesser Antilles: 82 species. Guianas: 16 species, 5 described here.

Notes: For the Guianas, ten Memora species have been re-classified as Adenocalymma species (Lohmann 2006). Adenocalymma and Memora were previously kept separate due to the pinnate leaves of Memora and 2-3-foliolate leaves of Adenocalymma. Adenocalymma tanaeciicarpum has intermediate characteristics.

1.1. Adenocalymma impressum (Rusby) Sandw.

Woody liana. Stems: branchlets round, grey-white, lenticels present or absent; nodes with **awl-shaped prophylls,** short-lived; older stems bark dark brown. **Leaves: 2-3-foliolate, tendrils simple;** leaflets elliptic-ovate to ovate, 4.5-17 x 2.2-8.8 cm, base rounded to cuneate, apex acuminate to acute; subleathery, light-grey-scaly, **margins thicker than blade and white outlined upon drying;** veins impressed above, prominent below; leaflet stalks dark red and swollen near apex. **Inflorescences:** axillary; bracts present (ovate, ± leafy, to 0.7 cm long); **calyx bell-shaped, > 1.3 cm long, with fine marginal teeth and glands;** corolla light to dark yellow, hairless outside, 6-7 cm long. **Fruits: capsules narrowly oblong,** ± 27 x 2.5 cm, surface smooth; seeds thick-bodied, wings narrow.

Uncommon; old-growth forest. Fr. Guiana, Suriname; also greater Amazonia.

1.2. Adenocalymma inundatum Mart. ex DC.

Common name: arawata kariwë (Tr); gaán háti (Sa); hehebena (Ar), kamoro (Ar); kuraiweimë (Tr); yapepuku (Ca).

Woody liana. Stems: branchlets **remotely 4-sided, bright green, smooth,** lenticels present; nodes with **awl-shaped prophylls,** short-lived; older stems variable to 8 cm diameter, **with conspicuous joint-like swollen nodes** and whitish-grey bark. **Leaves: 2-3-foliolate; tendrils simple;** leaflets elliptic-ovate to ovate, ± 10 x 5 cm, base rounded to subcordate, apex acuminate; subleathery, **margins with white outlines upon drying;** leaflet stalks green and not swollen near apex. **Inflorescences:** axillary, infl. stalks < 2 cm long; **calyx bell-shaped, 0.5-0.8 cm long, with conspicuous marginal teeth and glands;** corolla yellow, 5-7 cm long, hairy outside. **Fruits: capsules oblong, flattened,** to 20 x 4.5 cm, central rib slightly raised, surface wrinkled, green with white or grey specks, bearing lenticels; **seeds thick-bodied, brown, wings corky or absent. Notes:** Seed dispersal via water. A. var. surinamense with leaflets longer and more strongly leathery than var. inundatum.

Common; seasonally-flooded and non-flooded old-growth forest, secondary forest, riverbanks; Guianas; also widespread from coastal Brazil to Mexico.

BIGNONIACEAE

1.3. Adenocalymma schomburgkii (DC.) L.G.Lohmann

[Syn: Memora schomburgkii (DC.) Miers]

Common name: gaán háti (Sa); hehebena (Ar); kashiparaballi (Ar); kuraiweimë (Tr); piupiu (Ca).

Woody liana. Stems: branchlets round, surface greyish-white, striated or ridged; nodes with **conspicuous, leafy prophylls;** older stems round, bark greyish-white, soft, cracked horizontally, slash not changing color, wood white, soft; stem x-s arms less that ¼ distance to center. **Leaves: pinnate (5-7-foliolate)** (or 2-palmately compound, with 9 leaflets), **tendrils simple;** leaflets ovate to broadly ovate, 7-16 x 3-7 cm, base rounded, apex acuminate. **Inflorescences:** axillary, to 5 cm long, grey-hairy; leafy bracts present; **calyx cupular, 2.5 cm long, yellowish-green, margin fine-toothed, volcano-crater glands present;** corolla yellow, tubular, 7-8 cm long, lobes ± 3 cm long. **Fruits: capsules linear,** 15-65 x 2-4 cm, constricted at irregular intervals, midrib and two faint parallel ribs present; surface smooth and hairyless, greenish-blue, ± bumpy [*tuberculate*]; **seeds brown, corky,** to 3 x 5.5 x 0.4 cm. **Ecology:** Seed dispersal via water.

Common; seasonally-flooded old-growth forest, riparian and creek forest. Guianas; also greater Amazonia from Brazil, Peru to Colombia.

---

26. Adenocalymma schomburgkii, close-up [BH]
27. Adenocalymma impressum, woody stem [FRD]

1.4. Adenocalymma tanaeciicarpum (A.H. Gentry) L.G.Lohmann

[Syn: Memora tanaeciicarpa A.H.Gentry]

Woody liana. Stems: branchlets round; nodes with **small, leafy prophylls;** IP ridge present. **Leaves: 2-3-foliolate to pinnate-compound; tendrils simple;** leaflets large, elliptic to ovate, 4-15 x 1.5-7 cm, base cuneate to truncate, apex acuminate; leathery, lightly hairy to scaly [*lepidote*]. **Inflorescences:** axillary racemes, axes densely lepidote; **calyx bell-shaped, 1-1.3 x 0.6-0.9 cm, reddish-brown hairy, margin truncate, 5-toothed,** volcano-crater glands present; corolla yellow, large, scaly, to 4 cm long, hairless. **Fruits: capsules elliptic,** 9-10 x 5 cm, densely hairy; seeds winged.

Common; seasonally-flooded old-growth forest, riparian and creek forest. Guianas; also greater Amazonia - Brazil, Peru to Colombia.

1.5. Adenocalymma validum (K.Schum) L.G.Lohmann

[Syn: Memora flavida (DC.) Bur. & K.Schum.; Adenocalymma neoflavidum (DC.) L.G.Lohmann]

Common name: arawata kariwë (Tr); gaán háti (Sa); përaimë (Tr).

Woody liana, sometimes shrubby. **Stems:** branchlets round and reddish-purple; nodes with **small, leafy prophylls;** older stems round, ± 2.5 cm diameter, **bark corky and rough, flaky, brown,** inner dead bark chocolate-brown, **living bark slash white, turning orange;** stem x-s arms ½ distance to center. **Leaves: bipinnate** (or 2-palmately compound, with 9 leaflets), **tendrils 3-forked;** leaflets large, elliptic-ovate, to 13.5 x 4.5 cm, base rounded, apex acute; ± papery, hairless. **Inflorescences:** axillary and terminal; bracts and bracteoles absent or very small; **calyx tubular to cupular, margin truncate, volcano-crater glands absent; corolla yellow, large, scaly, to 6.5 cm long,** hairless. **Fruits: capsules cylindrical-compressed,** 10-15 x 4 cm, apex acuminate; surface greyish-green, with dense cream-colored lenticels, drying black; seeds narrow, wings brown, thin and soft. **Use:** bark, wood, leaves for aching body, joints, or face muscles by Suriname Akuriyos and Trios.

Uncommon; non-flooded and flooded old-growth forest, riverbanks. Guianas; also greater Amazonia and coastal Brazil.

2. Amphilophium Kunth

Woody lianas and vines; shrubs; distinctive juvenile shoots present in some species, with small, flattened leaves and aerial roots. Climbing via leaf tendrils.

General: Often conspicuously hairy, **hairs simple, forked** [*dendroid*] **or star-shaped** [*stellate*]; inner bark slash changing color (oxidizing) in some species.

Stems: Branchlets **round or 6-8-sided and ribbed** (ribs often detachable), often hairy; nodes often with **leafy prophylls, short-lived; IP glands and ridges typically absent.** Older stems ± round to 6-sided, often with knobby swellings at nodes, bark dark to white-grey, slash changing color to orange in one species; **stem x-s arms multiples of 4.**

Leaves: **2-3 foliolate, tendrils 3-forked or multiple-branched with terminal adhesive discs;** petiole round (x-s); leaflets often ovate; leathery; glands sparse or clustered in nerve axils, hairy below.

Inflorescences: Inflorescences terminal or axillary, showy; **calyx cup-shaped, truncate, shortly 5-lobed, toothed, or frilly-double-layered** (A. paniculatum), glands present or absent; corolla tubular or funnel-shaped, 3-10 cm long, rarely bent at 90° angle (A. crucigerum), leathery, 5-lobed, sometimes 2-lipped; **white, bluish-purple to pinkish-purple, throat sometimes yellow,** nectar guides absent, often hairy outside and with glands at apex; stamens held within corolla tube; ovary surface smooth, hairy outside [*ovules in multiple series per chamber, stigma rhombic or spear-shaped*].

Fruits: **Capsules oblong or elliptic,** thick, to 15 cm long, **surface often prickly, scaly, or bumpy** [*tuberculate*], sometimes hairy, lenticels absent, with or without glands, valves convex; seeds ± winged (wings reduced), **with a hairy or bumpy** [*papillate*] **seed coat.**

Ecology: Pollination often by bees. Seed dispersal via wind or water.

Distribution: Diverse vegetation types, including wet and dry old-growth forests, savanna, woodlands. Neotropics 47 species; Guianas 11 species, 4 described here.

Notes: Amphilophium now includes species of the genera Distictis, Distictella, and Pithecoctenium in the Guianas. Many Amphilophium have forked [*dendritic*] and/or star-shaped [*stellate*] hairs and 3-branched tendrils. A. crucigerum has simple hairs and multiple-branched tendrils with adhesive discs.

28. Adenocalymma impressum, branch node [FRD]

29. Adenocalymma inundatum, fruit [BH]

30. Adenocalymma schomburgkii, pinnately-compound leaves and flower [BH]

31. Amphilophium crucigerum, multi-branched tendrils [BH]

32. Amphilophium crucigerum, woody stems [BH]

33. Amphilophium crucigerum fruit [PT]

34. Amphilophium paniculatum, 6-sided branchlet [BH]

35. Amphilophium elongatum in flower [OG]

2.1. Amphilophium crucigerum (L.) L.G.Lohmann

[Syn: Pithecoctenium crucigerum (L.) A.H.Gentry]

Common name: baradaballi mibikoro, barudaballi (Ar); keskes kankan (Sr); monkey comb (GU).

Robust liana; hairs simple. **Stems: branchlets sharply 6-sided; prophylls leafy,** oval or shovel-shaped; older stems with outer bark brown and papery, inner bark green; stem x-s arms 4. **Leaves:** 2-3-foliolate, **tendrils many branched, ultimate units 3-forked, often with adhesive discs;** leaflets broadly ovate to round, to 20 x 20 cm, base cordate-truncate or rounded, apex acuminate; papery to thin; ± hairy above and below. **Inflorescences:** terminal to axillary, to 15 cm long, usually narrow, hairy to wooly; **calyx cupular, remotely 5-toothed,** 0.8-1.2 cm long, leathery, scaly and hairy; **corolla tube strongly curved, 5-lobed (not 2-lipped), cream-colored,** with musk scent. **Fruits: capsules compressed ellipsoid to oblong,** thick, 10-20 x 4.5-6.5 cm, **surface densely prickly or bumpy** [*tuberculate*]; seeds with wings, 2.5-3 long x 6-8 cm wide.

Common; Old-growth forest, ridges, riversides, and forest edges in coastal area. Guianas; widespread from Argentina and Uruguay to Mexico.

2.2. Amphilophium granulosum (Klotzsch) L.G.Lohmann

[Syn: Distictis granulosa (Klotzsch) Bureau & K.Schum.]

Woody lianas; hairs simple, forked or stellate. **Stems:** branchlets **6-sided with raised ridges; prophylls leafy;** stem x-s arms 4. **Leaves:** 2-3-foliolate, **tendrils 3-forked (without adhesive discs);** petiole 2-4 cm long; leaflets ovate, to 15 x 10 cm, base rounded, apex acuminate; surface slightly domed, 2° veins deeply sunken above; **leaflet stalks flexed. Inflorescences:** terminal racemes, scaly or hairy; **calyx cup-like, leathery, untoothed,** to 1.3 cm long, with long-vertical glands; **corolla tubular or bell-shaped,** 5.5-8.5 cm long, **tube white, with 5 flat, spreading lavender lobes and deep yellow throat,** linear glands present outside. **Fruits: capsules compressed-ellipsoid,** to 12.5 x 4 cm, finely bumpy [*tuberculate*], with orangish-brown scales and stellate hairs; seeds with broad wings, 1.2 cm long x 3 cm wide, darkly veined.

Uncommon; riversides, non-flooded old-growth forest, forest margins. Guianas; widespread in the Amazon basin.

2.3. Amphilophium magnoliifolium (Kunth) L.G.Lohmann

[Syn: Distictella magnoliifolia (Kunth) Sandwith]

Common name: kamoro (Ar)

Woody liana; reddish- to orangish-brown hairy. Stems: branchlets **round (not 6-sided),** bark reddish-brown and peeling; nodes flattened, prophylls leafy; older stems round and twisted, to 5 cm diameter, bark dark red-brown with raised lenticels, striated; wood white and soft; stem x-s arms 4. **Leaves:** 2-3-foliolate, **leaf tendrils 3-forked (without adhesive discs);** petiole 1.8-6 cm long; leaflets elliptic to oval, 8-15 x 6-11, base cuneate, apex acuminate; leathery, mostly hairless, glands sparse; veins impressed above. **Inflorescences:** terminal thyrses or racemes, slender, ultimate units 1-3-flowered; **calyx cupular, 1-1.4 cm long, whitish-green, hairy, often with conspicuous gland fields; corolla tubular,** 4.5-6 cm long, **5-lobed, white with yellow throat. Fruits: capsules woody, narrowly ovate to oblong,** 10-14 x 4-6 cm, rough, hairy, with sunken lenticels, each side with a faint midrib and 2 broad, lateral ribs; seeds with broad wings, to 2.4 cm long x 4 cm wide. **Use:** documented as an curare arrow poison ingredient outside of the Guianas (Schultes 1970).

Common; non-flooded old-growth forest and edges, also found in forest along rivers. Guianas; Brazil to Costa Rica.

2.4. Amphilophium paniculatum (L.) Kunth

Woody lianas; hairs (simple, forked or stellate) and peltate scales [*lepidote*] often present. **Stems: branchlets 6-sided with raised ridges; prophylls leafy,** to 0.3-1 cm long, short-lived; IP ridge present; older stems to 10 cm diameter, bark soft, fibrous, striated; **stem x-s arms 4. Leaves:** 2-3-foliolate, **tendrils 4-6-sided, 3-forked (without adhesive discs);** petiole 2-5 cm long, 4-6-sided; leaflets ovate to cordate, 4-9 x 2.5-8 cm, base truncate to cordate, apex acuminate; leathery, discolorous, lepidote above, with forked hairs below; 2° veins palmate at base, pinnate towards apex. **Inflorescences:** terminal racemose panicles, hairy; **calyx double-layered (outer layer frilly, inner layer 2-lipped), 0.9-1.2 cm long,** green, glands absent; **corolla tubular,** 2.3-3 cm long, straight in tube, 5-lobed, strongly 2-lipped, remaining closed even at maturity, **white, bluish-purple, or pinkish-purple,** nectar guides absent. **Fruits: capsules elliptic-flattened,** to 10-12.5 x 4-5 cm, woody, hairless and wrinkled, without wings or ridges; seeds broadly winged, 1.5 cm long x 4.5 cm wide. **Ecology:** Flowers open during daytime, closed or withered at night. Seed dispersal via wind.

Common; old-growth forest, xerophytic forest. Guianas; widespread in Neotropics and subtropics from Bolivia to Mexico and Trinidad.

3. Anemopaegma Mart. ex Meisn.

Woody lianas and herbaceous vines. Climbing via leaf tendrils.

General: Hairs simple, usually sparse, sometimes orangish-brown; peltate scales [*lepidote*] and/or lenticels often present. Mostly without conspicuous vegetative characters; inner bark slash changing color (oxidizing) in some species.

Stems: Branchlets round or 4- to 6-sided and ribbed; nodes with diverse prophylls (broadly leafy, awl-shaped [*subulate*], tiny and triangular,

absent); IP glands mostly absent, partial IP ridge often present. Older stems round or remotely 4-sided (A. parkeri), occasionally hollow; **stem x-s arms 4 or multiples of 4.**

Leaves: 2-3-foliolate; tendril 3-forked (simple); leaflets narrow-elliptic to ovate, thin to leathery, sometimes rough-surfaced [*asperous*].

36. Anemopaegma chamberlaynii, inflorescence [PT]

Inflorescences: Axillary or terminal, few-flowered racemes or solitary; calyx cupular or bell-shaped, margin truncate or 5-lobed, < 1.3 cm long, glands present or absent; **corolla** tubular to bell-shaped, large, straight in tube, 5-lobed, **creamy-white or pale yellow,** nectar guides present, lobes with basal glands in lines; stamens held within corolla tube; **ovary stalked** (rare in the family), smooth and scaly [*lepidote*] outside.

Fruits: Capsules woody, mostly elliptic in outline, flattened, stalked, calyx persistent, **surfaces** smooth (lenticels and glands absent) or hairy, midrib very thin; seeds flat and winged to thick and wingless.

Ecology: Seeds dispersed via wind or water.

Distribution: Widespread in South America, from Argentina to Mexico: 45 species. Guianas: 13 species, 3 described here.

Notes: A genus with few changes, except that Anemopaegma foetidum Bureau & K. Schum.

is now a synonym of Anemopaegma maguirei Sandw. Callichlamys fruits are similar to Anemopaegma fruits but woodier and broader.

3.1. Anemopaegma chrysoleucum (Kunth) Sandw.

Woody liana; mostly hairless. **Stems: branchlets round,** nodes with **prophylls large and leafy, 4 per node,** 0.3-1.3 cm long; IP ridge present; older stems round to ovate, bark smooth, finely striate; **stem x-s arms 8. Leaves:** 2-3-foliolate; **tendrils simple or 3-forked; leaflets narrow-elliptic,** ± 10 x 3.5 cm, base cuneate, apex acute; thin, with scales or tiny bumps [*papillose*]. **Inflorescences:** axillary racemes, often few flowered; calyx cup-shaped, truncate, persistent; **corolla > 6 cm long, mostly yellow with cream-colored lobes,** 4 dark yellow strips within. **Fruits:** capsules elliptic, 10-12 cm long, surface lightly scaly, shiny; seeds **1-4, thick, corky,** wings reduced or absent.

Uncommon; riparian forest. Guianas; also Peru, Venezuela, N Brazil to Central America.

3.2. Anemopaegma oligoneuron (Sprague & Sandw.) A.H.Gentry

Woody liana; orangish-brown hairy. **Stems: branchlets round or 4-sided,** striate; nodes with prophylls awl-shaped; older stems grayish-brown, **slash with cream-colored inner bark, changing color (oxidizing) to orange; stem x-s arms 8. Leaves:** 2-3-foliolate; **tendrils 3-forked; leaflets ovate to suborbicular,** 5-21 x 2.8-15 cm, base truncate to cordate, apex sub-acuminate; rough-surfaced [*asperous*] above and below, strongly domed [*bullate*], lightly hairy, glands clustered at leaf base. **Inflorescences:** axillary or terminal racemes; **calyx bell-shaped,** 1.2-1.5 cm long, **apex 2-3 parted,** hairless and without glands; **corolla** 5-6.5 cm long, hairless below, **creamy-white with yellow mouth,** glands absent; ovary scaly outside. **Fruits:** capsules oblong-elliptic, flattened, 13-14 x 5.5-6 cm, densely velvet-hairy.

Uncommon; non-flooded old-growth forest, xerophytic forest on sand. Guianas; also Peru, N Brazil, Venezuela, and Colombia.

3.3. Anemopaegma parkeri Sprague

Common name: onseballi (Ar); simio (Ca)

Woody liana. Stems: branchlets round or 4-sided, striate, surface splitting and peeling away; nodes with **prophylls** small, stumpy, and short-lived; older stems remotely 4-sided; stem x-s arms 4. **Leaves:** 2- foliolate; **tendrils simple or 3-forked; leaflets ovate,** 4-16 x 2.2-9 cm, base rounded to subcordate, apex acute; leathery, margin ± cartilaginous and curled. **Inflorescences:** axillary racemes, short, 3-flowered, hairy; calyx bell-shaped, truncate, green, 0.3-0.7 cm long; corolla tubular/bell-shaped, narrow at base, 4.5-6 cm long, **pale yellow with white lobes, throat deep yellow,** waxy, densely scaly [*leidote*] outside, with conspicuous glands at base of corolla lobes. **Fruits:** capsules ovoid-elliptic, flattened, 7-10 x 4-5 cm, stalked, green and drying yellowish-tan; seeds winged, 2.5-3.2 cm long x 4-4.5 cm wide.

Common; non-flooded old-growth, montane forest to 900 m, low stature shrubby forest, secondary vegetation on white sand. Guianas; also Amazonian Brazil and Venezuela.

4. Bignonia L.

Woody lianas or shrubs. Climbing via leaf tendrils.

General: Surfaces mostly hairless, often with peltate glands/scales [*lepidote*]; inner bark slash changing color (oxidizing) in some species, scent sometimes conspicuous (e.g., almond), sap sometimes sticky.

Stems: **Branchlets round to sharply 4-sided (even on same individual),** lenticels absent or few, bark green to reddish-brown; nodes with prophylls leafy or miniature-bromeliad-like (absent in B. nocturna), glands present or absent; IP glands typically absent, IP ridge partial or absent; **sometimes hollow** (B. prieurii). **Older stems round to 4-sided,** often with swollen nodes; bark reddish-brown, grey, or white, often softly woody; **stem x-s arms 8-16 (multiples of 4), reaching less than ½ distance from outer edge to center.**

Leaves: **2-3-foliolate, tendrils simple (3-forked to multi-forked); petiole round;** leaflets variable, papery to leathery, glands clustered in vein axils or scattered across blade.

Inflorescences: Axillary or terminal, racemes or bunches of cymes; **calyx cupular or tubular, margin 5-lobed, 5-needled, or 2-lipped,** leathery, hairless to densely hairy, glands sometimes present; **corolla tubular or trumpet-shaped,** 5-lobed, often flattened in middle, thin, **magenta (yellow, white, red),** nectar guides present or absent, glands present or absent; nectar disc usually absent; stamens held within corolla tube; ovary surface smooth or bumpy. [*Ovules in 1-series per chamber*].

Fruits: Capsules woody or leathery, highly variable (long and narrow, short and squat, flattened to inflated), margins flat, raised, ribbed or winged, midrib raised or furrowed; surface smooth, spiny or bumpy [*tuberculate*]; seeds flat and winged to thick and wingless.

Ecology: Seed dispersal via wind and water. The flowers in most species deceive pollinators with nectar guide markings on the corolla but no nectar-bearing disc within.

Distribution: Occurring across a wide range of tropical and subtropical forest types. Widespread in the Americas, from Argentina to the southern USA: 28 species; Guianas: 9 species, 6 described here.

Notes: This genus is new for the Guianas, with species transferred from the genera Cydista, Clytostoma, Mussatia, Potamoganos, Roentgenia and Tanaecium. Members of the former Mussatia (Bignonia prieurei, fruits linear and smooth or bumpy) and Clytostoma (Bignonia binata, fruits elliptic and spiny) in particular are similar. Bignonia nocturnum was previously placed in Tanaecium based upon the long-tubular hawkmoth pollinated flowers in common with other Tanaecium species.

4.1. Bignonia aequinoctialis (L.) L.G.Lohmann

[Syn: Cydista aequinoctialis (L.) Miers]

Common name: arawata kariwa (Tr); baskita titei, witte (Sr); gaán háti (Sa); haiariballi diamaru (Ar); japepuku (Ca); kuraiweimë (Tr); liane panier (FG); oepretete (Sr); sjimio (Ca); yapepuku (Ca).

Woody liana; mostly hairless and scaly. **Stems: branchlets 4-sided, ribbed;** nodes with leafy prophylls; older stems remotely 4-sided, 6-9 cm diameter, bark dark red-brown or greyish-brown, cracked and peeling in short strips, densely lenticellate; **inner bark slash bright red and light orange, changing (oxidizing) to dark red, exuding a sticky sap; stem x-s with 8 arms (4 longer, 4 shorter). Leaves:** 2-foliolate; tendrils simple, robust; petiole 1-4 cm long; leaflets ovate, to 12.5 x 8 cm, base cuneate to truncate, apex acuminate to obtuse; thin to papery, glossy; mid-nerve off center. **Inflorescences:** few-flowered, axes to 15 cm long; calyx bell-shaped, 0.7-0.9 cm long, truncate, glandular; **corolla bell-shaped,** 4.5-7 cm long, **tube white outside, lobes pink or magenta, with burgundy nectar guides inside yellow throat. Fruits: capsules linear,** 25-41 x 1.8-2.5 cm, apex with prickle, thickened midrib present, margins obtuse; seeds flattened, to 2 x 4.5 cm, brown. **Use:** stems and leaves for knee pain (Wayãpi).

Common. Old-growth forest, xerophytic forest on sand, and especially riverine forest. Guianas; also Brazil to the West Indies and Mexico.

4.2. Bignonia binata Thunb

[Syn: Clytostoma binatum (Thunb.) Sandwith]

Common name: fremusu wiri (Sr); tapanapïïmë (Tr); yapepuku (Ca)

Woody liana. Stems: branchlets round to remotely 4-sided, sometimes hollow, bark grey, smooth; nodes with **prophylls** resembling **miniature bromeliads;** older stems round to remotely 4-sided, sometimes hollow; **stem x-s arms 8. Leaves:** 2-foliolate; **tendrils simple;** leaflets elliptic, 7-18 x 3-8 cm, base obtuse, apex acuminate to acute; thick-leathery, wrinkled, scaly; glandular along midvein, especially at base. **Inflorescences:** axillary, few-flowered; calyx cup-shaped, 0.5-0.6 cm long, margin truncate or sharply 5-toothed, glandular; **corolla tubular/funnel-shaped,** 5-8.5 cm long, regularly 5-lobed (not 2-lipped), **purplish-pink (magenta) with white throat, nectar guides present. Fruits: capsules ellipsoid, short and squat,** 4.5-6.5 x 3-4.5 cm, **winged and with a densely bristly surface** (bristles ± 1 cm long); seeds corky, winged, 1.7 cm long x 2.5 cm wide. **Ecology:** Seed dispersal via water.

Common; riverine or creek forest, especially on white sand. Guianas; Argentina, S Brazil to S Mexico.

4.3. Bignonia microcalyx G.Mey.

[Syn: Potamoganos microcalyx (G.Mey.) Sandwith]

Woody liana; mostly hairless. **Stems: branchlets oval;** nodes flattened, with prophylls minute or absent; older stems oval; **stem x-s arms 8. Leaves:** 2-3-foliolate; **tendrils 3-forked;** petioles 3-4 cm long; leaflets elliptic to elliptic-elongate, 5-11 x 3-6 cm, base rounded, apex cuspidate-acuminate; papery, drying olive-green. **Inflorescences:** terminal racemes; calyx cup-shaped, less than 1 cm long, margin truncate or fine-toothed, finely scaly towards apex; **corolla tubular/funnel-shaped with 5 spreading lobes,** unconstricted at base, **purplish-pink. Fruits:** unknown. **Notes:** Flowers and inflorescences similar to Mansoa or Bignonia sordida.

Common; Non-flooded old-growth forest. Guyana, Suriname; endemic to the Guiana Shield, including areas of Brazil, Venezuela, and Colombia.

37. Bignonia aequinoctialis, in flower [PT]
38. B. aequinoctialis, branchlet node [FRD]
39. B. aequinoctialis, woody stem [FRD]
40. B. aequinoctialis, fruit capsules [PT]

41

4.4. Bignonia nocturna (Barb.Rodr.) L.G.Lohmann

[Syn: Tanaecium nocturnum (Barb.Rodr.) Bureau & K.Schum.]

Common names: amaretto liaan (SD); ëmëli (Way); ëmorijatë (Tr); ihipkunau (Pal); liane noyo (FG); wátawenú (Sa); watrawenu (Sr)

High climbing woody liana; all plant parts strongly almond-scented when cut or crushed. Stems: branchlets round or 4-sided, finely striate, **nodes naked** – without prophylls, IP ridge, or IP gland fields; older stems round, to 5 cm diameter, with swollen joints at nodes, bark whitish-grey with lines and scattered lenticels, **stem x-s arms 8-16. Leaves:** 2-3-foliolate; **tendril simple,** 7-18 cm long; petiole 4-7 cm long, scaly; leaflets widely ovate, 8-15 x 5-11 cm, base rounded to truncate, apex acuminate to acute, **margin wavy;** papery, hairless to scaly-hairy above and below; **3-veined-palmate from blade base,** vein axils glandular. **Inflorescences:** terminal thyrses; calyx cup-shaped, 5-toothed, green, marginal glands present; **corolla trumpet-shaped, long-tubular with flared lobes, not 2-lipped or lobed, conspicuously curved, to 14 cm long, white, throat green to pale yellow,** glands present; stamens even with or exceeding corolla mouth. **Fruits:** capsules woody, sub-round to round, 20-25 x 4.5 cm, convex with 4 lengthwise furrows; surface green, hairless, smooth; seeds oblong, thick, corky, wings as thick as seed body. **Ecology:** Hawk moth pollinated. Seed dispersal via animal or water. Almond-scented plant parts contain cyanogenetic heterosides (cyanide). **Use:** Leaves for herbal baths and post pregnancy health (Saramacca). The Saramacca name, 'watawenu' indicates the aboma (Anaconda) water spirit (van Andel & Ruysschaert 2011). Bark and leaves - for headaches, skin ailments, fevers, to make bees "sleepy" for successful honey hunting (Wayãpi) - and for colds, coughs and throat pain (Trio, Wayana). Indigenous groups in Colombia and Brazil make a hallocinogenic snuff from the leaves or bark and a narcotic tea from the roots (Gentry 1992, Prance et. al 1977).

Uncommon; non-flooded old-growth forest, riparian forest and low shrubby forest on granite or bauxite. Fr. Guiana, Suriname; also N Amazonia.

4.5. Bignonia prieurei DC.

[Syn: Mussatia prieurei (DC.) Bur. ex K.Schum.]

Common name: arawata kariwa (Tr); baskitatitei (Sr); gaán háti (Sa); kamoro (Ar)

Woody liana; reddish-brown hairy. **Stems: branchlets sharply 4-sided, ribbed, hairy, often hollow; nodes with prophylls leafy,** ovate to lanceolate with swollen base, to 2 cm long; **older stems sharply 4-sided, with raised ribbed angles, solid or hollow,** bark whitish-grey to light brown, flaky, striate, inner bark cream, wood soft; **stem x-s arms 8 (16). Leaves:** 2-foliolate, **tendrils simple (rarely 2-3-forked);** petiole ± 2 cm long; leaflets elliptic-oblong to oblong-obovate, 12-23 x 7-12 cm, base rounded, apex obtuse to short-acuminate; papery; 2° vein axils hairy. **Inflorescences:** terminal, narrow thyrses/panicles, 10-15 cm long, many-flowered, hairless or scaly; calyx cup-shaped, short, truncate, green; **corolla** tubular/funnel-form, 5-lobed and **2-lipped, to 3.5 cm long, white with some pink outside, yellow with dark burgundy stripes within,** strong-scented, glandular-scaly outside; ovary smooth. **Fruits: capsules irregularly shaped, ovoid,** to 20 x 5 x 3-4 cm, truncate at both ends, **surfaces very rough, bumpy** [*tuberculate*], yellowish-brown; seeds with brown wings, to 2.7 x 7.5 cm, firm, hairless. **Ecology:** Seed dispersal via wind.

Uncommon; old-growth forest, especially near rivers and creeks. Guianas; also the Colombian Amazon.

4.6. Bignonia sordida (Bureau & K.Schum.) L.G.Lohmann

[Syn: Roentgenia sordida (Bur. & K.Schum.) Sprague & Sandwith]

Common name: baskita-titei (Sr)

Liana; mostly hairless. **Stems: branchlets round to remote-4-sided,** unribbed; prophylls tiny and leafy, often absent; older stems round to remotely-4-sided, unribbed; **stem x-s arms 8. Leaves:** 2-3-foliolate; **tendril 3-forked (simple);** petiole 0.7-3.5 cm long; leaflets elliptic to elliptic-ovate, 6-18 x 3-9 cm, base unevenly rounded, apex acuminate; papery, drying olive green to reddish green with darker veins. **Inflorescences:** racemose panicles or racemes, to 25 cm long, narrow; calyx cup-shaped, small (0.4-0.5 cm long), green, margin truncate, glands present; **corolla** tubular/funnel-form, 4.5-5.5 cm long, tube ± curved and flattened, **pale violet or white outside, with dark red (magenta) nectar guides. Fruits: capsules long and narrow, with obscure midrib,** to 34 x 1.6-2 cm; pale brown, hairless, smooth; seeds 1.8 cm long x 5-6 cm wide, brown, seed body small, wings translucent. **Notes:** Similar to Bignonia (Cydista) aequinoctialis and Bignonia (Cydista) liliacina.

Uncommon; non-flooded old-growth forest, riparian forest, and savanna forest. Guianas; also E Venezuela and lower Brazilian Amazon.

41. Bignonia prieurii (baskita titei, Sr), flowering branch [BH]

42. B. prieurii, dried fruit [BH]

43. B. prieurii, woody stem with slash [BH]

5. Callichlamys Miq.

Woody lianas. Climbing via leaf tendrils. Monotypic genus.

5.1. Callichlamys latifolia (L.C.Rich.) K.Schum.

Common name: kalapasapoá (Wpi); kuraiwe-imë (Tr); parii (Tr)

Woody liana. General: mostly hairless, at least some hairs forked [*dendritic*]. **Stems: branchlets round to remotely 4-sided, bark light brown, lenticels abundant;** nodes with **prophylls triangular and tiny** (absent, short-lived); **IP glands absent,** IP ridge partial; **hollow chamber present; older stems round to remotely 4-sided,** observed to 8 cm diameter, bark reddish- or greyish-brown, with a fine network of ridges, lenticels abundant, **inner bark slash orange, changing color (oxidizing) to red, stem x-s arms 4. Leaves:** 2-3-foliolate, tendrils simple; petiole 2.5-18 cm long, sometimes with swollen apex and base [*pulvinus*]; **leaflets large, ovate, 7-35 x 4-20 cm,** base obtuse to rounded, apex acuminate; leathery, glands scattered, hairs present in vein axils. **Inflorescences:** axillary, racemes to 12 cm long, hairy; **calyx tubular to bell-shaped, spongy, large** (2-6 x 1-4 cm), **margin irregular to 2-lipped, solid to pale yellow,** enclosing corolla when young, glands scattered; **corolla funnel-shaped, 5-10 cm long,** tube straight, 5-lobed (not-2-lipped), solid yellow or yellow with burgundy nectar guides, fragrant, hairless outside; stamens held within corolla tube; ovary surface smooth. **Fruits: capsules elliptic to ovate, large,** 12-22 x 7-9 cm, unwinged, midrib rounded; surface bright green and hairless when young, finely mottled, glands scattered, base 2-lobed; seeds thin, ovate, larger than most of family, 2.5-4.3 cm long x 6.7-13 cm wide, wings translucent. **Use:** Grated bark as a top leishmaniasis remedy (Wayãpi); cited as curare arrow poison in Colombia (Gentry 1992). **Distribution:** Common; non-flooded, seasonally-flooded and riparian old-growth forest; lowland to pre-montane. Guianas; widespread in the neotropics, Bolivia and Brazil to Mexico.

Notes: One of the most easily recognized Bignoniaceae lianas in flower or fruit. Callichlamys fruits are similar to Anemopaegma fruits but woodier and broader.

44. Callichlamys latifolia, flowering shoot with spongy calyx [BH]

45. C. latifolia, node [BH]

46. C. latifolia, fruit capsule [FRD]

BIGNONIACEAE

6. Cuspidaria DC.

Woody lianas. Climbing via leaf tendrils or axillary tendrils.

General: Sometimes conspicuously hairy, hairs simple; without conspicuous color, sap, or scent.

Stems: Branchlets round, lightly hairy, lenticels few; nodes with **prophylls triangular and tiny** (absent, short-lived); **IP glands present,** partial IP ridge present; older stems round, lenticels scattered; **stem x-s arms 4.**

Leaves: 2-3-foliolate (C. subincanum) **to 2-3-pal-mately-compound** (C. inaequalis), **tendril simple;** leaflets ± 10 cm long, ovate to elliptic; leathery to papery, undersides sometimes white-grey with black veins (C. subincana); **glands clustered in vein axils;** 2° veins strongly ascending, **3° veins both parallel and perpendicular (ladder-like) to 2° veins.**

Inflorescences: Terminal, many-flowered thyrse; **calyx cup-shaped, shortly lobed or fine-toothed,** not-split down side, glands clustered near margins, ± hairy outside; corolla funnel-shaped, straight in tube, to 4 cm long, **5-lobed (not 2-lipped), light purple, magenta, pink, or red, often with white throat,** hairy outside;

stamens held within corolla mouth, **anthers sacs curving in U-shape;** ovary smooth and scaly outside.

Fruits: Capsules linear to oblong, up to 40 cm long, with or without raised lateral ridges, (sub) woody, smooth or minute-scaly, lenticels present, with or without saucer-like glands; seeds 2-winged or wingless, wings translucent.

Distribution: Neotropics and subtropics from Argentina to Mexico: 19 species; Guianas: 2 species

Notes: Without flowers or fruits, Cuspidaria might be confused with Fridericia, Lundia, Tynanthus and Xylophragma.

47. Cuspidaria subincana, leaves and fruits [BH]
48. C. subincana, leaves 3-palmately veined at base [FRD]
49. C. subincana, node, square stem [BH]
50. Cuspidaria inaequalis, flowers [OG]

BIGNONIACEAE

6.1. Cuspidaria inaequalis (DC. ex Splitg.) L.G.Lohmann.

[Syn: Arrabidaea inaequalis (DC. ex Splitg.) K.Schum.]

Common name: arawone simiori (Ca); gaán háti (Sa); kamoroballi (Ar); onseballi (Ar)

Woody lianas. Leaves: 2-3-palmately-compound (9-27 leaflets, terminal leaflets often replaced by tendril); petiole 4-6 cm long; leaflets ovate, to 10 cm long, base obtuse to rounded, apex acuminate; **papery, mostly hairless and green above and below** [*concolorous*]. **Inflorescences:** brown-hairy [*tomentose*]; **corolla light purple;** lobes hairy outside and inside. **Fruits:** capsules linear, thin, 7-40 x 1-1.6 x 0.2-0.25 cm, hairless, minutely scaly, saucer-like glands present; seeds oblong, to 1.3 cm long x 4.5 cm wide, winged.

Uncommon; old-growth seasonally-flooded forest, savanna-forest edges, secondary forest. Guianas; also Brazilian Amazon to Trinidad and Tobago, Guatemala, Belize.

6.2. Cuspidaria subincana A.H.Gentry

Woody lianas. Stems: branchlets dark-green; bark light brown with white lenticels, **Leaves: mostly 2-foliolate;** petiole 2-4 cm long; **leaflets ovate,** 6-10 cm long, base obtuse to rounded, apex abruptly cuspidate; **often white-grey below** [*discolorous*]. **Inflorescences:** terminal, pyramidal, lightly hairy; **corolla pink, reddish-pink, or light purple. Fruits:** capsules linear or oblong, with conspicuous central furrow and raised lateral ridges, hairless, minutely scaly, saucer-like glands present; seeds winged.

Uncommon; old-growth forest. Guyana, Suriname; also Colombia, Venezuela, Brazil.

7. Dolichandra Cham.

Woody lianas and vines. **Juvenile climbing shoots with flattened, small leaves and aerial roots** (similar to Markgravia species). Adult shoots climbing via leaf tendrils.

Common names: cat's claw vine (En); fowru-futu titei (Sr); nere amoi (Tr)

General: Sometimes hairy, hairs simple; vegetation without conspicuous color, sap, or scent.

Stems: Branchlets round or 4-sided, lenticels few or many; nodes with **prophylls awl-shaped, spear-shaped, leafy and ovate, or absent,** glands absent; IP glands and IP ridge present or absent. Older stems round or 4-sided, deeply furrowed; bark often whitish-grey, papery, inner bark dark red; **stem x-s arms many (irregular) or 4-8.**

Leaves: 2-foliolate; **tendrils 3-forked and clawed, resembling cat claws or bird talons** [*uncinate*]; leaflets elliptic to broadly ovate, > 5 cm long (D. uncata), < 5 cm long (D. unguis-cati), or > 4 cm long D. steyermarkii, thin or papery, glands scattered sparsely across blade.

Inflorescences: Axillary, dense cymes or panicles, few to many flowered; calyx cup-like to bell-shaped, margin irregularly lobed, split or truncate, glands absent or few; **corolla tubular/ bell-shaped, straight in tube, regularly 5-lobed (not 2-lipped),** 4.2 – 10 cm long, **bright yellow, yellow-orange, purplish-white or magenta,** nectar guides sometimes present, glands present or absent; **stamens held within corolla tube;** ovary surface smooth or minutely scaly [*ovules in multiple series*].

Fruits: Woody or leathery **capsules, long, straight and thin-walled** (D. unguis-cati to 95 cm long), without wings or ridges, persistent calyx present, **splitting open along 4 vertical lines**

(unique in tribe), either parallel to inner wall (D. uncata, D. unguis-cati) or both perpendicular and parallel to inner wall (D. steyermarkii); surface smooth or lightly scaly; seeds with body not well distinguished from wings, flat with thin wings or thicker with reduced wings.

Ecology: Pollination mostly by bees. Seed dispersal via water or wind. Flowers appear before new leaves have expanded (Mori 2002). D. uncata is more common in swampy areas and D. unguis-cati prefers well-drained, upland sites.

Distribution: Widespread from Argentina to the southeastern USA, West Indies: 9 species; Guianas: 3 species.

Notes: In the Guianas, several Bignoniaceae climbing taxa have 3-clawed tendrils resembling cat claws or bird talons: Dolichandra (former Macfadyena, Parabignonia), Pleonotoma clematis, and (sometimes) Tynanthus pubescens. Pleonotoma clematis is easily distinguished by sharply 4-sided stems and twice-palmately-compound leaves. Tynanthus pubescens is distinguished by clove-scent in cut parts and conspicuous hairs. With fertile material, Dolichandra species are distinguished by a bright yellow corolla (dark-red-purple in D. steyermarkii) and a capsule with 4 valves. To distinguish between D. uncata and D. unguis-cati, characters of the interpetiolar 'gland fields', calyx, and seeds are useful. D. unguis-cati is more variable than D. uncata.

7.1. Dolichandra steyermarkii (Sandwith) L.G.Lohmann

[Syn: Parabignonia steyermarkii Sandwith]

Woody liana or vine. **Stems:** branchlets round, bark brown and bearing many lenticels; **nodes with prophylls inconspicuous, short-lived; IP glands and IP ridge present; stem x-s arms 8-many. Leaves:** petioles 0.7-2 cm long, woody, leaflet stalks green; **leaflets oval to acuminate, 4-8 x 2-4.5 cm,** base rounded, apex obtuse to acute, margins sometimes toothed when young; leathery to papery. **Inflorescences: calyx tubu-**

lar/bell-shaped, 0.8-1 cm long, green, lobes short-pointed [*mucronate*], bearing glands; corolla 4.5-8.8 cm long, tube and lobes pale pink-orange. **Fruits:** capsules linear, 10-20 x 2 x 0.5 cm, valves splitting lengthwise and horizontally at mid-section; seeds 2-winged.

Uncommon. Old-growth forest and cloud forest. Guianas; Ecuador, Brazil, Venezuela, Panama, Costa Rica.

7.2. Dolichandra uncata (Andrews) L.G.Lohmann

[Syn: Macfadyena uncata (Andr.) Sprague & Sandw.]

Common name: fowrufutu titei (Sr); kamoro (Ar); lele simiole (Ca); nere amoi (Tr); prasara-titei (Sr); wakagamu (Way).

Woody liana or vine. Stems: nodes with **prophylls awl- to spear-shaped; IP glands conspicuous on younger shoots,** IP ridge present or absent; older stems round, to 2.5 cm diameter, furrowed, bark white-grey, papery, densely lenticellate, inner bark dark red, pith red; **stem x-s arms multiple, irregular. Leaves:** petioles 0.8-4 cm long; juvenile leaflets 1.5-3 cm long; mature leaflets elliptic to ovate, 5-18 x 2-9 cm, base cuneate to rounded, apex long-acuminate; membranous, minutely bumpy [*papillate*] or scaly [*lepidote*] on both sides, reddish or blackish below. **Inflorescences: calyx spathe-like and deeply split (not truncate), 1.5-3 cm long,** lobes short-pointed [*mucronate*]; glands absent; **corolla** 4.2-8.8 cm long, **pale yellow to orange-yellow,** sometimes with darker yellow nectar guides, hairless outside, with scattered glands and scales inside. **Fruits: capsules linear, to 16-26 x 1.5-2 cm, dark brown, valves splitting only lengthwise;** seeds to 3 x 1.5 cm, wings reduced, woody but delicate. **Use:** whole plant to treat aching joints, fevers (Wayana); avoid miscarriage, genital steambath, keep a man from cheating (Saramacca).

Uncommon. Riparian forest and swamps. Guianas; widespread in tropical and sub-tropical America.

7.3. Dolichandra unguis-cati (L.) L.G.Lohmann

[Syn: Macfadyena unguis-cati (L) A.H.Gentry]

Common names: awawe ansa (Sa); fowrufutu titei (Sr); griffes de chat (FG); kamoro (Ar); kwaitaka nangra (Au); lele simiole (Ca); msibiu awak (Pal); nere amoi (Tr); ongles de chat (FG); prasara-titei (Sr); wakagamu (Wa); wajamaka finga (Sa)

Woody liana or vine. Stems: nodes with **prophylls leafy, narrow to oval,** IP glands absent, IP ridge present or absent; older stems round, to 6 cm diameter; bark dark, softly fibrous, striate; stem x-s arms multiple, irregular. **Leaves:** petioles 0.5-3 cm long; juvenile leaflets 1-2 cm long; mature leaflets ovate, 5-16 x 1.2-7 cm, base cuneate to truncate, apex long-acuminate; thin, minutely scaly [*lepidote*], hairy, or with a few saucer-shaped glands. **Inflorescences: calyx cup-like, with flared, wavy margin, 0.5-1.8 cm long,** hairless to densely scaly, often with scattered glands; **corolla** 4.5-10 cm long, **pale yel-**

low to orange-yellow, sometimes with darker (orange) nectar guides, hairless outside, with scattered glands and scales inside. **Fruits: capsules linear, to 26-95 x 1-2 cm, dark brown, finely-scaly, valves splitting only lengthwise;** seeds flat, 1-1.8 cm long x 4.2-5.8 cm wide, wings thin, translucent, reddish-brown. **Use:** stem and leaves used to treat female sterility, hemorrages, and provide poison resistance (NW Guyana).

Uncommon. Non-flooded old growth forest. Guianas; widespread in tropical and sub-tropical America.

---

51. Dolichandra steyermarkii, flowering shoot, 3-clawed tendrils [BH]

52. Dolichandra uncata, 3-clawed tendrils, IP glands [BH]

53. Dolichandra unguis-cati, showy flowers, calyces saucer-like with wavy margins [BH]

BIGNONIACEAE

8. Fridericia Mart.

Woody lianas or shrubs. Climbing via leaf tendrils.

General: Vegetatively mostly non-descript; surfaces hairless, lightly hairy, with large scattered lenticels, with scattered scaly [*peltate*] hairs or layers of scaly hairs [*lepidote*], often drying conspicuously black or brown, inner bark sometimes brightly colored.

Stems: Branchlets round, lenticels abundant or few; **nodes with tiny, triangular prophylls,** short-lived; IP glands typically present, IP ridge present or absent. **Older stems round to remotely 4-sided;** bark white, grey to reddish-brown, or black, inner bark slash occasionally orange or pink and changing (oxidizing) to a darker color; **stem x-s arms 4.**

Leaves: 2-3-foliolate (often 2-foliolate and without a tendril); **tendrils simple;** petioles round (x-s), sometimes modified into a large, solitary tendril without leaflets; leaflet shape variable, margins entire (juvenile leaves rarely toothed), glands scattered across blade, clustered in vein axils, or absent.

Inflorescences: Terminal or axillary, (compound) thyrse/panicle; **calyx cup-shaped, tubular or urn-shaped,** thin or leathery, often colorful (not green), tightly wrapped around corolla or wide-mouthed and only distantly encircling corolla, **margin truncate, 5-lobed, 5-toothed, or 2-lipped,** glands absent or few; corolla tubular/funnel-shaped, 5-lobed (rarely 2-lipped), **medium- to small-sized (< 10 cm long), typically pink, magenta, or purple** (red, white), nectar guides and glands absent; stamens held within corolla tube, **anther sacs U-shaped or straight;** ovary smooth and scaly outside [*ovules in 1 series per chamber*].

Fruits: Capsules linear, straight, flat, mostly unwinged and with a prominent midrib, leathery, typically bearing lenticels and scales [*lepidote*]; seeds flat, oblong, wings 2, mostly translucent, either distinct or not distinct from seed body.

Ecology: Seeds dispersed via wind.

Distribution: Widespread in the American tropics and subtropics, from Argentina to Mexico and the West Indies: 67 species total; Guianas: 21 species, 6 described here.

Notes: Arrabidaea has long been recognized as a poorly-defined genus in the Bignoniaceae family. Following Lohmann (2006, 2014), most Arrabidaeae of the Guianas are now included within Fridericea. The remaining Arrabidaea species are included within the genera, Cuspidaria, Tanaecium, and Xylophragma.

8.1. Fridericia candicans (Rich.) L.G.Lohmann

[Syn: Arrabidaea candicans (Rich.) DC.]

Common names: kalayulu (Wpi); moussi (Ca)

Woody liana. Stems: branchlets oval, with raised lengthwise ridges; **IP glands present,** IP ridge absent; older stems oval, observed to 5 cm diameter, bark soft, white, lenticellate. **Leaves: leaflets ovate,** ± 9 x 6.5 cm; papery, discolorous, **white short-haired [*canescent*] below,** primary veins dark. **Inflorescences:** thyrses, with fine, long gray hairs; calyx cup-shaped, 0.3-0.6 cm long, purple, truncate, bearing glands; **corolla tubular/bell-shaped, 2.2-4 cm long, light purple or magenta with white throat,** lobes hairy, tube hairless; fragrant. **Fruits:** linear, flat capsules, midrib acute, 12-27 x 1-1.5 cm, brown, seeds 0.5-1 cm long x 2-3.7 cm wide, body dark, wings translucent, distinct from seed body. **Use:** Leaves to extract a purple dye (Trio); bark and sap to treat fever, headache, burns (Fr. Guiana).

Common; non-flooded and seasonally-flooded old-growth forest, low scrubby forest on white

54

55

56

57

58

sand. Guianas; widespread from Bolivia and Brazil to S Mexico.

8.2. Fridericia chica (Bonpl.) L.G.Lohmann

[Syn: Arrabidaea chica (Bonpl.) Verl.]

Common name: calajourou (FG); karawiru (Ca); karajura (Pal); tapanapi (Tr)

High climbing liana. Stems: branchlets oval, young axes smooth and green, bark with raised lenticels, striate, green to light grey; **IP glands present,** IP ridge small or absent; older stems **remotely 4-sided,** observed to 6 cm diameter (Gentry 1982), shallow-furrowed, bark whitish-grey and soft. **Leaves:** petiole 1.5-7 cm long; leaflets ovate, terminal leaflet 5-12 x 2-7 cm, base cuneate to truncate, apex acuminate to acute; thin, hairless, sparsely glandular. **Inflorescences:** terminal; calyx 0.3-0.5 cm long, truncate, minutely toothed, reddish-pink, usually non-glandula; **corolla** tubular/funnel-shaped, 5-lobed, **1.6-3 cm long, magenta with white throat,** fragrant, tube hairy outside and hairless inside. **Fruits:** capsules linear, 12-23 x 1 cm, midrib wavy and slightly prominent, surface purplish-red upon drying,

hairless, finely veined; seeds 0.7-0.9 cm long x 2.3-2.6 cm wide, wings translucent, distinct from seed body. **Uses:** Leaves to extract a red dye (Macushi, Trio).

Common; old-growth forest, low scrubby forest on white sand, and riparian forest. Fr. Guiana, Suriname; widespread from Argentina to Mexico.

8.3. Fridericia nigrescens (Sandwith) L.G.Lohmann

[Syn: Arrabidaea nigrescens Sandwith]

Woody liana; with forked [*dendroid*] and star-shaped [*stellate*] hairs. **Stems:** branchlets oval, grey, finely-hairy, with large, regular, elongate lenticels; **nodes with clusters of sharp-pointed bracts; IP glands absent, IP ridge conspicuous.** **Leaves:** petioles 4-7.5 cm long; leaflets oblong-ovate, 4-16 x 2-10 cm, base rounded to truncate, apex acute or obtuse; hairless or with white, forked hairs below. **Inflorescences:** terminal, many-branched, axes with forked hairs; calyx cup-like, truncate, 5-toothed, 0.3-0.6 cm long, with stellate hairs and saucer-shaped glands; **corolla** tubular/funnel-shaped, **pale purplish-**

pink, 2-3 cm long, mostly hairless within; ovary densely scaly. **Fruits: capsules linear, thin, 15-30 x 1.5-2 cm, margins sharply 2-winged,** apex acuminate, median nerve inconspicuous, green, hairless; seeds 1 cm long x 3.5-5 cm wide, wings translucent, distinct from seed body. Notes: stems and leaves drying dark brown to black. **Notes:** leaflets drying blackish above, olive-green below; calyx drying dark brown to black

Uncommon; non-flooded old-growth forest, low scrubby forest on white sand, montane forest to 800 m, secondary forest and forest edges. Guianas; also N and W Amazonian Brazil and Venezuela.

8.4. Fridericia oligantha L.G.Lohmann.

[Syn: Arrabidaea oligantha Bureau & K.Schum.]

Common names: sikime (Tr)

Woody liana. Stems: branchlets round, dark, with sparse peltate scales and lenticels; **IP glands present,** IP ridge absent; older stems round to remotely 4-sided, densely lenticellate, mostly hairless, inner bark orange. **Leaves:** petiole 1-2 cm long, lenticellate; leaflets elliptic to broadly ovate, to 12-20 x 5.5-12 cm, base obtuse to rounded, apex acuminate; papery to leathery, hairless; **commonly 3-palmately veined at base,** pinnate towards apex. **Inflorescences:** axillary, few-flowered; **calyx tubular, long** (± 2.5 cm long), **2-lipped, tightly encircling corolla base,** lightly hairy, glands absent; **corolla narrowly funnel-form, to 6 cm long, maroon to pink,** white spot on upper lobe, lower lobe of the throat folded. **Fruits:** straight capsules, 12-23 x 1.5 cm, with a prominent midrib disappearing ¼ distance from apex, surface dark brown; seeds 1.3 cm long x 4.5 cm wide, light yellow, translucent. **Uses:** Leaves to extract a black dye (Trio); bark and leaves for knee pain (Wayãpi).

Uncommon; old-growth forest, especially riparian forest, occurring on fertile loamy soils, brown sand, and white sand. Guianas; widespread in N Amazonia (Brazil, Colombia, Ecuador, Colombia).

8.5. Fridericia patellifera (Schlecht.) L.G.Lohmann

[Syn: Arrabidaea patellifera (Schlecht.) Sandwith]

Common names: ariminimë (Tr)

High-climbing woody liana. Stems: branchlets round to remotely 4-sided, bark light colored, with large, raised lenticels, scaly [*lepidote*]; **IP glands absent (present),** IP ridge absent; older stems round, observed to 4.5 cm diameter, bark silverish-grey, densely corky, brown-lenticellate, inner bark slash tannish-pink, changing (oxidizing) to dark orange, wood white. **Leaves:** mostly 2-foliolate; petiole 0.7-1.5 cm long; leaflets elliptic, broadly-elliptic to ovate, 5-10 x 3-6 cm, base cuneate to cordate, apex acuminate to cuspidate; papery with peltate scales above and below; veins pinnate (sometimes palmate at base); leaflet stalks flexed. **Inflorescences:** axillary or terminal, thyrses/panicles; **calyx unique, a broad cup or saucer, only distantly circling corolla tube base,** 0.4-0.5 cm long x 0.7-0.8 cm wide, truncate or shortly-lobed, deep pink to purple, without conspicuous glands; corolla tubular/funnel-shaped **with a very narrow base, 2.5-3 cm long, deep pink to blue-violet with white throat, conspicuously two-tone in bud** (dull pink to purple with contrasting white velvet-haired apex). **Fruits:** capsules woody, linear, 15-35 x 1.2-1.3 cm, midrib inconspicuous, surface hairless, lenticels numerous; seeds to 1 x 3.5 cm, wings with translucent-whitish margins. **Notes:** Specimens have been observed in Suriname with older stems tangles and young shoots growing across the forest floor. **Use:** leaves for treating sprains, rheumatism, muscular pains and contusions (Trio).

Uncommon; non-flooded old-growth forest, riparian forest, shrubby, low-stature forest on granite, and secondary forest. Guianas; also Amazonian Brazil to S Mexico.

8.6. Fridericia trailii (Sprague) L.G.Lohmann

[Syn: Arrabidaea trailii Sprague]

Woody liana. Stems: branchlets round, bark hairless to scaly [*lepidote*], smooth; **IP glands and ridge absent;** older stems round, bark light-brown, peeling in platelets, inner bark slash cream-colored, changing (oxidizing) orange. **Leaves:** typically 2-foliolate; petiole 3-3.5 cm long; **leaflets** large, ovate, to 18.5 x 10.5 cm, base obtuse to rounded, apex acute to long acuminate; papery, discolorous, scaly [*lepidote*] above, **white, short-hairy below;** venation dark-brown. **Inflorescences:** terminal panicles up to 10 cm long and many-branched; **calyx tubular,** tightly encircling corolla base, to 1 cm long, **pink,** margin ribbed, bluntly toothed, glands absent; **corolla tubular, ± 2 x 0.4 cm, red to reddish-pink with white throat,** lobes to 0.3 cm long, hairy outside. **Fruits:** capsules straight, flat, to 23 x 1.5 cm, surface scaly [*lepidote*], green, drying uniformly brown or black; seeds 1-1.3 cm long x 2.7-4 cm wide, wings translucent, strongly differentiated from seed body.

Rare; non-flooded and seasonally-flooded old-growth forest and swamps; disturbed areas. Guianas; also Bolivia, Peru, Brazil to S Venezuela.

9. Lundia DC.

Woody lianas or shrubs. Climbing via leaf tendrils.

General: Sometimes conspicuously hairy, mostly without conspicuous color, sap/water, or scent.

Stems: **Branchlets round,** bark densely lenticellate, lightly to densely hairy; nodes with tiny, triangular prophylls, without glands; **IP glands usually conspicuous,** IP ridge partial or absent, **stem x-s arms 4.**

Leaves: 2-foliolate; tendril 3-forked (L. corymbifera) or simple (other 3 Guianas species); petiole round, 2-7 cm long; leaflets elliptic to widely-ovate, base cordate to rounded, margin entire or ciliate, glands sparse, domatia present, blades conspicuously hairy on both sides; **3-5-palmately veined at lamina base, pinnately-veined above, 3° veins ladder-like, perpendicular to 2° veins.** Leaves drying dark green to brown above and grayish-green below.

Inflorescences: Axillary or terminal panicles (thyrses); **calyx closed and pointed in bud [*calyptrate*]** (L. corymbifera) or open, cupular at maturity, margin truncate or finely toothed, hairy outside, glands absent; corolla funnel-form to broadly tubular, 5-lobed, < 8 cm long, color variable (white, orange, pink, or dark red), throat yellow, lightly hairy outside, with or without nectar guides; nectar disc absent; **stamens held withing corolla tube, anthers covered with dense, long hairs [*villose*]; ovary smooth, hairy** [*ovules in 2 series per locule, stigma slanted-rectangular*].

Fruits: **Capsules long, narrow and flat, to 40 cm long,** usually hairy (non-glandular), **midrib blunt (L. densiflora) or prominently raised,** calyx not persisting; seeds narrowly oblong, flat, seed body thin, wings broad, translucent.

Ecology: Seed dispersal via wind. Flowers pollinated by Euglossa bees, possibly through deception because nectar disc is not present. (Alcantara & Lohmann, 2010)

Distribution: Neotropics from Brazil and Bolivia to S Mexico: 13 species; Guianas: 3 species, 2 described here.

9.1. Lundia corymbifera (Vahl) Sandwith

Woody liana or climbing shrub; not densely hairy. **Stems:** branchlets round, striate, bearing lenticels, dark brown. **Leaves: tendril simple;** petiole 3-5.5 cm long; leaflets elliptic to ovate, large, 6.5-11.5 x 3.5-7.5 cm, base truncate to cordate, apex acute to acuminate, margins ciliate; papery, discolorous. **Inflorescences:** calyx calyptrate in bud, cup-like at maturity, truncate, 0.3-0.6 cm long, drying light brown, hairy [*puberulous*]; corolla 2.5-4.3 cm long, white; **stamen filaments hairless except at extreme base or with a few scattered hairs. Fruits:** capsules linear, with prominent, sharp midrib on both sides, 30-60 x 1.5-2 cm.

Uncommon; riparian forest, seasonally-flooded old growth. Guyana, Suriname; also Bolivia, Peru, Ecuador, Brazil, Colombia and S Venezuela.

9.2. Lundia densiflora DC.

High climbing woody liana or climbing shrub; conspicuously brown or reddish-brown hairy [*tomentellous*]. **Stems:** older stems round, bark dark brown; slash cream-colored, not changing color. **Leaves: tendril 3-forked;** peti-ole 3-7 cm long; leaflets ovate, large, 7-18 x 4-12 cm, base round to subcordate, apex acuminate, margins slightly recurved. **Inflorescences:** Axillary thryses, 5-12 cm long, axes hairy; calyx truncate, yellowish-white, hairy [*tomentous*]; corolla showy, 5.5-8 cm long, white or cream-colored with yellow bands in throat, lobes 1.5-3 cm long, hairy; stamen filaments densely **hairy in upper half. Fruits:** capsules to 40 x 2 cm, margins thickly-wrinkled, **midrib blunt,** wings absent, hairy, numerous glands present, especially near midrib; seeds flat, 1 cm long x 3.5 cm wide, with 2 yellowish-brown wings.

Common; non-flooded old-growth forest, forest edges, riparian forest, submontane forest. Guyana, Suriname; also Bolivia, Peru, Ecuador, Brazil, Colombia and S Venezuela.

54. Fridericia candicans, habit [OG]
55. Fridericia chica, interpetiolar glands conspicuous [BH]
56. Fridericia sp., white flower [OG]
57. Fridericia candicans, leaves white below [OG]
58. Fridericia sp., calyx dark purple (not green) [PT]
59. Lundia densiflora, habit [CB]
60. L. densiflora, branchlet node [FRD]
61. L. densiflora, hairy leaf margin [FRD]
62. L. densiflora, habit [PT]

63   64   65   66   67

10. Mansoa DC.

Woody lianas; distinctive juvenile shoots rarely present, with small, flattened leaves and aerial roots. Climbing via leaf tendrils.

General: **Strong garlic or onion scent** in cut or crushed plant parts (M. alliacea and M. standleyi only); conspicuous sap, hairs and colors mostly absent.

Stems: **Branchlets round, becoming angled (4-6-sided) and ribbed with age,** hairless, grayish, lenticels few or absent; nodes with prophylls small and inconspicuous (conical, awl-like, or small-bromeliad-like); **IP glands present** (M. standleyi) **or absent; IP ridge present** (M. alliacea) **or absent** (M. standleyi); older stems angled, becoming deeply fissured; bark smooth, white with green (lichen) patches, inner bark slash white, wood cream-colored; **stem x-s arms 'multiples of 4'** (4-16).

Leaves: 2-3-foliolate, **tendrils 3-forked** (most Mansoa) **or simple** (M. standleyi); **petioles with glands at apex** (M. standleyi) or without (other Mansoa); leaflets elliptic to broadly ovate; papery to leathery, lightly hairy above and below, with glands clustered in vein axils; 2° veins pinnate (M. alliacea, M. onohualcoides) **or 3-palmately veined at base, pinnate towards apex** (M. standleyi, M. verrucifera); leaf stalks sometimes swollen at base and apex [*pulvinulate*].

Inflorescences: Axillary, raceme or panicle/thyrse; calyx cup-shaped to bell-shaped, margin wavy (M. standleyi) or fine-toothed (other Mansoa), lightly hairy, glands sometimes present; corolla tubular/funnel-shaped, straight in tube, **5-lobed (not 2-lipped), pale to dark blue or violet, magenta, red, or reddish-purple;** nectar guides absent, hairy present on lobes, glands absent; nectar disc absent; stamens held within corolla tube (or 1-2 exserted); ovary smooth and hairless to finely bumpy outside [*ovules in 2-4 series per locule, stigma rhombic*].

Fruits: **Capsules woody, linear to oblong,** flattened to thickened, **midrib strongly raised or flat, surface smooth** (M. standleyi) **to bumpy** (other Mansoa), surface gland-dotted, calyx falling away quickly, margins separating with age and leaving threadlike outlines; seeds thin, broad, and 2-winged or thick and ± wingless.

Ecology: Seed dispersal via water or wind.

Distribution: Widespread in the Americas from Argentina to Mexico: 13 species; Guianas: 4 species, 2 described here.

Use: The garlic lianas are very important ethnobotanical species in the Guianas. They are used by diverse cultures in traditional medicine, rituals, and to repel pests (see below).

Notes: The currently accepted name for Mansoa kerere (Aubl.) A.H.Gentry is Pachyptera kerere (Aubl.) Sandwith (see Pachyptera).

BIGNONIACEAE

10.1. Mansoa alliacea (Lam.) A.H.Gentry

Common name: abonenge tatai (Sa); abuya mibia (Ar); akapore (Tr); ayun tatai (Sa, Au); boningre uwii (Sa); bos gonofru (Sr); bos kno-flook (Du); cipo d'alho (BP); douvant-douvant (FG); garlic rope (GU); gonofru-titei (Sr); ilay kamwi (Pal); knoflookliaan (SD); knoflookwiri (Sr); kunofruktutitei (Sr); kwipokan (Aku); li-ane-ail (FG); tingititei (Sr); wïpore (Ca); wit bawang (Ja); wo'pole (Cr); wuporëng (Cr)

Woody liana; all parts with garlic or onion scent. **Stems:** IP glands present or absent; partial IP ridge present; older stems. **Leaves: tendrils 3-forked; petiole 1-1.3 cm long;** leaflets elliptic (broadly ovate), 9-11(20) x 4.5-5.5 cm, base rounded to cuneate, apex curved and asymmetrical; **2° veins pinnate. Inflorescences:** Axillary or terminal, open raceme; calyx cup-shaped, 1-2 cm long, 5-lobed at apex, with peltate scales, simple hairs, scattered glands; **corolla to 4.3 cm long, yellow/pink tube with pink to red lobes or white tube with blue-white lobes. Fruits:** capsules linear, 25-40(75) x 1-2.5(4) cm, **with a prominent, ± winged midrib** and 2 lateral ribs, base rounded and tapering; surface smooth, green; seed wings broad, translucent. **Use:** whole plant, stem, and/or leaf (external) for fever, flu, head cold, rheumatism, fatigue, laziness, and to repel bats and other pests.

Uncommon; non-flooded and seasonally-flooded old-growth forest; often naturalized or cultivated. Guianas; widespread, from Bolivia to Nicaragua and Lesser Antilles.

10.2. Mansoa standleyi (Steyerm.) A.H. Gentry

Common name: akapore (Tr); bos knoflook (Du); cipo d'alho (BP); knoflookliaan (SD); knoflookwiri (Sr); kunofruktutitei (Sr).

Woody liana; all parts with garlic or onion scent. **Stems:** IP glands present; IP ridge absent. **Leaves: tendrils simple; petioles 3-3.5 cm**

long, with glands at apex; leaflets elliptic, 22-24 x 10-12 cm, base acute, apex acute and acuminate; **palmately 3-veined at base. Inflorescences:** Axillary or terminal, open racemes; each flower subtended by 2 large, reddish-purple bracts, ovoid, 1-2 cm long; calyx large, cup-shaped, 1.3 cm long, truncate/lobed at apex, with peltate scales, simple hairs, glands absent; **corolla 3.5-6 cm long, tube white tinged with purple, lobes purplish, hairless. Fruits:** capsules linear, 35-75 x 2.6-3.8 cm, **with midrib inconspicuous;** surface smooth, green, hairless, wrinkled, gland-dotted; seed wings broad, translucent. **Use:** whole plant, stem, and/or leaf (external) for fever, blood pressure, stomach pain, stress, laxative and sweating, pregnancy steam bath, winti rituals and to repel bats and leaf-cutter ants.

Uncommon; non-flooded and seasonally-flooded old-growth forest; naturalized or cultivated within villages and cities. Guianas; widespread from Bolivia and Brazil to Guatemala.

---

63. Mansoa alliacea, cultivated in Saramacca village of Ston-huku (Suriname) [BH]

64. M. alliacea, branchlet 8 sided in cross-section, IP ridge conspicuous [BH]

65. M. alliacea, woody stem [FRD]

66. M. alliacea, stem x-s arms 16 [NBC]

67. M. alliacea, woody stem with deep fissures, x-s arms many, irregular [BH]

---

11. Martinella Baill.

Woody lianas. Climbing via leaf tendrils.

Distribution: Neotropics from Bolivia and Brazil to S Mexico: 2 species; Guianas: 2 species; 1 described here.

11.1. Martinella obovata (Kunth) Bur. & K.Schum.

Common name: akoacorollii (Ar); eye vine (GU); kamoro (Ar); kuraiweimë (Tr); once-a-mile (GU); uquilla (Ar)

Woody liana. General: Surfaces smooth and green, sometimes with **star-shaped [*stellate*] hairs;** mostly non-descript. **Stems:** branchlets round (4-sided), bark green, striate; nodes smooth, with tiny, triangular, short-lived prophylls, IP glands absent, **IP ridges conspicuous, forming a continuous ring** (or scar); **older woody stems round to remotely 4-sided** with shallow furrows, to 7 cm diameter, bark greyish-white, softly fibrous, lenticels present (circular to oblong); **stem x-s arms 4,** extending full length from outer edge to center. **Leaves: 2-3-foliolate; tendrils 3-forked;** petioles round, 3-3.5 cm long; leaflets ovate to elliptic, 9-10.5 x 4-4.7 cm, base rounded, apex acuminate; papery, with scattered glands below; leaflet stalks thickened and twisted. **Inflorescences:** axillary racemes or thyrses; 15-28 cm long, slender, flexuous, flowers opposite; **calyx tubular, 2-lobed or irregularly 3- to 4-lobed,** with scattered glands, hairy outside, light green, margin entire; **corolla funnelform,** 5-lobed, ± 4.5 cm long, magenta, pink, or wine-red, often white in mouth, hairless outside, without glands; stamens held within corolla tube; ovary smooth and scaly outside [*ovules in 1 series per chamber*]. **Fruits: Capsules long, narrow and flat, 46-80 (130) x 1.2-1.8 cm,** ridges/wings absent or irregular, pointed at both ends, midrib usually prominent; surface smooth, lenticels absent, green to brown, hairless to lightly hairy; seeds broad, 1-1.4 cm long x 4-5 cm wide, body smooth, wings ± wrinkled, opaque.

Ecology: pollinated via hummingbirds. **Uses:** Widely used in most of South America, including the Guianas, for conjunctivitis and to relive sore or irritated eyes. Also recorded as a curare arrow poison ingredient by the Barasana of Colombia (Schultes 1970).

Common; riparian forest. Guianas; widespread from S Amazon to Belize.

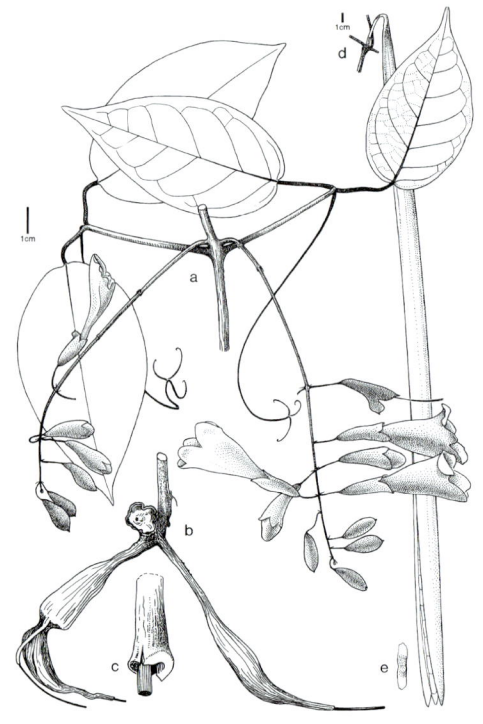

68. Martinella obovata a. flowering shoot; b., c. tubers and stem base; d. capsule; e. winged seed [NBC]

BIGNONIACEAE

12. Pachyptera

Woody lianas. Climbing via leaf tendrils. Bolivia and Brazil to Central America: 4 species; Guianas: 1 species.

Notes: Pachyptera kerere was previously included in the Mansoa genus. Pachyptera is distinguished by the stem x-s of 4 arms (vs. 8-16 arms in Mansoa) and 3-vertical-stacked prophylls (vs. small and inconspicuous prophylls in Mansoa).

12.1. Pachyptera kerere (Aubl.) Sandwith

[Syn: *Mansoa kerere* (Aubl.) A.H.Gentry]

Common name: kamoroballi (Ar); yapepuku (Ca)

Woody liana. Stems: branchlets remotely 4-sided, striate, hairless, bark soft and peeling when older, nodes with **prophylls unique - formed by 3 vertical-stacked, sword-shaped segments, both IP glands and IP ridge present and conspicuous;** older stems remotely 4-sided, bark smooth, grey, without lenticels, **stem x-s arms 4. Leaves:** 2-3-foliolate; tendril minutely-3-forked, 8-16 cm long; **petiole with apical glands;** leaflets ovate, terminal leaflet 9-24 x 3.4-10 cm, lateral leaflets 7.8-22 x 2.4-9.2 cm, base widely cuneate to narrowly subcordate, apex long-acuminate to acute; papery to leathery, bullate, scaly, glands sparse on blade. **Inflorescences:** axillary or terminal, racemes to 5 cm long, branches covered with fine, short hairs, flowers clustered towards apex; **calyx tubular, truncate or minutely 5-lobed, hairy outside, with saucer-shaped glands in lines at margins;** corolla narrowly tubular, straight in tube, **5-lobed (not 2-lipped), white,** up to 7.5 cm long, hairy outside, glands in lines towards apex; **stamens extending beyond corolla mouth,** anthers densely hairy [*villous*]; ovary smooth and scaly outside [*ovules in 1 series per locule, stigma spear-shaped*]. **Fruits: Capsules linear, flat, 12-25 x 3.2-3.8 cm,** margins deeply furrowed, surfaces green, finely hairy with scattered glands, bumpy [*tuberculate*] along sides, calyx not persisting; seeds to 2 x 6 cm, pale brown, **wings corky or wingless. Ecology:** Pollination via moths; seed dispersal via water.

Distribution: Uncommon; riparian forest. Guianas; also ranging from Central Amazon to Belize.

69. Pachyptera kerere, habit, nodes with three pointed, vertical-stacked prophylls [BH]

70. P. kerere node, with conspicuous IP glands and prophylls [BH]

BIGNONIACEAE

13. Pleonotoma Miers.

Woody lianas. Climbing via leaf tendrils.

General: Mostly hairless, rarely with conspicuous forked [*dendritic*] hairs (P. dendrotricha).

Stems: Branchlets ± 4-sided and winged, bark and wings peeling, lenticels absent; nodes with leafy, stalked prophylls present or absent, bearing glands when present, IP gland fields absent, partial IP ridge present. **Older stems conspicuously 4-sided, bark blackish-green to reddish-brown,** becoming loose and papery, slash regular or changing color from cream to orange (P. albiflora); **stem x-s arms 4.**

Leaves: Conspicuously 2-3-times compound – either palmately-compound (3-leaflet subunits) or pinnately compound (5-9-leaflet subunits); **tendrils 3-forked** (3-clawed); **petiole 4-sided;** leaflets elliptic-ovate; papery to leathery, thicker at base, glands scattered; veins pinnate; blade drying olive-green.

Inflorescences: Axillary or terminal racemes, narrow to wide; **calyx cup-shaped or tubular,** margin truncate, short-lobed or minutely toothed, often with immersed saucer-shaped glands (volcano-crater-shaped glands absent); **corolla tubular/bell-shaped, straight in tube,** ± flattened in mid-section, base often narrow, **5-lobed (not 2-lipped),** glands sometimes present, **white, cream, suffused with yellow, pink or tan, throat often yellow** (solid yellow in P. variabilis); ovary smooth and scaly outside [*ovules in 2 series per locule, stigma rhombic*].

Fruits: Capsules woody, linear to oblong-linear, flat, 10-50 cm long, central rib usually obscure, surface smooth, green, glands absent, lenticels and hairs few or absent; seeds broad, body compact, wings broad and translucent.

Ecology: Pollination via bees and hawkmoths. Seed dispersal via wind.

Distribution: Widespread from Argentina to Nicaragua and Honduras: 17 species; Guianas: 5 species; 2 described here.

Notes: Distinctive vegetative characters include: sharply 4-sided, ± winged branchlets and stems with detachable corners; papery, peeling bark; and compound leaves.

13.1. Pleonotoma albiflora A.H.Gentry

Woody liana. Stems: prophylls mostly absent; slash changing color from cream to orange. **Leaves: Twice-palmately compound, tendrils robustly 3-forked;** leaflets elliptic-ovate, 2.5-14 x 1.5-7.5 cm, base round, apex shortly acuminate; leathery. **Inflorescences:** open, lax racemes; calyx 2-lipped, 1 cm long, glands absent; **corolla 4.5 cm long, white or cream,** lobes with cilia. **Fruits: capsules 35-50 x 2-3 cm,** drying yellowish-green; **seeds to 7.5 cm wide.**

Common; non-flooded old-growth forest. Guianas; also E and S Venezuela, NE Brazil.

13.2. Pleonotoma clematis (Kunth) Miers

Woody liana. Stems: prophylls leafy (stalked, to 0.5 cm long) or inconspicuous; slash not changing color. **Leaves: Twice-palmately or -pinnately compound, tendrils 3-forked or 3-clawed;** leaflets elliptic-ovate, 1.7-9 x 0.8-3.8 cm, base asymmetrical, cuneate, apex long-acute; papery. **Inflorescences:** narrow racemes; calyx truncate or toothed, to 0.5 cm long; corolla, white, ± pinkish towards apex. **Fruits: capsules 10-40 cm,** margins thickened, drying white to brown; **seeds to 4 cm wide. Notes:** P. echitidea is similar to P. clematis, but fruits are smaller (capsules to 25 cm long, seeds to 2.5 cm wide).

Common; riparian and seasonal (deciduous and semi-deciduous) forest. Guianas; also Colombia, Venezuela and N Brazil.

72 | 73

14. Pyrostegia C.Presl.

Woody lianas. Climbing via leaf tendrils. Widely but unevenly distributed across tropical South America: 2 species; Guianas: 1 species.

14.1. Pyrostegia venusta (Ker Gawl.) Miers

[Syn: Pyrostegia dichotoma Miers ex K.Schum]

Common name: arimi arokë (Tr); tukuinetë (Tr)

High-climbing liana. General: Hairless or hairs simple, sometimes scaly [*lepidote*], surfaces mostly without conspicuous color, sap, or scent. **Stems: branchlets 6-8-sided, green, striate to ribbed,** bark not peeling, lenticels absent; **nodes with tiny, triangular prophylls,** IP glands absent, partial IP ridge present. **Older stems remotely 6-8-sided,** bark reddish-brown, flaky, with length-wise ridges; **stem x-s pattern with (phloem) arms in multiples of 4. Leaves:** 2-3-foliolate, tendril 3-forked (rarely 1-forked). Leaflets ovate, medium-small (to 10 cm long), base truncate to rounded, apex acuminate to acute; subleathery to papery, **translucent gland dots present,** sometimes pitted below, glands clustered in vein axils; veins often yellow, sunken above, prominent below, 2⁰ veins broadly looping. **Inflorescences:** Terminal or axillary, many-flowered, flat-topped panicles [*corymbose cymes*]; calyx **cupular and conspicuously flared towards apex,** truncate, minutely 5-toothed, scaly and bearing glands, yellowish-green; **corolla narrowly tubular,** straight in tube, to 8 cm long, **bright orange to orangish-red,** with 4-5 lobes near apex, lobes closed in bud and eventually curling back; stamens white, extending beyond corolla mouth; ovary smooth and scaly outside [*ovules in 1 series per chamber, stigma rhombic*]. **Fruits:** Capsules leathery, linear, flattened, to 20 x 1 cm, midvein raised, surface smooth, hairless; calyx persistent; seeds winged, body smooth, wings translucent. **Uses:** pantropical ornamental; water from stem for coughs (Trio).

Common; old-growth forest and secondary forest. Guyana, Suriname; widespread in the Americas, Argentina to Mexico.

---

71. Pleonotoma clematis, pinnately- compound leaves, 3-forked tendrils [SI]

72. Pleonotoma venusta, 6-sided branchlet with raised ridges and minute prophylls [FRD]

73. Pleonotoma albiflora woody stem with fibrous bark [FRD]

BIGNONIACEAE

74   75   76   77

15. Stizophyllum Miers

Woody lianas. Climbing via leaf tendrils.

General: Surfaces densely hairy (S. inaequi-laterum) to finely hairy (S. riparium), hairs simple but sometimes with a visible ring at the base; peltate scales and glands often present; without conspicuous scent or sap.

Stems: Branchlets round, hollow; nodes with **prophylls large and leafy** (S. riparium) or inconspicuous, glands absent; IP glands and IP ridge absent. **Older stems round,** 2-5 cm diameter, **with central hollow chamber,** bark softly fibrous, whitish; **stem x-s arms 4.**

Leaves: 2-3-foliolate, **tendrils simple or 3-forked; petioles often long;** leaflets ovate to elliptic, sometimes toothed when young; papery, ± hairy above and below, **translucent [pellucid-punctate] gland dots present below;** 2° veins pinnate, sometimes palmately-veined at base.

Inflorescences: Axillary or terminal racemes; calyx urn-shaped [urceolate] or funnel-form, 2-5 lobed, glands present or absent; **corolla tubular/funnel-form, 5-lobed (not 2-lipped), 3-5 cm long, white, pink or magenta,** without nectar guides; anthers held within corolla tube; ovary smooth and scaly outside [ovules in 2 series per locule; stigmas elliptic].

Fruits: Capsules pencil-like, narrowly elongate-linear, up to 45 cm long, obscurely ribbed, wings and ridges absent, calyx persistent; surface smooth or densely hairy [villose]; seeds thin, with broad, translucent wings.

Ecology: Pioneers that often occupy disturbed areas; hollow stems may shelter Crematogaster stinging ants; seed dispersal via wind.

Distribution: Widespread in tropical and subtropical Americas from Bolivia to S Mexico: 3 species; Guianas: 3 species, 2 described here (S. perforatum is a third species occurring in Guyana).

Notes: The most important identification characters for the genus are the hollow stems, papery leaves, and translucent gland dots.

15.1. Stizophyllum inaequilaterum Bur. & K.Schum.

Liana; densely brown-hairy. **Stems: prophylls inconspicuous. Leaves: tendrils simple to 3-forked, conspicuously hairy, petioles 2-7 cm long;** leaflets elliptic to ovate, 4-13 x 2-6 cm, base unevenly truncate to subcordate, apex acute to obtuse; **gland dots difficult to see through hairs;** 2° veins palmate at base. **Inflorescences:** axillary or terminal racemes; calyx urn-shaped, inflated, 0.6-1.2 cm long, irregularly 5-lobed, glands present; **corolla** 3.5-4.5 cm long, **whitish-pink with dark pink lobes. Fruits:** capsules 30-33 x 0.45 cm.

Uncommon; non-flooded old-growth forest, disturbed forest, xerophytic, low forest on bauxite or granite. Suriname, French Guiana; widespread, Bolivia to Panama.

15.2. Stizophyllum riparium (Kunth) Sandwith

Common name: ruwïïmë

Liana; shortly hairy. **Stems: prophylls large and leafy when present. Leaves:** tendrils simple, hairless; petioles 2-11 cm long; leaflets oblong-elliptic to ovate, 3-20 x 2-12.5 cm, base unevenly truncate to rounded, apex acute to acuminate; **gland dots easily seen; 2° veins pinnate throughout blade. Inflorescences:** axillary racemes; calyx funnel-form, 0.5-1.2 cm long, 2-5-lobed, peltate scales and glands present; corolla 3.2-5 cm long, **solid white to** cream-colored. **Fruits:** capsules 24-45 x 0.6 cm. **Use:** sap from cut stem for ear infections (Wayana) and irritated eyes (NW Guyana).

Common; non-flooded and flooded old-growth forest, disturbed forest. Suriname, French Guiana; widespread, Bolivia to Panama.

---

74. Stizophyllum riparium, juvenile shoot, 3-foliolate, leaflets with pinnate venation [BH]

75. S. riparium, growing towards the light [BH]

76. S. inaequilaterum, close-up of hairs [BH]

77. S. riparium, branchlets soft, hollow, often occupied by ants [FRD]

---

16. Tanaecium Sw.

Woody lianas. Climbing via leaf tendrils.

General: Vegetatively mostly non-descript; surfaces not densely hairy, sometimes scaly [*lepidote*] or densely lenticellate, rarely with sweet scented leaves (T. pyramidatum).

Stems: Branchlets round (4-sided); lenticels sparse, hairless to lightly hairy, bark not peeling; nodes with conical, awl-shaped and/or bromeliad-like prophylls, these with or without glands; IP glands and IP ridge present or absent. Older stems round to remotely 4-sided, robust, sometimes with joint-like swellings at nodes, bark dull white to yellow-white, softly woody, with vertical furrows; x-s arms 4 or 8.

Leaves: 2-3-foliolate, **tendril simple** (T. bilabiata) **or 2-3-forked;** petiole round, sometimes with apical glands; leaflets size and shape variable, papery to subleathery, glands present, scattered; margins entire, **young leaves sometimes toothed; secondary veins straight, ladder-like.**

Inflorescences: Terminal (axillary), thyrse or cyme; **calyx cup- to bell-shaped, 2-lipped,** leathery, hairy outside, glands present or absent; corolla tubular or funnel-form, **straight in tube, 5-lobed (not 2-lipped),** medium to large; **yellow, white, pink, or magenta, without nectar guides,** hairy outside [*villose*], without glands; stamens held within corolla tube; ovary smooth and scaly outside [*ovules in 1(2) series per locule, stigma shape elliptic*].

Fruits: Capsules woody or leathery, **linear, straight, flattened or inflated,** unwinged, surface ± hairy, typically bearing lenticels, glands absent; seeds thin and winged to thick and wingless.

Ecology: Seed dispersal via wind or water.

Distribution: Widespread in the American tropics and subtropics, from Argentina to Mexico, extending into the West Indies: 17 species total; Guianas: 5 species, 4 described here. Wet to dry forest, cerrado, and caatinga vegetation.

Notes: Four species occurring in the Guianas have recently been transferred to the Tanaecium genus, including Arrabidaea bilabiata, Arrabidaea revillae, Ceratophytum tetragonolobum, and Paragonia pyramidata.

79

80

16.1. Tanaecium bilabiatum (Sprague) L.G.Lohmann

[Syn: Arrabidaea bilabiata (Sprague) Sandwith]

Common name: bokotaballi (Ar); bokota firoberu (Ar); duldul (Sr); duludulu, duruduru (Au, Sr); prasara titei (Sr); tiapotano (Ca).

Woody liana; lightly hairy. **Stems:** branchlets round; nodes often with **tiny, awl-shaped prophylls;** IP gland fields and IP ridge absent; older stems round, bark whitish-grey, smooth; **stem x-s arms 4. Leaves:** 2-3-foliolate, **tendrils simple, robust;** leaflets elliptic to narrowly ovate, 3.5-23 x 2-10 cm, base rounded, apex obtuse to acute. **Inflorescences:** calyx bell-shaped, 2-lipped, > 1.5 cm long, with vertical glands, white or pinkish; corolla tubular, 4.2-7 cm long, **white or cream, slightly pink outside, throat yellow. Fruits: capsules flattened, 13-33 x 1.5-1.8 cm,** white-dotted, margins ± thickened; seeds woody, flat, unwinged. **Use:** stem and bark as a general tonic (tea), to stimulate urine production, clean the uterus, clean the blood, lose weight after pregnancy and as a component of male aphrodisiac potions (van Andel & Ruysschaert 2011); leaves (tea) to treat muscle pain, rheumatism and sprains (Trio).

Uncommon; old-growth forest, riparian forest. Guianas; Brazil and Venezuela.

16.2. Tanaecium pyramidatum (L.C.Rich.) L.G.Lohmann

[Syn: Paragonia pyramidata (L.C. Rich.) Bureau]

Common name: (rode) baskita-titei (Sr); gaán háti (Sa); kuraiweimë (Tr); opu (Pa); watratitei (Pa); yapepuku (Ca).

High-climbing, robust liana; lenticels abundant. **General:** cut plants parts smelling "sweet". **Stems:** branchlets round; **nodes with 2-3 conical, vertical-stacked prophylls** (see also Pachyptera kerekere); IP glands absent, partial IP ridge present; **older stems remotely 4-sided,** with joint-like swellings and white-yellow bark; stem x-s arms 8. **Leaves:** 2-foliolate; tendrils **2-3-forked; petiole bearing apical glands;** leaflets ovate, ± 12 x 6 cm, base rounded to cuneate, apex acute; mostly leathery; 2° veins strongly ascending. **Inflorescences:** terminal or axillary, many-flowered; calyx cup-shaped, to 1 cm long, flared, lobed or entire, pinkish-green, scales and glands present; **corolla tubular/ funnel-form, 5-lobed, to 7 cm long, pink or magenta with white throat. Fruits: capsules oval-inflated, 60 x 1.3-2 cm,** sides convex, valves with one obscure central furrow; surface green, rough; seeds with broad, papery wings (to 3.5 cm wide).

Common. Riparian forest, non-flooded old-growth forest, xerophytic forest on lateritic soil. Guianas; widespread in neotropics.

16.3. Tanaecium tetragonolobum (Jacq.) L.G.Lohmann

[Syn: Ceratophytum tetragonolobum (Jacq.) Sprague]

High-climbing, robust liana; lenticels abundant. **Stems: branchlets round to 4-sided,** bark with length-wise ridges and large, sparse lenticels; nodes **with narrow-pointed, awl-shaped prophylls; IP glands present,** IP ridge absent; **stem x-s arms 4-8. Leaves:** 2-3-foliolate, tendrils 3-forked; leaflets ovate, 8-14 x 6-10 cm, base rounded, uneven, apex acuminate to acute. **Inflorescences:** terminal, few-flowered panicles; calyx narrowly cup-shaped, to 1 cm long, margin truncate, yellow-green, glands present; **corolla tubular/funnel-form,** tube ± flattened, straight in tube, **5-lobed (not 2-lipped),** thick, **tube pale yellow outside lobes and throat cream to white. Fruits:** capsules 20-30 x 0.5-0.7 cm, remotely 4-sided with a central furrow on each valve, narrowing to a fine tip at apex, orangish-brown at maturity, densely lenticellate; seeds with broad, papery wings.

Uncommon. Old-growth forest. Guianas; widespread, Amazon Basin (all countries), Venezuela, C America, S Mexico including the Yucatan Peninsula.

---

78. Tanaecium pyramidatum, flowering shoot [BH]

79. T. pyramidatum, branchlet 4-sided, node with conspicuous leafy prophylls bearing glands, slightly raised IP ridge [BH]

80. T. pyramidatum, stem x-s arms ± 20 [NBC]

---

17. Tynanthus Miers

Woody lianas. Climbing via leaf tendrils.

Common name: "cipó-cravo" (Brazil)

General: Vegetative parts releasing strong scent of clove (Dutch: kruidnagel; French: clou de girofle) upon cutting; lightly to densely hairy (hairs simple); commonly bearing glands - shield-like [*peltate*] or saucer-like [*patelliform*].

Stems: Branchlets round to 4-sided, woody bark and lenticels present or absent, often finely striate, hairs and glands few to many; nodes with **tiny, triangular prophylls and IP glands present (T. pubescens) or leafy prophylls and IP glands absent (T. polyanthus, T. sastrei)** (bromeliad-like prophylls only in Tynanthus species outside of the Guianas), IP ridge partial or absent. Older stems round to remotely 4-sided, bark smooth, brown, striated; stem x-s arms 4.

Leaves: 2-3-foliolate, tendrils simple (T. polyanthus) or 3-forked/3-clawed, petiole sometimes grooved above; leaflets elliptic to obovate, margin entire (rarely toothed when young); leathery to papery, colors above and below contrasting [*discolorous*] or similar [*concolorous*], often with glands and hairs above and (especially) below; 2° veins pinnate.

Inflorescences: Axillary or terminal, panicles/thyrses with **small flowers (< 2 cm long),** axes often hairy; calyx cup-shaped, 0.1-0.45 cm long, margin truncate or finely toothed, glands absent; corolla funnel-shaped, **tube angled (not straight), conspicuously two-lipped** (2 lobes above, 3 lobes below), white, cream or pale yellow (rarely bluish or reddish), nectar guides present (T. pubescens) or absent, lower lobes usually hairy, glands absent; nectar disc poorly developed; stamens barely exceeding corolla tube, anther sacs U-shaped; ovary conspicuously hairy [*ovaries in 2 series per locule, stigma rhombic-shaped*].

Fruits: Capsules linear, flattened, 10-55 x 0.4-4.2 cm, with raised or winged margins, with 1-2 central vertical ridges; rough-surfaced, ± hairy, glands present or absent; seeds broad, flat, with smooth body and translucent wings.

Ecology: Flowers open during the day and close at night [*diurnal*]. Seed dispersal via wind.

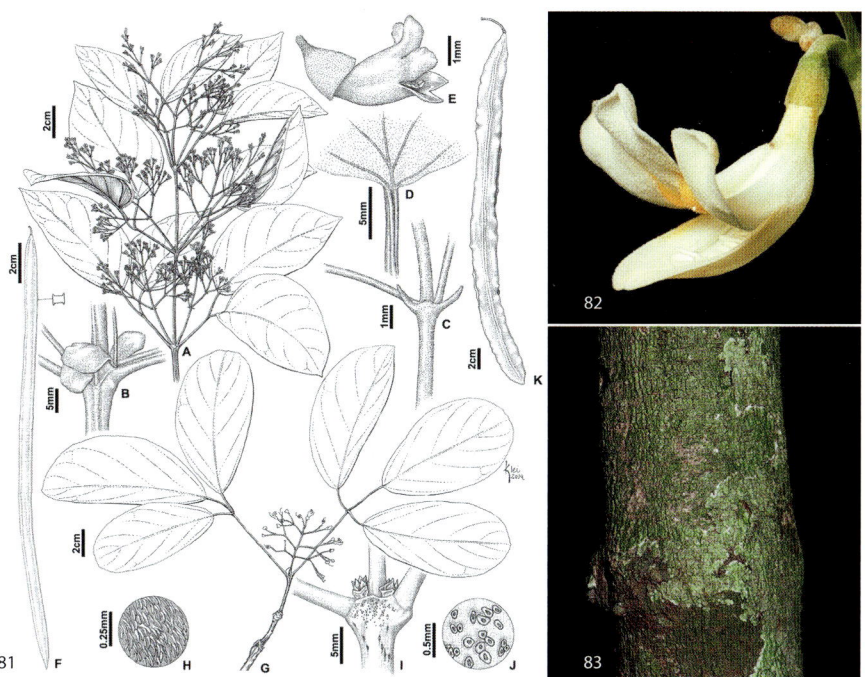

Distribution: Widespread in the neotropics: 14 species; Guianas: 3 species.

Notes: The most important identification characters for the genus are the clove scent, small, bent 2-lipped flowers, and fruits with marginal wings or ribs.

17.1. Tynanthus polyanthus (Bureau) Sandwith

Woody liana; with fine or scattered long hairs. **Stems:** branchlets with few to many lenticels, **nodes with leafy prophylls up to 2.5 cm long, IP glands absent. Leaves: tendrils simple,** petioles 1-5 cm long, bearing glands; leaflets elliptic to ovate, 4–16.5 cm x 2.5–12.6 cm, base cuneate to subcordate, apex caudate-mucronate, blades mostly thin, discolorous or concolorous. **Inflorescences:** calyx 1-2.7 cm long; **corolla short, 0.4-0.8 cm long, nectar guides absent. Fruits: capsules 10.5–25 x 0.4–0.9 cm wide,** leathery to woody (rough surfaced), lenticels present or absent, hairs and glands present, **margins slightly raised, unwinged, central ridge single;** seed body to 1.5 x 0.9 cm; wings 0.4–1.2 cm long.

Uncommon; non-flooded old-growth forest. Guyana (Upper Essequibo); Amazonian Bolivia to Venezuela and the Greater Antilles.

17.2. Tynanthus pubescens A.H.Gentry

Woody liana; with dense hairs. Stems: branchlets with many lenticels; **nodes with tiny, triangular prophylls, IP glands present;** older stem 3-4 cm diameter, inner bark pink, wood white. **Leaves: tendrils 3-forked or 3-clawed,** petioles 1-7.5 cm long, bearing glands; leaflets elliptic to obovate, 5.9–15.4 x 3.1–11.3 cm, base cuneate or obtuse, apex acuminate or obtuse; papery to leathery, **densely white-hairy below, discolorous. Inflorescences:** calyx 0.3-0.45 cm long; corolla 1-1.6 cm long, nectar guides present (yellow), hairs present within tube at base and on lower lobes. **Fruits: capsules 15-30 x 0.5-0.8 cm,** with double central ridge,. **Fruits:** capsules 20-55 x 2.3–4.2 cm, woody (rough surfaced), lenticels usually dense, hairs and glands present, **margins prominently raised (winged)**

and with a vertical double ridge at center; seed body to 2.8 x 1.3 cm; wings 1-1.8 cm long.

Uncommon; non-flooded old-growth forest, brown sand. Suriname, French Guiana; also N Brazil, Peru, Venezuela.

17.3. Tynanthus sastrei A.H.Gentry

Woody liana; with fine or regular hairs, **sunken-cup [patelliform] glands mostly absent.** **Stems:** branchlets with few to many lenticels, nodes with leafy prophylls 0.35-0.65 cm long, IP glands absent. **Leaves: tendrils 3-forked,** petioles 2.3-6.6 cm long, bearing hairs and glands; leaflets elliptic, 6–16.5 cm x 3.6–9.5 cm, base cuneate to obtuse, apex acuminate-mucronate,

blades papery to leathery, concolorous, shield-like [peltate] glands more abundant above than below. **Inflorescences:** calyx 0.1 – 0.2 cm long; **corolla short, 0.6-0.8 cm long, nectar guides absent, hairs absent within tube, dense on lower corolla lobes. Fruits:** Not seen.

Rare and apparently endemic to the region; non-flooded old-growth forest, xerophytic forest on white sand, open areas. Suriname (Suriname record, B.Hoffman coll no. 5302, Brokopondo Lake area) and French Guiana only.

85. Tynanthus polyanthus: a. flowering branch; b. node with leafy prophylls; c. inflorescence axis showing bracts; d. grooved leaf stalk; e. flower; f. flattened fruit. g–k. T. pubescens: g. flowering branch; h.branchlet hairs i-j. IP glands; k. fruit. [KS] (pub. in Madeiros & Lohmann 2015)

86. Tynnanthus-panurensis, flower (Brazil species) [FRD]

87. Tynnanthus panurensis, woody stem (Brazil species) [FRD]

18. Xylophragma Sprague

High-climbing lianas or shrubs; climbing via leaf tendrils. Widespread in the neotropics: 4 species; Guianas: 1 species.

18.1. Xylophragma seemannianum (Kuntze) Sandwith

High climbing liana or shrub. General: surfaces densely hairy (hairs simple, forked or stellate), often appearing scaly. **Stems: branchlets round to ± 4-sided,** dark green and bearing sparse lenticels; nodes with **prophylls small, conical or miniature-bromeliad-like; IP glands present and conspicuous,** IP ridge partial. Older stems round, 3-7(-10) cm diameter; bark white, soft, densely lenticellate; stem x-s arms 4. **Leaves:** 2-3-(rarely 5)-foliolate; tendrils simple; petioles round, 6-12 cm long; leaflets rhomboid-ovate to -obovate, 9-13 x 6-9 cm, base truncate to wide-cuneate, apex acuminate to acute; **thin to papery, stellate-lepidote above and below, especially below; 3-5-palmately-veined;** blades drying olive-tan. **Inflorescences:** terminal and axillary, panicles/thyrses at leafless nodes, large,

hairy; linear bract pairs present; **calyx** cup-like, small, **5-ribbed and toothed, bearing glands,** pale green, hairy; corolla tubular/funnel-form, straight in tube, 5-lobed (not 2-lipped), **3.5-6.2 cm long, pink to magenta with a white throat,** nectar guides absent, with a fragrant musky scent; stamens held within corolla tube; ovary smooth and scaly outside [ovules in multiple series per chamber]. **Fruits: Capsules oblong-elliptic, 5.5-16 x 3.5-5 cm, resembling fruit of Jacaranda,** wingless, without persistent calyx, surface smooth or with a raised zone of glands, blackish or brown; seeds 1.5-3 cm long x 3-5 cm wide, wings translucent.

Distribution: Rare; non-flooded old-growth forest, granite flats, xerophytic, deciduous, low forest. Guyana, Suriname; widely distributed from Bolivia, N Brazil to Mexico.

BORAGINACEAE

BORAGINACEAE

Occasional lianas, weakly-climbing shrubs, and vines; mostly herbs, shrubs or trees. Climbing via twining stems and elbow-like branchlets and petioles. Leaves are alternate, simple, with margins entire, wavy or finely toothed. Stipules are absent. Inflorescences are spirals, spikes or heads of tubular 5-parted flowers with 5 stamens and a superior ovary. Fruits are fleshy or dry drupes.

General: Few distinctive vegetative characters. **Surfaces often rough** [*asperous*] **and/or with erect and bristly hairs** (hardened with mineral deposits); lacking unique sap, scent, or taste.

Stems: Branchlets round, smooth, nodes without swellings or horizontal line. Older stems oval to round, sometimes flat-sided, slender to medium diameter, bark thick and fibrous (Tournefortia) to thin (Varronia); stem x-s pattern simple.

Leaves: Large and thick (Tournefortia) to medium-sized and stiff-asperous (Varronia); margin entire, wavy or toothed; 2° veins pinnate and curving to edge, not forming marginal loops; **petiole often flexed, twisted or jutting outwards at 90° angle.**

Inflorescences: Terminal or axillary, dense heads, spikes, or spiral cymes [*helicoid, scorpioid*] **calyx 5-lobed, persistent; corolla tubular, trumpet- or funnel-shaped,** greenish-white or yellowish-white; bisexual or unisexual; stamens 5, borne on tube and alternating with corolla lobes; ovary 4-lobed or unlobed, style 1, stigma 1 (Tournefortia) or 4 (Varronia).

Fruits: Climbing species with drupes, either naked (Tournefortia) or enclosed within a persistent, inflated calyx (Varronia).

Ecology: Seed dispersal via animal (birds).

Distribution: Cosmopolitan, mostly herbs and small shrubs outside of the tropics. Neotropics: 23 genera/500 species; Guianas: 2 genera /10 species of climbers.

Notes: Cordia is a common Boraginaceae genus that has been split up in recent years. The Varronia species treated here were previously known as Cordia species.

1. Tournefortia L.

Woody canopy lianas, climbing shrubs, and vines; also erect shrubs. Climbing via twining stems and twisted or flexed petiole bases.

General: **Surfaces not rough-surfaced;** vegetation densely hairy to hairless.

Stems: Older stem oval or with flat sides, furrowed and 'bumpy' [*tuberculate*]; bark thick and fibrous, horizontal lenticels often present.

Leaves: Elliptic to narrow; margins entire to wavy, **not toothed;** often with flattened hairs [*sericeous*] below; veins strongly raised below, often yellow; 3° veins parallel and ± 90° to midvein; petiole often twisted or flexed near leaf axil.

Inflorescences: Terminal, flowering axes straight or spiral; calyx 5-lobed, 0.2 cm long, persistent; corolla tube to 1 cm long, white to greenish-white, trumpet-shaped with recurved lobes – mouth of tube inflated, normal, or flattened; **ovary 4-lobed or unlobed, with 1 stigma/style.**

Fruits: Globose drupes, white, translucent or pale yellow-orange at maturity, 0.6-0.8 cm diameter, not enclosed by persistent calyx, 4-lobed or unlobed; nutlets 2-4.

Ecology: Seed dispersal via animal (bird).

Distribution: Pantropical distribution, ca. 150 species; Guianas: 6 climbers, 4 common species described here.

Notes: Black-drying leaves are one of the few reliable Tournefortia characters for sterile plants (Gentry 1995).

1.1 Tournefortia bicolor Sw.

Common name: baskita-titei (Sr)

Lianas or scandent shrubs; lightly hairy. **Stems:** older stems round or remotely 4-sided and deeply grooved, to 3 cm diameter; bark whitish-yellow, thick. **Leaves:** lanceolate to elliptic, 5-14 x 3-9 cm, base obtuse to rounded, apex acute; sub-leathery, shiny above, with flattened hairs above and below; petiole 0.5-1.5 cm long. **Inflorescences:** densely branched, spiral [*scorpioid*] racemes 1-4 cm long; corolla tube 0.5 cm long — constricted at throat — lobes broad. **Fruits:** ovoid, 0.8 cm long, maturing translucent green to white; **unlobed.**

Common; creek and riverine forests, and open areas in non-flooded old-growth forest. Guianas, widespread in the neotropics.

1.2 Tournefortia cuspidata Kunth

Lianas or scandent shrubs; densely hairy [*hirsute*] **Leaves:** lanceolate-ovate to lanceolate, 7-15 x 3-6 cm, base rounded or obtuse, apex acuminate; **with flattened hairs above and below;** petiole 0.5-1.3 cm long, twisted at base. Inflorescences: crowded straight spikes, 1-3 cm long, on long branches; corolla tube 1 cm long — neither constricted nor inflated – lobes broad. Fruits: ovoid, 0.8 cm long, maturing translucent green to white; **unlobed.**

Common; non-flooded old-growth forest, especially on slopes along creeks and rivers. Guianas; N South America to Central America and the West Indies.

1.3 Tournefortia maculata Lam.

Common name: awari imopi tëkënë (Tr); sekre-patu-bita (Sr); wayamu eresitjuru (Ca)

Lianas or scandent shrubs; mostly hairless. **Leaves:** ovate to elliptic, 5-10 x 2.5-5 cm, base acute to rounded, apex acuminate; robust, **with tiny bumps above and below;** petiole 0.7-1.5

cm long. **Inflorescences:** panicles slender, many-branched, 5-15 cm wide; corolla tube 0.4-0.6 cm long — neither constricted nor inflated – lobes long and narrow. **Fruits:** drupes ovoid, yellow, fleshy, sometimes black-dotted; **4-lobed.**

Uncommon; riparian forest, creek forest, and non-flooded old-growth forest. Guianas; also Brazil and Peru to Mexico.

1.4. Tournefortia ulei Vaupel

Lianas or scandent shrubs; lightly hairy. **Leaves:** broadly lanceolate to ovate, 6-17 x 3-8 cm, base obtuse, apex acuminate; thin, **with tiny bumps above and below;** petiole 1-1.5 cm long. **Inflorescences:** loosely-branched racemes, 2-10 cm long; corolla tube 0.7-0.8 cm long — inflated at throat — lobes broad, style well-developed,

0.2-0.3 cm long. **Fruits:** drupe ovoid, ca. 0.5 cm diam., maturing translucent green to white; **obscurely 4-lobed or unlobed.**

Common; non-flooded old-growth forest edges and secondary forest. Guianas; also Amazonian Brazil to Bolivia.

2. Varronia P.Browne (= some Cordia L.)

Some climbing shrubs; mostly erect shrubs and trees. Climbing weakly via angled branches or petioles.

General: Surfaces rough [*asperous*]; with short, curved bristly hairs intermingled with long or short hairs.

Stems: Branchlets round or flat-sided, center often hollow and occupied by ants or filled with spongy white material. Older stems round, bark thin, easily stripped, with fine networked ridges running along stem – not 'bumpy' or furrowed; wood soft.

Leaves: Lanceolate to ovate, **margins evenly-toothed (entire); surface rough above [*asperous*],** lightly to densely hairy below; 2° veins extending to margin, not looping; **petiole straight or elbow-like, at 90°angle from stem.**

Inflorescences: Terminal or axillary, in **spikes or heads [*not spiral*];** calyx 5-toothed or lobed, to 0.5 cm long, persistent; corolla trumpet- or funnel-shaped, 0.5-1 cm long, white or white-yellow; **ovary 4-lobed with a 4-branched style.**

Fruits: Fleshy transparent drupes, surrounded by an inflated calyx, maturing white to red, pulp sticky; 1-4-chambered, with a hard, bony center.

Ecology: Seed dispersal via animal (bird).

Distribution: Pantropical: ca. 100 species; Guianas: 2 weakly climbing species.

Notes: Varronia is a sister group to Cordia.

2.1. Varronia polycephala Lam.

[Syn: Cordia polycephala (Lam.) I.M.Johnst.]

Common name: man-blaka-uma (Sr)

Climbing or erect shrubs. Stems: branchlets with short bristles and matted long hairs intermixed. **Leaves:** 2-12 x 1-5 cm, base acute to obtuse, apex acute; with bumps [*tuberculate*] and lightly hairy above, densely brown-hairy

BORAGINACEAE

[*tomentose*] below; petiole straight (not elbow-like), 0.3-1 cm long, base extending onto stem [*decurrent*]. Inflorescences: axillary and terminal, branched, dense heads, 0.5-1.5 cm wide. Fruits: subglobose drupes, 0.5 cm diam., red at maturity.

Common; disturbed areas, marsh forest along creeks and rivers in coastal area. Guianas; also N South America, West Indies, and coastal Brazil.

1. Varronia schomburgkii in fruit [BH]
2. Tournefortia bicolor, branchlet and node [RA]
3. Varronia schomburgkii, leaf backside with marginal teeth and elbow-like petiole [BH]
4. Tournefortia sp., translucent berries without persistent calyx [PT]
5. Varronia schomburgkii, inflorescence with persistent calyces [CF]
6. Tournefortia bicolor with a spiral (helicoid) inflorescence [RA]
7. Tournefortia paniculata, robust woody stem with fibrous bark and horizontal lenticels [BH]

2.2. Varronia schomburgkii (DC.) Borhidi

[Syn: Cordia schomburgkii A.DC.]

Common name: blaka-uma (Sr); blaka wintje (Pa); danpuwi (Pa); diamarpe (Ar); fowru-loso wiwiri (Sr); kabujakoro makuja pipá (Ca); par-akatai (Ca); wayamaka erepatï (Ca)

Climbing or erect shrubs. Stems: branchlets with short bristles and fine short hairs inter-mixed. **Leaves:** 5-10 x 2-7 cm., base obtuse or ± rounded, apex acute; with flattened hairs above; petiole elbow-like, 0.6-0.9 cm long. **Inflorescences:** axillary, unbranched, loosely-flowered spikes to 15 cm long. **Fruits:** ovoid drupes, 0.5 cm diam., red at maturity.

Common; forest margin, secondary forest, and savanna. Guianas; also Trinidad and Tobago.

CANNABACEAE

CANNABACEAE

Rarely lianas or scandent shrubs; commonly erect shrubs and trees. Widespread in tropical and temperate zones. Neotropics: 4 genera/20 species; Guianas: one genus with a climbing species. Most neotropical, woody Cannabaceae were formerly placed in the Ulmaceae (Elm family).

1. Celtis L.

Lianas or climbing shrubs; mostly erect shrubs or treelets. Climbing via twining stems, angled branchlets, and spines. Stipules are present. Leaves are simple, alternate, palmate-3-veined. Inflorescences with tiny flowers clustered in leaf axils; stamens 5; ovary superior. Fruits are yellow-orange drupes. Neotropics: 15 species; Guianas: 2 species, one climbing.

1.1. Celtis iguanaea (Jacq.) Sarg.

Common name: busi-lemki-maka (Sr)

Shrubby liana or climbing shrub; armed. **General:** surfaces rough to touch [*asperous*], hairless [*glabrous*] or lightly hairy, cystoliths (calcium carbonate) present in leaves, notable scent or sap absent. **Stems: branchlets alternate, ± zigzag on main stem; stipules inconspicuous in leaf axils** (0.2-0.5 cm long) **or modified as 1-2 spines** (recurved on climbing branchlets, straight on normal branchlets), spines thick at base, to 1.5 cm long; older stems perfectly round, 3-8 cm diameter, with paired spines; bark light-colored, ± smooth, bearing lenticels; stem x-s simple. **Leaves: 2-ranked** [*distichous*]; ovate to elliptic, 3-14 x 2.6-6.5 cm, **base unequal** [*assymmetrical*], rounded, truncate, or subcordate; apex acuminate to acute; margins entire, roughly toothed, or with a few teeth at tip; papery and rough; **veins strongly palmate at base, pinnate towards apex,** 2° veins looping at margins, raised below, tiny pockets [*domatia*] present in 2° vein axils; petiole standard, without swelling [*pulvinus*] at apex, to 0.8 cm long. **Inflorescences:** axillary, many-flowered (male only) or few-flowered (female + bisexual), bracts large, to 4 cm long; **flowers radial, tiny (0.1 cm diameter), green;** (4-)5-parted, margins frilly, corolla (petals) absent; stamens 5, free; ovary bottle-shaped, 2-chambered;

styles 2, white and hairy. **Fruits: ovoid drupes,** 0.6-1.5 x 1-1.4 cm, maturing bright orange or yellow-orange, **single-seeded, styles persistent. Ecology:** Seed dispersal via animal (birds). **Notes:** Byttneria (Malvaceae) and Gouania (Rhamnaceae) also have zigzag stems, alternate 3-palmate-veined leaves, toothed to wavy leaf margins, and stipules. Byttneria is distinguished by star-shaped [*stellate*] hairs, a glandular zone at the leaf-base, and spiny fruits. Gouania is distinguished by coiled 'butterfly tongue' tendrils, paired conical glands at the leaf-base, and propeller-like fruits.

Uncommon; secondary forest, dry forest, forest clearings, and riversides. Guianas; widespread from C & S America to Caribbean.

1. Celtis iguanae, leafy branch [BH]
2. C. iguanaea. a. fertile branch; b. stipular spines; c. inflorescence with immature fruits. d. staminate flower, entire and longit. section; e. stamens, dorsal and lateral view; e. immature fruit, entire and longit. section; g, mature fruit [BA]
3. C. iguanaea, leafy branch [RA]
4. C. iguanaea, leaf underside [RA]
5. Immature fruit (drupe) of C. iguanaea [RA]
6. Inflorescence of C. iguanaea, female flowers [BH]
7. C. iguanaea, older woody stem with paired axillary spines [BH]

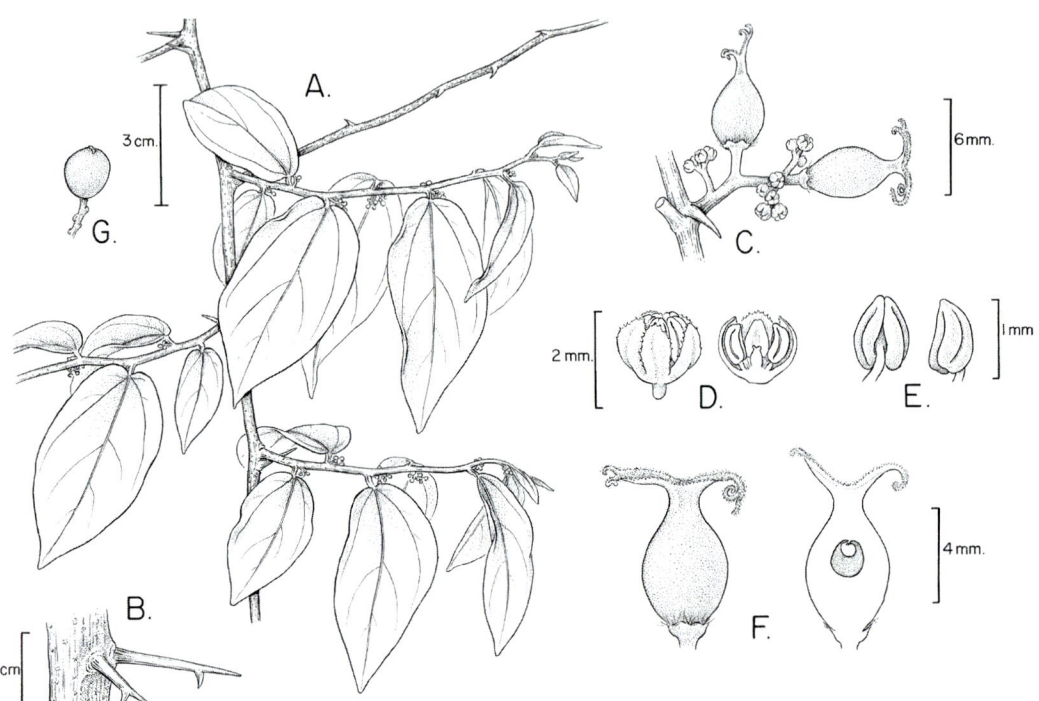

2   Published in Acevedo-Rodriguez et al. 1996, drawing by Bobbi Angel.

CELASTRACEAE

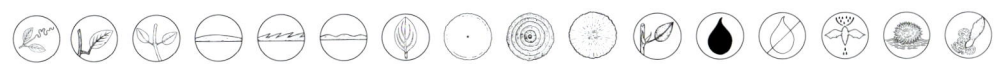

CELASTRACEAE

including Hippocrateaceae

A diverse family of robust woody canopy lianas, and climbing shrubs; also erect shrubs and trees. Climbing via tendril-like terminal shoots and recurved branchlets. Leaves are opposite to sub-opposite, simple, large, leathery and with margins toothed, wavy or entire. Stipules are present, but short-lived. Flowers are small, radial, with 5 free sepals and petals, a nectar disc, 3(5) stamens, and a superior ovary with 3(5) stigmas. Fruits are either dry capsules with winged seeds or fleshy berries with bulky seeds. The Celastraceae family includes two large subfamilies of lianas with similar flowers but very different stem and fruit traits: **Hippocratea group** (Hippocrateoideae-HG) includes Anthodon, Cuervea, Elachyptera, Hippocratea, Hylenaeae, Prionostemma, and Pristimera. **Salacia group** (Salacioideae-SG) includes Cheiloclinium, Peritassa, Salacia, and Tontalea.

General: Hairs largely absent, simple when present (Hippocratea, Peritassa); surfaces mostly smooth, sometimes rough [*asperous*]; **brightly colored rubbery or watery sap sometimes present (yellow, orange or red)**, in some cases visible as 'latex threads' when leaves or twigs are torn apart; raised gland-pairs absent, small flat glands or black dots sometimes present.

Stems: Branchlets mostly opposite, conspicuously smooth and green to densely woody, often with lenticels, spines, or tiny [**papillae**] to large [*tubercles*] bumps; nodes swollen or flat, stipules present (interpetiolar, tiny, usually short-lived) and sometimes leaving horizontal scar. Older woody stems often robust (> 20 cm diameter) and with thick, brown bark. **Hippocratea group:** outer bark variable, inner bark never orange, **stem x-s pattern radial** or simple. **Salacia group:** outer bark smooth, thin, grey or orangish-brown, inner bark orange, slash often with bands of red or orange, **stem x-s pattern concentric**, with narrowly- or widely-spaced rings. In species with wide rings, horizontal bands often connect adjacent rings (Cheiloclinium anomalum, C. hippocrateoides, Salacia kanukuensis, S. miqueliana, S. multiflora, Tontelea nectandrifolia).

Leaves: Highly variable in shape and size, mostly elliptic to ovate, 5-50 cm long, **margins entire, wavy or toothed;** tiny dots (black or clear) or pits sometimes present; **venation pinnate, 2° veins typically widely-spaced and curving steeply upwards**, often yellow, 3° veins net-veined or parallel; **petioles often thick**, emerging at 90° to branchlets or extending onto branchlets [*decurrent*].

Inflorescences: Axillary, between nodes or terminal upon short shoots; branching equally 1-many times [*dichotomous*], in loose or dense clusters [*fascicles*] upon a cushion-like base, in tight broom-like clusters or single-flowered; small bracts usually present; **flowers radial, tiny to small, often flattened at maturity,** yellowish-green, cream, white or pink; sepals 5, free, overlapping; petals 5, free, fleshy to papery, margins sometimes fringed, toothed or appearing irregularly torn [*erose*]; nectar disc conspicuous, flat or tubular, fleshy, thin, or cushion-like; stamens 3(5), distinct, erect or bent, sometimes inserted within holes in nectar disc; ovary superior, often sunken into disc, 3(5)-chambered, stigmas 3(5), linear or forming a flat 'stigmatic shield'.

Fruits: **Hippocratea group: propeller-like dry capsules** with 3 (1-2) flat to ovoid 'blades', separate or partially fused, usually splitting down middle on both sides to release seeds; seeds winged and attached to fruit wall by the wing tip, 6 to many. **Salacia group: berries** with 3-6 oblong, 3-angled seeds immersed in pulp.

Ecology: Seed dispersal via water (Cuervea, Hylenaea), wind, or animal (monkey gut).

Distribution: Pantropical: ± 350 species; Guianas: 11 genera, ± 41 species, 25 species described here.

Use: Common uses documented in the Guiana Shield region include medicine, dye, food (fruits) and drinking water.

Notes: The family is relatively easy to distinguish among opposite-leaved climbers with pinnate venation. Top sterile identification characters: **leaf margins wavy to shallow-toothed; tendril-like green terminal shoots; red to yellow latex or watery sap; skin-like bark, orange sub-bark; cross-section with spoke-like rays or concentric rings.** Celastraceae with entire leaf margins are more difficult to identify; compare with Apocynaceae (white milky sap), Malpighiaceae (silky hairs, raised leaf glands), Rubiaceae (interpetiolar stipules).

1. Anthodon Ruiz. & Pav.

Lianas, also erect shrubs and treelets. Tropical America: 1 species.

1.1. Anthodon decussatum Ruiz. & Pav.

Liana; shrub or treelet. **General:** colorful sap absent, hairless. **Stems:** branchlets round to 4-sided; older stems round; **stem x-s pattern radial. Leaves:** elliptic to ovate, 6-15 x 2-5.5 cm, **apex and margins toothed;** papery, all veins slightly raised above and below. **Inflorescences:** axillary, densely-flowered, flowers 0.5-1.2 cm diameter; petals greenish to whitish-yellow, **long and narrow with finely-toothed** margins; nectar disc short-tubular, tightly enclosing 3 large,

kidney-shaped stamens; stigmas 3, fused into tiny shield. **Fruits: dry, 3-lobed capsules to 18 cm diameter,** lobes emerging only at halfway point from center to outer edge (in contrast to propeller-like fruits with free lobes or 'blades'); seeds with large spreading wing, 3-3.5 x 1.5-2 cm. **Ecology:** Seed dispersal via wind.

Uncommon. Guyana, French Guiana, expected in Surinam; widespread, from Brazil to C America.

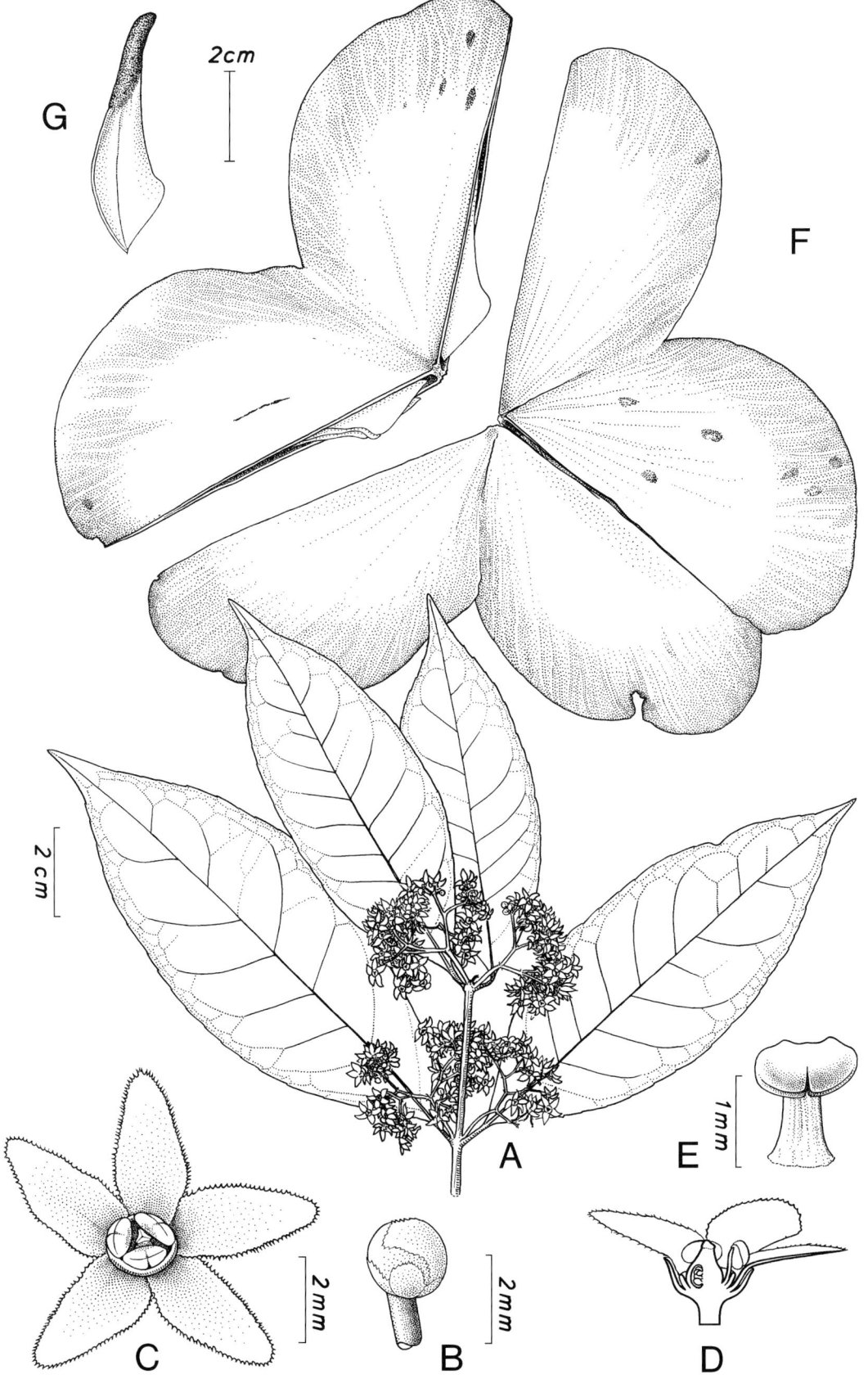

G

2cm

F

2 cm

A

C

2mm

B

2mm

E

1mm

D

2. Cheiloclinium Miers

Robust lianas or climbing shrubs; also erect shrubs and treelets.

General: Bright yellow to orange rubbery sap often present, visible on undamaged leaf and shoot surfaces and/or as 'latex threads' when torn and pulled apart.

Stems: Shoot tip with a spike-like bud emerging from between ultimate leaf pair; branchlets round or 4-sided, bright green and hairless. Older stems with **smooth, skin-like outer bark** and dull-orange inner-bark; slash red-white banded, looking like raw meat; stem x-s pattern **off-center concentric, rings widely-spaced, with irregular radial cross-links.**

Leaves: Shape variable, margins wavy to obscurely toothed; **black gland dots often present;** 1° and 2° veins conspicuously raised below, 3° venation usually parallel and 90° to midvein; petiole short and wrinkled.

Inflorescences: Axillary, not in broom-like clusters; flowers tiny, 0.2-0.5 cm diameter; petals usually long, thin and fleshy, margins entire (rarely fringed, barbed or irregular); stamens 3 (5), attached within 'pockets' in nectar disc; stigmas 3 (5), free and opposite to stamens.

Fruits: Sub-globose to ovoid berries to 4 cm diameter, with a tough, leathery fruit wall and lengthwise bands or ribs near base; seeds 2-6, oblong, angled, embedded in pulp.

Ecology: Seeds dispersed via animal (through gut of spider and howler monkeys).

Distribution: Tropical America: 15 species; Guianas: 6 species, 4 described here.

2.1. Cheiloclinium anomalum Miers

Robust canopy liana; colorful sap absent. Stems: branchlets slender, green to grey; bearing lenticels. **Leaves:** elliptic, 5-15 x 2-5.8 cm, margins wavy to toothed; papery to subleathery. **Inflorescences:** panicles 2-5 cm long on short stalks; flowers tiny, ± 0.25 cm diameter; petals greenish-white, fleshy; nectar disc yellow dotted with 5 pockets; **stamens and stigmas 5. Fruits:** berries to 3.5 x 3.2 cm, green to yellow, skin white-dusted, hairless, base broadly 3-ribbed; seeds 6, 1.5-2 x 1 cm.

Uncommon; non-flooded old-growth forest. Guyana, Suriname; also N Brazil to C America.

2.2. Cheiloclinium cognatum (Miers) A.C.Sm.

Common name: aretepe (Ca); bofru tiki (Sr); grootblad kapi, hoasoropan (Sr), mamuare (Sr), soutmeti-udu (Sr); urarian (Ar)

Robust liana, more commonly an erect shrub or tree; **external yellow sap and yellow latex threads present. Stems:** branchlets green with lenticels; older stem bark skin-like with shallow lengthwise cracks. **Leaves:** narrow to broad, 8-20(-28) x 2.5-7(-9) cm, margins wavy or inconspicuously toothed; papery, black gland dots present. **Inflorescences:** 2-10 cm long with short stalks; flowers 0.5 cm diameter; petals fleshy or thin, reddish-brown or yellow with reddish-brown edges, margins entire; **stamens and stigmas 3. Fruits:** berries to 3 x 3.5 cm, maturing greenish-yellow to orange, skin smooth and hairless, often bearing lenticels, base with 3 broad and slightly raised ribs; seeds 4-5(-6), angled, ovoid, 1.5 x 1.1.

Common; non-flooded old-growth forest. Guianas; widespread, Bolivia to C America.

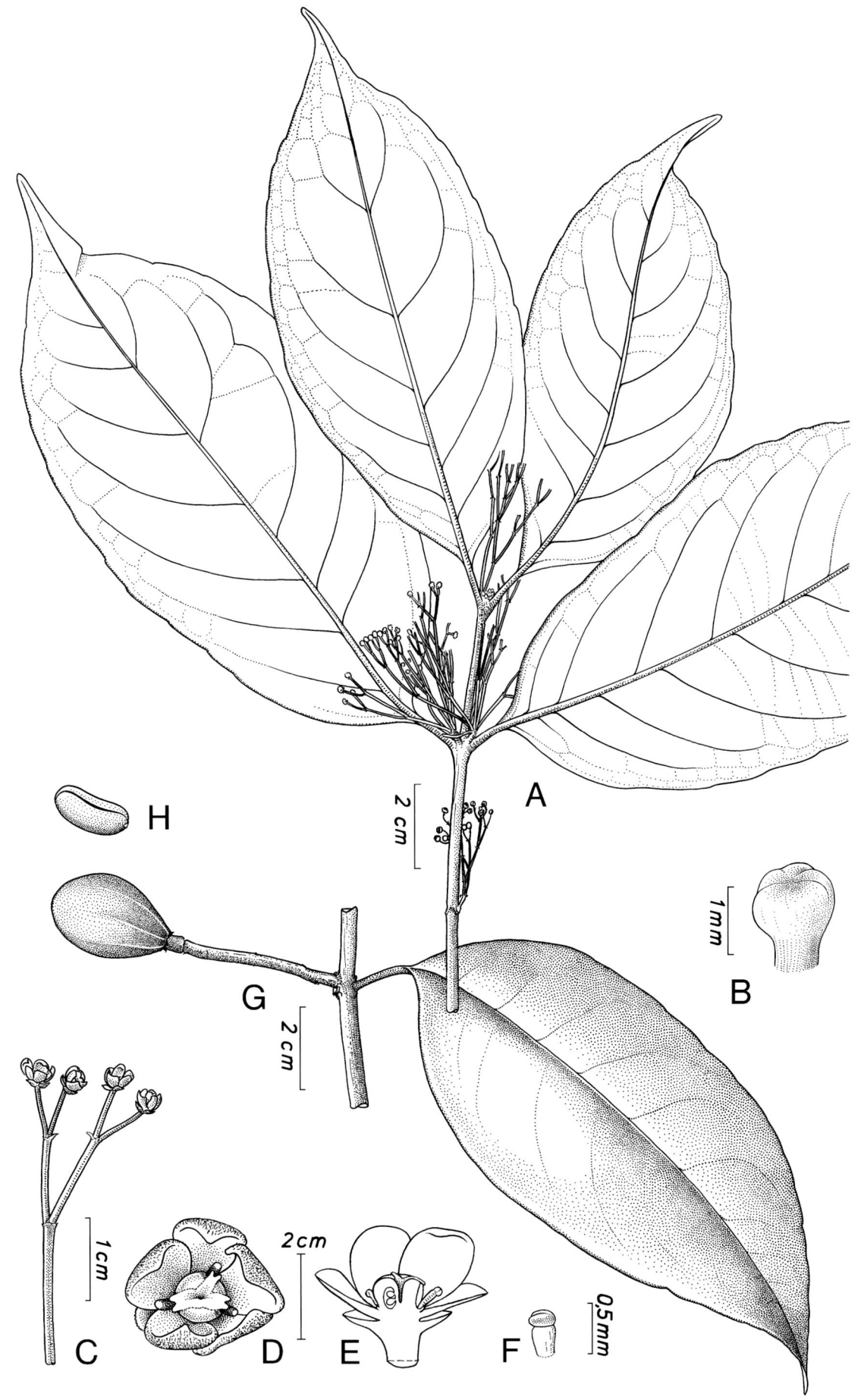

2.3. Cheiloclinium diffusiflorum (Miers) A.C.Sm.

Robust liana or treelet; external yellow or orange sap present, internal latex threads absent. **Stems:** branchlets round or remotely 4-sided, ± flattened at nodes, with lengthwise lines. **Leaves:** elliptic-oblong, 4-11 x 2-5 cm, margins variable; papery, shiny above; 3° veins obscure, parallel. **Inflorescences:** panicles compact and head-like, 2-4 cm long with short stalks; flowers tiny, ± 0.2 cm diameter; petals fleshy, yellow, margins entire; **stamens and stigmas 3. Fruits:** berries round, to 2 cm diameter, maturing orange-yellow, skin smooth; seeds unknown.

Uncommon; non-flooded old-growth forest. Guianas; also Venezuela, Brazil, and Peru.

2.4. Cheiloclinium hippocrateoides (Peyr. ex Mart.) A.C.Sm.

Common name: sipun (Tr)

Robust liana; external yellow sap sometimes present, internal latex threads absent. Stems: branchlets thin, 4-sided with lengthwise grooves, lenticels absent; older woody stems robust, ± 20 cm diameter, fluted and twisted. **Leaves:** broadly elliptic, 7-20 x 3-11 cm, black gland dots present. **Inflorescences:** panicles to 5.5 cm long with short stalks, flowers paired and tiny, ± 0.25 cm diameter; petals greenish-white to yellow or orange, ± fleshy, margins fringed; **stamens and stigmas 3. Fruits:** berries 4-5 cm diameter, maturing yellow or orange, fruit wall leathery, sharply 10-ribbed; seeds 3-6, to 2.2 x 1 cm.

Uncommon; old-growth forest, laterite hills, riparian forest. Guianas; widespread in Amazonia.

3. Cuervea Triana ex Miers

Lianas; also shrubs or treelets. Tropical America and Africa, West Indies: 5 species; Guianas: 1 species.

3.1. Cuervea kappleriana (Miq.) A.C.Sm.

Common name: karoshiri (Ar)

Robust liana or shrub. General: colorful sap absent; mostly hairless. Stems: branchlets round; bark smooth, **greyish to purplish-brown (never shiny green);** older stems round, bark thick and soft, inner bark layer dull orangish-brown; **stem x-s pattern radial. Leaves:** nearly opposite, elliptic to ovate, 15-23 x 8-11 cm, **margins entire or obscurely toothed;** leathery, all veins fine and slightly raised above and below (as in Hylenaea). **Inflorescences:** axillary, few-flowered panicles, each 5-12 cm long with stalk to 7 cm; flower large, 1-1.5 cm diameter; petals white, cream, or yellow, round, margins entire or irregular; nectar disc thin, margin wavy; stamens and stigmas 3. **Fruits: dry propeller-like capsules with 3 free, thick blades;** blades elliptic to obovate, 5-9 x 5.5-10 cm, **seeds 4-6, corky and unwinged. Ecology:** Seed dispersal via water. **Notes:** Sterile Cuervea and Hylenaea are very similar.

Rarely collected; riverine forest. Guianas; widespread, S Brazil to S Mexico.

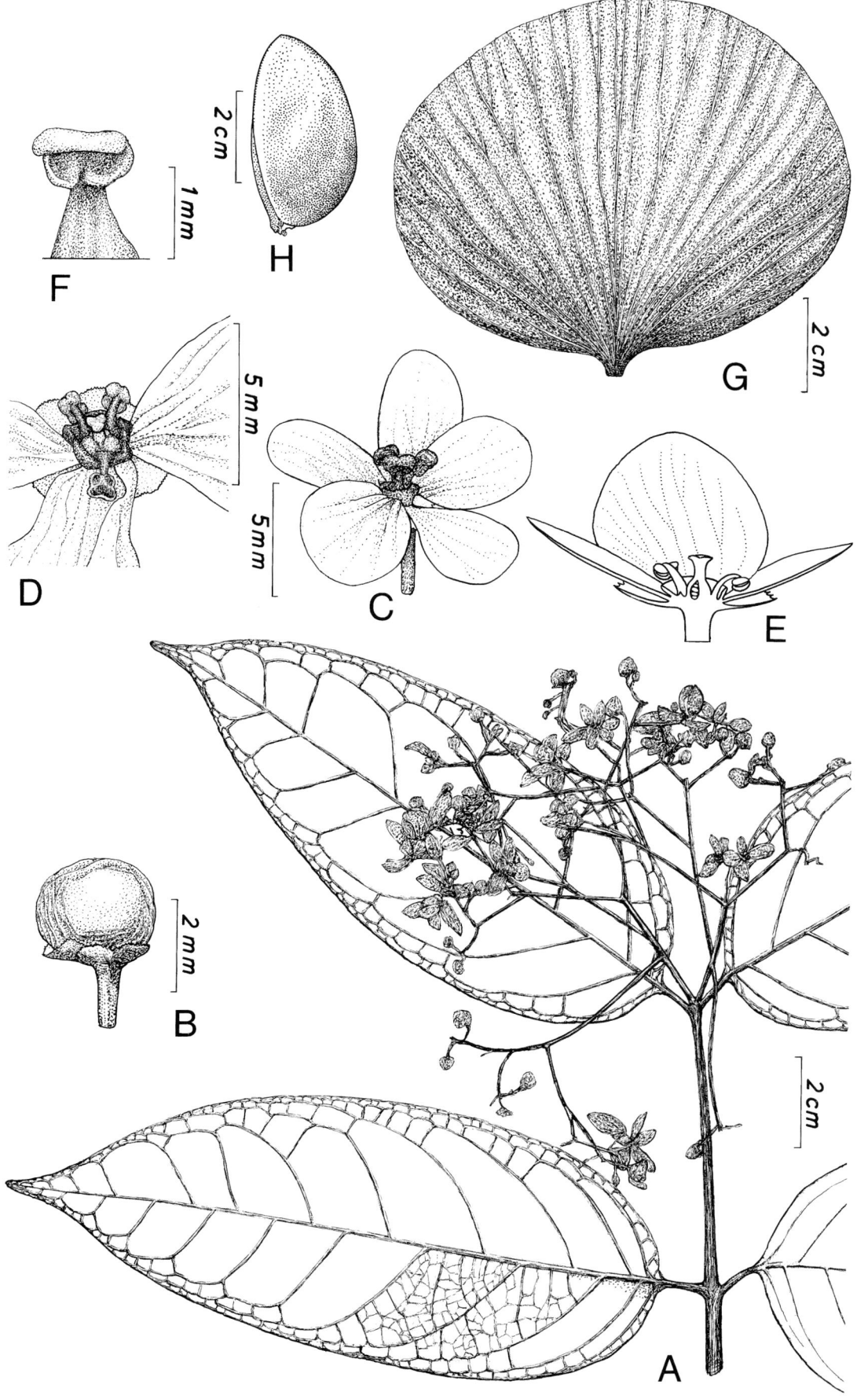

4. Elachyptera A.C.Sm.

Woody lianas and climbing shrubs; also erect shrubs and treelets. Tropical America and Africa, Madagascar: 8 species; Guianas: 1 species.

4.1. Elachyptera floribunda (Benth.) A.C. Sm.

Robust liana or climbing shrub; erect shrub. **General: colorful sap absent; mostly hairless. Stems:** branchlets 4-sided, **bark grey to purplish-brown (never shiny green and smooth);** older woody stems round with swollen nodes, bark thick, bumpy and bearing lenticels, subbark layer orange-brown; stem **x-s pattern radial. Leaves:** elliptic to broadly ovate, 10-15 x 5-7.5 cm, margins wavy or obscurely toothed; papery or thin-leathery, shiny above; 1° & 2° veins conspicuously raised below; 3° veins netveined, slightly raised below. **Inflorescences:** many-flowered panicles, solitary or paired, 2-11 cm long with stalks to 3 cm long, twigs sharply 4-sided; **flowers tiny, 0.15-0.3 cm diameter; petals white, cream or purplish,** fleshy, margins entire; **stamens and stigmas 3, forming a triangular shield. Fruits: dry propeller-like capsules with 3 free, flat blades;** seeds ellipsoid, **wing usually inconspicuous. Ecology:** Seed dispersal via water. **Use:** Reportedly used in Venezuela as a purple dye.

Uncommon; non-flooded and seasonally flooded old-growth forest. Guianas; widespread, S Brazil to S Mexico.

5. Hippocratea L.

Woody lianas; also shrubs or treelets. Tropical America and Africa, West Indies: 3 species; Guianas: 1 species.

5.1. Hippocratea volubilis L.

Robust liana. General: colorful sap absent; soft hairs present [*pilose*]. Stems: branchlets opposite, often 4-sided, **bark with orangish-brown hairs and tiny bumps** [*papillae*]; older stems often fluted, bark fibrous, sub-bark layer orangish-brown, **stem x-s pattern radial. Leaves:** elliptic to ovate, 3-14 x 2-6 cm, **margins wavy to obscurely toothed;** nearly leathery, hairless; 2° veins prominent above and below. **Inflorescences:** axillary or terminal, panicles with **conspicuous orange-brown hairs,** many-flowered, 3-15 cm long including stalk to 5.5 cm long; **flowers small, ± 0.5 cm diameter; petals greenish-white or yellow,** inner surface with a hairy horizontal band, margins entire or with hooked bristles [*barbellate*]; nectar disc conspicuous and fleshy; stamens and stigmas 3 (but inconspicuous). **Fruits: dry propeller-like capsules with 3 free, flat blades,** 3.5-8 x 1.5-5 cm, papery to sub-leathery, ribbed; seeds with a conspicuous basal wing, 2-4 x 1-2.5 cm. **Ecology:** seeds dispersed via wind.

Common; non-flooded old-growth forest, riparian areas, ridges. Guianas; widespread, SE Brazil to subtropical U.S.

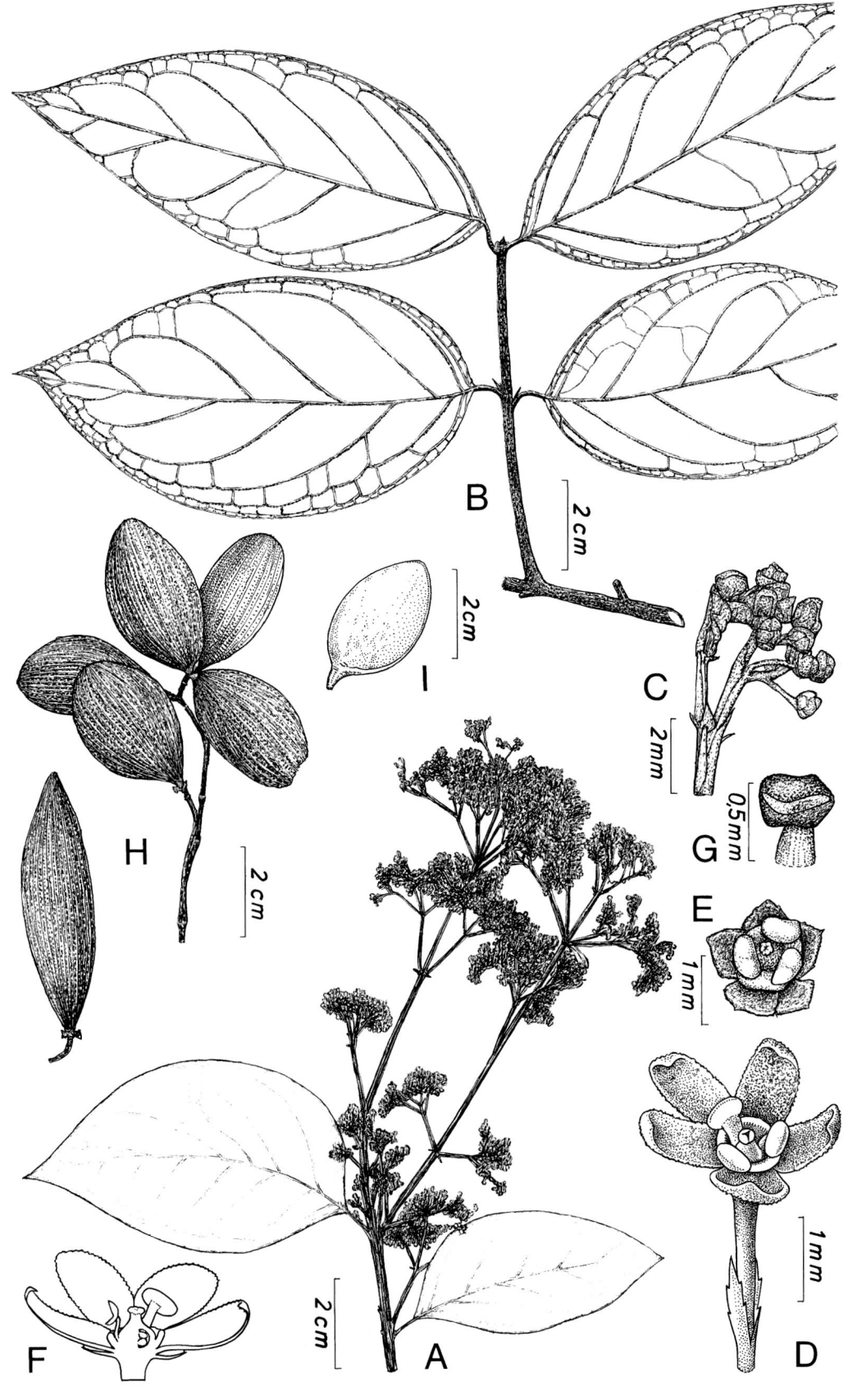

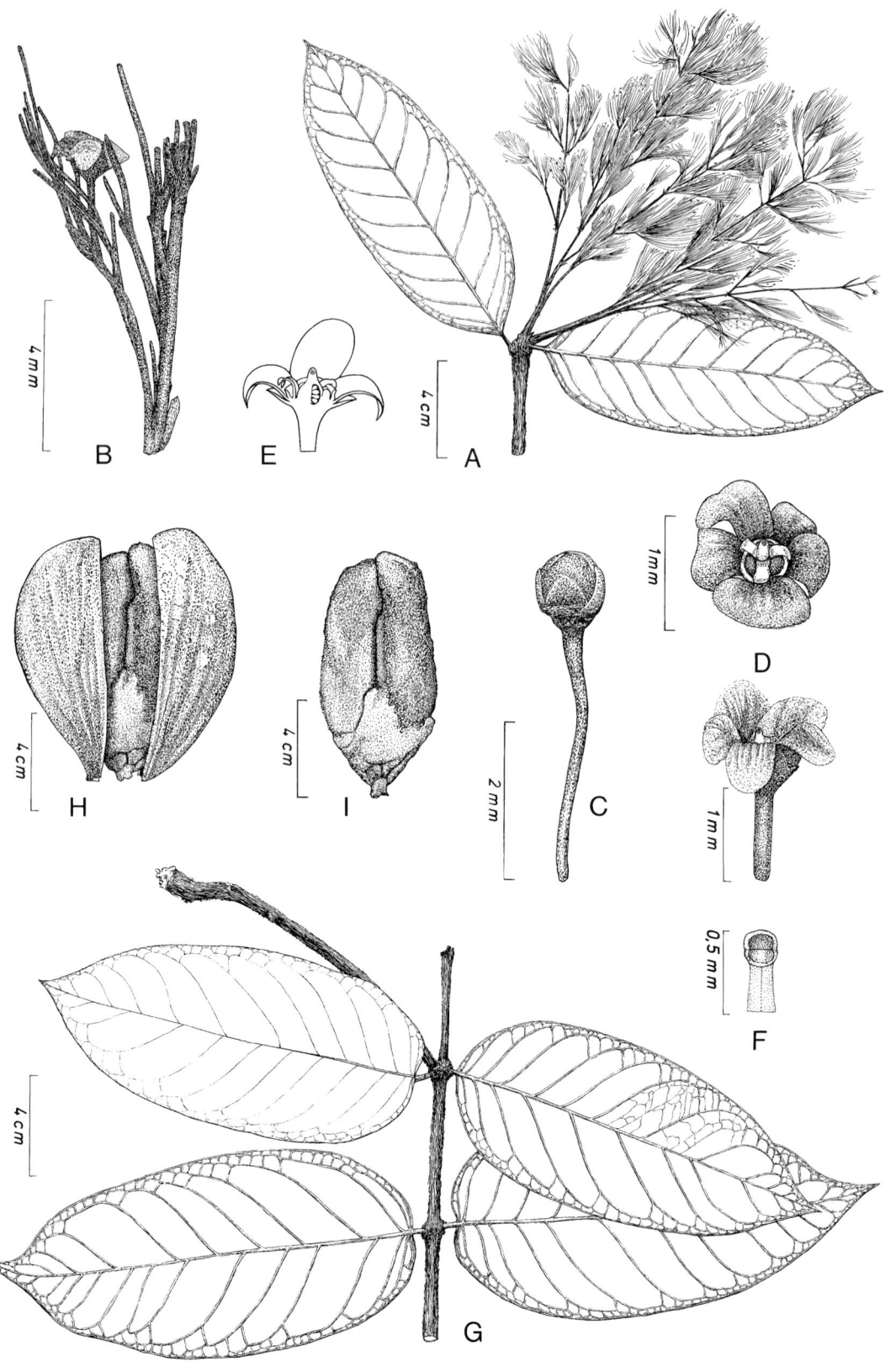

6. Hylenaea Miers

Robust lianas.

General: Colorful sap absent; hairless.

Stems: Branchlets often 4-sided, bark not bright green and smooth; older stem often fluted; **bark purplish- to reddish-brown and peeling; stem x-s pattern radial.**

Leaves: Sub-opposite, elliptic to ovate, medium-sized, **margins slightly wavy to entire (not toothed);** thin-leathery to papery; venation raised above and below, 3° veins densely to loosely net-veined; petiole to 1.5 cm long.

Inflorescences: Axillary or terminal, many-flowered panicles up to 20 cm long, **axes evenly-branching [*dichotomous*] or in broom- or hair-like clusters; flowers tiny,** 0.15-0.3 cm diameter; petals thin and fleshy, reflexed at maturity, yellow or greenish-yellow, margins entire, with or without a narrow stalk at base [*clawed*]; nectar disc slightly raised to tubular, margin entire; stamens 3, often reflexed, stigmas 3, often inconspicuous.

Fruits: **Dry propeller-like capsules with 3 free, flat to thick blades,** bulging outwards, ribbed or wrinkled, leathery to woody; **seeds 2-6, corky, wings often reduced.**

Ecology: Seeds dispersed via water.

Distribution: Tropical and subtropical America: 3 species; Guianas: 2 species + 1 species expected in Guyana; 2 described here.

Notes: At maturity, swollen, single 'blades' [*mericarps*] of Hylenaea species (to 10 x 6 x 5 cm) may fall to the forest floor separately from the other two segments (propeller-blades). At first glance, these propagules may be confused with Carapa guianensis seed pods and seeds. Sterile Cuervea and Hylenaea are easily confused.

6.1. Hylenaea comosa (Swartz) Miers

Common name: karoshiri (Ar)

Robust liana. Leaves: elliptic to oblong-ovate, 9-26 x 5-12 cm, base rounded or nearly heart-shaped [*subcordate*], thin-leathery. **Inflorescences:** axillary or terminal, with **3-8 broom- or hair-like clusters of fine branchlets,** each 4-14 cm long; flowers ± 0.15 cm diameter; petals greenish-yellow, without a narrow stalk at base, margins entire. **Fruits:** blades (nearly) ovate, 8-10 x 4.5-7.5 cm; leathery, green, ribbed; **seed body large, corky,** 4-8.5 x 1.5-2 cm, with an oblong basal wing to 5 x 3.5 cm.

Uncommon; Seasonally-flooded old-growth forest and swamps. Guianas (more likely to be seen than other Hylenaea species); widespread, S & C America.

6.2. Hylenaea unguiculata A.M.Mennega

Robust liana. Leaves: ovate to elliptic-ovate, 12-16 x 7-10 cm, base obtuse to heart-shaped, apex blunt [*obtuse*] or rounded; thin-leathery. **Inflorescences: branched and not in broom- or hair-like clusters;** petals yellowish, with a narrow stalk at base, gland-dotted. **Fruits:** blades ovoid or oblong, 7-10 x 4-8 cm, pale brown to grey; seed body large, corky, to 3 cm long, with a narrow wing to 4 x 1 cm.

Rare; Suriname: 2 collections; Brazil (Pará): 1 collection.

7. Peritassa Miers

Lianas or climbing shrubs.

General: Colorful sap absent; hairless or with few brown or grey hairs [*puberulous*].

Stems: Branchlets round, slender; nodes swollen or flattened. Older stems round, robust, furrowed; bark thick, fibrous, light brown to ash-grey, bearing lenticels; slash with burgundy red bands and white wood; **stem x-s pattern concentric,** with or without short, radial cross-links between rings.

Leaves: Elliptic-obovate, to 21 x 9.5 cm, **margins entire, wavy or toothed;** papery to leathery, darker green above, with scattered black dots below; midvein and 2° veins raised above and below, 3° veins net-veined, sometimes inconspicuous; petioles thick and wrinkled.

Inflorescences: Axillary or terminal, axes branching evenly 1-2 times; **flowers tiny to small,** 0.15-0.4 cm in diameter; sepals often cup-shaped; petals yellowish-green to cream-colored, thin and fleshy, reflexed or erect, margins often fringed; nectar disk short-tubular with wavy margin; stamens 3, erect, stigmas 3, inconspicuous.

Fruits: **Sub-globose to apple-shaped berries, 2-6 cm diameter,** fruit wall tough, skin wrinkled and often pitted; seeds 2-6, oblong, angled, embedded in pulp.

Ecology: Seeds dispersed via animal (through gut of large monkeys).

Distribution: Tropical and subtropical America: 19 species; Guianas: 5-6 uncommon species, 3 described here.

7.1. Peritassa glabra (A.C.Sm) Lombardi

High climbing robust liana, shrub or treelet. **Stems:** branchlets round, brown or grey, lenti-

cels present or absent. **Leaves:** narrowly elliptic to obovate, 11-15.5 x 2.5-6 cm; apex with narrow tip to 1.2 cm long, margins entire; papery to thin; **2° and 3° veins slightly raised below.** **Inflorescences:** axillary, panicles short, 1.5-3 cm long; flowers 0.2 cm diameter; petals oblong, thin to fleshy; stamens erect; stigmas inconspicuous. **Fruits:** **berries nearly round** (not apple-shaped), obscurely 3-sided, 2-2.5 cm diameter, maturing yellowish-orange with white dots, fruit wall leathery, smooth; inner pulp white or yellow and sweet-tasting; seeds 3-6, oblong, 2 x 1 cm, strongly attached to fruit by hair-like 'strings'.

Uncommon; non-flooded old-growth forest, especially on granite flats. Guianas; also NE and NW Amazonian Brazil.

7.2. Peritassa huanucana (Loes.) A.C.Sm.

Woody liana, climbing shrub or erect shrub/treelet. **Leaves:** nearly opposite, elliptic, 8-20 x 3.5-7 cm; 2° veins sunken above, raised below, **3° veins barely visible below. Inflorescences:** panicles hairy, short, ± 2 cm long; flowers ca. 0.15 cm diameter; petals greenish-yellow, slightly hairy, recurved, margins entire. **Fruits:** **berries egg-shaped,** to 4.5 x 3 x 3 cm, skin waxy, greyish-green to yellow or orange; seeds 0.15 x 0.8 x 0.5 cm.

Uncommon; old-growth non-flooded forest, riparian forest, ferrobauxite hills, disturbed areas. Guianas; also NW Amazon.

7.3. Peritassa laevigata (Hoffmanns. ex Link) A.C.Sm.

Common name: attakari (GU); kréré (Ca)

Liana or climbing shrub or erect shrub/treelet. Leaves: strictly opposite, narrowly elliptic to

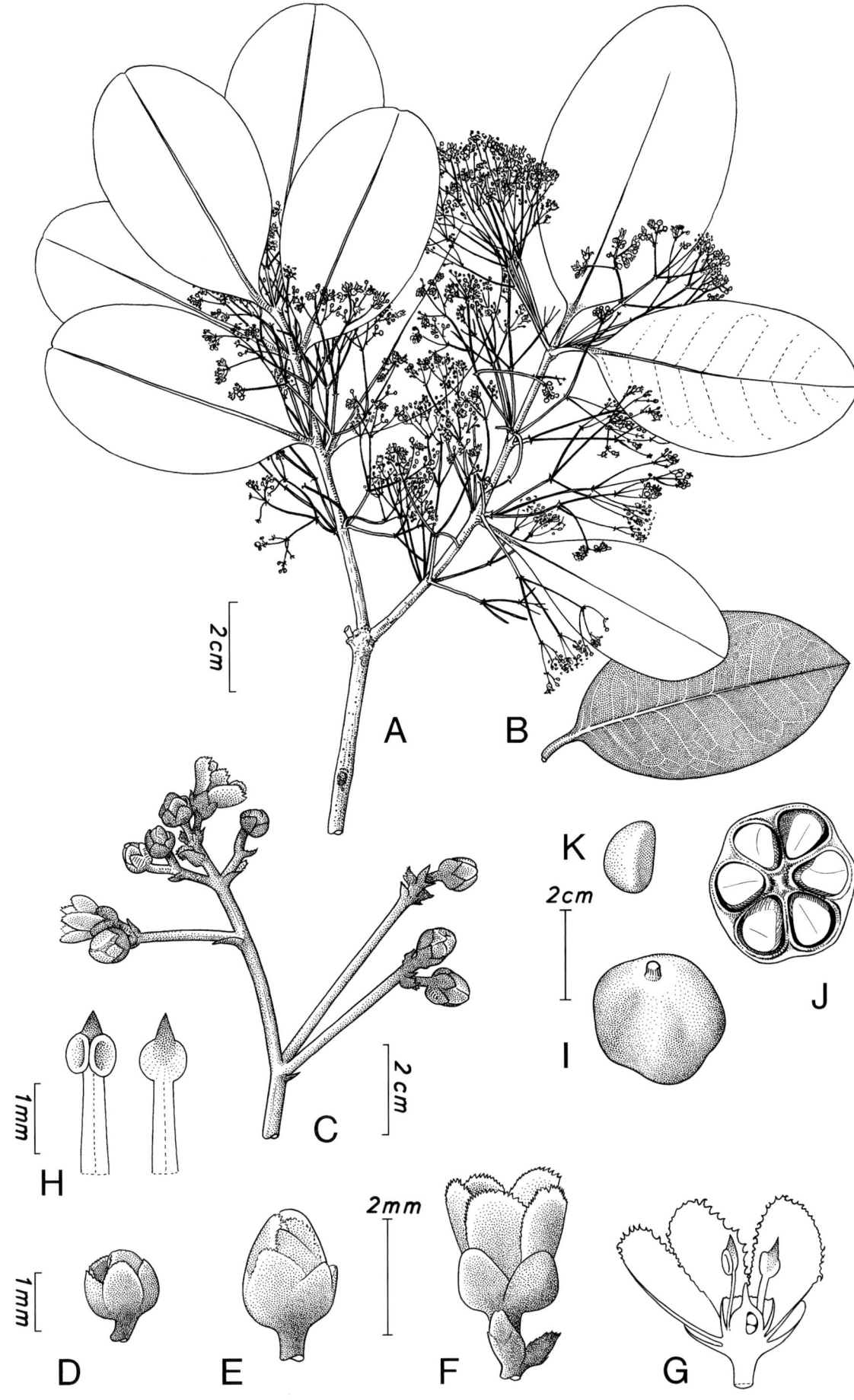

nearly oval, 5-13 x 2.5-5 cm, apex rounded, margins entire; midrib and 2° veins immersed above, prominent below, **3° veins barely visible below;** petiole to 2.5 cm long. **Inflorescences:** panicles hairless, to 8 cm long; flowers 0.25-0.4 cm diameter; petals nearly erect, creamish-white to yellow, hairless; margins entire or fringed. **Fruits:** **berries nearly round or apple-shaped,** to 5 cm diameter, skin blue-green to yellow or orange, waxy, wrinkled; seeds 4-6, 1.8 x 1 x 1 cm.

Rarely collected; non-flooded and seasonally-flooded old-growth forest. Guianas; also NW Amazon.

8. Prionostemma Miers

Woody lianas or climbing shrubs. Pantropical: 5 species; Guianas: one species.

8.1. Prionostemma aspera (Lam.) Miers

Common name: konikoni titei (Sr); kapi (Tr)

Robust liana or clambering shrub. General: sap watery, red to reddish-orange, torn parts often with red latex-threads; hairless to lightly hairy; **surfaces rough, sandpaper-like** [*asperous*]. **Stems:** branchlets round to flattened, slender, lenticels absent; older stems round, bumpy or deeply furrowed, robust, 5-15 cm diameter, swollen at nodes, bark light brown to grey, fibrous with cracked platelets; **slash with burgundy-red specks on white background, stem x-s pattern radial. Leaves:** ovate to oblong-elliptic, 5-10 x 2.5-4.5 cm, margin wavy or slightly toothed; leathery with age, surfaces rough, with clear dots below; 3° veins parallel, raised below. **Inflorescences:** often flowering profusely, flowers large for family, 0.9-1.2 cm diameter; petals round with a narrow stalk [*clawed*], thin, yellowish-green; nectar disc fleshy (similar to many Salacia); stamens and stigmas 3, orange and knoblike. **Fruits: dry propeller-like capsules with 3 free blades;** blades flat and broadly ovate, 7 x 5.5 cm, 0.8 cm thick; nearly woody, green or brown, surface rough, with many weak, longitudinal ribs; seeds numerous, 2.5-6 cm long, each with a broad apical wing. **Ecology:** Seed dispersal via wind. **Use:** Young shoots used in treating burns, woody stems used as a source of medicine and drinking water in the forest (Trios).

Uncommon but locally abundant; non-flooded old-growth forest, highland savanna, and secondary forest. Guianas; widespread in the American tropics.

9. Pristimera Miers

Robust lianas or climbing shrubs.

General: **Colorful sap absent;** mostly hairless.

Stems: Branchlets opposite, round or 4-sided, **usually grey, green or purple (never bright green),** lenticels present, smooth to minutely bumpy; nodes swollen; older stems round, robust, to 25 cm diameter, swollen at nodes, bark thick, often bumpy; **stem x-s pattern radial.**

Leaves: Narrowly elliptic to broadly ovate, < 20 cm long, **margins entire, wavy or obscurely toothed;** papery or thin, blade drying olive-greenish (not brown as in Hippocratea), scattered black gland dots often present below; **all veins ± raised below** (except 3° veins inconspicuous in P. holdeniana), 3° veins mostly net-veined; veins (drying) yellowish.

Inflorescences: Axillary, axes slender, branching evenly 1-2 times, with 2-5 flowers per ultimate

14

15

6

17

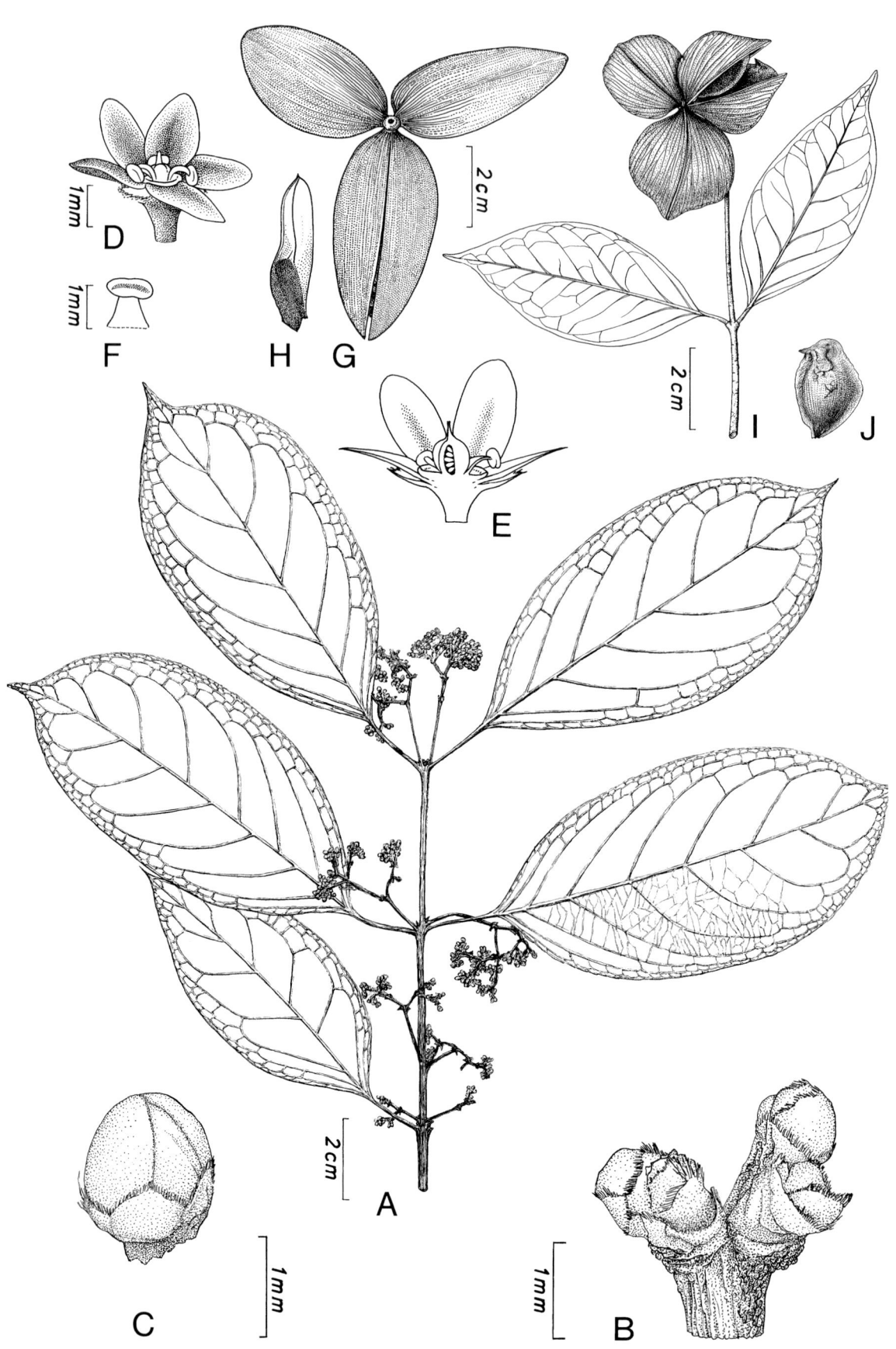

branchlet; **flowers tiny, 0.2-0.5 cm diameter;** petals yellowish-green or cream, thin, elliptic to oblong; nectar disk low, cushion-like; stamens 3, reflexed; ovary depressed, 3-lobed, stigmas either 3 and tiny or fused to form a shield; margins of bracts, sepals and petals entire, irregularly 'torn' or 'fringed'.

Fruits: Dry, propeller-like capsules with 3 free, flat blades; blades narrowly elliptic to almost round, 3-5 cm long, splitting down middle; leathery, greenish-brown, with lengthwise lines or ridges; seeds 2-8, conspicuously-winged at base, wing papery or thin, longer than seed body (unlike Hippocratea, both wing margins thickened).

Ecology: Seeds dispersed via wind or water.

Distribution: Pantropical: ± 38 species. Guianas: 5 species, 3 described here.

9.1. Pristimera nervosa (Miers) A.C.Sm.

Liana or climbing shrub; erect treelet. **Stems:** branchlets 4-sided, smooth, grey to purple; **older stem bark corky. Leaves:** 7-13 x 4-8 cm, margin wavy or obscurely toothed; papery to nearly leathery, black gland dots conspicuous. **Inflorescences:** 1-4 cm long with 0.5-2.5 cm stalk, densely-flowered, axes 4-sided; flowers ± 0.25 cm diameter; sepal margins strongly fringed; petals greenish-white to yellow, margins entire or 'torn'; stigmas 3. **Fruits: blades narrowly elliptic to ovate,** 5 x 2 cm, rounded at apex, often appearing winged; seeds 5 x 0.5 cm.

Uncommon; non-flooded and seasonally-flooded old-growth forest (including Mora forest). Guianas; northern S America to Panama.

9.2. Pristimera tenuiflora (Mart. ex Peyr.) A.C.Sm.
Liana or climbing shrub. Stems: branchlets round to 4-sided, minutely bumpy, grey, reddish-brown or purple, black-dotted. Leaves: narrowly

elliptic to ovate, 6.5-15 x 2.5-5.5 cm, apex long-acuminate to 0.12 cm, margin wavy or obscurely toothed; thin and papery. **Inflorescences:** compact, branching 1-2 times, 2 cm long with 0.5-1.5 cm stalk; flowers ± 0.25 cm diameter; sepal margins 'torn'; petals whitish, pinkish or green, margins 'torn' or entire; stigmatic shield present. **Fruits: blades ± round,** 3-3.5 cm diameter, apex short-acuminate, conspicuous veins present.

Rarely collected; non-flooded and seasonally-flooded old-growth forest. Fr. Guiana, Suriname; Amazonia to C America.

9.3. Pristimera verrucosa Kunth Miers

Liana or climbing shrub; erect treelet. **Stems:** branchlets round, densely lenticellate, ash-grey. **Leaves:** 5-11 x 2.5-6 cm, margin sub-entire, wavy to obscurely toothed; papery to nearly leathery. **Inflorescences:** branching > 2 times, 1.5-7 cm long including 2-3.5 cm long stalk, densely-flowered, twigs round; flowers 0.15-0.2 cm diameter; sepal margins 'torn'; petals greenish-white to yellow, margins entire. **Fruits: blades elliptic-ovate,** 4.5-5.5 x 2-3 cm, delicate, papery, with conspicuous lengthwise ridges.

Rarely collected; non-flooded and seasonally-flooded old-growth forest; gallery forest in savanna. Guyana; also N Colombia and Venezuela.

10. Salacia L.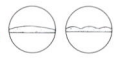

Robust lianas, also erect shrubs or treelets.

Common name: sipun, sipunuimë (Tr).

General: Colorful sap occasionally present; mostly hairless except for inflorescences.

Stems: Branchlets alternate or opposite, often 90° to stem, round to fluted, nodes flattened or flared, **bark mostly smooth and green, often waxy,** lenticels usually present. Older stems robust, **swollen at nodes, bark commonly smooth, skin-like and orange-brown;** slash with alternating red-orange and white bands; **stem x-s pattern concentric, rings usually widely-spaced and with short, radial crosslinks.**

Leaves: Size and shape variable, elliptic to obovate, **margins mostly entire, rarely wavy or irregularly toothed;** mostly thick and leathery, sometimes with black gland dots or tiny pits, dried leaves greyish- to brownish-olive; 1° vein usually conspicuously raised below, 2° veins raised above, flat or faint below, 3° veins usually faint above and below, reticulate or parallel; petioles often short, thick and wrinkled.

Inflorescences: Evenly-branching panicles or densely clustered stalks emerging from a cushion-like base; bracts present, shape variable; **flowers tiny (0.2 cm) to large (3 cm) diameter;** sepals often uneven; **petals overlapping, greenish-white to reddish,** margins entire, fringed or appearing 'torn' [erose], sometimes gland-dotted; nectar disc flat and short, cushion-like or tubular, margin entire or variable; **stamens 3, usually spreading widely, ovary sunken, with a single thick, blunt style** and inconspicuous stigmas or rarely with a miniature 'stigmatic shield'.

Fruits: globose to ovoid berries, 1.5-7 cm diameter; bluish-green to yellowish-orange; fruit wall leathery, skin smooth or wrinkled, often with specks, pock-marks or tubercles; seeds 6 to many, angled, embedded in a jelly-like edible pulp. Fruit stalk often becoming thick and woody.

Ecology: Seed dispersal via animal (large-bodied monkeys).

Distribution: Pantropical and subtropical genus: ca. 200 species (>100 in Africa); Guianas: ± 10 woody climbers, 5 described here.

Use: The fruit pulp of many species is edible. The stem sap of some species is used as a medicine (Tr).

Notes: Salacia and Tontelea are difficult to distinguish even in fruit. The leaves of Salacia are larger, duller, and less conspicuously-veined than Tontelea leaves. Salacia duckei A.C. Sm. is an unresolved name and is not presented here.

10.1. Salacia elliptica (Mart.) G.Don

Liana or climbing treelet; bright yellow rubbery sap sometimes present on surfaces. Stems: branchlets round, lenticellate. **Leaves:** variable, 10-17 x 4-8.5 cm, margins entire or slightly wavy; leathery, smooth above, 2° and 3° veins faint above and below. **Inflorescences:** hairless, in few-flowered clusters upon cushion-like base; flowers 0.5-1 cm diameter; petals greenish, yellowish-green, or orange, thin to fleshy, gland-dotted, margins often appearing "torn"; nectar disc short, with yellow rubbery sap visible on surface; stigmas inconspicuous. **Fruits:** berries ± globose, 1.5-4 cm diameter, waxy, blue-green maturing yellow-orange, fruit wall leathery, skin wrinkled; seeds ellipsoid, to 0.7 cm long.

Uncommon; old-growth forest, gallery forest, swamp. Guianas; widespread, from Bolivia to S Mexico.

10.2. Salacia impressifolia (Miers) A.C.Sm.

Liana or climbing treelet; colorful sap absent.
Stems: branchlets ± round with a central furrow, bark smooth with a few elliptic lenticels. **Leaves:** narrowly ovate to ovate, 9-20 x 4-8 cm, margins entire; (nearly) leathery; 2° veins sunken or flat on both sides, 3° veins mostly inconspicuous. **Inflorescences:** hairless, in few-flowered clusters upon cushion-like base; flowers 1-2.2 cm diameter; petals yellowish-green, pale brown or orange, fleshy to papery, thin to ± fleshy, margins entire; nectar disc short; stigmas inconspicuous. **Fruits:** berries ± globose, 3-4 cm diameter, maturing from green with grey-white powder to light tan or orange, fruit wall leathery, skin with thick wrinkles; seeds angular, 1.5 x 0.9 x 0.7 cm. **Notes:** Salacia impressifolia is similar to S. elliptica and S. juruana. S. impressifolia has oval leaves (S. elliptica: leaves narrowly elliptic; S. juruana: leaves elliptic-oblong) and larger flowers.

Uncommon; non-flooded old-growth forest. Guianas; S Amazonia to Central America.

10.3. Salacia insignis A.C.Sm.

Woody canopy liana; shrub or treelet; **colorful sap absent. Stems:** branchlets round, often reddish-brown, with lengthwise lines and lenticels. **Leaves:** elliptic-oblong, 25-28 x 4-7.5 cm, apex with 'drip tip' to 1 cm long; papery, smooth; 2° veins raised above and below, 3° veins flat above, slightly raised below, parallel & 90° to midvein. **Inflorescences:** hairless, panicles branching with few flowers at tips, 20-25 cm long with 3-5 cm stalk; flowers 0.6-1 cm diameter; petals yellowish-white, greenish-yellow or orange, thin; nectar disc cushion-like; stigmas inconspicuous. **Fruits:** berries ± globose, 2-3 cm diameter, greyish-green to yellowish-orange, fruit wall leathery, later woody, skin smooth or slightly wrinkled; pulp whitish, sweet-tasting; seeds 2(-3), 2 x 1 cm, base and apex obtuse.

Uncommon; non-flooded old-growth forest. Fr. Guiana, Suriname; also widespread in Amazon Basin.

10.4. Salacia juruana Loes.

Liana or tree to 15 m tall; colorful sap absent. Leaves: elliptic-oblong, 12-29 x 7-15 cm, margins entire; leathery; 2° veins curving to ± straight, flat or slightly raised above, raised below, 3° veins faint, parallel & 90° to midvein. **Inflorescences:** hairless, in few-flowered clusters upon cushion-like base; flowers ± 1.3 cm diameter; petals greenish-yellow or yellowish, often with dark spot at base, fleshy or thin; nectar disc cushion-like; stigmas inconspicuous. **Fruits:** berries ± globose, to 6 cm diameter, maturing bluish-green to yellow, skin with tubercles.

Uncommon; non-flooded old-growth forest. Guianas; also Venezuela, Ecuador, Peru, and Amazonian Brazil.

10.5. Salacia multiflora (Lam.) D.C.

Liana or shrub; colorful sap absent. Stems: branchlets 90° to main stem, bark smooth, greyish-brown, lenticels present. **Leaves:** obovate to elliptic-oblong, 6-19 x 3.5-9 cm, apex often with apical gland; sub-leathery. **Inflorescences:** sometimes red and hairy, panicles branching evenly, 1-4 cm long with stalk < 0.5 cm; flowers 0.7-1 cm diameter; petals green or cream; nectar disc tubular or conical; stigmas inconspicuous. **Fruits:** berries ellipsoid to nearly egg-shaped, 3.5-7 x 2.5-4 cm, yellow-green with greyish-white specks, faintly 3-furrowed, fruit wall leathery; seeds 4-6, angled, orange.

Uncommon; non-flooded old-growth forest, riverbanks. Guianas; widespread from Nicaragua to Bolivia.

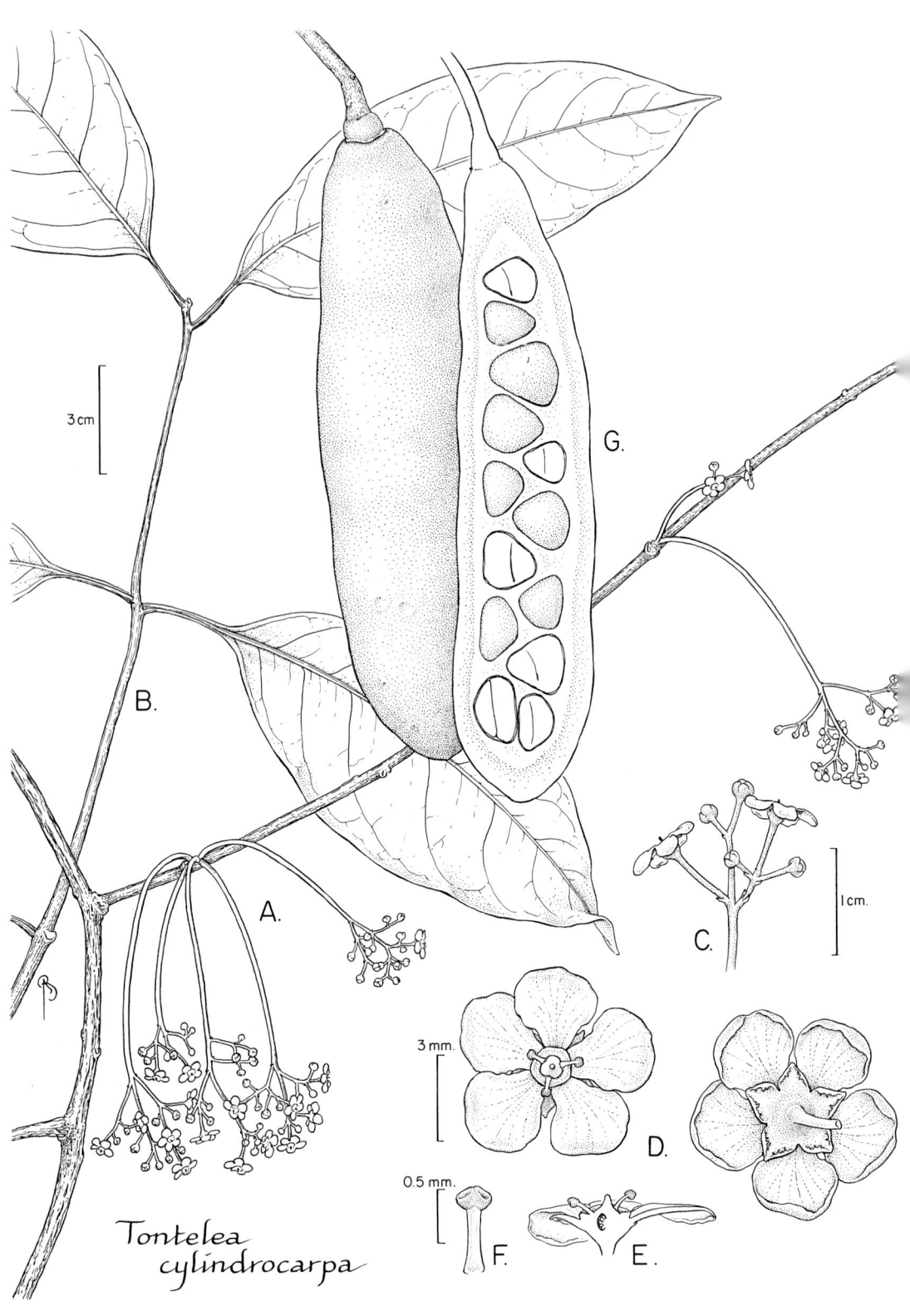

Tontelea
cylindrocarpa

11. Tontelea Miers

Robust lianas; occasional shrubs, treelets, or medium-sized trees.

General: **Colorful sap absent;** mostly hairless.

Stems: Branchlets round with swollen or flattened nodes. Older stems round, large, bark thick and fibrous to smooth, with lengthwise furrows and lenticels; slash with red and white alternating bands; **stem x-s pattern concentric, rings usually widely-spaced and with short, radial cross-links.**

Leaves: Often nearly opposite, size and shape variable, **margins entire or rarely wavy; mostly leathery, dark green and shiny; often with black gland dots below;** midrib raised above and below, 2° veins flat or raised, sometimes with parallel inter-secondary veins, 3° veins netveined or parallel; petioles wrinkled, often angular or canal-shaped.

Inflorescences: **Axillary or in-between nodes, branching evenly, loosely-flowered on short, stiff branchlets or on fine, long branchlets; flowers small, 0.2-0.4 (-0.7) cm diameter; petals greenish-white or yellow,** mostly thin and ovate, margins often fringed; nectar disc short-tubular with margin entire or wavy; **stamens 3, erect or reflexed; stigmas 3 and conspicuously spreading or inconspicuous as in most Salacia.**

Fruits: ± globose to elliptic berries, up to 5 cm diameter, or a long-cylindrical berry, fruit wall thick and leathery to corky, blue or bluish-green when young, maturing yellow-orange and speckled; seeds few, angled, embedded in soft, edible pulp; fruit stalks often stout, up to 0.5 cm diameter.

Ecology: Seed dispersal via animal (through gut of large-bodied monkeys). Some species leafless when flowering.

Distribution: Widespread in the Neotropics: 17 species; Guianas: 6 species, 4 described here.

Notes: Tontelea leaves are generally smaller, shinier, and with more conspicuous venation than Salacia leaves. Tontelea and Salacia fruits are similar.

11.1. Tontelea attenuata Miers

Liana; shrub or treelet. Stems: branchlets grey, with lengthwise lines or ridges, lenticels present or absent. **Leaves:** narrowly elliptic to oblong-elliptic, 8-19 x 2.5-7 cm, papery to thin-leathery; 2° and 3° veins flattened above, flattened to slightly raised below. **Inflorescences:** axillary, compact panicles, 2-3 cm long; **flowers 0.3-0.4 cm diameter;** petals broadly oblong-elliptic, fleshy; stamens green; stigmas 3, spreading. **Fruit: berries to 4.5 x 2 cm,** maturing orange, skin finely wrinkled. **Notes:** leaves and shoots similar to T. laxiflora.

Rarely collected; non-flooded and seasonally-flooded old-growth forest. Guianas; widespread in N Amazonia.

11.2. Tontelea cylindrocarpa (A.C.Sm.) A.C.Sm.

High-climbing robust liana. Stems: branchlets slender, brown or grey, lenticels present. **Leaves:** often nearly opposite, nearly oval to elliptic, to 10-30 x 4-14 cm, margin slightly wavy; subleathery to papery; 2° veins flat above, raised below, 3° veins parallel, visible below. **Inflorescences:** in-between nodes, **long, loose panicles,** 7-20 cm long with stalks 4-16 cm; **flowers 0.6-0.8 cm diameter;** petals ovate, thin, recurved; nectar disc fleshy; stamens erect; stigmas inconspicuous. **Fruits: cylindrical berries, 8-10 (-16) x 3-5 cm,** obscurely 3-sided, brown-purplish, hairless, fruit wall woody, smooth; seeds 10-20, angular-ovoid, 2-2.5 x 1.5-1.8 cm, embedded in pulp. **Notes:** often leafless when flowering.

Uncommon; non-flooded old-growth forest, especially on granite flats and savanna edges. French Guiana, Suriname; also N Amazonian Brazil and Venezuela.

11.3. Tontelea laxiflora (Benth.) A.C.Sm.

Robust liana, shrub or treelet. Stems: branchlets purplish or grey, with lengthwise lines, lenticels present or absent. **Leaves:** narrowly elliptic to nearly ovate, 8-17 x 2.5-6 cm, apex with narrow 'drip tip' to 1.5 cm long, margins entire; papery to sub-leathery, 2° veins flattened above and raised below, 3° veins densely net-veined, slightly raised below. **Inflorescences:** axillary, **long, loose panicles,** 3-9 cm long, with very slender twigs; **flowers 0.4-0.5 cm diameter;** petals reflexed; stamens erect; stigmas 3, each 2-lobed [*bifid*]. **Fruits: berries to 2 cm diameter,** green to yellow-orange or red, smooth, fruit wall leathery to corky; pulp thin, whitish; seeds 1, ellipsoid, to 1.7 x 1 cm. **Notes:** shoots and leaves similar to T. attenuata.

Uncommon; non-flooded old-growth forest, slopes and riverbanks. Guianas; also Venezuela and N. Brazil.

11.4. Tontelea passiflora (Vell.) Lombardi
[Syn: Tontelea ovalifolia (A.C.Sm.) A.C.Sm.]

Robust liana. Stems: branchlets greyish, with some lenticels. **Leaves:** elliptic to nearly ovate, 9-22 x 4-11 cm, margins entire, wavy, or toothed; leathery, 2° veins flat above, slightly raised below, 3° veins flat above, inconspicuous below. **Inflorescences:** axillary, in **dense branching clusters,** to 4 cm long; **flowers 0.4-0.7 cm diameter;** petals fleshy, margins irregular; stamens nearly erect; stigmas 3, spreading. **Fruits: berries to 5 x 3 cm,** fruit wall leathery, skin smooth or wrinkled; seeds angular.

Uncommon; non-flooded old-growth forest. Guyana, French Guiana; also adjacent Amazonian Brazil to Panama and Bolivia.

1. Salacia impressifolia, robust woody stem and leaves with wavy to toothed margins [BH]
2. Cheiloclinium cognatum, yellow latex threads in broken leaf [BH]
3. Salacia sp., robust stem with skin-like bark, bright orange sub-bark [BH]
4. Salacia mucronata, x-s pattern concentric with cross-links [BH]
5. Salacia impressifolia, yellowish-white, thick-walled fruit and seeds [BH]
6. Salacia cordata, x-s pattern concentric without cross-links between rings [BH]
7. Anthodon decussatum: a. habit; b. bud; c. open flower; d. flower longit. section; e. stamen; f. fruit; g. seed [NBC]
8. Cheiloclinium hippocrateoides: a. habit; b. bud; c. part of infl.; d. open flower; e. longit. section of flower; f. stamen; g. fruiting branch; h. seed [NBC]
9. Cuervea kappleriana. a. habit; b. bud; c,d. open flowers; e. longit. section of flower; f. stamen; g. fruit 'blade' (mericarp); h. seed [NBC]
10. Elachyptera floribunda: a. flowering branch; b. sterile branch; c. part of inflorescence; d. open flower; e. open flower; f. longit. section of flower; g. stamen; h. fruiting branchlet and solitary mericarp; i. seed [NBC]
11. Hylenaea comosa: a. habit; b. part of infl. with fertile flower; c. bud; d. flowers; e. longit. section of flower; f.stamen; g. fruiting branch; h. opening 'blade' (mericarp); i. seed [NBC]
12. Peritassa compta: a. habit; b. leaf; c. part of infl.; d,e. buds; f. open flower; g. longit. section of flower; h. stamens; i. fruit; j. section of fruit; k. seed [NBC]
13. Prionostemma aspera, flowering branch [BH]
14. Prionostemma aspera, propeller-like fruit [BH]
15. Elachyptera floribunda, radial x-s pattern [BH]
16. Prionostemma aspera, woody stem slash with watery red sap [BH]
17. Prionostemma aspera, woody robust stem [BH]
18. Pristimera nervosa (Miers) A.C. Smith: a. habit; b. part of infl. with buds; c. bud; d. open flower; e. longit. section of flower; f. stamen; g. mericarps; h. seed. Pristimera tenuiflora (Mart.) A.C. Sm.; i. fruit; j. seed [NBC]
19. Tontelea cylindrocarpa. a. part of stem with inflorescences; b. apex of stem with leaves; c. part of inflorescence; d. apical (left) and basal (right) views of flower; e. medial section of flower; f. stamen; g. medial section of fruit showing embedded seeds [BA]

COMBRETACEAE

COMBRETACEAE

Robust canopy lianas, shrubs, treelets and trees including a mangrove genus. Pantropical, especially Africa and South America, 20 genera/500 species. Only one genus with liana species in the neotropics.

1. Combretum Loefl.

Robust canopy lianas, weakly climbing shrubs, erect shrubs and trees. Climbing via clockwise twining shoots and opposite branchlets. Leaves are opposite, simple, and entire. Stipules are absent. Inflorescences are showy, with flowers white to red-orange-yellow; the stamens are often longer than the corolla and the ovary is inferior. Fruits are 4(5) winged or ridged.

General: Surfaces often scaly [*lepidote*], **reddish-brown to golden-yellow;** lightly hairy to hairless.

Stems: Branchlets often with shedding reddish-brown bark; nodes simple, stipules absent. Older stems robustly woody, 5 to 30 cm diameter, round, oval (sometimes multiple-cabled); outer bark thick, grey to reddish-brown, inner bark green or red, wood hard, with **central canal slowly releasing a colorless, thick gum when cut;** stem x-s pattern simple.

Leaves: Elliptic to broadly ovate, base ± rounded; leathery to papery, conspicuously smooth, sometimes pitted; leaf pockets [*domatia*] sometimes present in vein axils below; **2° veins widely spaced, gently curved to straight,** usually form-ing marginal loops, 3° venation usually parallel & slanted towards midvein [*oblique*], all veins flat or sunken on upper surface, midvein raised below, 2° and 3° veins raised or flat below; **petiole short, dark, often scaly.**

Inflorescences: Terminal or axillary, with **panicles of small** (0.1-0.5 cm long) whitish-cream flowers or spikes of large (2-3 cm long) **red-orange-yellow flowers, opposite-paired at nodes,** usually with conspicuous bracts, buds and branchlets, often scaly; flowers irregular-zygomorphic and 5-parted in C. cacoucia, otherwise radial and 4-parted; receptable [*hypanthium*] present, with persistent calyx lobes; petals free or absent; **stamens** (4)-8-(10), **often long and extending well beyond petals** [*exserted*]; **ovary inferior,** 1-locular, style and stigma 1.

Fruits: Two basic models: i) dry, papery, 4-5 winged samaras; or ii) spongy, fat, 4-5 ridged pseudo-drupes; 1 seed per fruit.

Ecology: Shoots sometimes growing far across the forest floor. Fruits of several species have conspicuous insect galls. Seed dispersal via wind (samara) or water (drupe).

Distribution: Pantropical: 255 species; neotropics: 31 species; Guianas: 8-9 species; 5 described here.

Notes: Besides flaking bark and gummy sap, Combretaceae lianas are distinguished largely by the absence of characters (Gentry 1993). The leaves are less rough than many Verbenaceae (Petrea). Flat leaf glands are not present as in Lamiaceae (Aegiphila). Compared to Malpighiaceae, there are no T-shaped hairs, lenticels, raised or flat leaf glands, or notable stem x-s pattern. Compared to Celastraceae, brightly-colored sap, toothed/wavy leaf margins, or unique stem anatomy are absent.

1.1. Combretum cacoucia (Baill.) Exell ex Sandw.

Common name: fremusu-wiri (Sr), jekara (Tr); kararawa andïkïrï (Ca); karu arib (Pal); puspustere (Sr); rabu-de-arara (BP); sekema (Ca); yariman (Ar); yoyoca (BP).

Liana or shrub; minutely hairy or felty, **scales absent. Stems:** older woody stem round, observed to 3 cm diameter, inner bark green. Leaves: elliptic to broadly ovate, 15-19 cm x 8-10 cm, base ± heart-shaped, apex acuminate; sub-leathery, hairy domatia present in 2° leaf axils below. **Inflorescences: hanging spikes to 50 cm long with large spear-shaped bracts** (to 2 x 0.5 cm) **below each flower;** flowers 5-parted; corolla irregular [*zygomorphic*], large, 2.2-3.2 cm long, bright orange-red, hairy; stamens 10, dark red, extending 0.5-1.5 cm beyond petals. **Fruits: five-sided pear-shaped drupes with 5 hard ribs, to 8 x 2.5 cm,** larger than other

Combretum fruits in the region, green or yellow, minutely hairy. **Uses:** dry seeds burned to repel bats from homes. Fruits said to be poisonous. Triterpine saponins are present (Grenand 1987).

Common; especially in secondary vegetation and along lower rivers. Guianas; widespread from Brazil to C America.

1.2. Combretum fruticosum (Loefl.) Stuntz

Common name: Kadiamen (GU).

Liana or shrub; scales golden-yellow. Leaves: elliptic to narrowly elliptic, 3.5-19 x 1.5-10 cm, base cuneate to rounded or ± heart-shaped, apex variable; papery to sub-leathery, **conspicuously scaly and pitted below,** hairless to densely hairy below; petiole 0.6-1.5 cm long, scaly. **Inflorescences:** showy spikes, flowers pointing vertically upwards, each spike 4-15 cm long; flowers variable, radial, 4-parted, 1-2 cm long; receptacle 4-angled, scaly; petals rarely absent; stamens 8, extending beyond petals 2-3 cm. **Fruits: samara-like with 4 wings,** ovate, 1.2-3 x 1-3 cm, becoming dark red and scaly, especially on body.

Rare in region; wide habitat range, including old-growth, seasonal, and disturbed forests. Guianas; widely distributed across tropical America from Argentina to Mexico.

1.3. Combretum laxum Jacq.

Common names: kamooraballi (GU); kupirisi (SA); supple jack (GU)

Robust liana or shrub, observed with leafless twining shoot to 1 m long (BH5217); hairless to densely hairy; **scales white or dark, small and inconspicuous. Stems:** older woody stems recorded to 14 cm diameter. **Leaves:** variable, broadly ovate to elliptic, to 4-25 x 1.4-13 cm, base broadly cuneate to rounded, apex variable; papery to leathery, scaly below; petiole < 1.1

COMBRETACEAE

cm long. **Inflorescences:** narrow, erect panicles, 3-17 cm long, axes reddish-brown, hairy; **flowers radial, 4-parted, cream to yellow,** tiny, 0.23-0.38(0.5) cm long; petals broadly spathe-shaped; stamens 8, white, **extending beyond petals 0.3-0.4 cm** (longer than C. pyramidatum). **Fruits: highly variable,** either spongy with 4 thin, sharp ribs, 1.4-3 x 0.6-1.2 or samara-like with 4 large papery wings, 1.4-3.8 x 1.3-3.1.

Very common; flooded and non-flooded old-growth forest, secondary forest and along rivers. Guianas; widely distributed in the neotropics.

1.4. Combretum pyramidatum Desv. ex Hamilton

Lianas or treelets; appearing hairless, **scales white-clear or dark, inconspicuous. Stems:** older stems oval with central hollow chamber; outer bark dark brown and slightly exfoliating, inner bark green. **Leaves:** narrowly oblong-ovate to widely ovate, 4-20 x 2-12.5 cm, base rounded to cordate, apex acuminate to rounded; thick and leathery, scaly below; 2° veins looping, midvein and 2° veins raised below, 3° veins flat; petiole 0.3-1.4 cm long. **Inflorescences:** panicles showy, 2-7 cm long, branchlets hairless; flowers radial, 4-parted, yellow or cream, tiny, 0.2-0.3 cm long; petals broadly spathulate; **stamens 8, white, extending beyond petals 0.15-0.3 cm (shorter than C. laxum). Fruits: drupe-like with 4 hard wings or ridges,** body 1.5-4.5 x 0.7-1.7 cm, wings 0-0.3 cm wide, greenish-yellow, minutely scaly.

Common; esp. lowland riparian forest but also in drier secondary forest. Guianas; also W Amazonia.

1.5. Combretum rotundifolium L.C. Rich.

Common name: fremusu noto (Sr); jekara (Tr); keskes bosro (Sr); keskes kankan (Sr), meu abesu (Pal); meyu akánta (Wpi); peigne singe rouge (FG); sekema (Ca); yariman (Ar).

Liana or shrub; scales green, reddish-brown, or golden. Stems: older stems oval, observed to 5 cm diam., outer bark grey-brown, thick, cracked in platelets, inner bark slash with a network of reddish lines. **Leaves:** ovate to elliptic, 4-17 x 2-11 cm, base rounded to ± heart-shaped, apex acute to acuminate; papery to leathery, scaly and pitted below; petiole 0.3-1 cm long, scaly. **Inflorescences: showy 'bottle-brush' spikes,** each 5-12 cm long, flowers pointing upwards; flowers regular, 4-parted, large, 1.2-2 cm long, **flower buds long and pointed (not round);** calyx bright red; receptacle [*hypanthium*] 4-angled; stamens 8, extending beyond petals 2-3 cm, bright yellow or red. **Fruits: samara-like with 4 papery wings,** ovate, 3 x 1.8-3 cm, green to greenish-red scaly and reddish-brown-dotted. **Uses:** Medicinal — fevers (stem), headaches (gum); common ornamental in urban areas. **Notes:** C. rohrii is similar but rare (Fr. Guiana and Brazil).

Common; flooded and non-flooded old-growth forest, rivers and creeks, to 230 m. Guianas; also C & W Amazonia.

1. Combretum cacoucia inflorescence [PT]
2. C. cacoucia habit [PT]
3. C. cacoucia in fruit [PT]
4. Combretum pyramidatum in fruit with stem [BH]
5. Combretum laxum in flower [FRD]
6. Combretum rotundifolium in flower [PT]
7. C. rotundifolium branchlet with flaking red-brown bark [BH]
8. C. rotundifolium, robust woody stem [BH]
9. C. rotundifolium in fruit [PT]

COMBRETACEAE

CONNARACEAE

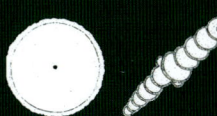

CONNARACEAE

A family of mostly woody lianas and scrambling shrubs with a few erect shrubs and treelets. Climbing via twining branchlets and recurved petioles. Stipules are absent. Leaves are alternate and compound with a terminal leaflet. Flowers are small and radial (not pea-like), with a 5-lobed calyx, 5 petals, 10 stamens, and a superior ovary. Fruits are 1-seeded pods [*follicles*], splitting along one side; seeds often have a colorful aril.

General: Surfaces densely to finely reddish-brown hairy – hairs simple (Cnestidium & Rourea), dendroid (Connarus & Pseudoconnarus), or glandular (Connarus).

Stems: Branchlets hairless to conspicuously hairy, lenticels often present; nodes smooth, without a horizontal line or ridge; **stipules absent.** Older stems round to oval, small to medium in diameter; bark smooth, brown, yellowish, or grey; **slash sometimes with watery red sap;** stem x-s pattern almost always simple, 'eccentric' in a few Rourea.

Leaves: Leaf stalk [*base*] swollen, round and wrinkled; leaflets 3-21, mostly sub-opposite to alternate on rachis, terminal leaflet larger than others; blades obovate to lanceolate, margins entire or wavy, sometimes curled [*revolute*], **dark green and smooth above, lighter-colored/ hairy below,** papery to leathery, sometimes with fine bumps [*papillae*] below; **main veins pinnate or 3-nerved (Pseudoconnarus),** sunken above, raised below, 2° veins with marginal loops; leaflet stalks [*base*] thick, cylindrical, wrinkled with age.

Inflorescences: axillary, terminal or on stem [*cauliflorous*], with small radial flowers (< 1 cm long), small bracts present; flower stalks [*pedicels*] jointed; calyx fused at base, sepals overlapping [*imbricate*] or not [*valvate*], sometimes persistent (Rourea); petals 5, free or fused at the base, white, yellowish or light pink, with dark gland dots (Connarus); stamens 10 (5 long, 5 short); ovary with free segments [*carpels*] 5 or 1 (Connarus), stigma(s) simple or two-lobed.

Fruits: One-seeded pods [*follicles*] splitting on one side, 1-5 per flower, with or without a stalk, Rourea fruits with basal cup [*persistent calyx*]; seeds 1, black, with a yellow, white, orange or red aril.

Ecology: Seeds dispersed via animal (birds, through gut).

Distribution: Pantropical: 16 genera/300-350 species; Neotropics: 5 genera/110 species; Guianas: 4 genera/17 species of woody climbers.

Notes: Connaraceae species are most similar to Leguminosae-Papilionoideae (legume) climbers.

Both families have alternate, simple, compound leaves with swollen, cylindrical leaf and leaflet stalks. Connaraceae can be distinguished by the lack of stipules, spines, and a "green bean" scent.

1. Cnestidium Planch.

Lianas, scandent shrubs, and treelets. Widespread from N South America to Mexico; 2 species, 1 known in the Guianas.

1.1. Cnestidium guianense (Schellenb.) Schellenb.

Common Name: inekuipë (Tr); rode man-neku (Sr)

Canopy liana or scandent shrub; densely hairy when young. **General:** hairs reddish-brown [*pilose, tomentose*]. **Stems:** branchlets oval, angled; older stems oval, 2-3 cm diameter, reddish-brown to grey; **slash with some red sap;** stem x-s pattern simple. **Leaves:** leaflets **5-7(-9) foliolate,** leaflets ovate-lanceolate, 4.5-11 x 2-6 cm, base attenuate to rounded, drip tip to 0.5 cm long; papery or subleathery; stalks 0.2-0.5 cm long. **Inflorescences:** panicles to 20 cm long, densely hairy [*pilose*]; flowers 2.5-3 cm long, white; sepals not overlapping; gland dots absent; **ovary with 5 free segments. Fruits:** follicles ± flattened, ovoid, 1.2-2 x 0.4-0.7 cm, often >1 per flower, ripening bright orange, hairy, persistent calyx cup absent; seeds ovoid, 1.2 x 0.7 cm., black, shiny with a yellow aril.

Uncommon; non-flooded old growth forest, esp. on lateritic soils. Guianas; also Venezuela.

1.   Connarus patrisii [PT]
2.   Connarus cf. perrottetii, woody stem and leafy shoot with orangish-red hairs [BH]

2. Connarus L.

Large lianas, sprawling shrubs and treelets; also erect shrubs and treelets.

General: Most surfaces densely reddish-brown hairy [*tomentose to puberulous*]; hairs simple, dendroid (with tree-like branching), or glandular.

Stems: Branchlets oval, hairy, bearing small lenticels. Older stems round to oval; bark fibrous, not peeling away, reddish-brown or grey, often bearing lenticels; **slash with clear red sap in one species** (C. perrottetii); stem x-s pattern simple.

Leaves: Leaflets 3 (C. coriaceous, C. punctatus), 5-9 (C. patrisii, C. perrottetii, C. punctatus), or > 11 (C. fasciculatus); leaflets hairless above and ± hairy below, tiny bumps [*papillae*] absent below.

Inflorescences: Axillary, terminal, or pseudo-terminal panicles; dark gland dots or glandular hairs present; flowers white or yellow; sepals small, overlapping; **ovary 1.**

Fruits: Follicles elliptic to round, 1 per flower, woody or leathery, stalked or unstalked, persistent calyx cup absent; seeds 1, aril usually present.

Distribution: Pantropical; Neotropics: 54 species; Guianas: 6 climbing species known, 5 described here.

2.1. Connarus coriaceous Schellenb.

Common name: akareowoi, akareroai, arakaituran (Ca); sukerdyap (Sr);

Liana or a tree to 5 m. **Stems:** bark brownish-grey, lenticels dense and small. **Leaves: 3-foliolate;** leaflets elliptic, 7-16 x 3-7 cm, ± rounded at base, apex long and acuminate; **leathery, hairy below;** leaflet-stalks 0.2-0.5 cm long. **Inflorescences:** panicles to 22 cm long, densely hairy [*puberulous*]; **flowers white, 0.45-0.6 cm long,** **fragrant, gland dots present. Fruits:** follicles to 2.5 x 1.2 cm, orange or red, hairless; seeds black.

Uncommon; riverine and creek forests, savanna borders. Guianas; also Venezuela, Brazil, and Peru.

2.2. Connarus fasciculatus (DC.) Planch.

Usually a small tree, sometimes climbing. **Leaves: 11-21 foliolate;** leaflets 9-20 x 3-6 cm, **papery, hairless below. Inflorescences:** borne on stem in clusters, racemes up to 2 cm long, reddish-brown hairy [*tomentose*]; **flowers white, 0.3 cm long, gland dots present. Fruits:** follicles to 2.5 x 1.2 cm, orange, hairless; seeds black, aril white.

Rare; non-flooded old growth forest. French Guiana, Suriname; also Amazonian Brazil.

2.3. Connarus patrisii (DC.) Planch.

Liana. Stems: bark brownish-grey, lenticels dense and small. **Leaves: 7-9 foliolate;** leaflets elliptic to oblong-lanceolate, 9-13 x 3-6 cm, base rounded to subcordate; **subleathery, hairy below;** leaflet-stalks 0.5-0.7 cm long. **Inflorescences:** axillary or pseudo-terminal, panicles dense and wide, to 15 cm long; **flowers white, 0.4 cm long, gland dots present but inconspicuous. Fruits:** follicles to 1.5 x 1.2 cm, orange to red, hairy or hairless; seeds oblong, black, aril yellow.

Not common; seasonally-flooded old growth forest. Guianas; Venezuela and adjacent Brazil.

2.4. Connarus perrottetii (DC.) Planch. var.

Common name: maneko (Pa); seweyuballi (Ar); wakapuimë (Tr)

Liana or tree to 8 m tall. **Stems:** bark with small, sparse lenticels; **slash with ± sticky, clear red sap. Leaves: 5-7 foliolate;** leaflets elliptic to long and narrow, 8-16 x 2.5-5 cm, base round to subcordate, drip tip to 1 cm long; **papery, hairy below;** leaflet-stalks 0.6-0.8 cm long. **Inflorescences:** axillary or terminal, panicles to 30 cm long; **flowers white, 0.3 cm long, gland dots few. Fruits:** follicles to 2 x 1.3 cm, hairy, leathery; seeds reddish-black, shiny.

Common; non-flooded old growth and riverine forests. Guianas; adjoining N. Brazil.

2.5. Connarus punctatus Planch.

Common name: basterd tingimoni (Sr)

Shrubby liana or treelet. Stems: bark greyish-brown with small lenticels. **Leaves: (3)-5-7 foliolate,** leaflets elliptic to oblong-lanceolate, 8-16 x 2.5-7 cm, base round to subcordate, drip tip to 1 cm long; **subleathery, hairy below;** leaflet-stalks 0.5-1 cm long. **Inflorescences:** axillary or pseudo-terminal, panicles dense, to 15 cm long, hairy; **flowers white, 4.5-6 cm long, gland dots many. Fruits:** follicles ± pyriform, 2.1-2.5 x 1-1.2 cm, red, leathery, finely ribbed or veined; seeds black, smooth, shiny.

Uncommon. Swamp forest, seasonally-flooded and non-flooded old growth forest. Guianas; also widespread in Amazonia.

3. Pseudoconnarus Radlk.

Lianas. N South America: 5 species; Guianas: 2 species, one described here.

3.1. Pseudoconnarus subtriplinervis (Radlk.) Schellenb.

Woody liana. General: densely reddish-brown hairy when young [*tomentose*], hairs simple or dendroid. **Stems:** older stems round to oval, lenticels small and few; bark fibrous, not peeling away; **slash without red sap;** stem x-s pattern simple. **Leaves: always 3-foliolate, entire leaf 5.5-10 cm long; all leaflets equal in length, lateral leaflets distinctly 3-nerved at base,** blades ovate to obovate, 4.5-9.5 (-12) x 2.2-5(-9) cm; subleathery, lightly hairy to hairless and shiny above, **with fine bumps [*papillae*] below;** leaflet-stalks thick, wrinkled, 0.2-0.3 cm long. **Inflorescences:** axillary, panicles to 8 cm long, densely hairy; sepals densely hairy outside; corolla without hairs or gland dots; **ovary with 5 free segments. Fruits:** follicles 1-several per flower, without persistent calyx cup.

Uncommon; upland forests. Guianas; also Venezuela and adjacent Brazil.

---

3.   Connarus patrisii [CB]

4.   Connarus perrottetii, wrinkled, round leaf stalk [FRD]

5.   C. perrottetii, slash with red sap [FRD]

6.   C. perrottetii in fruit [BH]

4. Rourea Aubl.

Lianas, climbing shrubs, erect shrubs, and treelets.

General: With dense reddish-brown hairs [*tomentose*] or hairless; hairs simple.

Stems: Older stems round to oval; bark smooth-fibrous, w/o furrows, lenticels present or absent; inner bark reddish, yellow, or white, **slash without red sap;** stem x-s pattern simple, rarely 'eccentric' (= off-center-concentric).

Leaves: **3-5 (7-9)-foliolate, lower leaflet pairs alternate to subopposite;** leaflets hairy or hairless, tiny bumps [*papillae*] rarely present when leaflets > 5; leaflet-stalks sometimes absent.

Inflorescences: Axillary, terminal, or pseudo-terminal panicles; sepals overlapping, hairy or glabrous; petals hairless, **gland dots and glandular hairs absent; ovary with 5 free segments.**

Fruits: Follicles 1(2) per flower, ellipsoid to ovoid, **persistent calyx cup present;** seed 1.

Distribution: Pantropical; Neotropics: 48 species; Guianas: 8 species known, 2 common species described in detail here.

4.1. Rourea frutescens Aubl.

Liana or treelet; mostly hairy [*tomentose*]. **Leaves:** 3-5(-9)-foliolate; leaflets obovate, 3-12 x 2-7 cm, base rounded to attenuate, apex acuminate; **papery to subleathery, hairy below;** leaflet-stalks 0.1-0.3 cm long. **Inflorescences:** axillary or terminal, panicles to 10 cm long; flowers white; sepals densely hairy within. **Fruits:** follicles narrow-ovoid, slightly bent, 1.4-2 x 0.6-0.7 cm, red, hairless, wrinkled along axis.

Common; old-growth forest. Guianas; also Venezuela, adjacent Brazil, and Trinidad.

4.2. Rourea surinamensis Miq.

Common name: broin-ati-titei (Sr); bruinhart-titei (SD)

Woody liana, climbing shrub, or treelet; mostly hairless. Stems: older stem round, with scattered knobs, observed at 3.5 cm diameter. **Leaves:** **3-foliolate, rarely 1-to-5-foliolate;** leaflets ovate or elliptic, 4-21 x 1.5-8.5 cm, base rounded, apex acuminate; **leathery, hairless, shiny above and below;** leaflet-stalks thick, black, 0.4-0.6 cm long. **Inflorescences:** axillary or pseudo-terminal, panicles to 12 cm long; flowers white, 0.35-5 cm long, sepals and petals hairless. **Fruits:** follicles narrow-ovoid to -oblong, bent, 1.1-1.3 x 0.4-0.5 cm, maturing reddish, hairless, wrinkled along axis.

Common; old growth forest, savanna forest, secondary forest on sand. Guianas; also E Venezuela and W Indies.

---

7.   Connarus species, close-up of white flowers [PT]

CONVOLVULACEAE

CONVOLVULACEAE

Robust canopy lianas, subwoody lianas and herbaceous vines; less commonly herbs, shrubs or trees; often producing tubers. Climbing via twining shoots, tendril-like shoots, inflorescence axes, and clambering branchlets. Stipules are absent. Leaves are alternate and simple, ovate to heart-shaped, and pinnate-veined. Flowers vary from small to large and showy, 5-parted, with a persistent calyx, tubular to bell-shaped corolla, 5 stamens and superior ovary with 1-2 stigmas. Fruits are capsules, berries or nuts, with one-to-four seeds.

General: Bald or hairy, **hairs often silky** [*sericeous*] **when present,** simple, 2-branched (T-shaped) or rarely star-shaped [*stellate*]; **milky white sap present in vines, usually absent in canopy lianas.**

Stems: Branchlets round to irregular; nodes smooth, without interpetiolar ridge or line, stipules absent; **vines often woody only at base** [*suffrutescent*]. Older woody stems round to irregularly grooved (in cross-section), sometimes very large in diameter; bark softly woody, yellow or brown, with lengthwise platelets or wavy lines, pith often large and soft; stem x-s pattern simple (most vines, some Dicranostyles), **narrow-banded-concentric** (Dicranostyles), **broad-banded-concentric** (Maripa), or likely foraminate in one species (Colycobolus glaber).

Leaves: Elliptic to ovate-oblong, ± leathery, hairy in lianas, heart-shaped [*cordate*] in vines, margins entire or sometimes lobed (vines); **venation always pinnate; petioles normal length**

in lianas, exceptionally long in some vines (3-10 cm), often canal-shaped above; always without a swelling [*pulvinus*] or conspicuous gland-pairs.

Inflorescences: Axillary or terminal, flowers solitary or in racemes, umbels, dichasia, panicles; flowers **often tiny to small in canopy lianas** (exc. some Maripa), **mostly large and showy in vines,** radial, 5-parted; sepals free, unequal or equal, often expanding and persistent in fruit, cup-like in Maripa; **corolla fused, narrow-tubular, funnel-form with folded pleats, or bell-shaped** [*campanulate*], lobed or entire; stamens 5, alternate with corolla lobes, often of different lengths; nectar disc usually present; ovary superior, 2-chambered; style and stigmas 1 or 2.

Fruits: Nut-like, fleshy, indehiscent **berries** (Dicranostyles, Lysiostyles, Maripa) or dry capsules (vines, Bonamia, Calycobolus), 1-4 chambered; seeds 1-4, often 3, hairy or bald.

Ecology: Insect pollinated (bees and moths); Seed dispersal via animal or wind.

Distribution: Cosmopolitan, mostly (sub)tropics: ca. 55 genera/1200 species; Guianas: 5 genera/14 species of robust lianas, 6 genera/54 species of subwoody, basally woody or herbaceous vines.

Notes: Small vines are easily recognized to family, with mostly heart-shaped leaves, long petioles, showy "morning-glory" flowers, and white sap. The robust liana species are more difficult to recognize, with alternate, short-stalked leaves, smaller flowers and without white sap.

1. Bonamia Thouars

Large lianas or vines; rarely small shrubs. Climbing via twining shoots. Tropics and subtropics of New and Old World: 45 species; Amazonia: 6 species; Guianas: 1 species.

1.1. Bonamia maripoides Hallier f.

Common names: satijnblad liaan (SD)

Woody canopy liana; hairs **golden-yellow silky [*sericeous*]; white sap absent. Stems:** branchlets sometimes twisted; older stems round, softly woody; bark grey with large scattered lenticels; stem x-s pattern simple. **Leaves:** ovate to elliptic, 6-14 x 3-8 cm, margins entire; subleathery, **dark green above, hairy below; petiole short** (< 2 cm). **Inflorescences:** axillary cymes, often compound (umbella-shaped); leafy bracts present on infl. stalk; calyx with 2 hairy, larger outer sepals and 3 bald, smaller inner sepals; corolla funnel-form, white, 2-2.5 cm long. **Fruits:** fruiting calyx expanding and persistent; with **woody, egg-shaped capsules,** 0.8 x 0.6 cm, 8-valved, 2-chambered, densely hairy when young; seeds (2-)4, triangular, black, hairless. **Ecology:** Seed dispersal via wind.

Uncommon; old-growth non-flooded and riverine forest, forest edges, savanna woodland. Guianas; also Venezuela and N Brazil.

CONVOLVULACEAE

2. Calycobolus Willd. ex Roem. & Schult.

Woody canopy lianas. Climbing via twining or tendril-like shoots. Tropical America (3 species) and Africa (25 species); Guyana: 1 species.

2.1 Calycobolus glaber (Kunth) House

Woody canopy liana; mostly hairless, white sap absent. **Stems:** older stems round, stem x-s pattern possibly foraminate (with 2D 'islands' inside - see icon and Acanthaceae chapter). **Leaves:** elliptic to broadly elliptic, 5-17 x 3-6 cm, base obtuse or rounded, apex obtuse to acuminate, mucronulate; papery to leathery, **hairless above and below; petiole short** (< 1.6 cm). **Inflorescences:** axillary or terminal on lateral stems, 3-11-flowered umbels; leafy bracts present on infl. stalk; calyx with 2 outer sepals much larger (ca. 2 cm long) than 3 inner sepals (ca. 0.5 cm long); **corolla funnel-form, white (pale yellow-** ish, bluish), **2 cm long,** corolla lobes obtuse; stamens not extending beyond corolla mouth; ovary ovoid, styles 2. **Fruits: fruiting calyx expanding and persistent,** papery, ovoid, 2.3-2.8 cm long, cupping fruit tightly at base; fruits are bladder-like sacs hidden within the calyx [*utricle-like*], brown to dark red, 0.6-0.7 cm long; seeds 4 or less, ovoid to ellipsoid, black, hairless, to 0.3 cm long. **Ecology:** Seed dispersal via wind.

Rare; old-growth non-flooded and riverine forest, forest edges, savanna woodland. Guyana; also tropical South America from Bolivia to Colombia.

3. Dicranostyles Benth

Woody canopy lianas. Climbing via twining or tendril-like shoots.

General: **surfaces with silky, sericeous hairs** (not golden, often rust-colored).

Stems: Older woody stems round or irregularly grooved in cross-section, often large (> 15 cm diameter), bark smooth, with lengthwise line or channels, yellowish-brown; **stem x-s pattern simple or concentric (rings very close).**

Leaves: Elliptic-oval to oblong; usually softly hairy below; petiole short (< 2 cm long), with canal above.

Inflorescences: Axillary, racemose or thyrsiform, rachis ± hairy, infl. stalk bearing small leafy bracts; **flowers tiny, 0.5-0.8 cm long, white or light pink;** sepals free, nearly equal; corolla 5-lobed, lobes not ending in thread-like extensions (vs. Lysiostyles), with a 2-lobed stigma protruding beyond the corolla mouth.

Fruits: **nut-like berries without calyx cupule,** nearly cylindrical, thick-leathery to woody, with juicy, transparent pulp; seeds 1(-4), often 3-sided, hairless.

Ecology: Seeds dispersed via animal.

Distribution: South America to Costa Rica: 15 species. Guianas: 4 species; 2 described here.

3.1. Dicranostyles guianensis Mennega

Woody canopy liana. Stems: branchlets with rusty, sericeous hairs, lost with age. **Leaves:** elliptic to ovate, 6-11 cm x 3-7 cm, **white- to reddish-silky-hairy below; petiole 1-1.7 cm long. Inflorescences:** axillary, cymes on lateral branches, 10-15-flowered, 1-2 cm long, with small bracts; flower stalks hairless; sepals hairy;

corolla nearly disk-shaped, hairy throughout. **Fruits:** ± ellipse-shaped, to 3.8 x 2.5 cm, leathery, greyish-green, light green dotted; seeds 1, to 2.2 x 1.8 cm.

Rare; old-growth, non-flooded forest, granite flats. Fr. Guiana, Suriname; also E Venezuela.

3.2. Dicranostyles villosus Ducke

Liana; sometimes shrub; mostly hairless. **Stems:** branchlets greyish-brown. **Leaves:** elliptic to ob-

long, 4-10 cm x 2-4 cm, papery, **lightly hairy above and below; petiole < 1 cm long. Inflo-**rescences: axillary racemes or thyrses, 1.5-3 cm long, with ovate, thin bracts; flower stalks hairy [*villose*]; corolla disk-shaped, white, hairy at apex but not within; ovary bald. **Fruits:** oval to pear-shaped, to 2.5 x 1.4 cm, leathery to woody, smooth or rough, glossy green to black, persistent calyx reflexed; seeds 1, to 1.5 x 1 cm.

Uncommon; old-growth non-flooded forest, mountain savanna forest. Fr. Guiana; Suriname.

4. Ipomoea L.

Occasional subwoody lianas; often perennial or annual vines, shrubs and trees; often producing tubers. Climbing via twining shoots or clambering stems.

General: Hairless to lightly hairy; cut parts **often with milky sap.**

Stems: Branchlets round or angled, twisted. Older stems round, generally less than 2-cm diameter; bark smooth with fine, lengthwise lines; stem x-s pattern simple.

Leaves: Broadly ovate, **heart-shaped,** deeply-lobed, or rarely palmate-compound; **pinnate-veined; petiole often long (> 3-10 cm).**

Inflorescences: Axillary; sepals unequal, persistent, enlarged in fruit; corolla large and showy, funnel-form or bell-shaped, commonly white, pink or purple; ovary 2-4(3)-chambered.

Fruits: Capsules, elliptical or globose, 4-6-valved; seeds 4-6, hairy.

Ecology: Seed dispersal via wind.

Distribution: Pantropical, occas. temperate; ca. 600 species, one of the most species rich vine genera in the tropics; Guianas: ± 34 species; very few species possess woody stems beyond the base.

Uses: Ipomoea batatas (L.) Lam., the 'sweet po-

tato', is an important food plant throughout the tropics and frequently escapes to grow as a semi-wild species.

4.1. Ipomoea batatoides Choisy

Liana or perennial vine; lightly hairy or hairless. **Stems:** stems round, herbaceous with a woody base, 0.5-2 cm diameter, becoming entirely woody with age. **Leaves:** broadly ovate to nearly round, 3-10 x 3-9 cm; thin and soft, hairless above, soft hairy below; petiole to 7 cm long. **Inflorescences:** axillary or terminal, 1 to several-flowered; sepals nearly equal; **corolla funnel-form, 4-5 cm long, deep to light pink to white,** hairless. **Fruits:** almost globose, 1-1.2 x 0.8-1 cm, hairless; seeds ca. 1 cm long, covered with long, brown wooly hairs.

Common. Non-flooded old-growth forest, secondary forest, open areas. Guianas; also widely distributed from S. America to Mexico.

4.2. Ipomoea carnea Jacq.

Liana or perennial vine; also shrub or treelet;

lightly hairy or hairless. **Stems:** stems round, herbaceous with a woody base, 1-3 cm diameter, becoming entirely woody with age. **Leaves:** broadly heart-shaped to oval, 10-25 x 5-8 cm; thin and soft, hairless above, softly hairy to hairless below; petiole to 5 cm long. **Inflorescences:** terminal, several-flowered; sepals nearly equal; **corolla funnel-shaped, to 10 cm long, pink or rose-red,** with hairy bands. **Fruits:** ovoid to round, 2 x 1-1.5 cm, hairless; seeds 1-1.5 cm long, covered with long, brown wooly hairs.

Uncommon. Riversides and disturbed places. Guianas; also Venezuela, Central America, & the West Indies.

5. Lysiostyles Benth.

Monospecific genus. Climbing via twining or tendril-like shoots.

5.1. Lysiostyles scandens Benth.

High climbing woody lianas; hairs red- to yellowish-brown, silky [*sericeous*]. **Stems:** older stems unevenly angular (x-s), > 4 cm diameter, bark scaly, greyish-brown; stem x-s pattern simple. **Leaves:** oblong to elliptic-ovate, 7.5-14 x 5-10 cm; leathery, densely silky-haired below; petiole often > 2 cm long, with canal above. **Inflorescences:** axillary, cylindrical, 2.5-8 cm long; flowers fragrant, small, tubular, 0.2-0.8 cm long; sepals nearly equal, oval or round; **petal lobes ending in fine thread-like extensions. Fruits:** nut-like berries, ± globose, 2.7 x 2.3 cm;, white or buff-gray, 1-chambered, with juicy, transparent pulp; seeds 1, curved, 1.6-1.8 cm long. **Ecology:** Insect-pollinated (bees); seeds dispersed via animal (through gut of monkeys).

Rare; flooded and non-flooded old-growth forest, riparian forest. Guianas; Venezuela and adjoining Brazil.

CONVOLVULACEAE

6. Maripa Aubl.

Woody canopy lianas. Climbing via twining shoots and stout, recurved, tendril-like branchlets.

General: Hairless or occasionally with star-shaped [*stellate*] hairs; tiny black gland dots present or absent (visible with 10x hand lens).

Stems: Branchlets round or unevenly angular in cross-section, slender, hollow, greyish-brown; nodes smooth, without a swelling. Older woody stems cylindrical, oval or angled, > 5 cm diameter; bark light yellow-brown to grey, soft and fibrous, often peeling away in platelets, **stem x-s pattern off-center concentric, with conspicuous yellow and white rings.**

Leaves: **Large, ovate to elliptic, margins entire; conspicuously leathery,** with or without numerous, tiny black gland dots on upper side, sometimes hairy below; veins strongly impressed above, often prominent and yellow below; **petiole length mostly standard or short,** 1-4 cm long.

Inflorescences: Axillary or terminal on lateral branches, panicles or racemes, axes usually hairy; **flowers medium- to large-sized (1.5-6 cm long), calyx persistent and forming a cupule in fruit, sepals equal or unequal in size; corolla funnel-form to bell-shaped,** 5-lobed, violet, rose or white, fragrant, lobes softly hairy outside; ovary usually hairless.

Fruits: Calyx cupule tightly or loosely attached to fruit, sometimes reflexed away from fruit; **Berries nut-like,** ± globose to ellipsoid (flattened globose); firm to hard; pulp gelatinous, often sweet, fragrant; seeds 1(-4), hairless.

Ecology: Insect-pollinated; Seeds dispersed via animal (mostly through gut of spider monkeys).

Distribution: Widespread in the neotropics: 20 species; Guianas: 7 species; 5 species described here.

Uses: Fruits edible.

Notes: Maripa is a very distinctive genus. The **concentric stem x-s pattern of Maripa has broad bands, whereas Dicranostyles stems have very narrow, crowded concentric rings when present.** Dicranostyles has similar fruits but without the persistent calyx cupule of Maripa. Other robust liana genera in the family differ by having much smaller flowers and/or capsular, wind-dispersed fruits. Maripa and Ipomoea flowers are often similar, but the leaf and growth forms are very different.

6.1. Maripa densiflora Benth.

Liana. Leaves: ovate to elliptic, 6.5-20 x 4-11 cm; gland dots present or absent above. **Inflorescences:** terminal, in dense heads; **sepals unequal, 0.4-0.5 cm long,** with small hairs [*ciliate*], external sepals with glandular scales; **corolla white- to rose-colored,** funnel-form, 1.5-1.8 cm long. **Fruits:** ± ellipsoid, 2.7-3.0 x 1-1.6 cm, dark brown, often softly hairy at tip, **calyx strongly connected to fruit;** seed 1, 2-2.5 cm long.

Common; seasonally-flooded forest and swamps; Guianas; also Brazil, Venezuela.

6.2. Maripa glabra Choisy

High-climbing liana or erect tree. Leaves: elliptic to ovate, 8-17 x 3-10 cm; **gland dots present above;** petiole stout, 1.5-2 cm long. **Inflorescences:** terminal, in heads to 10 cm long; **sepals equal, nearly round, 0.4-0.5 cm long,** external sepals hairless; **corolla pink or white,** funnel-form, 1.5-2.3 cm long. **Fruits:** cylindrical to ellipsoid, 2-2.7 x 1.2-1.8 cm, light green to brown, **persistent calyx loosely connected to fruit;** seeds 1, ca. 2 x 1-1.5 cm, brown.

Uncommon; non-flooded forest, seasonally-

CONVOLVULACEAE

5

6

7

1. Maripa cf. scandens [CB]
2. Bonamia maripoides, flowering branch with golden-yellow silky hairs on sepals [BH]
3. Ipomoea batatoides, habit with heart-shaped flowers and large tubular flowers [STRI]
4. Maripa reticulata, flowering shoot with silky hairs on leaves and branchlets, sepals oval and equal in size [PT]
5. M. reticulata, robust woody stem [FRD]
6. M. reticulata, close-up of flowers [PT]
7. Maripa cf. scandens, close-up of flowers [CB]
8. Maripa violacea, fruiting branch [BH]
9. M. violacea with large violet flowers [PT]
10. M. violacea, mature fruits containing two large seeds [PT]
11. M. violacea, immature fruit [PT]
12. M. violacea, mature fruit bursting open with two large seeds [PT]

CONVOLVULACEAE

flooded forest and riverine forest. Guianas; also
N Amazonian Brazil.

6.3. Maripa reticulata Ducke

Common names: akawe pomëdë (Ca); babun-
malasi (Sr)

**Lianas or scandent shrubs; star-shaped [*stel-
late*] hairs present. Leaves:** elliptic to ovate, 8.5-
15 x 3.5-12 cm; **gland dots present above, yellowish-brown hairs present above and below;**
petiole 1-2 cm long. **Inflorescences:** axillary,
3-7-flowered racemes, 2.5-3.5 cm long, hairless or with stellate hairs; flowers 1.5-2 cm long,
fragrant; sepals ± equal, oblong; **corolla white,**
bell- or funnel-shaped, 1-2 cm long. **Fruits:** globose or ellipsoid, to 2 cm long, brownish-black,
shiny, slightly wrinkled, **persistent calyx lobes
reflexed;** seeds 1.

Uncommon; old-growth lowland forest, forest
edges, and along water courses. Guianas; also
Amazonian Brazil.

6.4. Maripa scandens Aubl.

Common names: hoasoropan (Ar); monkey
syrup (GU).

Woody canopy liana. Leaves: broadly ovate, 10-
18 x 4-9 cm, **gland dots absent above;** petiole
0.8-2.5 cm. **Inflorescences:** terminal, in compound heads; sepals equal, ovate-round, large, to
0.5-0.9 cm long, external sepals with flattened
hairs; **corolla lilac, pink or white,** tubular or
funnel-shaped, 2-3.5 cm long. **Fruits:** ellipsoid-
oval, 2-3 x 1.3-2 cm, greenish-brown, **persistent
calyx tightly appressed to fruit;** seeds 1(-2), to
2.2 x 1.5 cm, black.

Common; old-growth lowland forest, forest edges, and along water courses. Guianas; Venezuela
and N Amazonian Brazil.

6.5. Maripa violacea (Aubl.) Ooststr. ex Lanj. & Uittien

Common names: onseballi (Ar); patawana-titei
(Sr)

Woody canopy liana. Leaves: elliptic to ovate,
8.5-20 x 4-8.5 cm, **gland dots absent above;**
petiole 1.5-3.5 cm. **Inflorescences:** terminal, in
compound heads, 15-30 cm long; sepals ± equal,
ovate, large, 1-1.6 cm long, external sepals hairless, inner sepals purple-haired; **corolla violet
or dark blue, broadly bell-shaped, very large,
3.5-6 cm. Fruits:** globose to ellipsoid, 3.5 x 2.5-
3.5 cm, green to dark brown, with 0.2 cm long
tip; **persistent calyx lobes reflexed;** seeds (1-)2-
4, to 1.5 x 1.6 cm. **Notes:** M. violacea is distinguished from other Maripa species in Suriname
by the very large flowers and calyx.

Common; old-growth lowland and secondary forest, forest edges, especially slopes, also in swampy
areas. Guianas; also N Amazonian Brazil.

CUCURBITACEAE

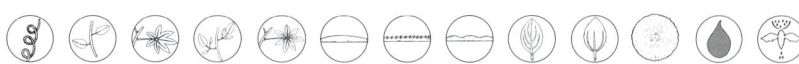

CUCURBITACEAE

Herbaceous or sub-woody vines; rarely shrubs or trees. Climbing via tendrils. Leaves are alternate and of diverse forms — palmately-compound, deeply lobed, entire – even on the same plant. Stipules are absent. Flowers are unisexual, 5-parted, with a conspicuous hypanthium, small sepals and fused petals, 5 stamens and an inferior ovary, with one chamber and one stigma. Fruits are dry capsules or fleshy berries.

General: Entire plant green, surfaces often rough [*asperous*]; watery sap often present; cut parts sometimes with an unpleasant 'rank' or cucumber scent.

Stems: Branchlets round or oval, slender and with vertical stripes or ridges, often hairless; nodes smooth, without an interpetiolar ridge or stipules; **tendrils emerging on the node at 90° from the petiole, simple or 2-7-forked,** usually spiral-coiled. Older stems round to oval, softly woody, medium-size in diameter (2-5 cm); **stem x-s pattern simple or 'radial', with conspicuous spoke-like rays.**

Leaves: Palmately-compound, palmately-lobed, or rarely entire; margins often minutely or **obscurely toothed,** round toothed or entire; thin to leathery; **flat, disc-shaped glands sometimes present** as a row on the petiole or leaf base; petiole often long.

Inflorescences: Axillary, in racemes, umbels, panicles, or solitary; **flowers usually unisexual,** with separate male and female inflorescences on the same plant, or with populations of separate male and female plants [*dioecious*]; **female flowers larger,** solitary, axillary, **male flower smaller and grouped,** often short-lived; all flowers (3)5-parted, radial; floral cup [*hypanthium*] conspicuous, greenish-white to orange or red; sepals small, lobes fused or free; petals usually united, sometimes free, white, yellow, red or orange; stamens 5, 3 or less; **ovary inferior,** 1-chambered; style 1 with a 2-lobed stigma.

Fruits: Hard-skinned berries [*pepo*] or dry capsules, often with spots or stripes; smooth or with spines, bumps, ridges; seeds 1-many, of diverse shapes and sizes.

Ecology: Insect-pollinated; seed dispersal via animal (through gut of birds or monkeys).

Distribution: Cosmopolitan, primarily in the (sub)tropics: 120 genera/760 species; Neotropics: 30 genera/400 species; Guianas: 19 genera/46 species, many of these cultivated and/or naturalized; 3 genera with sub-woody climbers.

Notes: Distinguished from other tendrillate climbers (Vitaceae, Sapindaceae, Passifloraceae) by the distinctive cucurbit tendrils, leaf shapes, stem cross-section, unisexual flowers and the inferior ovary. The family is relatively easy to recognize but distinguishing between its genera can be challenging.

1. Cayaponia A.SilvaManso

Sub-woody lianas and herbaceous vines with perennial roots. Climbing via tendrils.

General: Vegetation hairless or stiff-haired and asperous; **cucumber scent present.**

Stems: Branchlets round or oval, slender and with vertical stripes or ridges, often hairless and green; nodes with simple or 2-5-branched tendrils. Older stems round or 6-sided, softly woody; stem x-s pattern simple or radial.

Leaves: Entire to deeply 3-5-lobed, or 3-5-foliolate, asperous, **often with disc-like glands near base below.**

Inflorescences: Solitary or in groups, racemes or panicles; **hypanthium, sepals and corolla green-white;** corolla 5-lobed, sometimes yellow; male flowers with hypanthium shortly bell-shaped to deeply cylindrical, stamens usually 3; female flowers with hypanthium shallowly funnel-form to cylindrical, usually smaller than the male, staminodes (infertile stamens) 3, style stout, stigmas 3.

Fruits: Berries round or oval, smooth, not opening with valves, leathery, maturing red; calyx leaving a circular scar at fruit apex; seeds 1-30, irregularly ovate, slightly compressed, with thick, hard teeth.

Ecology: Flowers often opening at night.

Distribution: Mostly American (sub)tropics, from southern Brazil to southern U.S. and the Caribbean: ± 60 species; Guianas: 10 species, 3 described here.

1.1. Cayaponia cruegeri (Naudin) Cogn.

Common name: liba-pongo (Sa)

High-climbing herbaceous or subwoody vine; fragrant. **Stems:** ± stout, sparsely shortly hairy or hairless. **Leaves:** entire or deeply 3- or 5-lobed, 8-14 x 6-12 cm, membranous, asperous on both sides; tendrils simple or 2-fid. **Inflorescences:** axillary, solitary flowers on lax shoots; male flower hypanthium green, 1.0-1.3 cm long, hairless outside, long-hairy within, corolla lobes 1.6-1.8 cm long; female flower hypanthium green, 0.3-0.5 cm long, corolla lobes 2-2.5 cm long. **Fruits:** berries oblong-ellipsoid, 2.5-4.8 cm diam., light green to reddish-brown or black, dull, smooth; seeds 10-30, broadly elliptic, ± compressed and angular.

Common; edges of non-flooded, swamp and riverbank forest. Guianas, also NE and NW Amazonia & Trinidad.

1.2. Cayaponia ophthalmica R.E.Schult.

High-climbing sub-woody vine; hairless or with short hairs. **Stems:** nodes with very robust tendrils, 4-5-fid; stems cylindrical or 6-furrowed, stout. **Leaves:** broadly ovate-cordate in outline, 10-30 x 18-38 cm, **5-7-lobed, base with conspicuous ear-like lobes on both sides;** leathery. **Inflorescences:** axillary, numerous, crowded in stalkless clusters; male hypanthium long-hairy inside, to 2 cm long, corolla whitish, lobes to 1.0

cm long; female hypanthium bell-shaped, to 0.7 cm long, corolla lobes to 1.1 cm long. **Fruits:** fruiting clusters large, globose, with numerous bracts; berries nearly egg-shaped, to 3.5 cm diameter, pale green, white speckled, later dark red, fruit crowned by calyx ring; seeds 3, compressed, broadly ovate in outline, 2 x 1.3 cm.

Locally abundant; old-growth non-flooded forest. Guianas; also Venezuela, Brazil, Bolivia, Colombia.

1.3. Cayaponia rigida (Cogn.) Cogn.

Common names: aturaimë (Tr)

High-climbing sub-woody vine. Stems: branchlets smooth and hairless; older stems ± cylindrical, robust, to 5 cm diameter, soft and flexible, bark yellowish-brown, **slash with watery transparent sap and cucumber scent. Leaves:** 3 leaflets, 5-13 x 6-14 cm, **lateral leaflets conspicuously unequal at base;** hairless, **disc-shaped glands absent. Inflorescences:** axillary, in small panicles or racemes, greenish-yellow, male hypanthium hairless, to 0.8 cm long, corolla lobes oblong, 1 cm long; female flowers unknown. **Fruits:** berries globose, 1-1.5 cm diam, maturing red, crowned by calyx ring; seeds 3, compressed, < 2 cm, brownish-black. **Uses:** Medicinal plant of Trio Indians and Saramaccans.

Uncommon; old-growth non-flooded forest. Guianas.

2. Gurania (Schldl.) Cogn.

Subwoody or herbaceous vines with perennial roots; climbing via tendrils.

Stems: Nodes with **tendrils simple, often stout.** Older stems cylindrical or furrowed, often robust and softly woody, green; stem x-s pattern simple or with radial spoke-like rays.

Leaves: Entire to deeply 3-5-lobed or 3-5-foliolate, **disc-shaped glands absent near base below;** hairless to sparsely hairy, usually non-asperous.

Inflorescences: Male flowers in dense umbellate heads on long stalks, hypanthium and sepals showy, bright yellow to orange or red, petals bright yellow, shorter than sepals, stamens 2; female flowers in pendulous clusters on leafless branches.

Fruits: Berries, nearly elliptic, hanging, fleshy, many-seeded, with small spots; seeds smooth, slightly compressed.

Ecology: Male and female flowers on same plants; Bird- and butterfly-pollinated; Seed dispersal via animal (fruit-eating bats).

Distribution: Widely distributed in the neotropics: 30-35 species; Guianas: 11 species, 4 described here.

2.1. Gurania huberi Cogn.

Sub-woody herbaceous climber to 10 m. Stems: cylindrical, hairless, smooth. **Leaves:** 3(-5)-leaflets, lateral leaflets with unequal base; succulent, deep green, glossy above, paler beneath. **Inflorescences:** male flowers 15-30, in dense clusters on 1-21 cm stalk, hypanthium 10-15 cm long, bright orange-red; female flowers 1-3 on pendulous, leafless branches. **Fruits:** berries green with small pale spots.

Uncommon; secondary growth, edges of non-flooded old-growth forest and savanna forest. Guianas; Amazonian Brazil.

2.2. Gurania lobata (L.) Pruski

Common name: boskomkommer (SD)

Robust, sub-woody climber, to 6 m high.
Stems: soft, striate, densely hairy to ± hairless.
Leaves: ovate to broadly kidney-shaped, 10-30 x
10-30 cm, usually shallowly to deeply 3-5-lobed,
rugose with curled or stiff hairs. **Inflorescences:**
female flowers 2-11, in clusters on leafless, terminal branches, stalks 0.1-3 cm long. **Fruits:** berries ellipse-cylindrical, 4-7 x 1.5-2.3 cm, smooth,
hairless, with longitudinally lines of small, pale
flecks; seeds numerous, smooth, 0.5-0.8 cm
long, grey mottled.

Common; non-flooded old-growth forest, especially forest margins. Guianas; also Brazil, Venezuela, Ecuador, Bolivia, & Trinidad.

2.3. Gurania robusta Suessang.

Common name: bos pampoen (SD);
horotoballi (Ar)

Very robust herbaceous climber, to 20 m high.
Leaves: broadly kidney- to heart-shaped, 15-30
x 25-40cm, the central lobe largest. **Inflorescences:** male flowers 20-100 in dense, robust
spikes on 17-26 cm long stalks; hypanthium,
and calyx orange-red with white hairs; female
flowers unknown. **Fruits:** berries green.

Common; non-flooded old-growth and secondary forest. Fr. Guiana, Suriname.

2.4. Gurania subumbellata (Miq.) Cogn.

Common names: pongu (Sr); wilde sukwa (SD)

Soft-woody, high-climbing vine. Stems: ± angular, soft hairy. **Leaves:** ovate, 10-20 x 10-30
cm, 0-3-5-lobed, the central lobe largest; thin,
dull green with short stiff hairs above. **Inflorescences:** male flowers 50-100 in dense clusters
on 5-45 cm long stalks, hypanthium 0.4-0.7 cm
long; female flowers few on ± leafless, pendulous
branches. **Fruits:** berries with short apex, 3.5-4
cm long; seeds unknown.

Common; non-flooded old-growth forest. Guianas; also Brazil and Venezuela.

CUCURBITACEAE

3. Psiguria Neck. ex Arn.

Widespread in the neotropics: 12 species; Guianas: 1 species. Climbing via axillary tendrils.

3.1. Psiguria triphylla (Miq.) C. Jeffrey

Common name: boskomkomer (Sr)

High climbing, softly-woody vine. General: surfaces hairless to sparsely hairy, usually **non-asperous. Stems:** cylindrical, 2-10 cm diameter, deep green, ± hairless; nodes with **tendrils simple and often stout;** stem x-s pattern simple or radial. **Leaves:** extremely variable, entire to deeply 3-lobed, occasionally with 3-5 leaflets (more than one form on individual plants); **disc-shaped glands absent near base below;** petiole to 2 cm long. **Inflorescences:** Male and female flowers in separate inflorescences; with a greenish hypanthium; with sessile male flowers in extremely dense spikes, 20-250 flowers on 5-40 cm long infl. stalks; female flowers 1-5 clustered at each node on stout, hanging, leafless branches; hypanthia 0.5-1.4 cm long, sepals 5, greenish; petals red or orange; stamens 2. **Fruits:** berries, elliptical to cylindrical, 3.5-7 x 1.5-2.8 cm, green to yellowish with lengthwise stripes; calyx persistent, pulp whitish, edible; seeds numerous, 0.7-1 x 0.4-0.5 cm. **Ecology:** Animal-dispersed (parrots, small monkeys).

Common, with a wide ecological amplitude; non-flooded old-growth forest, mountain savanna, flooded grasslands, riverine forest, cloud forests, especially along edges. Guianas; widespread in the neotropics, up to 2,100 meters.

1. Gurania subumbellata, male plant [PT]
2. G. subumbellata, male inflorescence [PT]
3. G. subumbellata, leaves [PT]
4. Gurania reticulata, female inflorescence [DM]
5. Cayaponia sp., flower longitudinal section [BH]
6. Gurania reticulata, male inflorescence [PT]
7. Gurania lobata, female inflorescence [PT]
8. G. lobata, fruits [PT]
9. Cayaponia sp., stem x-s pattern 'radial' - with spoke-like radial rays [BH]
10. Cayaponia sp., young shoot with tendril and female flower [PT]
11. Psiguria umbrota, female flower [PT]
12. Cayaponia triangularis, flower with stamens attached to corolla [PT]
13. Cayaponia sp., older subwoody stem with tendril at node [BH]
14. Psiguria triphylla, fruits [PT]

CUCURBITACEAE

DICHAPETALACEAE

DICHAPETALACEAE

Lianas, shrubs, or trees. Pantropical: 3 genera/260 species. Guianas: one genus of trees (Tapura) and one genus of lianas.

1. Dichapetalum Thouars

Mostly canopy lianas; sometimes erect shrubs or treelets. Climbing via twining shoots and lateral branching. Stipules are present. Leaves are alternate, simple, entire, leathery to thin, with pinnate venation and gland dots; petioles thick. Inflorescences are uniquely borne on the petiole apex. Flowers are radial, white, with 5 stamens and a superior ovary. Fruits are small, dry drupes.

General: Surfaces finely to densely hairy [*tomentose, pilose, hirsute*]; notable sap or scent absent.

Stems: Branchlets oval, ± hairy, with raised lenticels; **nodes with paired linear stipules, persistent (D. pedunculatum) or soon falling away (D. rugosum).** Older stems round to irregular, **bark often fibrous, flaky and peeling away,** silver to reddish-brown, with raised lenticels or pustules; **stem x-s pattern simple, lobed or interrupted.**

Leaves: Often widely elliptic to ovate, margins entire; leathery to thin, blade bullate or flat; **dark gland dots present below;** venation pinnate with marginal loops, 1^0 and 2^0 veins commonly immersed above and prominent below; 3^0 veins net-like [*reticulate*].

Inflorescences: **Evenly-branching** [*dichotomous*], **flat-topped cymes borne on petiole apex;** flowers ± radial, 5-parted, small; sepals free, overlapping [*imbricate*]; **petals free,** often overlapping, white or whitish-yellow, **each petal split into 2 lobes; stamens 5, free;** nectar glands 5; ovary 1-3-chambered with 2(3) styles.

Fruits: **Dry, ellipsoid drupes,** tannish to red, hairy [*puberulous*]; inner wall [*endocarp*] hard, bony; seeds 1 per chamber.

Ecology: Plants dioecious or monoecious; insect-pollinated (bees); dispersed via animal (not through gut).

Distribution: Pantropical with greatest diversity in Africa and Malaysia; ± 250 species; Guianas: 3 species, 2 described here.

Notes: Dichapetalum schulzii Prance is a rare liana species known from remote southern forests in Suriname (voucher specimen: Irwin, Prance, Soderstrom & Holmgren 54582).

1.1. Dichapetalum pedunculatum (DC.) Baill.

Common name: awari imopi tëkënë (Tr); kanakudiballi (Ar)

Lianas, shrubs, or treelets; hairless to minutely hairy [*tomentose*]. **Stems:** branchlets zigzag, brown to tan, hairs lost with age, **nodes with persistent, linear stipules, 0.5-1 cm long;** older stems irregularly lobed, observed at 8 cm diameter; bark brown to grey, with large lenticels, cracked and peeling, sub-bark layer green; stem x-s pattern lobed or interrupted. **Leaves:** ovate-elliptic to oblong, 5.0-16.0 x 2.0-7.5 cm (smaller than D. rugosum), base subcordate to subcuneate and slightly unequal; petiole 0.4-1.2 cm long. **Inflorescences:** calyx with whitish-yellow hairs; corolla tan, petals deeply split, hairless. **Fruits:** drupes 0.8-1.1 x 0.6-0.7 cm, yellow with grayish-brown hairs, 1(2)-chambered.

Uncommon; non-flooded old-growth forest, riparian forest, secondary forest, and open areas. Guianas; also Amazonian Brazil, Venezuela, and Trinidad-Tobago.

1.2. Dichapetalum rugosum (Vahl) Prance

Common name: awari imopi tëkënë (Tr)

High-climbing lianas or shrubs; densely hairy [*tomentose*] when young. Stems: branchlets yellowish-brown hairy, **nodes with small, short-lived, lanceolate stipules, 0.2-0.4 cm long;** older branchlets with smooth, silvery bark and lenticels; older stems round, with large lenticels, bark brown, thin, flaking away, inner bark and wood yellow; stem x-s pattern unknown. **Leaves:** orbicular or ovate-elliptic, 6-32 x 3.5-21 cm, base rounded or subcordate; **leathery, bul-**late, densely yellowish-brown-hairy [*pilose or hirsute*] below;** petiole 2-3.5 cm long, densely hairy. **Inflorescences:** axillary or terminal; calyx with reddish-brown hairs; corolla tan, petals deeply split, hairless. **Fruits:** drupes 1.3-1.8 x 0.9-1.1 cm, dark brown with yellowish-brown hairs, 1(2)-chambered.

Uncommon; non-flooded old-growth forest, ridge forest and secondary forest. Guianas; also Amazonian regions of Colombia, Peru, and Brazil.

---

1. Dichapetalum rugosum, liana habit with dichotomous-branching inflorescence borne on the petiole [BH]
2. D. pedunculatum, with linear stipules, small white flowers and small dry drupes [BH]
3. D. pedunculatum, robust woody stem with irregular shape and lobed x-s pattern [BH]
4. D. rugosum, leaf underside with raised veins [FRD]
5. D. rugosum, dry drupes with dense reddish-brown hairs [FRD]
6. D. rugosum inflorescence with small, white 5-parted flowers (perianth and stamens free) [BH]

4

5

6

DILLENIACEAE

1

2

DILLENIACEAE

Robust canopy lianas and climbing shrubs; also erect shrubs and small trees. Climbing via twining shoots and clambering branches. Leaves are alternate (spiral), simple, with entire, wavy or toothed margins. Stipules are present but mostly inconspicuous. Inflorescences showy, flowers with 3-7 free sepals and petals, many stamens, and a superior ovary. Fruits dry capsules or fleshy berries, sometimes enclosed by persistent sepals.

General: Vegetation robust and often rough-surfaced or 'sand-papery' due to silica fibers [*asperous*]; hairs simple, in bundles or rarely stellate (Tetracera); colored sap absent, abundant water sometimes present.

Stems: Branchlets round to flat; bark in woody twigs red and papery; shoots sometimes with a 1-2 meter leafless zone below ultimate leaves; nodes with stipules small and short-lived or (rarely) winged and fused to leaf stalk. Older stems among the largest lianas occurring in the region (5-35 cm diameter), round to oval, often twisted and furrowed; **bark conspicuously dark red, orange-red to silver-grey, papery and often shedding;** the wood [*stele*] of larger stems in some species holding abundant pure water; **stem x-s pattern concentric or simple, often with obvious spoke-like rays.**

Leaves: Diverse, mostly large and broadly ovate (also lanceolate, elliptic, orbicular), base mostly cuneate, **margins entire or wavy, often toothed in upper half** (remotely-toothed with tiny spines, sharply toothed); blade rough-surfaced (Davilla, Tetracera) or smooth (Doliocarpus, Pinzona); **2° veins parallel, close-together, straight to slightly curved** (Davilla, Pinzona, Tetracera), or **far-apart and strongly to slightly curved** (Doliocarpus), **3° veins parallel** (forming ladder-like steps in-between 2° veins, slanted towards midvein) **or net-veined;** leaf stalk [*petiole*] variable — with a deep groove or channel on upper side [*canaliculate*], ± winged, often wider or thicker at base.

Inflorescences: Terminal, axillary, or in-between nodes [*cauliflorous*], in tight clusters, racemes or panicles; **flowers radial, medium-sized, showy,** yellow or white, 3-7-parted; sepals free, overlapping, sometimes persistent; petals free, crumpled in bud, short-lived; **stamens many** (15-200 per flower), often maturing from the center outwards; ovary superior, segments [*carpels*] 1-2 (Davilla, Doliocarpus, Pinzona) or ± 3-5 (Tetracera), stigma(s) terminal.

Fruits: Variable – fruit enclosed by two persistent yellow-orange, hardened sepals (Davilla); red leathery capsules enclosing seeds in a fleshy white aril (Doliocarpus); 2-lobed capsules (Pinzona); or 4-5 free follicles, each enclosing ± 4 arillate seeds (Tetracera).

Ecology: In some species, juvenile plants produce larger, toothed leaves, and adult plants produce smaller, un-toothed leaves. Many species flower simultaneously and abundantly. Seed dispersal via animal (birds, monkeys, ants).

Distribution: Mostly neotropical (exc. pantropical Tetracera): 6 genera/95 species; Guianas: 4 genera/24 species of woody climbers.

Use: Dilleniaceae liana stems are commonly used in the Guianas as components of herbal aphrodisiac concoctions and to make a chocolate-milk-like drink. The abundant fresh water from the stem can quench thirst in the forest, but is also used to treat heart disease and high blood pressure. Researchers have isolated more than 130 compounds, many with proven biological activity (Lima et al. 2014). Davilla species are important in genital steam-baths and rituals for Saramaccan and other Maroon peoples (van Andel & Ruysschaert 2011).

Notes: Easily recognized family due to the reddish-brown bark, straight 2° veins and leaf shape/margins. Compare to Cordia (Boraginaceae); Euphorbiaceae (Mabea); Piptocarpha (Asteraceae); and Solanaceae.

1. Davilla Vand.

Lianas or clambering shrubs.

General: Vegetation strongly to slightly rough-surfaced; hairs simple when present.

Stems: Branchlets round or irregularly flattened, papery or smooth, sometimes densely hairy; nodes marked by circular stipule scar (D. alata) or scars inconspicuous. Older stems mostly < 5 cm diameter, rarely to 15 cm diameter, bark red or grey, papery, peeling or not peeling easily; stem x-s pattern with radial spokes [*radial*]; abundant water sometimes present in stem.

Leaves: Highly variable shapes, lanceolate to ovate, base commonly rounded, **margin entire to remotely-toothed with tiny spines,** ± curled [*revolute*]; leathery (papery); **veins usually sunken or flat above, raised below;** 2° veins straight or curved, close-together, 3° veins parallel, forming ladder-like steps in-between 2° veins, slanted towards midvein; **leaf stalk [petiole] slightly to strongly winged,** always channeled above.

Inflorescences: Axillary or terminal panicles, hairy [*villous to sericeous*]; calyx 5-parted, un-equal, **the 2 inner sepals much larger than the outer 3 and becoming hardened; petals ± 5, yellow,** short-lived; stamens 50-70; ovary 1-chambered [*2 fused carpels*] with one style.

Fruits: Dry fruit with hardened, persistent dull-orange sepals, forming 2 half-globes that spread apart to display a fleshy white, round aril with 1-2 black seeds.

Ecology: Insect-pollinated; dispersal via animal (birds).

Distribution: Widespread in the Americas from Paraguay to subtropical Mexico: 30 species; Guianas: 5 species, 4 described here. Common in disturbed areas, savanna edges and secondary forest.

Notes: Species descriptions below do not include flower characters because petals are short-lived and sepals are persistent and best described in fruit.

1.1. Davilla alata (Vent.) Briq.

Woody liana or climbing shrub; hairy [*hirsute*], **surfaces rough. Stems:** branchlets brown-haired, with extended stipule scars encircling nodes. **Leaves:** elliptic, 7.5-17.5 x 5.5-11.5 cm, base and apex rounded, margin curled; leathery, densely golden-brown hairy above and below; **petiole 5-8 cm long, broadly-winged (± 0.75 cm wide). Inflorescences:** terminal panicles to 15 cm long. **Fruits:** few-fruited; follicles 2, free, hairless; calyx lobes (sepals) to 2 cm diameter, with flattened [*sericeous*] golden-brown hairs outside; aril white, seeds 2.

Uncommon; savanna and creek forests. Guianas only.

1.2. Davilla kunthii A.St.-Hil.

Common name: brontitei (Sr); diatetei (Au); diatitei (Sr); fajatatái (Sa); kapadula (GU); ka-buduli (Ar); kawtitei (Sr); sakëtaitu (Tr); schuur-papierliaan (SD); tameyu'u (Ca); watra tité (Sr).

Woody liana or climbing shrub; hairy, surfaces rough. Stems: branchlets white-haired, stipule scars inconspicuous; woody stems 3-15 cm diameter, bark reddish-brown with shallow furrows and bumps, **not easily peeling. Leaves:** variable, 5-17 x 3-11 cm, base rounded, apex acuminate, margin entire, curled; thick-leathery, shiny above, with white or brown hairs below; **petiole to 3 cm long, narrowly winged. Inflorescences:** panicles to 15 cm long. **Fruits:** many-fruited; follicles 2, free, yellow to reddish-brown; calyx lobes 0.5-0.6 cm diameter, hairs flattened; seeds 1. **Notes:** Abundant drinkable water in stem. Compared to D. rugosa, this species has coarser leaves and more conspicuous 3° veins.

Very common; savannah forest, shrub savannas and secondary forest. Guianas; widespread in tropical America.

1.3. Davilla nitida (Vahl) Kubitzki

Common name: cacaotitei (Sr); diatetei (Au); diatitei (Sr); fajatatái (Sa); kapadula (GU); kabuduli (Ar); sakëtaitu (Tr); schuurpapierliaan (SD); skraatititei (Sr); tameyu'u (Ca); watratité (Sr).

Woody liana or climbing shrub; mostly hairless, surfaces only slightly rough. Stems: branchlets smooth, stipule scars inconspicuous; woody stems with grey bark, **easily peeling. Leaves:** diverse, elliptic to ovate, 4-9(15) x 2.5-5(7) cm; base rounded, apex folded-obtuse, margin entire; leathery; **petiole to 2 cm long, narrowly winged. Inflorescences:** axillary, many-flowered panicles to 15 cm long. **Fruits:** many-fruited; follicles 2, free; calyx lobes 0.6-0.7 cm diameter, hairless; seeds 1, aril white.

Common; edges of gallery forests and savannas, disturbed areas. Guianas; widespread in tropical America.

1.4. Davilla rugosa Poir.

Woody liana or climbing shrub; reddish-brown hairy, surfaces rough. Stems: branchlets hairy, stipule scars inconspicuous; woody stems with brownish-grey bark, **easily peeling. Leaves:** variable, lanceolate to ovate, base rounded to cuneate, **margin entire or remotely toothed with tiny spines;** sub-leathery or papery, dense hairs on veins below; **petiole to 1.5 cm long, winged at apex,** densely hairy. **Inflorescences:** axillary panicles 2-10 cm long, axes occasionally with small leaves or bracts, densely hairy [*pilose to hirsute*]. **Fruits:** many-fruited; follicles 2, free; calyx lobes 0.8 cm diameter, yellowish-red to orange, densely hairy to hairless outside; seeds 1, aril white.

Uncommon; riverine, marsh forest, and savanna. Guianas; also NW South America, Central America, and Cuba.

2. Doliocarpus Roland.

Woody canopy lianas, climbing shrubs, and vines; also erect shrubs.

Common name: asrikatité (Sr); dialopu (Sr); dialopu tité (Pa); diatité (Sr); faja tatái (Sa); kapadula (GU); kabuduli (Ar); kawtité (Sr); sakëtaitu (Tr); tameyu urang (Ca); watratité (Sr).

General: Surfaces usually smooth, not rough-surfaced; hairs simple when present.

Stems: Branchlets round or irregularly flattened, papery or smooth, sometimes densely hairy; **nodes marked by circular stipule scars** or scars inconspicuous. Older stems round to flattened, **attaining very large diameters, bark reddish-brown, papery,** wood dark red; **stem x-s pattern always with concentric rings** [*concentric*]; abundant water usually present in stem.

Leaves: Variable but mostly large and broad, **margins entire to wavy, often becoming toothed/spine-tipped in upper half** (especially in juvenile plants); **leathery, smooth;** veins usually sunken/flat above and raised below (rarely 3° veins raised above); **2° veins not close-together, curved or straight to margin,** 3° veins parallel or net-veined; leaf stalk [*petiole*] variable — with a deep groove or channel on upper side [*canaliculate*], ± winged, often wider or thicker at base.

Inflorescences: Axillary or borne in-between nodes [*cauliflorous*] on woody stems, in dense clusters or long racemes, axes red, hairless to finely hairy; **calyx lobes 4-5, mostly equal** (vs. 2 sepals larger in Davilla), overlapping, often persistent; **petals 3-4, white or yellow,** overlapping, short-lived; **stamens 20-100;** ovary 1-chambered, stigma 1.

Fruits: Leathery capsules, globose, bright red, mostly splitting in half to display 1-2 round or flattened seeds, each embedded in a fleshy to thin white aril.

Ecology: Seed dispersal via animal (bird, monkey, through gut).

Distribution: Neotropics: 40 species; Guianas: 11 species, 5 described here.

2.1. Doliocarpus brevipedicellatus Garcke

Woody liana or clambering shrub; minutely hairy or hairless. **Stems:** branchlets round; **stipule scars conspicuous, encircling nodes. Leaves:** highly variable in shape, 5.5-14 x 3-5.5 cm, base obtuse, apex acuminate, **sharply to remotely-toothed in upper half;** leathery to papery; **2° veins curved upwards or almost straight; 3° veins sub-parallel,** visible above and below; petiole short, 0.5-1 cm long. **Inflorescences:** racemes, hairy; calyx lobes (4)5, subequal, minutely hairy; petals 3, white. **Fruits:** capsules to 0.6 cm diameter, with prickle at apex; seeds (1)2, aril white.

Uncommon; old-growth lowland forest, and upland savanna. Guianas; also Brazil, Venezuela, Bolivia, & Peru.

2.2. Doliocarpus dentatus (Aubl.) Standl.

Woody liana, scandent shrub, or treelet; lightly hairy to hairless. **Stems: branchlets round, densely silky-haired when young, stipule scars conspicuous, encircling nodes. Leaves:** ovate-elliptic to obovate, 7-20 x 2.5-10.5 cm, base acute, apex acuminate, **with spine-tipped teeth in upper half;** leathery to papery, smooth, paler below; **2° veins slightly curved upwards or straight, 3° veins parallel;** petiole 1-2.5 cm long, brown. **Inflorescences:** axillary or cauliflorous, in dense, 15-30-flowered clusters; calyx lobes 4(5); petals 2-3, about as long as the sepals. **Fruits:** capsules 0.6-0.7 cm diameter, cherry-red, hairless; persistent calyx to 0.5 cm long, fruiting style 0.1 cm long; seeds 1(2), blackish-brown, shiny, aril off-white. **Notes:** Doliocarpus dentatus subspecies esmeraldae (Steyerm.) Kubitzki has reddish-brown hairs.

Common; especially along creeks and rivers. Guianas; widespread in the neotropics.

2.3. Doliocarpus major J.F.Gmel.

Woody liana or climbing shrub; finely hairy [*pilose*] to hairless. **Stems:** branchlets round, stipule scars inconspicuous. **Leaves:** narrow to almost round, 5-13 x 2-6 cm; base acute or rounded, apex acuminate-acute, **with spine-tipped teeth in upper half;** leathery, violet-grey to blackish above, **white-scaly and with small gland dots below; 3° veins parallel;** petiole short, 0.4-1 cm long, hairy. **Inflorescences:** axillary, 2-6-flowered clusters, flower stalks hairy; calyx lobes 5, unequal, finely hairy; petals 3-4, pale yellow, ± 1 cm long. **Fruits:** capsules to 1.2 cm diameter, red, apical prickle 0.2 cm long, persistent calyx to 0.7 cm long; seeds (1)2, black, flattened, aril fleshy, off-white.

Common; especially at forest edges and along rivers. Guianas; also Venezuela, Brazil, Ecuador, Peru, and Bolivia.

2.4. Doliocarpus paraensis Sleumer

Woody liana or climbing shrub; mostly hairless. **Stems:** branchlets round or flattened, smooth, stipule scars inconspicuous. **Leaves:** narrowly to widely obovate, often asymmetrical from base to tip, 8-18 x 3-7.5 cm, base cuneate, apex short acuminate, **margin entire (not toothed), curled;** leathery, hairless, shiny above and below; **2° veins curving upwards with marginal loops, 3° veins net-veined, conspicuous above and below;** petiole 0.5-2 cm long, dark, abruptly swollen at base. **Inflorescences:** axillary or inter-nodal, dense clusters, calyx lobes 5, inner 3 oblong, minutely hairy, petals 3. **Fruits:** capsules large, to 3 cm diameter, red, apical prickle 0.4 cm long, persistent calyx to 0.6 cm long; seeds 2, blackish, kidney-shaped, aril very thin.

Common; old-growth forest, secondary forest, and savanna. Guianas; adjoining Brazil.

2.5. Doliocarpus spraguei Cheesman

Small liana; scandent or erect shrub; mostly hairless. **Stems:** branchlets round to oval, smooth, grey, hairless; stipule scars inconspicuous. **Leaves:** elliptic to obovate-elliptic, 5-12 x 2-6 cm, base acute, folded lengthwise, apex acuminate and acute, **rarely toothed in upper half;** thick-leathery, greyish-brown, **shiny above and below; 2° veins curving upwards, 3° veins net-veined;** petiole 0.5-1 cm long, dark. **Inflorescences:** axillary or inter-nodal, 10-20-flowered dense clusters (umbel-like); calyx lobes 5, unequal, obovate-elliptic, petals 3, ± round, to 0.3 cm long, yellow or white. **Fruits:** capsules 0.6-0.7 cm diameter, red, persistent calyx to 0.35 cm long; seeds 1(-2), black, ± kidney-shaped, aril off-white.

Common; white sand forest, savanna, scrub-savanna, and savanna forest. Guianas; also Brazil, Colombia, and Trinidad.

3. Pinzona Mart. & Zucc.

Monospecific genus.

3.1. Pinzona coriacea Mart. & Zucc.

Common name: dialopu (Sr); dialopu tité (Pa); diatité (Sr); ënkomikomi (Tr); faja tatái (Sa); sakëtaitu (Tr); watratité (Sr).

Woody canopy liana. General: surfaces usually finely hairy or hairless and smooth, **never rough to touch; hairs simple when present.** **Stems**: branchlets round, bark smooth and papery, greyish-green, stipule scars inconspicuous; older stems round and knobby, becoming large in diameter, bark thin, red or silvery, papery and peeling, slash red; **stem x-s pattern always with concentric rings** [*concentric*], visible even in small cuttings; abundant water usually present in stem. **Leaves:** elliptic to obovate, 9.5-16 x 5-12 cm, base acute to obtuse, apex rounded, **margin entire or ± toothed in upper half;** leathery, 2° veins often straight, sunken above, raised below; **3° veins rigidly parallel (closer together than in Doliocarpus),** slightly raised above and below; petiole 1.5-2.5 cm long, winged, with a groove on upper side [*canaliculate*]. **Inflorescences:** axillary, racemes or panicles of racemes, 3-7 cm long; sepals 3-4, ± equal, persistent; petals 2-3, obovate, white; stamens 25-50; ovary 2-chambered, styles 2. **Fruits: capsules 2-lobed, persistent calyx lobes (e.g., Davilla) absent,** each lobe globose, ca. 0.5 cm diameter, greenish to red, splitting to display 1-2 black seeds, each embedded in an orange aril. **Ecology:** Seeds dispersed via animal (birds, through gut; possibly ant-dispersed as well).

Uncommon; old-growth and semi-deciduous forest. Guianas; also Amazonian Brazil and Peru to Costa Rica and the Antilles.

4. Tetracera L.

Robust woody lianas or climbing shrubs.

General: Vegetation mostly rough-to-touch [*asperous*]; **hairs often star-shaped** [*stellate*].

Stems: Branchlets round or irregularly flattened, papery or smooth, sometimes densely hairy; nodes with inconspicuous stipule scars. Older stems round, attaining large diameters, **bark red or silver-red, papery, peeling easily,** slash dark red; **stem x-s pattern usually simple/radial, rarely with concentric rings;** abundant water usually present in stem.

Leaves: Oblong-elliptic to ovate, **margins entire;** leathery, **lightly hairy to densely stellate-hairy** [*tomentose*]; 2° veins curving upwards or straight; 3° venation mostly parallel, slanted towards midvein; petioles slightly winged or simple.

Inflorescences: Terminal or axillary, in loose racemes or panicles; flower-stalks jointed; bracts narrow and pointed; flowers bisexual or unisexual, **sepals usually 5** (3-15), equal or unequal, overlapping, **persistent but not enclosing fruit (e.g., Davilla); petals usually 5** (1-6), **white or yellow,** overlapping, obovate or oblong; **stamens 50-200;** ovary with 3-5 free segments [*carpels*].

Fruits: 4-5 pointed, leathery, free follicles per flower, spreading or erect, splitting open along the upper surface to display 1-4 black seeds with red arils.

Ecology: Seed dispersal via animal (through gut).

Distribution: Pantropical: 45 species; Neotropics: 20 species; Guianas: 7 species, 3 described here.

DILLENIACEAE

Pinzona coriacea

4.1. Tetracera asperula Miq.

Common name: sabana-kawtitei (Sr); watrat-itei (Sr)

Liana or climbing shrub; small tree. **Leaves:** oblong to elliptic-oblong, 5-16 x 2.5-6.5 cm, base rounded or subacute, apex rounded-acute with a fine tip, margins entire; leathery, dark green and glossy above, **surface turning from rough to smooth with age;** 2° veins curving upwards, raised on both sides, **3° veins net-veined;** petiole 0.8-2.5 cm long, unwinged. **Inflorescences:** terminal panicles, 8-20 cm long, most surfaces stellate-hairy, axes reddish-green; sepals 5, unequal in fruit; petals 5, double the length of the sepals. **Fruits: follicles 4-5, each 3-sided, dark brown or purplish-brown,** to 1.7 x 0.8 cm, smooth, shiny; seeds 2-3 per follicle, 0.5 x 0.4 cm, black, shiny, including irregularly torn aril longer than the seed.

Common; especially open areas, including white sand savannas and granite flats at higher elevations. Guianas; adjoining Brazil.

4.2. Tetracera costata Mart. ex Eichler

Liana or climbing shrub; small tree. **Leaves:** elliptic to oblong, 5.5-12.5 x 3.5-7.5 cm, base cordate or rounded, apex rounded or retuse, **margins entire;** leathery, smooth, greyish-green, hairless above, **densely stellate-hairy below; 2° veins straight,** sunken above, raised below, **3° veins parallel,** raised on both sides; petiole 1-2 cm long, slightly winged at apex, hairy. **Inflorescences:** terminal panicles, 15-25 cm long, greyish-brown tomentose with stellate hairs; sepals 5-6, orbicular, outside densely tomentose; petals 5-6, obovate. **Fruits:** follicles 4, purplish-brown at maturity, rough-surfaced.

Common; old-growth lowland forest, forest edges. Guianas; also Brazil (Mato Grosso) and Trinidad.

4.3. Tetracera surinamensis Miq.

Liana or climbing shrub; small tree. **Leaves:** broadly elliptic to ovate-elliptic, 6-11 x 4.5-7 cm, base rounded, apex obtuse, **margin entire;** leathery, smooth or slightly rough above, rough below; veins sunken above, raised below, **2° veins straight, 3° veins parallel;** petiole 1.5-2.5 cm long, unwinged. **Inflorescences:** terminal panicles, stellate-haired; sepals 5, unequal in fruit; petals 4-5, obovate, white, unequal, slightly longer than sepals. **Fruits:** follicles 4, purplish-brown at maturity, rough-surfaced.

Uncommon; old-growth lowland forest. Guianas; also N Amazonian Brazil and Ecuador.

---

1. Davilla kunthii, fruits of two hardened, yellow-orange sepals enclosing a black seed with fleshy white aril [BH]

2. D. kunthii, close-up of fruits [BH]

3. D. kunthii in flower [OG]

4. Doliocarpus dentatus, young leafy shoot, leaves with widely-spaced 2° veins [PT]

5. D. dentatus, leaf underside close-up, 3° veins parallel, slanted towards midvein [FRD]

6. Doliocarpus spraguei, flowering branch [PT]

7. Doliocarpus dentatus, leaf with widely-spaced 2° veins and spine-tipped teeth [RA]

8. D. dentatus, red capsule split to reveal black seed with white aril [RA]

9. D. dentatus, silvery-red bark with slash [BH]

10. Pinzona coriacea. a) flowering stem; b-d) bisexual flowers; e) stamens; f) two-chambered ovary, 2 seeds per chamber, styles 2; g) fruiting branch; h) bi-lobed fruit, persistent sepals; i) open fruit capsules; j) arils around seeds; k) seeds [BA]

11. P. coriacea, flower, ovary with two chambers and styles [FRD]

12. P. coriacea, stem with flaky reddish-brown bark [FRD]

13. P. coriacea, bi-lobed capsules with 2 orange-aril covered seeds [FRD]

14. Tetracera asperula, flowers in bud and open [PT]

15. Tetracera willdenowiana, fruiting branch [PT]

16. Tetracera volubilis. a) flowering branch; b) large leaf, remotely toothed; c) bumpy leaf underside [papillose]; d) flowers open and in bud; e) habit; f) 4-follicle fruit [NCB]

EUPHORBIACEAE

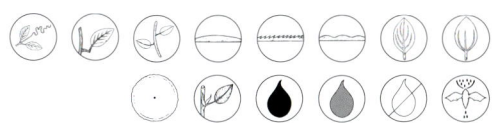

EUPHORBIACEAE

Occasional woody lianas, climbing shrubs and vines; usually erect trees, shrubs, herbs, or succulents. Climbing via twining shoots, tendril-like shoots, or weakly sprawling across vegetation. Leaves are alternate and simple, venation palmate and/or pinnate, often toothed and with diverse, conspicuous glands. Stipules are present. Inflorescence units usually with many tiny male flowers above and 1-2 larger female flowers below; ovary superior, 3(4) chambered. Fruits are 3-4 parted capsules, berries or drupes.

General: **Diverse sap colors in cut parts** – bright red, orange, watery (Croton), milky white latex (Mabea), white-translucent, slimy, turning [*oxidizing*] red, pink or blue (Omphalea). Diagnostic hairs [*stellate, dendroid*] **and scaly surfaces** [*peltate, lepidote*].

Stems: Branchlets round, hollow or soft-centered; nodes without a ridge or line, **stipules axillary** [*intrapetiolar*], **often inconspicuous,** short-lived or modified as hairs, glands, or scales. Older stems round, to 30 cm diameter; bark smooth or rough, **often with lengthwise "cuts"** and lenticels; stem x-s pattern simple (Keller 1996).

Leaves: **Blades and petioles of unequal size and shape** on one branch (equal in Mabea); margins wavy to finely toothed; leathery to papery; venation palmate, pinnate, or palmate-pinnate; **conspicuous paired glands at leaf base or petiole apex** [*basilaminar glands*], also with flat glands or tiny white spheres [*laminar glands*] and leaf pockets [*domatia*] below; 3° veins often parallel and 90° or slanted to midvein.

Inflorescences: Terminal or axillary, often in spikes; flowers unisexual, plants bisexual [*monoecious*]; perianth in 1-2 whorls or reduced/absent; male flowers — stamens 1-2-many; female flowers – mostly ovary 3-chambered with free stigmas or column; Plukenetia ovary 4-chambered with fused column.

Fruits: Capsules mostly 3-parted dry capsules (Plukenetia 4-parted), large indehiscent berries or drupes.

Ecology: Insect-pollinated; seed dispersal via animal (monkeys, birds, rodents, ants) or explosive fruits (Mabea).

Distribution: Widespread in tropical and temperate zones (218 genera/6745 species);

Guianas: 5 genera/10 species of woody climbers; 3 genera/16 species of herbaceous vines.

Notes: Two main (artificial) vegetative groups of Euphorbiaceae climbers in the Guianas: 1) Palmate venation, leaf-base gland pairs: Croton, Omphalea, and most Plukenetia. Climbers that might be confused with this group: Ampelozyziphus, Aristolochia, Byttneria, Celtis, Convolvulaceae, Heisteria, Menispermaceae, and Sparattanthelium. Among these, leaf-base gland pairs occur only in Byttneria. 2) Pinnate venation throughout, white milky sap (leaf-base gland pairs absent) – Mabea. Plukenetia supraglandulosa has pinnate-veined leaves but without white sap. A similar general appearance is evident in pinnate-veined Convolvulaceae, Cordia, Heisteria, and Piptocarpha (Asteraceae) – all of which lack milky sap.

1. Croton L.

Large and variable genus, including lianas, shrubs, trees, and a few herbs. Climbing via scrambling branchlets. Cosmopolitan; ca. 1200 species; Guianas: ca. 30 species, one liana.

1.1. Croton pullei Lanj.

Common name: katamïïmë (Tr)

Liana or scrambling shrub. General: cut parts with watery red-orange sap, clear resin and incense-like scent; surfaces of reddish-brown-scales mixed with white stellate hairs. **Stems:** older woody stems round, to 25 cm diameter, swollen at nodes; bark blackish-green to grey, with shallow lengthwise grooves or lines and raised lenticels; slash with dark orange-red living bark, red sap and white wood; pith hollow, exuding a resin or gum. **Leaves:** broadly ovate, 14-18 x 7-8 cm, base cordate to squarish, apex acuminate, margins finely-toothed to ± entire; papery, densely hairy or scaly below, basal glands crater-like or raised (not elongate); veins palmate at base, pinnate beyond base; 3° veins ± parallel; petioles of variable lengths, 2-7 cm long. **Inflorescences:** terminal, spikes 12-30 cm long; flowers scaly-stellate outside; sepals green and scaly; corolla yellowish-green or -white, with long yellow hairs within; stamens 5-many, yellow; ovary 3-chambered, styles free or fused. **Fruits:** weakly 3-lobed capsules with persistent styles, to 1 x 0.8 cm, splitting into three parts; seeds 3. **Ecology:** Seeds dispersed via animal. **Uses:** Latex and bark used to treat cuts, leishmaniasis, stomach aches, and ear infections (Tr).

Common; non-flooded old-growth forest, secondary forest and riverbank forest. Guianas.

2. Mabea Aubl.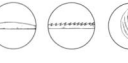

Rarely lianas or climbing shrubs; usually erect shrubs or trees. Climbing via scrambling branchlets.

General: Cut parts with **abundant white sap;** most surfaces reddish-brown hairy, with simple or **unique forked** [*dendroid*] **hairs.**

Stems: **Branchlets arranged in distinctive whorls** (5-8 branchlets meeting at one point), visible on ground and canopy individuals; nodes smooth, with stipules toothed at base, falling early. Older stems round or oval, 3-15 cm diameter, bark flaky or smooth, living bark thin, burgundy red, **slash often red-white banded** (appearance of raw beef).

Leaves: Elliptic-elongate, margins slightly wavy

(M. pulcherrima) to finely toothed (M. taquari); hairless and shiny above, dull below; venation pinnate; 2° veins curving upwards and forming marginal loops, 3° veins networked [*reticulate*]; with 40-50 globose, pearl-like glands along leaf margins; **laminar and leaf-base glands absent; petioles short, equal in length.**

Inflorescences: Showy, hanging "bottle-brush" panicle or raceme, with a mixture of yellow, brown, and burgundy colors, including male flower triads and a few longer female flowers at base; **male flowers with 5 sepals,** subtended by leafy bracts bearing 2-glands; **female flowers with 6 sepals, hairy,** subtended by leafy bracts without glands; ovary 3-chambered, stigmas 3 and coiled.

Fruits: Globose capsules, 3-furrowed, woody, velvet-hairy; seeds 3, ovoid, smooth, crowned by a small appendage [*caruncle*]. Fruits are not very useful in distinguishing between species.

Ecology: Bat, bird, and insect-pollinated; Seeds dispersed when capsules forcibly and audibly explode upon drying, perhaps secondarily via water.

Distribution: Widespread in the Neotropics: 38-40 species; Guianas: 11 species with 1 true liana and 1 shrubby climber.

Use: Inflorescences producing a very tasty, abundant, honey-like nectar.

2.1. Mabea pulcherrima Muell.Arg.

Liana to 25 meters tall. **Leaves:** elliptic, 6-13 (-18) x 2-5 cm, **margin entire or sub-entire,** curled, with marginal pearl-like glands; leathery, dark green above, white below; petiole grooved. **Inflorescences:** terminal or axillary, pendent "bottle-brush" panicles, to 12 cm diameter; flowers green to red-green or dull yellow; stamens up to 100 per flower. **Fruits:** capsules to 2.7 x 3.5 cm, green to reddish-yellow.

Uncommon; old-growth forest along water courses, lowland to sub-montane elevation. Guianas: also most of the Amazon Basin.

2.2. Mabea taquari Aubl.

Common name: dréwatra koko (Au); kotiki (Sr); mërii asoso (Tr).

Liana, climbing shrub; also small tree. **Leaves:** elliptic to narrowly-elliptic, 4-12 x 1.5-5 cm, margin finely toothed, with or without **marginal pearl-like glands;** hairless and shiny above; petiole densely hairy. **Inflorescences:** terminal or axillary, pendent "bottle-brush" panicles, 2-4 cm wide, reddish-brown hairy; flowers green to red-green. **Fruits:** capsules to 1.1 x 1.2 cm, green to reddish-yellow.

Common; old-growth forest near rivers, rock islands, rapids. Guianas; also Venezuela, Trinidad, E Amazon watershed.

1. Mabea pulcherrima [BH]
2. Croton pullei, flowering branch [BH]
3. C. pullei, female flowers close-up with stigmas [BH]
4. C. pullei, liana in host tree [BH]
5. Omphalia diandra, flowering branch [BH]
6. O. diandra, with robust tendril-like shoot [BH]
7. O. diandra, fruit, a large globose berry [PT]
8. O. diandra, ripe fruit and angular seeds [BH]
9. Mabea pulcherrima, unique whorled branching [BH]
10. O. diandra, thick, fleshy petiole with raised-gland pair at apex [BH]
11. Plukenetia polyadenia, leafy branch and woody stem [BH]
12. Mabea taquari, burgundy inner bark [BH]
13. Mabea pulcherrima, stem with white sap [BH]

3. Omphalea L.

Lianas; sometimes erect trees. **Climbing via conspicuous, thick, tendril-like shoots.** Neotropics, Australasia and Africa; 17 species, one species known in the Guianas.

3.1. Omphalea diandra L.

Common name: baboen-noto (Sr); babunnoot (SD); idaballi (Ar); meku kuware (Ca); sityò (Ca); warikë (Tr).

Robust, woody canopy liana. General: cut parts with **slimy, white-cloudy sap, quickly oxidizing blood red or pink-blue** and with a scent of passion fruit; hairless or sparsely orange-red hairy, hairs simple. **Stems:** branchlets round, ridged or striped, often green, hollow or with soft pith; stipules inconspicuous; older stems round, 10-20 cm diameter, bark blackish-grey with shallow lengthwise cuts or grooves; slash with dark red living bark and white-pinkish wood, sap slimy. **Leaves:** broadly ovate or elliptic-ovate, large, 8-27 x 5-16 cm, base rounded to subcordate, apex acuminate, **margins wavy, notched or few-toothed;** succulent and thick, dark green glossy above; orange-red hairy with scattered marginal glands below; **basal gland pair large, crater-like or globose; 3-veined at** base, **pinnately-veined above;** 2° veins curving and widely-spaced, terminating in marginal teeth, 3° veins parallel; petioles of different lengths. **Inflorescences:** axillary, large paniculate cymes, 15-70 cm long (often many male flowers and 1 large female flower); ultimate flowering units on rachis subtended by conspicuous, large, gland-bearing leafy bracts; male flowers with a distinctive mushroom-shaped cap; female flowers with 3-locular ovary and massive column of fused styles. **Fruits: large, round to ovoid berries,** to 11.5 x 8 cm, blackish-green to orange-yellow, with 3 blunt ribs, pulp yellow, thick, sap cloudy white and sticky, indehiscent; seeds 2-6, three-sided, to 5 cm long, with abundant milky sap. **Ecology:** Seed dispersal via animal (spider monkeys and rodents). **Uses:** Fruit and seeds edible as survival food. Bark used to heal skin cuts (Tr).

Common; riparian or swamp forest zones. Guianas; widespread in the Neotropics.

4. Plukenetia L.

Woody canopy lianas, slender lianas, and vines. Climbing via twining shoots.

General: Colored sap absent; vegetation lightly hairy to hairless, hairs mostly simple.

Stems: Branchlets round, length-wise striped or grooved; nodes with small inconspicuous stipules that often fall early. Older stems round, 0.5 to 5 cm diameter; bark thick and fibrous, greyish.

Leaves: Mostly ovate with round or squared base and 3-palmate veined (P. supraglandulosa entirely pinnate); margins toothed to wavy; **basal gland pair commonly flat and elongate;** 2° veins curving and widely-spaced, 3° veins ± parallel; laminar glands present above, below, or as "pearls" along margin; petioles short or long.

Inflorescences: Terminal or axillary, in spike-like simple racemes or panicles, sub-units with many male flowers above and 1-2 larger basal female flowers; stamens free, from 3 to many; ovary 4-chambered, **styles fused and forming a massive "stylar column".**

Fruits: i) **Capsules 4-sided or -parted, wider than long** < 1.5 cm diameter (P. supraglandulosa, P. verrucosa) or 2.5-6 cm diameter (P. volubilis); ii) **berries ± round,** 6-11 cm, indehiscent (P. polyadenia). In all species — seeds 4 per fruit, lens-shaped or globose, caruncle absent (fleshy structure attached to seed, usually attracting ants).

Ecology: Individual plants usually [*monecious*], rarely [*dioecious*]. Seed dispersal via animal.

Distribution: Pantropical: 17 species; Guianas: 5 species, 4 described here. Plukentia loretensis was documented only once, by the 1952 Guppy expedition in southern Guyana and Suriname.

4.1. Plukenetia polyadenia Muell.Arg.

High-climbing robust liana. Leaves: elliptic to ovate-elliptic, 7-11 x 4-7 cm, base rounded to obtuse, apex bluntly acuminate, with narrow tip to 1.5 cm long, **margin untoothed;** papery to sub-leathery; **3-palmate veined;** basilaminar glands 2-4, separated by a small knob; **laminar glands absent;** petiole 1-5 cm long. Inflorescences: axillary; male infl. 5-35 cm long, female infl. 1.5-7 cm long. **Fruits:** subglobose berry, obscurely 4-sided, to 11 cm diameter, indehiscent; seeds ovoid, to 5.6 cm long.

Uncommon; old-growth forest, especially at forest margins; lowland to sub-montane elevations. Guianas; also widespread in Amazonia except for Colombia.

4.2. Plukenetia supraglandulosa L.J. Gillespie

Slender liana. Leaves: elliptic or ovate-elliptic, 7-13 x 3-8 cm, base acute, with fine tip, **margin toothed;** papery; **venation entirely pinnate,** basilaminar glands 2-6; **scattered laminar glands present above and below;** petiole 0.5-2 cm long. **Inflorescences:** axillary, racemes 2-8 cm long; ovary 4-winged. **Fruits:** capsule with 4 thick wings, 0.7 x 1.5 cm., hanging, dehiscent.

Rare endemic in sub-montane forests near Brazil. Fr. Guiana, Suriname; also adjacent areas of Brazil (Amapá).

4.3. Plukenetia verrucosa J.E.Smith

Robust canopy liana. Leaves: oblong-ovate, 5-10 x 2.5-6.5 cm, base cordate or Plukentia loretensis was documented only once, by the 1952 Guppy expedition in southern Guyana and Suriname. truncate, apex with narrow tip 0.5-1.5 cm long, **margin untoothed; 3-palmate veined;** paired glands and **needle-like glands present** at petiole apex; pearl-like glands present at leaf margins; petiole 1-3 cm long. **Inflorescences:** terminal or axillary, racemes to 4 cm long; ovary and stigma deeply 4-lobed. **Fruits:** capsule flattened-globose with 4 horizontal projections [*tubercles*], 0.7 x 1.4-1.5 cm; stellate hairs present.

Uncommon; open areas and upland secondary forest. Guianas; also Trinidad and N Brazilian Amazon.

4.4. Plukenetia volubilis L.

Slender liana or vine. Leaves: triangular-ovate, 6-13 x 6-12 cm, base truncate to subcordate, margin toothed; thin, shining above, pale below; **3-palmate-veined;** basilaminar glands present, **laminar glands absent;** petiole 3-8 cm long. **Inflorescences:** axillary or terminal, racemes 5-18 cm long; ovary 4-winged, stigma thick, bilobed. **Fruits:** capsule deeply 4-lobed, 1.5-2.5 x 2.5-4 cm; seeds oblong, to 2 cm long, brown with brown lines. **Use:** Seeds reported to be edible when roasted, sometimes cultivated.

Uncommon; old-growth forest, disturbed areas, secondary forest, riparian forest; lowland to higher elevations; Guyana, Suriname; widespread in N & W tropical South America.

Additional climbing Euphorbiaceae species in the Guianas

Acalypha includes one slender, weakly climbing shrub in the Guianas, Acalypha scandens. There are three genera of small, herbaceous Euphorb vines in the region, including Dalechampia, Margaritaria, and Tragia.

GNETACEAE

GNETACEAE

A gymnosperm family with one genus.

1. Gnetum L.

Lianas, scandent shrubs, and a few trees. Climbing via opposite branching, jointed branchlets and twining shoots. Leaves are opposite, simple, and entire. Stipules are absent. Fertile branchlets in whorls and bearing whorls of tiny 'flowers' or large, oblong 'fruits'.

General: Colored, watery sap (orange-red, yellow-brown, or white) **and/or clear resin sometimes present in cut parts.**

Stems: Branchlets jointed and usually swollen at nodes, bark smooth and reddish- to greenish-brown, lenticels present. Older stems round to oval, observed up to 5 cm diameter; bark fibrous, furrowed, reddish-brown to dark brown or black, lenticels present; **stem x-s pattern concentric,** central chamber sometimes exuding jelly-like resin.

Leaves: Broadly-elliptic to ovate; leathery to papery, **blade either smooth and green or with silky-white reflective fibers above and fine lines below** (G. urens); veins sunken above, 2° veins pinnate, curving, often with marginal loops, 3° veins net-like [*reticulate*]. Drying black, tan, or greenish-yellow.

Fertile: Plants either male or female [*dioecious*]. On fertile branchlets, male or female 'flowers' [*sporangia*] occur in whorls at nodes, each whorl supported by a cup [*bract collar*]. 'Fruits' [*pseudodrupes*] are single seeds with a thin, tough covering, 1-many per whorl, broadly oblong, to 5 x 3 cm, bright orange, red, purple or black when ripe.

Ecology: Pollination likely by nocturnal insects (Vicentini 1999); seed dispersal via animal or water.

Distribution: Pantropical, 35 species; Neotropics, 8-10 species; Guianas: 3 species known, 2 additional species expected.

Uses: Edible seeds (Trio); medicinal bark and leaves; incense from resin; cordage from bark.

Notes: Gnetum is the only gymnosperm genus with pinnate-veined angiosperm-like leaves. **The presence of jointed stem branchlets is a good recognition character for sterile Gnetum species.** The genus might be confused with other opposite-leaved taxa that possess whitish or orange-red sap and a concentric stem x-s pattern (Apocynaceae, Celastraceae).

1.1. Gnetum nodiflorum Brongn.

Common name: kôsyiton (Pa); towa (Tr); towauri (Vz)

Slender to robust liana; sap watery orange-red. Stems: branchlets light-brown, smooth, with white, thick lenticels; jointed but without swollen nodes; older stem bark thick, furrowed, with scattered lenticels, inner bark orange. **Leaves:** elliptic, large, to 18 x 9 cm, base unequal [*and obtuse*], apex acuminate; **leathery, surfaces 'smooth' (without silky-fibers or fine lines),** green, drying greenish-yellow-tan. **Fertile:** uppermost whorls distant - 1-2 cm apart (male) or 1.5-2 cm apart (female). Pseudo-drupes broadly oblong, to 3.5 x 2 cm, maturing reddish-brown, faintly ribbed, base uneven.

Common; savanna-forest margins, riverbanks, open areas, ridges, and mountains, to 1800 m. Guianas; widespread across N. Amazonia.

1.2. Gnetum paniculatum Spruce ex Benth.

Robustly woody liana; sap white or cream. Stems: branchlets smooth; older stem bark fibrous, dark brown to black, densely covered with lenticels. **Leaves:** oblong-elliptic, to 17 x 8 cm, base unequal to obtuse, apex acuminate; leathery, **surfaces 'smooth' (without silky-fibers or fine lines),** green, drying black. **Fertile:** many branched, uppermost whorls crowded – 0.1-0.2 cm apart (male) or < 1 cm apart (female). Pseudo-drupes ellipsoid to ovoid, to 5 x 3 cm, maturing bright orange to red.

Rarely collected; evergreen flooded forest and river banks. French Guiana & Guyana, expected in Suriname; also Guiana Shield regions of Brazil and Venezuela.

1.3. Gnetum urens (Aubl.) Blume

Common name: towa (Tr); towauri (Vz)

Slender liana; sap white, cream, or yellow-brown. Stems: branchlets smooth, reddish-brown to light grey; nodes swollen; older stem bark fibrous, dark, densely covered with lenticels. **Leaves:** elliptic, medium-sized, 12 x 6 cm, base unequal, tip acuminate to acute; yellowish-green; thin-papery; **upper surface with white, silky, reflective fibers and lower surface with fine lines. Fertile:** uppermost whorls crowded - 0.1-0.2 cm apart (male) or < 1 cm apart (female). Pseudo-drupes broadly oblong, to 4 x 2 cm, maturing reddish-brown to bright red, faintly ribbed, base uneven.

Uncommon; evergreen forest, black water swamps, tepuis, to 1700 m. Guianas; also Guiana Shield regions of Brazil and Venezuela.

Additional Gnetum species

Gnetum camporum (Markgr.) D.W. Stev. & T. Zanoni
Small, very thick leaves, up to 6 x 4 cm, with silky fibers above; > 1000 m, Gran Sabana, Venezuela, higher elevations, expected in Guyana.

Gnetum leyboldii Tul.
Leaves large, very thick, up to 20 x 15 cm, without silky fibers above; Lowland N Brazil, Venezuela, expected in S Guianas.

1. Gnetum nodiflorum, habit and edible fruit (pseudodrupe) [BH]

2. G. urens node, leaf blade with silky fibers [FRD]

3. G. nodiflorum, woody stem [FRD]

4. G. nodiflorum, woody stem x-s pattern of concentric rings [BH]

5. G. nodiflorum, fertile male branchlets ('microsporangia') [FRD]

6. G. nodiflorum, whorls of male 'flowers' with 'bract collars' [FRD]

7. G. nodiflorum, woody stem bark and x-s pattern of concentric rings [BH]

HERNANDIACEAE

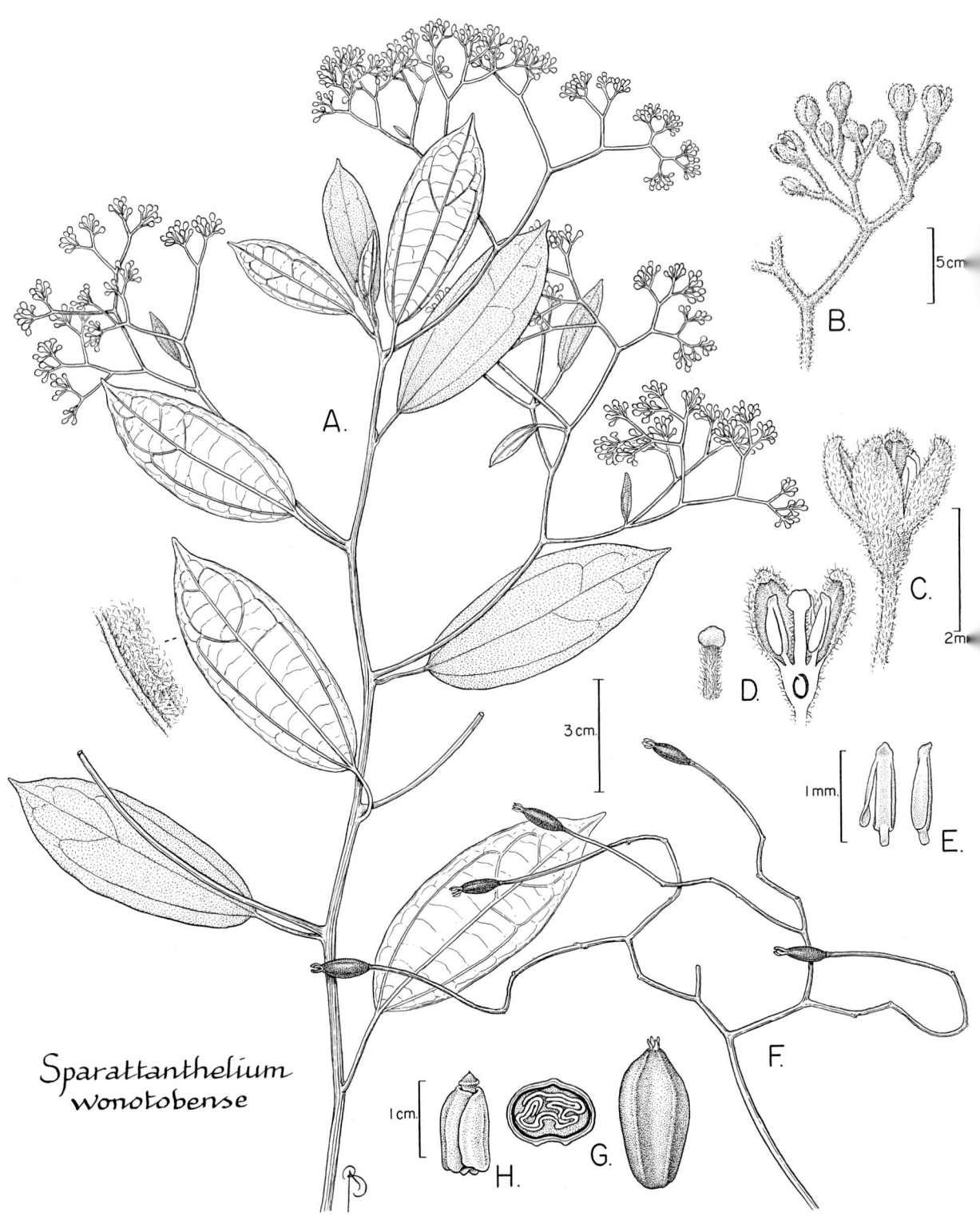

Sparattanthelium
wonotobense

Sparattanthelium wonotoboense. a) flowering shoot with old inflorescence stalks persisting as climbing hooks; b) inflorescence sub-unit; c) flower; d) flower long-section and style-stigma; e) stamens; f) fruiting branchlets; g) fruit; h) seed. [BA – Bobbi Angell]

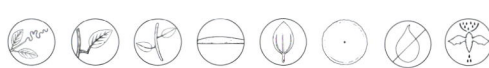

HERNANDIACEAE

A small family of lianas, shrubs, and trees. Pantropical: 5 genera/60 species; Guianas: 1 genus.

1. Sparattanthelium Mart.

Lianas or weakly climbing shrubs; aromatic. Climbing via unique branch-hooks in leaf axils (woody base of inflorescence stalks on older stems) and lateral branchlets. Stipules are absent. Leaves are alternate, simple, entire, 3-veined from below the leaf base, with ladder-like cross-veins. Inflorescences are long and widely-branched, bearing flat-topped clusters of tiny, 4-5 parted greyish-white flowers; ovary inferior, stigma 1. Fruits are ovoid, dry, greyish-white drupes (one-seeded).

General: Vegetation often in shades of black and dark green, **cut parts strongly aromatic** – with "wintergreen" scent of essential oils; young vegetation lightly hairy to densely hairy, hairs simple when present.

Stems: Branchlets wiry, slightly zigzag along axis, **bark black, smooth, with fine vertical lines,** lenticels absent; nodes smooth. Older stems round, to 5 cm diameter, bark becoming softly woody, brown, with vertical furrows and platelets; stem x-s pattern simple.

Leaves: Narrowly elliptic to ovate, never toothed, margins usually curled downwards [*revolute*], rarely flat (Sparattanthelium aruakorum); papery to leathery, hairless and smooth above, undersides lightly hairy or with a soft, dense, carpet-like layer of hairs; **3-palmately-veined from below leaf base,** veins raised below, **2° veins ladder-like,** 90° to midrib and palmate vein pairs [*scalariform*]; **petiole simple,** often long (not swollen at apex or base [*pulvinus*], not extending onto stem [*decurrent*], not grooved above).

Inflorescences Axillary or terminal, flat-topped panicles, with long infl. stalk, branches dividing widely and evenly [*dichotomous-branching*]; flowers radial, tiny (< 0.2 cm diameter), greyish-white, 4-5-parted, with one whorl [*tepals*]; bisexual; stamens 4-5; ovary inferior, 1-chambered; style 1, cylindrical.

Fruits: **Branching widely and often strongly zig-zag, axes often white or grey;** drupes ovoid to ellipsoid, dry, smooth, leathery to woody; seed 1.

Distribution: Generally rare. Neotropics: 14 species; Guianas: 5 species, 4 described below.

Ecology: Seed dispersal via animal (birds).

Notes: Compare with Ampelozizyphus (Rhamnaceae), Aristolochia, Dioscorea, many Menispermaceae.

1.1. Sparattanthelium aruakorum Tutin

Clambering shrub. Leaves: oblong-lanceolate to ovate, 6-15 x 2-4.5 cm, base obtuse to acute, often unequal, apex acuminate, **margins flat (not curled);** leathery, shiny above, **finely hairy below** when young, not felty; petiole 1-3.5 cm long. **Inflorescences:** axillary, panicles to 13 cm long, branchlets hairless, flowers (tepals) **5-parted,** with fine greyish-white hairs. **Fruits:** ovoid-ellipsoid drupes, ± 1.7 x 0.7 cm, silvery-grey with 6 faint ribs.

Rare; non-flooded old-growth forest. Guyana, Suriname.

1.2. Sparattanthelium tupiniquinorum Mart.

Clambering shrub. Leaves: lanceolate to elliptic, 5.5-16 x 2-6 cm, with drip-tip to 1.3 cm long, margins curled; leathery, shiny above, **finely hairy below, with whitish-grey hairs along nerves;** petiole 1.2-4 cm long. **Inflorescences:** axillary, panicles slender, hairy [*tomentose*], branchlets dark-red, flowers 4-5-parted. **Fruits:** ellipsoid drupes, 1-1.5 x 0.7-0.9 cm, silvery-grey with 5-7 ribs.

Rare; old-growth forest, lowland to lower montane, coastal. Guianas; Venezuela, Brazil.

1.3. Sparattanthelium uncigerum (Meissn.) Kubitzki

Common name: wanëkë (Tr)

Clambering shrub or liana. Leaves: lanceolate to obovate, 4-8 x 2-3.5 cm, drip-tip to 1.7 cm long, margins curled; **hairless and dull above with a dense, grey carpet-like layer below,** both sides with a similar shade of color; petioles 0.7-3.2 cm long. **Inflorescences:** axillary, panicles narrow, to 17 cm long, branches greyish-white hairy, with round knobs at branch junctions; **flowers 4-parted. Fruits:** depressed-ovoid drupes, 1.5-2 x 0.4-0.5 cm, white to silvery-grey, with 1-2 deep furrows on both sides.

Uncommon; lowland to lower montane old-growth forest, riverine forest. Guianas; Venezuela.

1.4. Sparattanthelium wonotoboense Kosterm.

Common Name: oneka (Ca); wanëkë (Tr)

Liana. Leaves: elliptic, 5-8.5 x 2.2-3.6 cm, margins curled; **hairless and shiny above with a dense, grey carpet-like layer below;** petiole 1-3 cm long. **Inflorescences:** axillary, panicles densely-flowered, infl. stalk to 10 cm long, branchlets brownish-purple to greyish-white, without knobs at branch junctions; flowers grey to orangish-purple. **Fruits:** resembling S. uncigerum, but smaller.

Rare; non-flooded and seasonally-flooded old-growth forest, riverine forest. Guianas only.

2. Sparattanthelium botocudorum, clambering liana in fruit (Brazil) [QP]

3. Sparattanthelium uncigerum, fruit close-up [BH]

4. S.uncigerum, fruiting branchlets, zig-zag with round knobs at junctions [BH]

5. S. aruakorum, stem cross-section [BH]

6. S. uncigerum, leaves dark green above, drying black, grey-green below [BH]

4

6

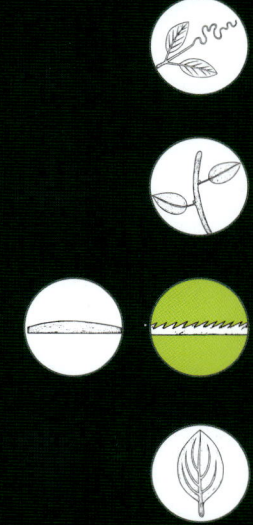

ICACINACEAE

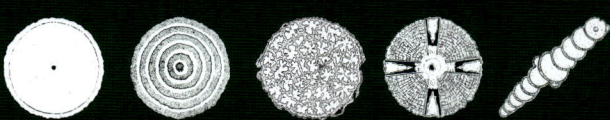

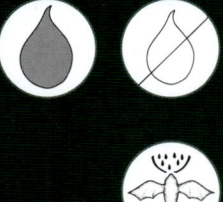

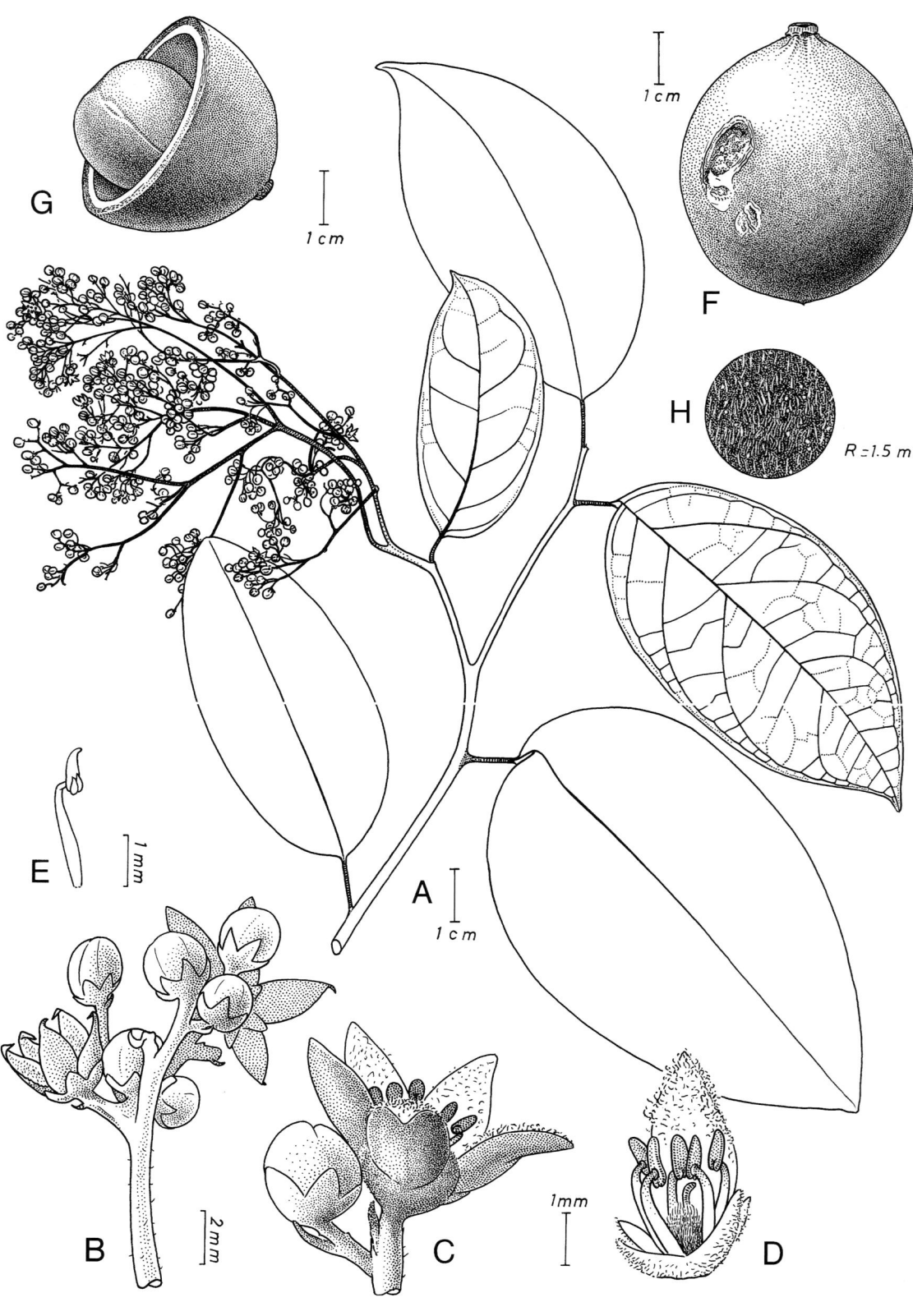

Casimirella ampla. a) flowering branch with terminal panicle and widely spaced secondary veins; b) inflorescence sub-unit;
c) open flower; d) petal, ovary and stamens; e) stamen; f) mature fruit (drupe); g) fruit with seed; h) hairs on leaf underside.

ICACINACEAE

Lianas and climbing shrubs; also erect shrubs and trees. Climbing via twining stems or branchlets. Stipules are absent. Leaves are alternate and simple with margins entire or ± toothed. Inflorescences in dense spike-like racemes or branching panicles, flowers small, free, fleshy and 4-5 parted; ovary superior, one-chambered. Fruits are one-seeded drupes with a hard inner wall.

General: Hairs variable – short and stiff, flattened, wooly, star-shaped [*stellate*], forked or see-saw-/t-shaped [*malpighiaceous hairs*]; cut parts often with a medicinal or rank scent.

Stems: Branchlets round or flat, bark with lengthwise furrows and lenticels. **Older stems oval or flat and with a central groove, up to 10 cm diameter;** bark corky or smooth; slash producing clear resin or gum and sometimes turning [*oxidixing*] orange (Leretia); **stem x-s pattern often complex and changing with age** – concentric, eccentric, interrupted or foraminate (see icons).

Leaves: Ovate to elliptic, robust, drying black to greenish-tan; veins pinnate, raised below; **2° veins widely-spaced,** looping or ending at marginal gland-teeth; 3° veins net-veined [*reticulate*] or parallel (Pleurisanthes); **petiole twisted and grooved above.**

Inflorescences: Wide-branching panicles or narrow, spike-like racemes; flowers very small, radial, free, 4-5 parted; sepals cup-shaped, fleshy; petals fleshy, not overlapping, white, green, or yellow; stamens 4-5, alternate with petals; ovary 1-chambered, stigma head-like.

Fruits: Ovoid or elongate drupes, 1-seeded, 2-5 cm long, sometimes hairy or rough-surfaced.

Ecology: Seeds dispersed via animal (through gut).

Distribution: Pantropical (temperate): ca. 60 genera/450 species; Guianas: 3 genera with woody climbers.

Use: Several species produce 5-20 kg edible tubers underground. The tubers require processing to remove toxins but are sometimes used to avoid starvation in the Guianas.

Notes: Similar to other lianas with alternate, pinnate-veined leaves without stipules. Compare with robust Convolvulaceae, Menispermaceae, Nyctaginaceae, Olacaceae, and Phytolaccaceae. Flattened stems also present in Bauhinia, Coccoloba, Machaerium, and Menispermaceae.

1. Casimirella Hassler

Only one species known to occur in the Guianas.

1.1. Casimirella ampla (Miers) R.A.Howard

Robust liana or climbing shrub. General: hairs simple, star-shaped [*stellate*], or forked (not see-saw/t-shaped). **Stems:** branchlets and twigs flat, furrowed, reddish-brown; **stem x-s pattern not observed. Leaves:** broadly lanceolate to elliptic, 8-20 x 3-10 cm, base acute or rounded, apex obtuse to acuminate; **margins entire;** papery to leathery, mostly hairless and smooth; venation strongly raised below; 3° veins net-veined to ± parallel; petiole 0.8-1 cm long. **Inflorescences:** terminal, branching panicles, densely reddish-brown-hairy; **flower stalks jointed;** petals white-yellow, narrow to ovate, 0.35-0.4.3 cm long, wooly-haired within; stamens 4; style slightly curved.

Uncommon; old-growth forest. Fr. Guiana; also NE Brazil.

2. Leretia Vell.

Monospecific genus.

2.1. Leretia cordata Vell.

Woody canopy liana or climbing shrub; also erect shrub or small tree. **General:** hairs diverse, including t-shaped hairs. **Stems:** branchlets flat, furrowed, ± zig-zag, slash with rank scent; **older stem x-s pattern complex,** changing from 'foraminate' to 'interrupted' with age. **Leaves:** ovate to elliptic, 10-30 x 3-12 cm; base acute to obtuse, apex acute; **margins entire;** papery to leathery, drying grey; **3° veins net-veined** (not parallel). **Inflorescences: branching panicles,** 10(-18) cm long, densely golden-hairy; **flower stalks jointed,** 0.1-0.4 cm long; petals white, oblong to narrow, **wooly-haired within; stamens 4;** style often curved. **Fruits:** drupes ellipsoid, 3-4.5 x 1.8-2.5 cm, green to dark red or purple, initially hairy; seed 2.5 x 2 cm, with vertical veins and ribs.

Uncommon; old-growth forest, creeks, slopes. Guianas; also Panama, Amazonia.

2. Pleurisanthes parviflora full leaf [AP]

3. P. parviflora, flattened, 2-lobed stem [FRD]

4. Pleurisanthes emarginata, flowers close-up [FRD]

5. Leretia cordata, stem with off-center concentric x-s pattern [BH]

6. Pleurisanthes emarginata, flattened inflorescences along one side of stem [FRD]

3. Pleurisanthes Baill.

Canopy lianas or climbing shrubs.

General: Hairs simple [*pilose, hispid*], without t-shaped hairs; strong scent absent.

Stems: Branchlets round, often densely hairy. **Older stems robust, ± flat, 2-cabled,** to > 10 cm diameter, bark thick, slash with or without a clear, sticky resin; stem x-s pattern concentric or eccentric.

Leaves: Oblong-elliptic, base ± rounded to cordate, **entire or with gland-tipped teeth** (P. emarginata); leathery to subleathery; hairs often stiff or flattened, rough surfaced below; 3° and 4° venation conspicuously parallel (P. parviflora) or net-like (P. emarginata); petiole channelled above, ± twisted; petiole ca. 1 cm long.

Inflorescences: Terminal or lateral, axes flattened with flowers (racemes) clustered to one side or simple axillary panicles; **flower stalks unjointed;** calyx cup-like, 5-lobed; flowers 4-5 parted, yellow or white, hairless within; ovary conical, 1-chambered.

Fruits: Ovoid to flattened drupes, ca. 2 cm diameter, surface with light brown hairs, rough-surfaced; seed 1.

Distribution: Widespread in low numbers in the South American tropics, from Bolivia to Colombia and the Guianas; Guianas: 4 species, 2 described here.

3.1. Pleurisanthes emarginata van Tiegh.

Liana or climbing shrub. Stems: slash with colorless resin. Leaves: 8-12 x 5-7 cm, young leaves larger than adult leaves, base ± rounded; margins with gland-tipped teeth; hairy above and below. **Inflorescences:** terminal, many narrow racemes along one side of branch, each 3-8 cm long; calyx lobes triangular; **corolla 4-parted,** white, petals oval, 0.1 cm long. **Fruits:** unknown.

Uncommon; old-growth forest. Fr. Guiana, Suriname; also Brazil.

3.2. Pleurisanthes parviflora (Ducke) R.A. Howard

Liana or climbing shrub. Stems: slash without resin. **Leaves:** 8-16 x 3-6 cm, base slightly heart-shaped, apex acuminate; margins entire; ± hairless above, hairs dense and flattened below. **Inflorescences:** axillary, panicles 5-15 cm long, flowers single or 2-6-clustered; calyx lobes triangular; corolla 5-parted, white, petals oblong to narrow, to 0.25 cm long. **Fruits:** Drupe orange with white speckles.

Uncommon; old-growth forest. Fr. Guiana, Suriname; N Amazonian Brazil.

6

LAMIACEAE

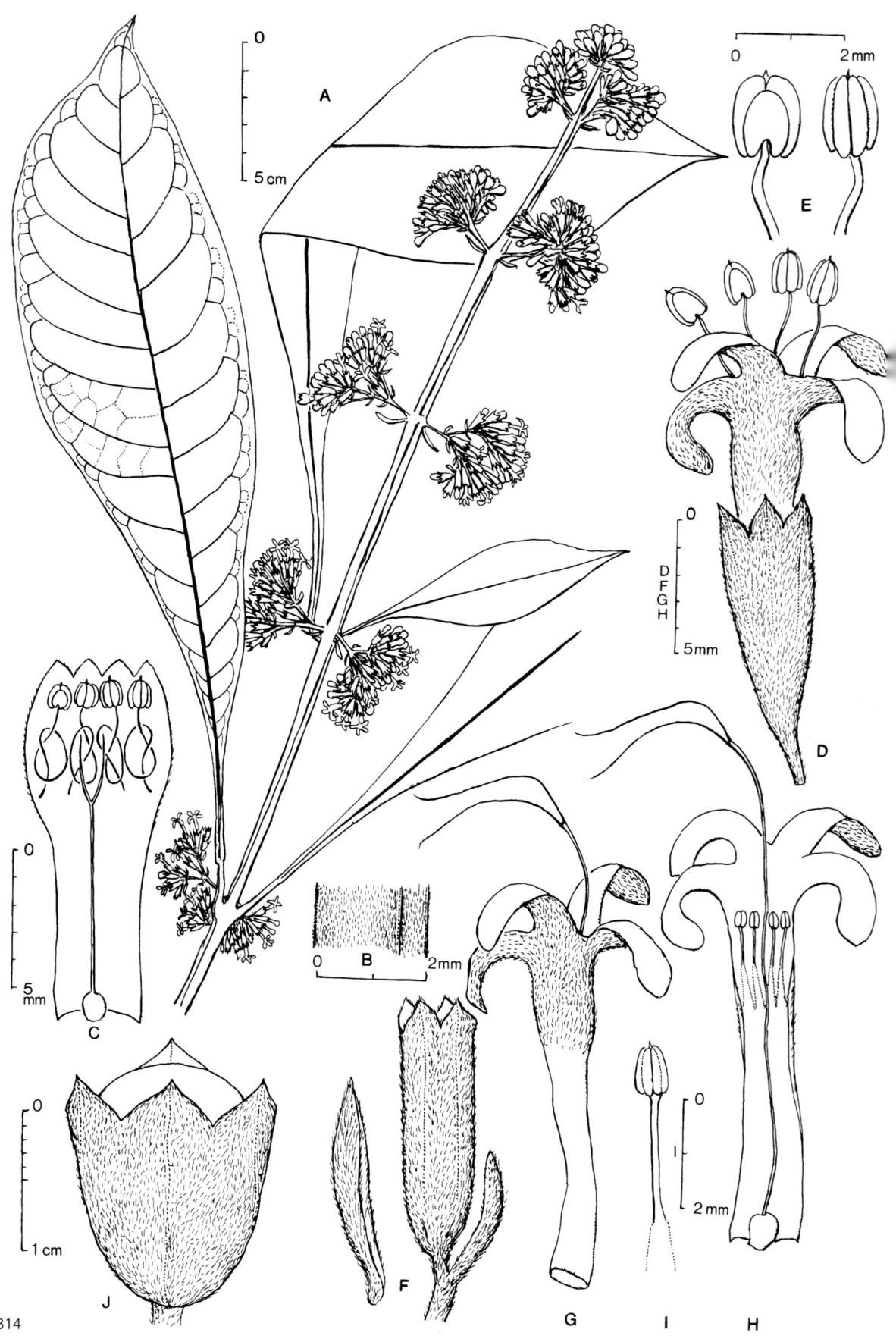

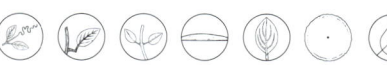

LAMIACEAE

Rarely lianas or climbing shrubs; commonly perennial or annual herbs, shrubs, and trees. Cosmopolitan, mostly tropical or subtropical; ca. 236 genera/7173 species; Guiana climbers represented by one native genus (Aegiphila).

1. Aegiphila Jacque.

Robust lianas and climbing shrubs; also erect shrubs and small trees. Climbing via twining stems, opposite branchlets, and angled petiole bases. Leaves are mostly opposite, simple, with entire margins in the Guianan species. Stipules are absent. Inflorescences are showy, with dense clusters of white or yellow, 4-parted tubular flowers; the ovary is superior. Fruits are ovoid drupes nested within a persistent calyx cup.

General: A genus with relatively few vegetative identification characters; most species lightly hairy.

Stems: Branchlets remotely 4-sided in cross-section, often bearing lenticels; nodes smooth; pith soft, white, & cottony. **Older stems 0.5-6 cm diameter, round to oval, without swollen nodes,** softly woody, bark yellow-brown to grey, thick; stem x-s pattern simple.

Leaves: Broadly ovate to elliptic, to 30 cm long, leaf bases often narrow, apex acuminate; **dark gland dots usually present above and below and flat circular glands usually present below near petiole,** surfaces smooth; **venation pin-**nate; 1° and 2° veins sunken above, raised below; 2° veins widely spaced, curving upwards and looping at margins; 3° veins net-like [*reticulate*]; petioles mostly 0.5-1 cm long.

Inflorescences: Axillary and/or terminal, densely-flowered; calyx tubular to bell-shaped; **corolla funnel-shaped and 4-lobed, cream to light yellow or green,** 0.5-2.5 cm long; **stamens 4,** inserted on corolla, to 1.5 cm long; ovary appearing undivided (4-chambered within), **style up to 2.5 cm long, forked.**

Fruits: Drupe-like, 1-3.5 cm diameter, yellow, orange-red, or black; fully to partially enclosed by **persistent calyx cup;** seeds 1-4.

Ecology: Insect- and bird-pollinated; seed dispersal via animal (bird); individuals often with 2 different flower forms – stamens short/style long or stamens long/style short. This pattern is likely to increase the chances of cross-pollination.

Distribution: Widespread in the neotropics from Brazil to Mexico: 156 species; Guianas: 14-15 species, 4 woody lianas introduced here.

Notes: Aegiphila was transferred from the Verbenaceae family. Another woody liana genus, Petrea, remains in the Verbenaceae family (in this volume). By comparison, Aegiphila has smooth leaves and robust drupe-like fruits, while Petrea has rough, pitted leaves and miniscule fruits that are dispersed via wind.

1.1. Aegiphila laevis (Aubl.) J.F. Gmel.

Common Name: awari imopi tëkënë, kumpuruni (Tr); manprasara (Sr)

Low climbing liana or shrub. Leaves: ovate-elliptic, 10-16 x 4-8 cm, base rounded or cuneate; leathery, shiny. **Inflorescences:** terminal, many-flowered panicles; **calyx bell-shaped,** green, margin entire or 4-lobed; **corolla to 1 cm long,** cream or yellow; style to 2.2 cm long. **Fruits: oblong-ellipsoid drupes,** 1.0-1.2 cm long, yellow; fruiting calyx 0.8 x 1 cm.

Uncommon; secondary forest and old-growth forest edges; to 450 m. Guianas; also Brazil and Venezuela.

1.2. Aegiphila macrantha (Ducke)

Robust liana. Leaves: elliptic, 7-14 x 4-8 cm, base broadly cuneate; thin to sub-leathery. **Inflorescences:** axillary, few- to many-flowered; **calyx cone-shaped,** 0.7-0.9 cm long, green, margin slightly 4-lobed or -toothed; **corolla large, 1.5-2.5 cm long,** white or cream; style to 2.5 cm long. **Fruits: subglobose, large drupes,** 2.5-3.5 cm long, smooth and dull, maturing yellow to brown; fruiting calyx large, 1.5-

2.5 cm long and 3-4 cm wide.

Uncommon; secondary forest and old-growth forest edges; to 800 m. Guianas; Amazonian Brazil, Venezuela, and Trinidad.

1.3. Aegiphila membranacea Turcz.

Liana, shrub, or treelet. Leaves: ovate-elliptic, large, 7-23 cm x 3-7 cm, base rounded to broadly cuneate; thin. **Inflorescences:** terminal, large panicles, 8-12 cm long; **calyx bell-shaped,** 0.2-0.4 cm long, margin entire, with 4 hairy zones; **corolla to 0.8 cm long,** pale yellow; styles to 1 cm long. **Fruits:** oblong-ellipsoid drupes, 0.6-1 cm long, yellow to orange; fruiting calyx 0.4 x 0.75 cm.

Uncommon; secondary forest and old-growth forest edges, to 700 m; Guianas; northern S America from Brazil to W Colombia.

1.4. Aegiphila racemosa Vell.

Common Name: awari imopi tëkënë, kumpuruni (Tr)

Liana or shrub; conspicuously hairy. Leaves: ovate-elliptic, 6-15 x 3-6 cm, base rounded to broadly cuneate; **leathery, golden-hairy below;** only flat circular glands present (no dark gland dots). **Inflorescences:** terminal, large panicles, 8-21 cm long; **calyx bell-shaped,** 0.3-0.5 cm long, with triangular lobes; corolla to 1 cm long, pale yellow; style to 1.5 cm long. **Fruits:** ellipsoid drupes, ca. 1.3 x 1 cm, yellow to orange-red or black, base and apex flattened; fruiting calyx 0.5 x 0.6 cm; pulp fleshy, mealy.

Uncommon; non-flooded old-growth forest, secondary forest, and riparian forest; to 350 m. Guianas; tropical S America to Peru, Colombia, and S Brazil.

1. Aegiphila bracteolosa, habit and inflorescences [NBC]
2. Aegiphila racemosa, inflorescence close-up [BH]
3. Aegiphila membranacea, mature woody stem [FRD]
4. A. membranacea, circular gland dots on leaf underside [BH]
5. Aegiphila racemosa, flowering stem [BH]

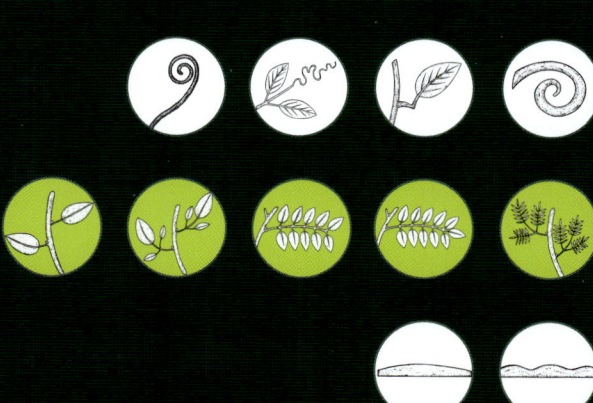

LEGUMINOSAE

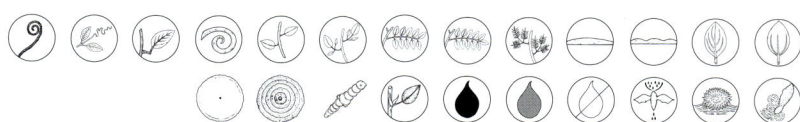

LEGUMINOSAE

also known as Fabaceae (the pea family)
(Sub-families: Caesalpinioideae, Mimosoideae, Papilionoideae)

Robust lianas, clambering shrubs, vines; also herbs, shrubs, and trees. Climbing via clockwise twining shoots, clambering branches, hooks, tendrils, or spiny tendril-like branchlets. Stipules are present, often inconspicuous or short-lived. Leaves are alternate and compound, 2-3-foliolate (Caesalpinioideae, Papilionoideae), 1-2-pinnate (typically odd-pinnate in Papilionoideae and even-pinnate in Caesalpinioideae and Mimosoideae), or 1-foliolate and partially to fully divided into two lobes (Caesalpinioideae, Bauhinia); always with a cylindrical swelling [*pulvinus*] at the base of leaf and leaflet stalks; leaf glands and leaflet stipules [*stipels*] present or absent. Inflorescences are of diverse arrangements and floral forms, with flowers pea- or butterfly-like (Papilionoideae, central 'flag' petal enclosing other petals), tubular with showy anthers (Mimosoideae, all petals similar) or radial with spreading petals (Caesalpinioideae, central 'flag' petal enclosed by other petals). The calyx and corolla are 4-5-parted, only partially fused or free, with 9-10 (>100) stamens and a 1-chambered, superior ovary with a 1-2 lobed style. Fruits are 1-many seeded legume pods that usually split open along two seams to release seeds [*dehiscent*] or, less commonly, remain closed [*indehiscent*].

Ecology: Seed dispersal primarily via wind and water, less often through self-dispersal (explosive release) or animals.

Distribution: Cosmopolitan, mostly in the Neo- and African tropics. The Caesalpinioideae and Mimosoideae are primarily (sub)tropical, while Papilionoideae include also many temperate representatives. Neotropics: 314 genera in total; Guianas: 14 genera/90 species of woody climbers.

Use: Globally, one of the most important plant families for humanity - providing protein, medicine, shelter (timber), crop ferti-

lizer, bio-fuels, industrial processes and much more. In the Guianas, Leguminosae lianas rarely serve as food sources but are important as medicines, ritual and craft plants, nitrogen-fixers, and fish poisons.

Notes: In agreement with views expressed by prominent experts (Lewis & Schrire 2005), the older family name, Leguminosae, is used here rather than Fabaceae, although both are valid. There are three major groups within the legume family that have been treated historically either as distinct families or subfamilies by different specialists. Here we follow the cur-

LEGUMINOSAE

rent APG classification, which considers the legumes as a single family made up of three subfamilies, the Caesalpinioideae, Mimosoideae, and Papilionoideae (Doyle et al., 2000; Wojciechowski, 2003).

Liana families in our region most likely to be confused with the Leguminosae include the Connaraceae (stipules and pulvinus absent); the Polygalaceae (stipules absent, leaves simple, flowers papilionoid-like but with only 3-petals); and the Sapindaceae (stipules and tendrils present, pulvinus absent, sap watery-white when present).

Leguminosae (Fabaceae) - Subfamily Caesalpinioideae

Lianas and clambering shrubs; also erect shrubs and trees. Climbing via twining shoots, hook-like tendrils (Bauhinia) or clambering branchlets (Senna).

General: **Hairs simple when present,** often silky [*sericeous*] (Bauhinia); **spines absent, prickles mostly absent (except some Senna); watery red sap absent,** resinous sap present in some Bauhinia.

Stems: Branchlets round, angled, or conspicuously flattened; nodes without interpetiolar ridge, stipules present, usually inconspicuous or short-lived. Older stems slender to robust, round (Bauhinia, Senna) to conspicuously flat and ribbon-like (Bauhinia), wood often soft; **stem x-s pattern simple, concentric, or eccentric** (flattened stem with growing point at one side).

Leaves: **Even-pinnate** (Senna) **or uniquely 1-2-foliolate** (Bauhinia); stipels usually absent; petiole and leaf rachis glands present (Senna) or absent; **always with a cylindrical swelling [*pulvinus*] at the base of leaf and leaflet stalks.** Erect trees and shrubs in the subfamily often have bipinnate leaves.

Inflorescences: Axillary or terminal, in racemes, spikes, cymes, or panicles; floral bracts usually absent; flowers bisexual, 5-parted, irregular [*partially zygomorphic*]; calyx with the upper 2 sepals ± fused (Bauhinia), or sepals free (Senna); **central 'flag' petal overlapped by 4 lateral petals;** stamens 10, usually free and in one whorl, sometimes fused at base.

Fruits: Dry legume pods, splitting at seams on 2 sides, **flattened** (Bauhinia) **or cylindrical** (Senna), many-seeded.

Ecology: Seed dispersal primarily via wind and water, occasionally by animal (birds, Senna) or explosive self-dispersal (some Bauhinia).

Distribution: Pan- and subtropical; 171 genera/2,250 species; Neotropics: 75 genera; Guianas: 2 genera with woody climbers.

1. Dioclea sp., inflorescences borne on stem [BH]
2. Dioclea sp., inflorescence [BH]

Leguminosae leaf diversity – compound, alternate leaves

3

4

5

6

7

8

3. Bauhinia cupreonitens, 1-foliolate leaf, split at apex [FRD]

4. Bauhinia guianensis, 2-foliolate, split to base [FRD]

5. Mucuna urens, 3-foliolate [FRD]

6. Deguelia amazonica, odd-pinnate (9 leaflets) [FRD]

7. Acacia altiscandens, even-bipinnate, each pinnae of 6-8-leaflets and opposite [FRD]

8. Entada polyphylla, even-bipinnate, each pinnae of 22-26 leaflets and opposite [FRD]

1. Bauhinia L. ⟲ ⬡ ⬡ ⬡ ⬡ ⬡ ⬡ ⬡ ⬡ ⬡

Woody lianas; also shrubs or small trees. Climbing via twining shoots, tendrils, or persistent leaf bases on old stems.

General: often with silky [*sericeous*] or felty [*velutinous*] hairs in all younger growth, lost with age; hairs simple; **spines and prickles absent; red sap absent.**

Stems: Branchlets round or flattened, lenticels present; stipules thin and soft, short-lived; **tendrils slender to woody, hook-like, occurring irregularly at nodes,** each subtended by a leaf scar. **Older stems often flattened and ribbon-like, reminding of staircases or ladders with parallel wavy ridges,** sometimes oval or contorted; bark often light yellowish-brown to grey; **slash often with clear, resinous gel or sap;** stem x-s pattern simple to eccentric (flattened with off-center concentric rings).

Leaves: 2-foliolate (with 2-leaflets, each with a basal hinge and brush-like remains of the rachis), **2-lobed** (divided ¼ to ¾ distance from the tip), or **1-foliolate** (undivided simple leaf); leathery to soft and thin, hairless above, with a dense, fine layer of hairs below; 3-6 palmate-veined.

Inflorescences: Terminal or axillary, racemes, panicles, or 1-2 flowered; hyranthium present (tube- or cup-shaped); calyx 5, entire, splitting into 2-5 parts when flower opens; petals 5, free, **central 'flag' petal enclosed by lateral petals;** stamens 10, free or fused; ovary 1-chambered, stigma head-like or uneven.

Fruits: Pods woody, flat, sometimes splitting or bursting open to release flat seeds [*dehiscent*], otherwise opening only through decomposition [*indehiscent*].

Ecology: Seed dispersal via water, animal (through gut), or mechanical.

Distribution: Pantropical: ± 300 species; Guianas: ± 20 species, with 10 woody climbers, 3 described here.

Use: The leaves and stems of Bauhinia species are commonly used for various medicinal purposes in the Guianas (diarrhea, venereal disease, broken bones, aphrodisiacs). Leaves are used in ritual and steam baths by Maroon groups in Suriname. Use as a fish poison has been documented in northwestern Guyana (van Andel 2000).

Notes: Local plant names commonly reflect traditional ecological knowledge. Most names for Bauhinia climbers in the Guianas indicate the mythical (or actual) use of the stems as ladders by forest animals (e.g., turtle ladder, echelle tortue, hikuritarafon, logoso sikada, sekrepatu trapu). Although it is difficult to imagine a turtle climbing a liana, monkeys and other canopy dwellers do use lianas as climbing ropes and bridges to move horizontally between trees. The Trio Indian name for Bauhinia lianas, mïrokoko ehe, indicates the resting place of a potoo bird species (Nyctibius griseus).

1.1. Bauhinia guianensis Aubl.

Common name: echelle tortue (FG); guayamu-frati (Ca); hikuritarafon (Ar); khawouieng gumapuiyik (Pat); kujule huhanukutpuh (Way); logoso sikada, bê (Sa); mïrokoko ehe, pipaman (Tr); monkey ladder (GU); sekrepatu trapu (Sr); turtle ladder (GU); yaoutimouta (Ca, Galibi).

High-climbing, robust liana; hairs reddish-brown. **Stems:** branchlets with tendrils at nodes; older stems flattened, wavy. **Leaves: 2-foliolate,** leaflets ± ovate, variable in size, 6-14 x 2-10.5 cm, base lobed, apex with short or longer narrow tip (to 3 cm); leathery, **hairless above, densely hairy below;** palmate-4-5 veined; petiole 2.5-3.5 cm long. **Inflorescences:** racemes to 16 cm long, densely hairy [*tomentose*]; flower stalk bracts 1-3, awl-shaped; calyx bell-shaped, 0.5-0.8 cm long; corolla white, pink, or pale yellow, petals 1-1.5

cm long, hairy outside, one petal recurved, narrower and shorter than others. **Fruits:** pod very flat, club- or egg-shaped in outline, 5-8.5 x 1.5-2.5 cm, hairy, obscurely veined, tapering to a 0.5 cm long stalk; seeds 4.

Common; non-flooded old-growth forest, riparian forest and secondary forest. Guianas; widespread from Bolivia to S Mexico and Trinidad.

1.2. Bauhinia smilacina (Schott) Steud.

Lianas; mostly hairless. **Stems:** branchlets with numerous tendrils at nodes; older stems ovoid to flattened. **Leaves: 1-foliolate,** (ovate)-oblong, 5-8 x 2.5-4 cm, base rounded to lobed; leathery, **hairless above and below;** palmately 5-veined; petiole 0.5-1 cm long. **Inflorescences:** bracts white, minute; calyx pink, bell-shaped and persistent, 0.4 cm long; corolla pink, petals usually hairless, ca. 0.5 cm long. **Fruits:** pod ovate in outline, 3-3.3 x 1.7-1.9 cm, red, slightly hairy to hairless, thin and soft, not opening at maturity; 1-seeded.

Uncommon; riverine forest and non-flooded old-growth forest slopes. Fr. Guiana, Suriname; also Brazil (Amapá, Pará, Atlantic Forest).

1.3. Bauhinia surinamensis Amshoff

Common name: khawouieng gumapuiyik (Pat); sekrepatu trapu (Sr); wayamu nepulu (Ca)

Liana; hairs reddish-brown [*pilose*]. **Stems:** branchlets with long soft hairs and numerous tendrils at nodes; older stems flat and wavy. **Leaves: 2-lobed, cut ½ to 1/3 length from apex,** lobes 7-10 x 5-7 cm, apex acuminate; leathery, **hairless above, shortly hairy below. Inflorescences:** racemes to 15 cm long, bracts very small; calyx bell-shaped with short hairs, 0.5-0.7 cm long; corolla white, petals ca. 1 cm long, brown-hairy outside. **Fruits:** elongate in outline, to 9 x 2.2 cm, with short hairs, opening at maturity.

Uncommon; old-growth forest, especially riverine forest and slopes. Guianas; adjoining NE Brazil.

2. Senna Miller

Softly-woody, high-climbing lianas or bushy, climbing shrubs; also trees, shrubs or herbs. Climbing via clambering branchlets.

General: Hairless or minutely hairy, hairs simple; **unarmed or armed, red sap absent.**

Stems: Branchlets round or 4-5-angled, furrowed, often softly woody; bark fibrous, peeling in small vertical strips; **stipules inconspicuous to large, leafy and persistent.** Older stems round or irregularly angled, lightly furrowed to deeply fissured; bark silver or dark brown, wood an extremely bright "dayglo" yellowish-orange to dull yellow or white; stem x-s pattern simple.

Leaves: **Even-pinnate, mostly 4-foliolate,** leaflets opposite, often diamond-shaped (2D),

nectar glands (club-shaped, cone-shaped, or flat) often present **on the pulvinus and/or in-between leaflets;** leaflet stipules [*stipels*] absent.

Inflorescences: Terminal or axillary, racemes or panicles, sometimes reduced to 1-2 flowers; **corolla irregular-radial;** flower stalks with bracts; cup-shaped hypanthium present; sepals 5, free; petals 5, yellow; stamens 10 (3 short-infertile, 7 long-fertile); ovary often off-center.

Fruits: Pods ± cylindrical, 4-angled, or 4-winged, > 7 cm long, splitting along 1 or 2 sides; many-seeded; seeds in 2 rows, embedded in pulp or gel.

Ecology: Seed dispersal via water, animal (through gut), or mechanical.

Distribution: Pantropical, center of diversity in the neotropics: 260 species; Guianas: 5 species of woody climbers, 3 described here.

Notes: Some Piptadenia (Leguminosae: Mimosoideae) species have similar leaves.

2.1. Senna chrysocarpa (Desv.) H.S.Irwin & Barneby

Common name: busipesi (Par); okó búka (Sa); pokopoko uru (Tr)

High-climbing lianas or bushy, climbing shrubs; also shrubs or treelets; hairs grey to golden-yellow. Stems: branchlet nodes with short-lived bristle-like stipules; older stems with bark deeply furrowed, greyish-silver to chocolate brown, wood very bright yellowish-orange. Leaves: 4-foliolate, 3-8 cm long, with 1 finger-like rachis gland per leaf, between lower leaflet pair; main leaf stalk 1-2.5 cm long; leaflets ovate-elliptic, upper pair 2-6 x 1-2.5 cm; blades leathery, hairs persistent below. Inflorescences: racemes or panicles crowded in uppermost shoots, 3-4 cm long; petals 1.3-2.2 cm long, with dark yellow hairs outside. Fruits: pods curved, 5-10 x 0.5-0.8 cm; glossy yellowish-green to black, sub-fleshy, with obscure cross-veins.

Common; riparian and secondary vegetation, savanna woodland, and white sand savanna. Guianas; also Venezuela and E Amazonian Brazil.

2.2. Senna latifolia (G.Mey.) H.S.Irwin & Barneby

Common name: legwana titei (Sr); lokonanjo (Ar)

Bushy, climbing shrubs; treelets; hairless or inconspicuously hairy. Stems: branchlet nodes with conspicuous, persistent leafy stipules, 0.5-3 cm long; older stems irregularly and deeply fur-rowed, bark brown to yellow, bearing lenticels, wood not bright yellowish-orange. Leaves: 4-foliolate, 12-18 cm long, 2 rachis glands per leaf (1 cone-shaped gland between basal leaflets and 1 flat gland between upper leaflets); main leaf stalk 3-10 cm long; leaflets ovate-oblong, upper pair 9-16 x 5-9 cm; blades nearly leathery, glossy above, dull below. Inflorescences: terminal, dense panicles, to 5.5 cm long, minutely hairy; petals 2-3.5 cm long, clawed, hairless, bright yellow or orange. Fruits: pods straight, hanging, to 30 x 1.2 cm, hairless, cross-veined.

Common; roadsides, sandy riverbanks, savanna, secondary forest. Guianas; widespread from N South America to Peru and Bolivia.

2.3. Senna quinquangulata (Rich.) H.S.Ir-win & Barneby

Common name: agaimargaiyik (Pat); arimi ka-rau (Tr); jorka pesi (Sr); lokonanjo (Ar); maloko pesi (Par); okó búka (Sa); punatawa (Way).

High-climbing lianas or bushy, climbing shrubs; hairs grey. Stems: branchlets 5-angled, soft; nodes with linear-sickle-shaped stipules, 0.3-1 cm long, short-lived; older stems 5-angled, knotted, wood orangish-yellow. Leaves: 4-fo-liolate, 10-22 cm long; with 2 rachis glands per leaf (club-shaped, cone-shaped, or slender); main leaf stalk 2-2.5 cm long; leaflets (ovate) lanceolate, upper pair 7-15 x 3-7 cm; blades leathery, glossy above, finely hairy below. Inflo-rescences: large terminal panicles, hairy; petals 1-1.6 cm long, hairy outside, bright yellow or golden. Fruits: pods straight, hanging, 15-20 x 1-1.5 cm, hairless, with winding, thick, promi-nent veins.

Common; open areas in old-growth and disturbed areas, riparian forest. Guianas; also widely distributed from N South America to S Mexico, Tobago.

Leguminosae (Fabaceae) - Subfamily Mimosoideae

Occasional lianas and clambering shrubs; rarely herbs, usually erect shrubs or trees. Climbing via twining shoots and clambering, reflexed or tendril-like branchlets.

General: Hairs fine to dense [*tomentose*], hairs simple when present; extra-floral nectar glands present; **nodes often armed; red, watery sap absent, drinkable stem water sometimes present.**

Stems: Branchlets round, oval to 5-ribbed, **bark red and flaky** with wavy lengthwise lines, often with **prickles, scales and/or glandular hairs;** nodes without interpetiolar ridge, stipules present, usually inconspicuous and short-lived in climbers, persisting as **projections or spines** in some Mimosa. Older stems large- to medium-sized, round to 5-sided, wood soft and white; **stem x-s pattern simple (not concentric or eccentric).**

Leaves: **Bi-pinnate with many tiny leaflets** (Acacia, Entada, Mimosa) **or fewer, larger leaflets** (some Mimosa, Piptadenia); pinnae (the 1-pinnate subunits of the leaf) usually opposite on the main leaf rachis, leaflets opposite or alternate; always with a cylindrical swelling [*pulvinus*] at the base of leaf and leaflet stalks; **(nectar) glands** (finger-, disc- or crater-like) **present on petiole base, along upper leaf rachis, or absent.**

Inflorescences: **Showy heads or spikes,** simple or compound; flowers tiny, tubular, and fragrant; commonly white, yellow, to pink/red; (4)-5-parted; flower stalks often with bracts; calyx fused at base; corolla free or basally fused; stamens 4-10 or > 100, **stamens** [*filaments*] free or tubular, usually **brightly-colored and extending far beyond the corolla.**

Fruits: **Dry legume pods, always strongly flattened** in woody climbers of the region, splitting vertically along 1-2 seams or horizontally into many 1-seeded segments [*articles*].

Ecology: Nectar glands on leaves attract ant species that protect plant species from herbivory. Seed dispersal via water, wind, animal, or explosive self-dispersal.

Distribution: Common in secondary forest and open places. Widespread in (sub)tropical, and warm temperate regions: 82 genera/>3000 species; Guianas: 30 genera/88 species, with 4 genera/12 species of woody climbers, 8 species described here.

Notes: Entada is distinguished from other Mimosoid climbers by the lack of both prickles and leaf glands and a unique broom-like inflorescence with hanging spikes of white flowers. Infertile Mimosa and Acacia species are difficult to distinguish. Mimosa species tend to have the first pair of leaflets per pinnae strongly reduced. Piptadenia floribunda is distinguished by larger leaflets and finger-like stalked glands at the petiole base and along the axes of both pinnae and leaflets.

3. Acacia Mill.

Occasional lianas and clambering, secondary forest shrubs; usually erect shrubs or trees. Climbing via twining apical shoots and clambering, reflexed or tendril-like branchlets.

Distribution: Pan- and subtropical, ca. 200 species; Guianas: 2 climbers, 1 described here.

3.1. Acacia tenuifolia (L.) Willd.

Common name: akalamaka (Sa); amosaitïrï (Ca); karaka (Way); kararaikyik (Pat); pïtïkïrïkï (Tr); wacht-een-beetje (SD); waktipikinso (Sr); wayamaka (Ca)

Liana, shrubby climber; tree to 12 m; leaves sensitive to touch. **Stems:** younger stem (2 cm diameter) round, bark smooth, black, with **sharp prickles in vertical rows;** older stem (5 cm diameter) obscurely angled, red with flaky bark, prickles absent, softly woody. **Leaves:** pinnae 20-35 paired, leaflets 40-100-paired; **nectar glands present at petiole base and on upper leaf rachis. Inflorescences:** axillary or terminal, **glo-bose heads in racemes,** lightly hairy; perianth < 0.3 cm long. **Fruits:** pods linear or oblong, 15-20 x 4 cm, surface with horizontal lines, but **not breaking into 1-seeded segments,** coarsely wrinkled; 10-15 seeded. **Notes:** A less common species in the Guianas, Acacia articulata Ducke, is distinguished by the absence of nectar glands on the upper leaf rachis, inflorescences in compound spikes and the fruit breaking horizontally into one-seeded segments.

Common; old-growth forest, secondary forest, especially along rivers. Guianas; also N South America and the West Indies.

4. Entada Adans.

Woody lianas only. Climbing via twining shoots and clambering, reflexed or tendril-like branchlets and leaf axes.

General: Hairless to finely hairy, hairs simple when present; ± unarmed.

Stems: Branchlets round, smooth, **unarmed or prickles few, small and insignificant;** older stems round, medium-sized, to ± 5 cm diameter, wood soft and white; stem x-s pattern simple.

Leaves: Bi-pinnate; pinnae opposite, **terminal pinnae sometimes tendril-like,** clasping; leaflets opposite, few-to-many, blades oblong; **leaf glands absent.**

Inflorescences: terminal, broom-like, often hanging, spikes solitary or densely crowded in panicles to one side of branch; flowers tiny, 5-parted, white, yellow (rarely reddish-brown); calyx fused and bell-shaped; petals ± free; **stamens** 10, free, **barely extending beyond corolla mouth.**

Fruits: Pods conspicuously wide and flat; straight or ± curved, soft and thin to woody with thickened, persistent margin, constricted at regular intervals, **breaking horizontally into 1-seeded segments** [*articles*]**;** seeds round, thick.

Ecology: Seeds dispersed via water and explosive self-dispersal.

Distribution: Neotropics and Africa: ca. 30 species, Guianas: 2 species.

Notes: The only mimosoid climber lacking armature and leaf glands. Similar to the Caesalpinioid legume genus Senna, except with bipinnate leaves.

4.1. Entada polyphylla Benth.

Common name: yuwana-hi (Ar)

Woody liana. Leaves: pinnae 4-8-paired; leaflets 20-40-paired, blades 0.8-1.6 cm long. **Inflorescences:** terminal, 20-25 cm long panicle, 40-50 pendant spikes, each 5-10 cm long; flowers yellowish or white. **Fruits:** pods 30-40 x 5-7 x 0.2 cm; seeds ± 16.

Rare; riverine forest and forest margins, flooded or non-flooded. Guianas; widespread in N Amazonia.

4.2. Entada polystachya (L.) DC.

Common name: yuwana-hi (Ar); sipokï (Ca)

Woody liana. Leaves: pinnae 2-6-paired, terminal pair occasionally converted into long tendrils; leaflets 6-8-paired; blades 2-3.5 cm long. **Inflorescences:** terminal, 30-60 cm long panicle, spikes numerous, each 5-7.5 cm long; flowers yellowish or reddish-brown. **Fruits:** pods 30-45 x 5-10 x 0.2 cm; seeds seeds ± 16.

Common; coastal area, on ridges and along deltas. Guianas; also widespread in the Neotropics, to Mexico and the Caribbean.

9. Bauhinia rubignosa, robust stem [BH]
10. Bauhinia guianensis. a. flowering stem with tendril; b. ribbon-like stem; c. variation in leaf shape; d. fruit pod, dehiscent [NCB-NHN]
11. Young leaves produced by buried woody stem of a Bauhinia species on forest floor [BH]
12. Bauhinia kunthiana, tendril [BH]
13. Winding stem of Bauhinia sp. [BH]
14. Bauhinia rubignosa, robust stem [BH]
15. Senna chrysocarpa, robust woody stem with bright yellow wood [BH]
16. Senna quinquangulata, flower up close [OG]
17. Senna quinquangulata, leaves 4-foliolate, each bearing 2 nectar glands (one flat, one finger-like) [OG]
18. Senna quinquangulata, raised nectar gland on 4-foliolate leaf [BH]
19. Senna latifolia with stipules and pulvinus (cylindrical swelling at petiole base) [OG]
20. Senna chrysocarpa, leaves 4-foliolate, each bearing one finger-like nectar gland [OG]
21. Recurved 'cat's claw' prickles (Senna chrysocarpa) [BH]
22. Acacia tenuifolia, with very small leaflets, flowers in heads [OG]
23. Acacia tenuifolia, branchlet with spines in lines. [OG]
24. Entada polyphylla, inflorescence with spikes clustered to one side [OG]
25. Entada polystachya, with very wide, flat fruit pods splitting horizontally into one-seeded segments [OG]

5. Mimosa L.

Woody lianas and shrubby climbers; also herbs, shrubs, and trees. Climbing via twining shoots and clambering, reflexed or tendril-like branchlets.

General: Sparsely to densely hairy [*tomentose*]; armed; nectar glands on leaves sometimes present.

Stems: Branchlets round, angled or ribbed, bark green or black, smooth or hairy, often armed with **recurved 'cat's claw' prickles,** scattered or in vertical lines, **stipules linear,** short-lived or persistent as **paired projections or spines.** Older stems observed to 15 cm diameter, round or remotely angled, bark soft, rusty red, flaky; wood soft and white; stem x-s pattern simple.

Leaves: Bi-pinnate, pinnae opposite, variable in size and number; leaflets opposite, usually many and small; **disc- or crater-like nectar glands sometimes present** (not finger-like).

Inflorescences: Axillary or terminal, usually showy heads, solitary or densely clustered; flowers 4-5-parted, white, yellow, or pink; **stamens always < 10,** free or fused at base, **extending far beyond corolla mouth.**

Fruits: Pods flattened, ovate or narrow, hairless to lightly hairy, soft and thin to woody with thickened, persistent margin, constricted at regular intervals, **breaking horizontally into 1-seeded, indehiscent segments [*articles*].** Pantropical, mostly neotropical: 450 species; Guianas: 25 species; 7 woody climbers with 4 described here.

5.1. Mimosa annularis Benth.

[Syn: M. paniculata Benth.]

High-climbing liana or shrub. Stems: 5-ribbed, with recurved prickles in lines. **Leaves:** pinnae in 2-7 pairs; pinnae leaflets 3-11 pairs, blades diamond-shaped, 1.5-2.5 x 1-2 cm; petiole 2-6 cm long; **leaf glands present at petiole base and on upper leaf rachis; stipules persistent, awl-shaped, to 1 cm long. Inflorescences:** axillary, round heads in large, fragrant panicles, > 30 cm long; flowers white. **Fruits:** pods linear, straight or ± recurved, 6-10 x 0.9; hairless or reddish-scaly; segments 0.4-0.8 cm long (lengthwise); seeds to 0.45 cm diam.

Uncommon; riverine forest. Guianas; also adjacent N Brazil (Amapá and Pará).

5.2. Mimosa guilandinae (DC.) Barneby

High-climbing liana; shrub. Leaves: pinnae in 2-3 pairs; pinnae leaflets 2-3 pairs, blades obovate, hairless but reddish-scaly below; **petiole with an urn-shaped nectar gland, between the pulvinus and first pair of leaflets. Inflorescences:** axillary or terminal, globose heads; **flowers 5-parted, white. Fruits:** pods Entada-like, wide and flat (7-10 cm wide), but armed with recurved prickles in lines.

Uncommon; old-growth forest, interior and coastal; French Guiana; Brazil (Amapá).

5.3. Mimosa microcephala Humb. & Bonpl. ex Willd.

Liana or shrubby climber; shrub; hairs yellowish-brown to red. **Stems:** round or ribbed, with recurved prickles in lines along ribs, or scattered; **stipules triangular or lance-shaped,** persistent, to 0.5 cm long. **Leaves:** pinnae 8-14-paired; pinnae leaflets 40-50-paired, blades (linear-)elliptic, to 0.65 x 0.9 cm; petiole 0.4-2.1 cm long; **leaf nectar glands absent. Inflorescences:** axillary, in small clustered heads (each. 0.4 cm diameter), or terminal, in compound heads, each infl. more than 30 cm long; flowers 4-parted, fragrant, **pinkish-brown, filaments white. Fruits:** pods linear; 1-6 x 0.4 cm, ± hairy; segments 0.4-0.7 cm long (lengthwise); seeds 3-10.

LEGUMINOSAE

Locally abundant; non-flooded riverine forest, rocky savanna. Fr. Guiana, Suriname; N Amazonian Brazil.

5.4. Mimosa myriadenia (Benth.) Benth.

Common name: akalamaka (Sa); kifundu maka (Pa); pïtïkïrïkï (Tr); wacht-een-beetje (SD); yu-wana-hi (Ar).

High-climbing liana or clambering shrub. Stems: branchlets 5-sided with prickles along ribs, minutely hairy; **stipules small, persistent, awl-shaped;** older stems 3-5 cm diameter, also armed. **Leaves:** pinnae 5-15-paired; pinnae leaflets 16-26-paired, blades linear to diamond-shaped, 0.3-1 x 0.1-0.3 cm; petiole 1.3-3 cm long, bearing prickles; **disc-shaped glands present, 1 on petiole base and 3-5 on upper leaf rachis. Inflorescences:** axillary, spikes in panicles to 7.5 cm long; **flowers 5-parted, white. Fruits:** pods linear-oblong, 7.5-9 x 1.2-2 x 0.1 cm, soft, gland-dotted, segments 0.4-0.8 cm long (lengthwise); seeds 8-15.

Common, often forming tangles in the forest understory; riverine forest, non-flooded old-growth, forest edges. Guianas; also N Brazil, Venezuela.

6. Piptadenia Benth.

Rarely lianas or scrambling shrubs; mostly erect shrubs and trees. Climbing via twining shoots and clambering, reflexed or tendril-like branchlets. Neotropics to S Brazil: 20 species; Guianas: 3-4 species, with one woody climber.

6.1. Piptadenia floribunda Kleinhoonte

Common name: boesi-branti (Sr); pïtïkïrïkï (Tr)

Liana or scrambling shrub; armed. **Stems:** branchlets 5-ribbed, **with recurved prickles in lines along ribs;** stipules inconspicuous, short-lived; older stems round to 5-ridged, to ca. 8 cm diameter, bark rusty red, thin and flaky; wood soft and white; stem x-s pattern simple. **Leaves:** bipinnate-compound; pinnae opposite, 2-paired; leaflets opposite, 3-4-paired; blades large, elliptic to slightly elongate, 4-6 x 2.3-3.7 cm, lightly hairy below; **with stalked, finger-like glands at petiole base and along leaf axes, armed with prickles. Inflorescences:** terminal panicle of spikes, > 30 cm long, spikes 2-3 per node, each spike 3-5 cm long; **flowers small, tubular, yellow;** stamens 10, free, extending far beyond corolla mouth. **Fruits:** pod broadly ovate, ca. 10 x 3 cm, flat, soft and thin to leathery, splitting along 2 seams; seeds compressed, ellipsoid in outline. **Ecology:** Seed dispersal via wind.

Uncommon. Non-flooded old-growth forest. Fr. Guiana, Suriname; adjoining Brazil.

26. Mimosa myriadenia, fruit pods [NH]
27. M. myriadenia, small leaflets, white flower spikes [NH]
28. M. myriadenia, disc-shaped nectar gland at petiole base, prickles on stem [BH]
29. M. guilandinae, large leaflets, flowers in heads with flat fruits [OG]
30. M. myriadenia, white flower spikes [NH]
31. M. pigra, small leaflets, hairy fruit pods [BH]
32. Piptadenia floribunda, with finger-like leaf rachis gland (top) and softly woody stem with spines (bottom) [BH]
33. P. floribunda, shrubby habit, with white flowers in spikes [OG]
34. P. floribunda, bi-pinnate leaves with large leaflets [OG]

Leguminosae (Fabaceae): Papilionoideae

Woody lianas, climbing shrubs, sub-woody and herbaceous vines; also erect herbs, shrubs, or trees. Climbing via twining shoots, clambering branchlets, and tendril-like branchlets.

General: hairs simple when present; sometimes armed; **red sap present or absent; cut parts often with a "green-bean" scent.**

Stems: Branchlets round, oval, flattened, or furrowed; bark thin and smooth to rough and fibrous, occasionally bearing lenticels; nodes without a horizontal ridge or line, **stipules present, often inconspicuous and short-lived, occasionally modified into spines.** Older stems of diverse shapes and sizes, robust and woody to slender and sub-woody; **stem slash sap red and watery when present; stem x-s pattern simple, eccentric or flattened-concentric** (Machaerium, Mucuna).

Leaves: Once-compound: i) odd-pinnate, leaflets 5-9 (leaflets alternate - Dalbergia, Machaerium; or opposite - Derris, Lonchocarpus); ii) even-pinnate (Abrus); or iii) palmately-compound (tri-foliolate - Clitoria, Dalbergia, Dioclea, Mucuna); **always with a cylindrical swelling [*pulvinus*] at base of leaf blade stalks,** sometimes both at base and apex [*double pulvinus*]; tiny stipule-like structures [*stipels*] usually present in leaflet axils, occasionally modified into spines (Machaerium); leaf and petiole glands absent (vs. present in Mimosoideae).

Inflorescences: Racemes, panicles or solitary; flowers bisexual, 5-parted, **pea- or butterfly-like** [*bilateral symmetrical*]; sepals partially fused, **corolla with a central 'flag' petal enclosing the other 4 petals, including 2 lateral petals [*wings*] and 2 lower petals [*keel*]**; the keel united and enclosing stamens and ovary; stamens 9-10; ovary superior, 1-chambered, style and stigma 1.

Fruits: **Legume pods with 1-seeded segments,** flattened, thick, or round; splitting length-wise, horizontally, or non-opening; **also samaras, nuts, and drupes.**

Distribution: Cosmopolitan: ca. 31 tribes, 500 genera/>10,000 species; Guianas: 52 genera/180 species; Guianas: 8 genera/63 species of woody climbers.

Ecology: Insect-, bird-, and bat-pollinated. Seeds dispersed via water, wind, and animal.

Uses: Important source of medicines and fish poisons.

7. Abrus Adans.

Slender lianas, climbing shrubs, or robust vines. Climbing via twining shoots. Pantropical, introduced from the paleotropics and often growing in disturbed places: 16 species; Guianas: 1 species naturalized.

7.1 Abrus precatorius L.

Common name: crab's eyes (GU); kokriki (Sr); lickrish (GU); paternosterboontje (SD); petit panacoco (FG)

Slender lianas, climbing shrubs, or robust vines. General: finely hairy; **unarmed; red sap absent.** Stems: branchlets round, winding, with flattened white hairs, **nodes with conspicuous, awl-shaped stipules;** older stems flattened to 2-cabled, small in diameter, bearing many lenticels; stem x-s pattern simple. Leaves: **even-pinnate (8-14-foliate),** 5-10 cm long, axes grooved, petiole >1 cm long; **leaflets opposite,** nearly ovate, 0.8-2 x 0.4-0.7 cm, base and apex rounded with a narrow extended tip; finely hairy below. **Inflorescences:** axillary, terminal, or stem-based racemes, to 8 cm long, thick-stalked; calyx bell-shaped; corolla <1 cm long, pink to white, flag petal 1 x 0.6 cm, wing petals shorter than keel; stamens 9. **Fruits:** nearly linear pods, 2-3.5 x 1-1.3 cm, ± flattened, woody, thickened over seeds, beaked, splitting along one side; seeds 3-5, to 0.6 cm long; scarlet red and black, aril-

late. **Ecology:** seed dispersal via animal (through bird gut). **Use:** Seeds poisonous. Roots have been used as a liquorice substitute. There are a wide variety of medicinal uses (dePhillips et al. 2004).

Common; growing in secondary forest and scrub, often in disturbed areas near coast. Guianas; pantropical.

8. Clitoria L.

Lianas, climbing bushy shrubs, subwoody vines; also erect herbs, shrubs, or small trees. Climbing via twining shoots.

General: Hairless or with tiny, hooked hairs, visible with a 10x hand lens (Berry et al. 2000); **red sap and spines absent.**

Stems: Branchlets round, bark grey to reddish-brown, smooth. Older stems round, less than 4 cm diameter, bark smooth or with lengthwise cracks.

Leaves: Palmately-compound with three leaflets [*trifoliolate*]. Leaflets ovate to elliptic, uneven at base, pinnate-veined, 2° veins close together.

Inflorescences: Axillary, terminal, or borne on the stem, often solitary, paired or few-flowered, flowers large and showy; flower stalks with 2 persistent bracts; calyx conspicuously toothed or lobed, persistent; **petals clawed, colored pink, blue or violet, flag petal round and notched at the top,** often wrinkled, **much longer than other petals (4.5-5 cm long); keel shorter than the wings;** stamens 10, 2-bundled, persistent; style elongate, curved, stigma unlobed.

Fruits: Woody or leathery pods, linear-oblong, ca. 20 cm, becoming twisted and explosive at maturity, surface smooth or with a long rib; stalk ± 2.5 cm long; seeds 2-10, flattened or round.

Distribution: Widespread in the American (sub)

tropics, also paleotropics: 60 species; Guianas: 5 species, 3 woodier species described here.

Ecology: Mostly insect-pollinated. Flowers "upside down" compared to the other papilionoids with pollen deposited on the back-side rather than front-side of pollinators. Seed dispersal via explosive release as mature capsules dry out.

Uses: Seeds of some species are used as fish poisons.

8.1. Clitoria arborescens R.Br.

Common name: inekuipë (Tr); kaw-ai (Sr); kurumu eneru (Ca); panapana (Ca)

Lianas, rarely shrubs or trees; hairy [*pilose to sericeous*]. **Stems:** stipules narrowly triangular, to 0.8 cm long. **Leaves:** leaflets ovate to oblong, 6-16 x 3-8 cm; leathery, hairless above, **densely golden- or reddish-hairy below;** veins parallel; petiole 2.5 cm long. **Inflorescences:** axillary, racemes 3-20 cm long, golden-hairy; flowers paired; calyx silky, petals violet and white, flag petal to 4 x 3.5 cm, lightly hairy outside. **Fruits:** pods to 20 x 2-3 cm; leathery, hairless.

Common; especially along riverbanks, sometimes cultivated. Guianas; also Venezuela and Colombia.

35

36

37

38

8.2. Clitoria pendens Fantz

Common name: gatè-titei (Mat); manneko (Par)

High-climbing lianas; mostly hairless. Stems: stipules lanceolate, 0.5 cm long. **Leaves:** leaflets ovate to oblong, 10-20 x 3-9 cm; **hairless above and below;** net-veined; petiole 5-10 cm long. **Inflorescences:** axillary or stem-borne, racemes to 80 cm long, finely hairy, flowers paired; calyx silky; petals violet or dark pink, flag petal to 5 x 3 cm, lightly hairy outside. **Fruits:** pods to 20 x 1.5 cm, leathery, lightly hairy.

Uncommon; along riverbanks, rapids, old-growth forest, forested slopes. Guyana, Suriname; also adjoining areas of N Brazil (Pará).

8.3. Clitoria sagotii Fantz

Lianas or trailing vines; mostly hairless. Stem: stipules triangular to spear-shaped, 0.2-0.5 cm long. **Leaves:** leaflets oblong, 10-20 x 5-8 cm; **hairless above, sparsely hairy below,** loosely net-veined; petiole 2-11 cm long. **Inflorescences:** racemes sometimes borne on stem below node, small, few-flowered; calyx lightly hairy; petals violet, flag petal to 5 x 3.5 cm, lightly hairy outside. **Fruits:** pods to 22 x 2 cm, leathery, lightly hairy.

Uncommon; along riverbanks, old-growth forest margins, open areas. Guianas; also Venezuela.

9. Dalbergia L.f.

Woody lianas, clambering shrubs; also erect shrubs and trees. Climbing via twining shoots and clasping, tendril-like branchlets.

General: Young vegetation and inflorescences green and hairless to softly hairy; **spines and red sap absent.**

Stems: Branchlets often slightly zigzag; bark green and smooth to brown and fibrous with lengthwise lines; **stipules small, inconspicuous, short-lived.** Older stems round to oval, to medium-size diameter; stem x-s pattern simple.

Leaves: Leaflets 1 (D. ecastophyllum), 3 (D. monetaria), or 5-15 odd-pinnate and alternate, leaflet stipules [*stipels*] absent; petiole 1-2 cm long.

Inflorescences: Terminal or axillary, often many-flowered branching racemes, flowers small; bracts small, lost early, bracteoles small, persistent on flower stalks; calyx bell-shaped, sometimes persisting in fruit, teeth short, unequal, upper 2 fused; **petals long-clawed, white or purple with a white patch,** hairless, flag petal < 1 cm long; stamens 9-10; style 1.

Fruits: Pods oblong-, liver-shaped, or circular-flattened, to 4.5 x 2.5 cm, leathery to thin and soft, indehiscent; seeds 1(2), kidney-shaped, compressed.

Ecology: Occurring as lianas in upland forest, as clambering shrubs along rivers. Fruits often adapted to fresh- or saltwater seed dispersal.

Distribution: Pantropical: 100 species; Guianas: ca. 10 species of lianas or climbing shrubs, 4 species described here.

Notes: Species of Dalbergia and Machaerium adapted for water dispersal often produce similar round to crescent-shaped, flattened green fruits. **In Machaerium, seeds are located at the base of the fruit, while in Dalbergia seeds are visible as a bump at the center of the fruit.**

35. Abrus precatorius, dehiscent pods revealing red and black seeds [RP]
36. Dalbergia ecastophyllum, fruiting branch [OG]
37. Dalbergia foliosa, robust leaves and crowded inflorescence [PT]
38. Dalbergia ecastophyllum, fruiting branch [OG]

9.1. Dalbergia ecastophyllum (L.) Taub.

Common name: akareowoi (Ca)

Clambering shrub or liana to 4 m; moderately hairy. Leaves: uni-foliolate, single leaflet ovate, 6-12 x 3-7 cm, base rounded, apex bluntly tapering; nearly leathery, **flattened-hairy beneath. Inflorescences:** axillary, panicles to 3 cm long, few branched; calyx persistent, hairy; corolla white, flag petal and wings each 0.6-0.8 cm long. **Fruits:** pods round to ovate, flat, kidney-shaped, constricted on one side, 2-2.5 x 2.5-3 cm; texture soft, strongly veined; fruit. **Ecology:** Seeds adapted to saltwater dispersal.

Common; coastal areas, mangroves, estuaries, and sandy banks along rivers. Guianas; (sub)tropical Americas, including West Indies; also in W Africa.

9.2. Dalbergia foliosa (Benth.) A.M.Carvalho

Common name: akarerowoi (Ca); kwata (Ar); sirito (Ca).

High-climbing lianas, erect shrubs or treelets; **hairy. Leaves: 7(5-15)-foliate;** leaflets ovate to oblong, 3-7 x 1-3 cm, apex rounded, occasionally notched; ± leathery, **yellow-hairy below;** 2° veins parallel. **Inflorescences:** axillary or stemborne, panicles short, flowers in clusters; calyx densely hairy, short-lived; corolla violet, with flag and wings 0.7-0.8 cm long, ± equal in length to calyx. **Fruits:** ± elliptic to sickle-shaped pods, 2-3 x 1-2 cm, green, leathery, bumpy; stalk short, 0.1-0.2 cm long. **Ecology:** seeds dispersed via water; leaves usually lost in the dry season.

Common in riparian areas. Guianas; widespread in tropical South America, extending to Argentina.

9.3. Dalbergia monetaria L.f.

Common name: aturia-rang (Ca), jenodeka (Way); karukwiakat (Pal); maneko (Par); mandeku (Sa); pitikiri (Tr).

Woody lianas, clambering shrubs; treelets; **mostly hairless. Stems:** branchlets with lenticels. **Leaves: 3(5)-foliate;** leaflets ovate to oblong, 8-15 x 5-8 cm, base rounded or obtuse, apex (long-)tapering; papery to nearly leathery, hairless above and below, shiny; 2° veins netveined, slightly raised. **Inflorescences:** axillary, panicles to 3 cm long, sparsely branched (resembling D. ecastaphyllum), flowers clustered; calyx persistent, shortly 5-toothed; corolla pink or white, flag and wings ca. 0.6 cm long. **Fruits:** pods round, 3-3.7 x 3 cm; leathery, hairless, smooth, green at maturity, veins conspicuous; stipe thin, 0.4-0.8 cm long.

Common; riparian and flooded forests. Guianas; also N South America, West Indies, and Central America.

39. Deguelia amazonica, habit [FRD]

40. D. amazonica, older stem with silver-grey bark [FRD]

41. D. amazonica, flower, flag petal with yellow-white central patch [FRD]

9.4. Dalbergia riedelii (Benth.) Sandwith

Robust woody lianas or climbing shrubs; reddish-brown hairy. Stems: branchlets often tendril-like; bark smooth and dark. **Leaves: 5-9-foliate or more;** leaflets variable, oval to lancet-shaped, 4-15 x 2.5-4(-7) cm, apex acute; hairless above, **softly hairy below. Inflorescences:** axillary, panicles short; axes softly brown-hairy (velvet- or felt-like); corolla white to cream, flag petal 0.5-0.7 cm long, notched at apex, wings nearly as long as the flag, keel slightly shorter. **Fruits: round pods with conspicuous hairs,** 3-3.7 x 2.5-2.8 cm, maturing green- to dark-brown; margin ribbed.

Common; along rivers and creeks, non-flooded old-growth forest. Guianas; widely distributed in the Amazon Basin.

10. Deguelia Aubl.

Lianas or clambering shrubs; less frequently trees. Climbing via twining shoots and tendril-like branchlets.

General: Spines absent; red sap often present.

Stems: Branchlets round, bark smooth; nodes with small, inconspicuous, short-lived stipules. Older stems round, attaining sizes of 10 cm diameter or more, bark brown and fibrous, with lengthwise lines or furrows; slash with blood-red, watery sap; stem x-s pattern simple.

Leaves: Odd-pinnate (3-17 leaflets); leaflets (nearly) opposite, with **leaflets increasing in size towards apex;** leaflet stipules [*stipels*] present or absent; hairless above, often silky [*sericeous*] or with short, flattened hairs [*strigose*] below; the main leaf stalk [*petiole*] up to 10 cm long.

Inflorescences: Axillary or terminal, racemes on short shoots, axes often densely reddish-brown hairy, flowers small, **in clusters of 6 or more; bracts short-lived;** bracteoles 2, usually directly under the calyx; calyx bell-shaped, 5-toothed to entire; **petals clawed** (narrow stalk, broad apex), **white, yellow, purple to magenta,** hairless; flag petal oval, notched at tip, reflexed; stamens 10; ovary hairy, style 1.

Fruits: Pod flattened, ± cylindrical, sometimes elongate, soft to leathery, rarely woody, the margin sometimes narrowly winged, mostly indehiscent; seeds 1–3, oblong to kidney-shaped.

Ecology: Seed dispersal via wind or water. Plants often sterile. Most are forest species, but D. nitidula is typically found in savannas.

Distribution: Neotropics, center of diversity in the N Amazon basin: 20 species; Guianas: 7 species of woody climbers, 2 described here.

Uses: Many species in this genus contain rotenone and are important throughout the Amazon as fish poisons. Deguelia utilis (A.C.Sm.) A.M.G. Azevedo is a species long used by indigenous groups in fish poisoning and has great cultural significance. It typically occurs in old indigenous plantations and dense secondary forests and rarely in undisturbed forests (Ducke 1949).

Notes: Duguelia species are part of a complex that has been repeatedly rearranged by taxonomists (Derris, Lonchocarpus). Many species are poorly collected and apparently rarely produce flowers and fruits. We follow **Camargo and de Azevedo Tozzi (2014)** in placing many species within a "resurrected" Duguelia genus.

10.1. Deguelia amazonica Killip

[Syn: Derris amazonica Killip]

Common name: inekuran (Ca); ineku (Tr); ingineko (Par)

High-climbing lianas. Stems: older stems round, bearing lenticels, slash with very little red sap. **Leaves:** 7-9-foliolate; leaflets large, oblong or elliptic, 6-12 x 3-5 cm; papery or subleathery, dark green, hairless, and shiny above, paler and sparsely reddish-brown hairy below. **Inflorescences:** axillary, racemes to 45 cm long, branchlets densely reddish-brown hairy [*tomentose*]; flower clusters distant; flower stalks slender, 0.5-0.8 cm long; calyx bell-shaped, persistent, brown-hairy, 0.4-0.6 cm long; corolla pink or lilac to whitish-yellow, **flag petal ± 1 cm diameter,** broadly obovate, with yellow spot at base. **Fruits: flattened, nearly cylindrical pods, 9-12 x 2.2-2.5 cm,** base and apex rounded, finely pointed; thin and soft, densely golden-hairy [*sericeous*]; wing thin, ± wavy, 0.2-0.6 cm wide; seeds 1-2.

Common; seasonally-flooded old-growth or swamp forest, river banks, secondary forest. Guianas; widespread in the Amazon basin.

10.2. Deguelia scandens Aubl.

[Syn: Derris guianensis Benth; Derris scandens Pittier]

Common name: haiariballi (Ar); inekuran (Ca); ingineko (Pa); ineku (Tr)

High-climbing lianas; occasional trees. Leaves: 7-9-foliolate leaflets (ovate-)oblong, large, 6-12 x 3-7 cm; leathery, hairless above, hairless or sparsely hairy below. **Inflorescences:** axillary, racemes 16-18 cm long, few-flowered, slender, branchlets densely reddish-brown-hairy [*tomentose*]; flower stalks slender, 0.4-0.5 cm long, hairy; calyx shortly bell-shaped, 0.3 cm long, sometimes falling early; corolla yellow-white, **flag petal ca. 0.8 x 0.4 cm,** elongate-ovate, notched above. **Fruits:** flattened, **linear oblong pods, 7-9 x 1-1.6 cm,** constricted between seeds; soft, hairless; strongly-veined; wing thin and soft, 0.2-0.3 cm wide, ± wavy; seeds 1-2.

Uncommon; riverine forest, secondary forests in upland areas. Guianas; also Venezuela and N Amazonian Brazil.

11. Dioclea

Woody lianas and vines; sometimes small shrubs. Climbing via twining shoots, branchlets often tendril-like, clasping.

General: Fine hairs usually present; **unarmed; red sap present.**

Stems: Branchlets round, bark smooth with lengthwise ridges; nodes without horizontal ridge, stipules free, small, sometimes with a "basal appendage". Older stems round, mostly < 7 cm diameter, bark dark red and fibrous, **sap blood-red and watery, drying "brick-red"** in some species **(Lewis and Owen 1989);** wood white; **stem x-s pattern simple or concentric.**

Leaves: 3-foliolate (palmately compound with three leaflets), leaflets broadly ovate to broadly spear-shaped, margins entire; venation pinnate, 2° veins curving upwards, prominent below, 3° veins parallel or net-veined; leaflet stipules [*stipels*] at base of swollen leaflet stalks usually present, thread-like or awl-shaped.

Inflorescences: Axillary or borne on stem [*cauliflorous*], **with showy, long, erect racemes of bluish- to reddish-purple flowers;** flower stalks often clustered on stalked knobs [*tubercles*] along the infl. stalk; bracts inconspicuous, bracteoles persistent or short-lived; calyx conspicuous, 4-5-toothed, colored; corolla 2-3 cm long, **flag petal reflexed,** sometimes with a yellow spot, wings free; stamens 10.

LEGUMINOSAE

Fruits: Pods elongate to ± round, straight or curved, flattened but often with thick side walls [sutures], unstalked, greenish-yellow to reddish-brown-hairy, dehiscent or indehiscent; seeds several per pod, variable in shape and size, compressed.

Ecology: Forest species often produce showy floral displays on robust stems at ground level, or on leafless smaller branches in the lower canopy. Seed dispersal via wind or water.

Distribution: Pantropical, mostly New World: ca. ± 55 species; Guianas: 13 species, 3 described here.

11.1. Dioclea guianensis Benth.

Common name: tamoko enu (Tr)

Robust lianas, subwoody lianas and slender vines; vegetation softly hairy. **Stems: stipules small, basal appendages present. Leaves:** leaflets (elongate) ovate, 5-10 x 3-6 cm, base rounded, apex with short tip; **thin, greenish-yellow, ± with flattened-silky hairs above and below. Inflorescences:** axillary, axes hairy; bracteoles ovate to spear-shaped; calyx silky hairy outside, persistent; corolla bluish- to reddish-purple, flag petal with yellow spot on inner face. **Fruits:** pods flattened, 6-7(-10) x 1-1.5 cm; subwoody, golden-soft hairy, one side wall thickened, ± winged, to 0.12 cm wide; apex pointed; dehiscent; seeds several, flat, to 1 cm diameter.

Common; open areas, along roadsides and riverbanks, savannahs. Guianas; also N Amazon basin, Mexico, and Central America.

11.2. Dioclea violacea Benth.

Common name: kokomandialu (Par); tamoko enu (Tr)

Woody lianas; most surfaces with soft, ± flattened hairs. **Stems: stipules, basal appendages**

conspicuous, large. Leaves: leaflets ovate, 6-15 x 4-10 cm; **leathery, dark green, mostly hairless above and below;** leaflet stalks soft golden-haired. **Inflorescences:** axillary, axes dark brown- to golden-hairy; bracteoles round, with marginal hairs; calyx with lowest lobe longest; corolla violet, flag with a central yellow spot. **Fruits:** flattened pods, 9-13.5 x 4-6 cm; woody, dark brown hairy, horizontal-veined; side walls [*sutures*] to 1 cm thick, 3-ribbed; dehiscent; seeds 2-4. **Ecology:** Water-dispersed, seeds floating due to internal chambers **(Lewis and Owen 1989).**

Common; riverine forest, especially in cultivated areas. Guianas; also Brazil, Trinidad and Tobago, Central America, and Madagascar.

11.3. Dioclea virgata (Rich.) Amshoff

Common name: amatobomiki (Par); kurumï enuru (Ca); pikin kaw-ai (Sr).

Woody lianas, scrambling shrubs, or sub-woody vines. **Stems: stipules small, basal appendages absent. Leaves:** leaflets ovate to pointed, 6-11.5 x 3.5-4.5 cm; **thin to nearly leathery, moderately hairy above and below. Inflorescences:** axillary or terminal, racemes single, sparsely hairy, tubercles elongate and upturned; bracteoles oval; calyx hairless outside, persistent or short-lived, purplish-red; corolla purple. **Fruits:** flattened pods, slightly constricted between seeds, 7-12 x 2 x 0.3 cm; leathery; one side wall thickened, to 0.3 cm thick, 3-ribbed; dehiscent; seeds 5-10, oblong, ca. 1 x 0.6 cm.

Common; in shrubby riverine vegetation and forest edges. Guianas; also adj. Brazil and Venezuela.

LEGUMINOSAE

12. Lonchocarpus

Robust lianas; shrubs and trees. Climbing via twining shoots and tendril-like branchlets. Neotropics, especially N South America: ± 120 species; Guianas: 5 species, 1 described here.

Notes: For decades, there has been much confusion about the poorly known genera Lonchocarpus, Deguelia and Derris. (See Deguelia in this chapter).

12. 1. Lonchocarpus chrysophyllus Klein-hoonte

Common name: haiari (Ar); ineku (Tr); ku-maruballi (Ar); kumata (Ca); ndeku (Sa) (man) neku (Sr); wakorokoda (Ar).

Woody canopy lianas. General: hairs golden [*tomentose, sericeous, strigose*]; **red sap present;** unarmed. **Stems:** branchlets round to oval, bark smooth, dark green to black; nodes with small, short-lived stipules; older stems round, to 25 cm diameter, bark brownish-grey with lenticels, **slash with blood-red watery sap and scent of cucumber;** stem x-s pattern simple. **Roots:** sometimes with watery, whitish sap. **Leaves: odd-pinnate** (5-9 leaflets); rachis to 25 cm long with a 10 cm long petiole; leaflets opposite, increasing in size towards the terminal leaflet; **blades oblong, large, to 17 x 7 cm, hairless and dark green above, golden below;** hairs silky-flattened [*sericeous*] or short-flattened [*strigose*]; leaflet stipules [*stipels*] usually absent. **Inflores-**

cences: axillary or terminal, flowers small and clustered in dense panicles, branchlets hairy; bracteoles 2, usually directly under the calyx; calyx bell-shaped, shortly 5-toothed to entire, ca. 0.6 cm long, silky hairy; corolla purple (white, yellow, to magenta in co-species of genus), **petals clawed,** ca. 1.2 cm long, hairy outside; flag petal reflexed, ovate, usually notched at apex, with a white center; stamens 10; ovary hairy, style 1. **Fruits:** pod flattened, ± round to (narrow-)oblong, soft to leathery, rarely woody, the margin often winged, indehiscent; seeds 1–12, oblong to kidney-shaped. **Ecology:** seed dispersal via wind or water. Plants often sterile. **Use:** Lonchocarpus species are still important throughout Amazonia as fish poisons. In northwestern Guyana, Lonchocarpus species were cited in local treatments for cancer and HIV infections (van Andel 2000).

Distribution: Uncommon; creek forest; rarely collected with flowers or fruits. Guianas; also N South America, Central America, and the West Indies.

13. Machaerium Pers.

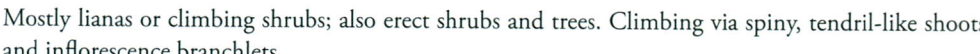

Mostly lianas or climbing shrubs; also erect shrubs and trees. Climbing via spiny, tendril-like shoots and inflorescence branchlets.

General: Often with dense golden-brown hairs; armed; **red sap present.**

Stems: Branchlets round, 3-5-sided, or conspicuously flattened, bark smooth or rough, densely hairy or hairless, **nodes usually with paired spines** (= modified stipules). Older stems round, conspicuously angled or flattened, 5-25

cm diameter, bark brown, grey, or silver-white, lenticels often present; **inner bark with blood-red, watery sap that oxidizes dark red; stem x-s pattern concentric (bull's eye) or eccentric (in flattened stems).**

Leaves: **Odd-pinnate,** leaflets alternate, **either 5-7-9-foliolate, leaflets large, ± ovate,** terminal leaflet usually largest, **or many-foliolate, with small, linear leaflets** (M. ferox, M. microphyllum, M. myrianthum); veins pinnate; leaflet stipules [*stipels*] absent.

Inflorescences: Axillary or terminal, in racemes or panicles, **axes often golden-hairy** [*strigose, pilose*], **tendril-like, spines present;** flowers small, 0.4 - 1.8 cm diameter, often inconspicuous; bracts spiny or soft; bracteoles 2 on flower stalks; calyx bell-shaped, toothed or entire; petals white, yellow, pink, blue, or purple, with a central color spot; stamens 10; ovary stalked, stigma 1.

Fruits: Samaras with 1 dorsal wing or crescent-shaped pods with a low crest, stalked; seeds 1, flat, kidney-shaped.

Distribution: (Sub)tropics of the New World (1 species W Africa): 130 species; Guianas: 18 species., 4 described here.

Notes: Species of Dalbergia and Machaerium adapted for water dispersal often produce similar round to crescent-shaped, flattened green fruits. **In Machaerium, seeds are located at the base of the fruit, while in Dalbergia seeds are visible as a bump at the center of the fruit.**

13.1. Machaerium inundatum (Benth.) Ducke

Common name: bohoribada (Ar); brantimaka (Sr); kromoko (Sr)

High-climbing, shrubby lianas; also erect shrubs and trees; unarmed. **Stems:** round or ± angled; **stipules short-lived, not modified into spines. Leaves:** 4-7-foliolate; leaflets 3-8 x 2-5 cm, **leathery, hairless above and below;** 3° veins net-veined. **Inflorescences:** axillary, panicles to 4 cm long, flowers unstalked, axes golden hairy; bracteoles nearly round, persistent; calyx ca. 0.3 cm long, shortly toothed; petals ± 0.8 cm long, flag petal densely hairy outside. **Fruits: crescent-shaped pods with medial prominent rib,** ca. 4 x 2-2.3 cm, base thick, apex with 0.4 cm long,

sharp wing-like crest; **nearly woody, green** with short, brown hairs when young.

Uncommon; marsh forest, along rivers. Guianas; also N South America including Peru.

13.2. Machaerium lunatum (L.f.) Ducke

Common name: anuana menepuru aturai (Ca); ate tatai (Sa); bariribada (Ar); brantimaka (Sr); bodori (Ar); bunduri pimple (GU); Chinese earring (GU); ëripoimë (Tr); mïkakijee (Tr); moon fruit (GU).

Shrubs or lianas. Stems: round, flattened, or angled; **spine-like, paired stipules present. Leaves:** (3)-5-11-foliolate; leaflets elongate, 2-5 x 0.5-1.5 cm, apex abruptly tipped; thin, **hairless above, finely hairy below. Inflorescences:** terminal, panicles to 15 cm long; bracts and bracteoles circular, 0.1-0.13 cm diameter; calyx persistent, ca. 0.4 cm long, with short hairs; corolla blue to violet, flag ca. 0.9 cm long, with few silk hairs outside. **Fruits: crescent-shaped pods,** 3-4 x 2 x 0.5 cm, not winged.

Common; in coastal swamps. Guianas; also coastal N South America, coastal Central America, Caribbean, and Africa.

13.3. Machaerium macrophyllum Benth.

High-climbing lianas; occasional trees. **Stems:** branchlets brown-haired when young; **stipules spine-like and persistent or short-lived;** slash with dark red sap. **Leaves:** 5-7-foliolate; leaflets nearly ovate, 8-15 x 4-8 cm; **hairless above, brown or rusty-haired below;** 2° veins prominent below. **Inflorescences:** terminal, panicles large, rusty-haired; bracteoles circular, 0.2-0.3 cm in diameter; calyx persistent, hairy, 0.2 cm long; corolla 1.0-1.2 cm long, flag petal circular, with few silky hairs outside. **Fruits:** pods 5.5 x 4 cm, apex rounded, with flattened hairs, unwinged.

Common; old-growth forest, creek forest, swamp forest. Guianas; also Amazonian Brazil, Venezuela, and Peru.

LEGUMINOSAE

42. Dioclea virgata [PT]
43. Mucuna sloanei, young fruit with horizontal ridges
44. M. sloanei, hanging inflorescence [PT]
45. Mucuna urens, mature fruit opening to display seeds for dispersal [BH]
46. Machaerium lunatum, in flower [PT]
47. Machaerium sp., stem cross-section with concentric rings and red sap [BH]
48. Machaerium sp., leaves with prickly branchlets [BH]
49. Machaerium lunatum, axillary spine [PT]

13.4. Machaerium quinatum (Aubl.) Sandwith

Common name: ëripoimë (Tr)

Climbing shrub; also erect shrub or treelet; unarmed. **Stems: stipules narrowly ovate,** 1 cm long, short-lived. **Leaves:** 7-15-foliolate, up to 40 cm long; leaflets largely ovate, 5-12 x 2.5-5 cm; leathery, ± **hairless above, rusty hairy below;** 2° nerves prominent below. **Inflorescences:** terminal, many-branched panicles; bracts and bracteoles ovate; calyx persistent, yellow-hairy [*tomentose*], 0.3-0.6 cm long, short-toothed; corolla 0.8-1.7 cm long, flag purple with central white stripe, reflexed, brown-silky hairy outside; keel white. **Fruits:** winged samaras, curved or straight, 5-9 x 2-3.2 cm, dark brown hairy; wing firm, broader toward tip.

Common; marsh and creek forests. Guianas; also N South America.

14. Mucuna Adans.

Woody vines; also herbs. Climbing via twining shoots and tendril-like branchlets.

General: **Stinging or itching hairs often present; spines absent; red sap present;** sometimes with drinkable water in older stems.

Stems: Branchlets round, slender; nodes with **stipules inconspicuous,** short-lived. Older stems round, small- to medium-sized, **slash with blood-red, watery sap; stem x-s pattern concentric.**

Leaves: 3-foliolate (palmately compound); terminal leaflet long-stalked, lateral leaflets often strongly asymmetrical; papery and soft; **leaflets 3-veined from base,** veins widely spaced; leaflet stipules [*stipels*] narrow and short-lived.

Inflorescences: Axillary pendent racemes or umbels, with stiff, brown hairs; flowers large and showy; bracts and bracteoles often large, soon falling; calyx bell-shaped, 4-5-toothed, sometimes partially fused and 2-lipped; corolla cream, yellow, or greenish-purple, **'flag' petal with ear-like lobes at base** [*auriculate*], usually much shorter than the wings; stamens 10, fused; ovary hairy.

Fruits: Pods linear to broadly elongate, thick and leathery, surface with horizontal ridges and often with stiff, brown, stinging hairs, opening with 2 valves; seeds thick-disc-shaped, with a marginal scar circling ¾ of the seed body.

Distribution: American tropics and W Africa: 120 species; Guianas: 3 native species; 2 described here. M. pruriens (L.) DC. var. utilis is a cultivated, naturalized vine from the Paleotropics.

Ecology: Bird- or bat-pollinated, pollen released explosively; seed dispersal via water (drift seeds) or possibly animal (bat-dispersed).

Uses: Widely used in traditional medicine. Seeds used in handicrafts and games.

14.1. Mucuna sloanei Fawc. & Rendle

Common name: graine tonnere (FG); kaw-ai (Sr); kuluway (Wpi); tamoko enu (Tr); urikti (Pal); zieu bourriqu (FG)

High climbing, woody lianas. Stems: branchlets hairless. **Leaves:** leaflets (elongate) ovate, 8-15 x 4-9 cm; thin, **hairless above, with silky silver hairs below** [*appressed-sericeous*]; petiole 10-15 cm long. **Inflorescences:** axillary, **compact umbels; stalk < 10 cm long;** calyx hairy, 1 cm long, teeth narrowly triangular; **corolla yellow,** 'flag' petal 4 cm long. **Fruits:** linear, oblong pods, 10-18(-20) x 4-6 x 2.5-3 cm, purplish-black, with fine hairs, with 2 lengthwise, wavy wings, to 0.7 cm wide; seeds 2-4.

Common; marsh forest along rivers and creeks. Guianas; N South American tropics and W Africa.

14.2. Mucuna urens (L.) Medik.

Common name: amogoyenuyik (Pat); cow itch (GU); horse eye (GU); kaw-ai (Sr); tamoko enu (Tr); tawaau (Way).

High climbing, woody liana. Stems: branchlets hairless; older stem round, often thick and soft, **containing drinkable water. Leaves:** leaflets (elongate) ovate, 8-12 x 4-7 cm, thin, **hairless or sparsely hairy above and below;** petioles 6-12 cm long. **Inflorescences:** axillary, **pendant racemes, stalk to 1 m long,** zig-zag, flowers in clusters of 2-3; calyx hairy, 1 cm long, teeth broad; **corolla pale greenish-white with violet,** 'flag' petal 3.5 cm long. **Fruits:** linear, oblong pods, 10-20 x 5 x 2-2.5 cm, reddish-brown, finely hairy, with 2 lengthwise, wavy wings, to 0.3 cm wide; seeds 1-4.

Common; marsh forest. Guianas; widespread in the neotropics.

LOGANIACEAE

1 2 3 4 5 6 7

LOGANIACEAE

A family of mostly woody canopy lianas with some herbs, shrubs, and trees. Distributed widely in the tropics and subtropics with a few temperate zone members, ± 15 genera/205 species; Guianas: 6 genera/35 species; with only one genus of climbers in the neotropics.

1. Strychnos L.

Robust woody lianas that grow initially as understory shrubs; rarely non-climbing shrubs. Climbing via robust hooks. Leaves are opposite, simple, entire, and palmately-3-5-veined. Stipules are absent in Guianan species. Inflorescences are terminal or axillary, with dense to open clusters of small, tubular, white or whitish-green flowers; ovary is superior. Fruits bright yellowish-orange berries with seeds immersed in pulp.

General: Vegetation shiny green and hairless to densely hairy; hairs simple, with mixed long and short hairs; an unpleasant or 'poisonous' scent in cut parts; colored sap absent.

Stems: Branchlets with a scar, ridge or ring between petioles, sometimes with knobs, spines or tiny paired leaves; leaf axils with 1-2 unmistakable, robust hooks (**1D – coiled in one plane**). **Older stems** robustly woody and often **almost perfectly round,** 2-30 cm diameter, **ringed at nodes;** bark commonly orange-brown, rough and with thick orange-brown lenticels (also white-grey, brown, reddish-brown, without lenticels); wood hard, white or yellow; **stem x-s pattern 'foraminate' - with scattered, tiny dark specks** [*secondary phloem islands*], in dry stems appearing as light-colored specks or holes.

Leaves: 3-5-palmate venation, veins splitting at 0-3 cm from the leaf base and often joining again near the tip (often appearing as a 'leaf within a leaf'); finer (3°) veins ± parallel and slanted [*oblique*] or net-veined; petioles often hairy and short, 0.5-2.5 cm long. Blade sometimes with tiny white to clear gland dots, underside often with tiny hairy pockets in vein axils [*domatia*] or warty bumps [*tubercles*]. Without 1-2 raised glands at leaf-base or on petiole.

LOGANIACEAE

Inflorescences: Axillary to terminal, few-flowered clusters [*cymes, thyrses*]; **flowers tubular, narrow, 4-5-parted, white, greenish-white, or yellow, usually < 1 cm long;** calyx fused, tiny (0.1-0.2 cm long); corolla often hairy inside; often with tiny bumps [*papillae*] outside; stamens 4-5, attached to corolla, extending beyond throat; corona (ring) sometimes present; ovary superior, usually 2-chambered [*2-locular*], stigma 1.

Fruits: ± **round, indehiscent berries that ripen bright orange-yellow to orange,** rarely grey or black, 2-3 (10) cm diameter, leathery or corky, shell [*pericarp*] smooth and shiny; 1-2-chambered with white to orange pulp; **seeds disc-shaped,** few to many.

Ecology: Some species are giant-sized lianas that occupy the upper forest canopy or occur in dense tangles on the forest floor. Seed dispersal via animal (often ingested by monkeys).

Distribution: Pantropical: ± 100 species; Guianas: ± 22 species, 7 described below.

Use: The wood and bark of Strychnos species, is a main ingredient of traditional paralyzing hunting poison recipes ('curare') across tropical South America (**Biset 2002, Krukoff 1972, Roth 1922-1923, Schultes & Raufauff 1990).** In the Guianas, many indigenous groups use the same or similar common name (e.g., oerari, ulali, urari, wirari) for the plant and the potion. Guianan species with widespread historical use and documented potency include Strychnos glabra, S. guianensis, S. tomentosa, and S. toxifera (**Krukoff 1972**). Strychnos species little used for curare are often given 'look-like' names such as orari-dan, urari-balli, and wirari-më (wirari-muh). Due to the introduction of guns for hunting, traditional knowledge of curare preparation is held only by a few remaining elder men in the Guianas.

Some Strychnos species are a main ingredient of aphrodisiac 'man batra' and other bitter medicinal tonics in and outside the Guianas (**van Andel et al., 2012**). To prepare such tonics, the wood, bark, and seeds of a variety of forest trees and lianas are soaked in alcohol. Strychnos melinoniana and other species used for this purpose are commonly known as devildoer in Guyana, dobrudua in Suriname, and dobouldoi in French Guiana. Many of the 'tonic' Strychnos species are not mentioned as curare plants in ethnobotanical accounts. **Fanshawe (1954)** noted for Strychnos erichsonii that "the total alkaloids have a convulsant, strychnine-like action; hence its use as an aphrodisiac."

Notes: The opposite, 3-5-palmate-veined leaves and axillary hooks together distinguish the genus well, although the hooks may be absent. Compare with Fabaceae-Bauhinia (2-foliolate, tendrils coiled, axillary stipule), Melastomataceae (opposite leaves, ladder-like cross-veins) and Rubiaceae-Uncaria (opposite leaves, conspicuous stipule, pinnate veins, short coiled hooks or spines in leaf axils).

1.1. Strychnos erichsonii M.R.Schomb. ex Progel

Common name: kumarawa (Ca-GU); kwabanero (Ar); ledi dobuldua (Sa); oraridan (Ca-GU); pulewinah (Way); red devildoer (GU); rouge dobouldoi (FG); urari-balli (Ar).

Very large woody canopy liana; very hairy to lightly hairy. Stems: branchlets often densely short-hairy; older stems > 25 cm diameter; **bark dark reddish-brown or orange,** rough, with large orange-brown lenticels; wood slash pale-red to yellow. **Leaves:** shape variable, **medium- to large-sized 7-26 x 3.5-13.5 cm,** base rounded to obtuse, apex acuminate; thin to leathery, two-tone, with short hairs and tiny dots below, mostly hairless above; 3- to 5-palmately-veined, **inner vein pair splitting 0-1 cm above base; 3° veins conspicuous;** petiole 0.2-1.3 cm long, hairy. **Inflorescences:** axillary or terminal, dense clusters to 2.5 cm long, hairs brownish-yellow woolly; corolla tube to 0.8 cm long; stamens barely extending out of throat. **Fruits: berries sub-round, 2-3.5 cm diameter,** wall 0.2

cm thick; seeds (1)3-5, 1.7 x 1.3 x 0.4 cm. **Use:** Aphrodisiac and medicinal tonic in the Guianas. Curare arrow poison (Way).

Common; non-flooded or seasonally-flooded old-growth forest. Guianas; also Venezuela, Brazil (Amapá and Pará), Colombia, and Peru.

1.2. Strychnos guianensis (Aubl.) Mart.

Common name: barauitu (Ww-GU); barawito (Ww-SU); orariyik (Pat); orari/urari (Ca-many); rouamon (Ca-SU); wilali (Way); wïrarï (Tr).

Medium-sized woody liana; very hairy to lightly hairy. Stems: branchlets felty; older stems large, > 25 cm diameter; **bark dark brown with large, scattered lenticels, sub-bark dull orange;** wood orange-white, hard; slash scent strong, unpleasant. **Leaves:** shape variable, **small- to medium-sized, 3-9 x 1.5-4.5 cm;** lightly hairy to densely felt-like below, hairs straight and long or curved and short; palmate veins 3-5, **inner vein pair alternate, splitting at 0-1.3 cm from base, not meeting at apex,** outer vein pair sub-marginal and faint, 3° veins faint above and below; petiole 0.2-0.6 cm long, hairy. **Inflorescences:** few-flowered, loose clusters, with short, rough hairs [*hirsute*]; calyx lobes to 0.15 cm long, corolla greenish-white, tube equal in length to corolla lobes; stamens and style extending beyond throat. **Fruits: berries oblong or ovate, 2-3.5 x 1.7-3 cm,** with a fine point at apex, shell very thin; seeds 1-2, disc-shaped, ca. 0.4 x 1.6 cm. **Use:** Principal ingredient in the traditional curare arrow poison recipes of indigenous groups throughout tropical South America and the Guianas.

Very common; non-flooded and flooded old-growth forest, mountain savanna forest, secondary forest, white water river basins. Guianas; also widespread in the South American tropics.

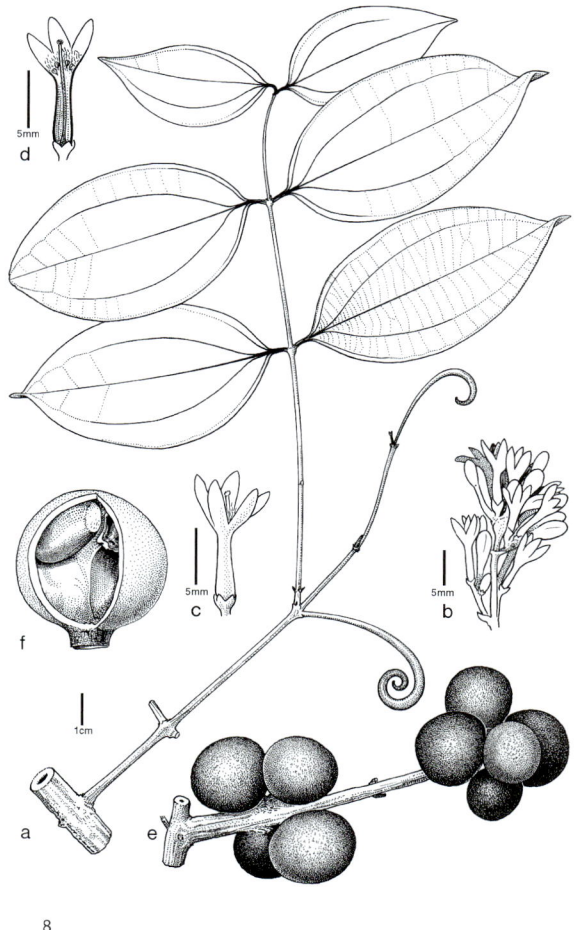

1. Strychnos melinoniana, young leafy branch with hook-tendril [PT]
2. S. melinoniana, leafy branch with immature fruits [BH]
3. S. erichsonii, 'foraminate' pattern in stem with scattered black dots when fresh, tiny holes when dry [BH]
4. S. melinoniana, bark and lenticels close-up [BH]
5. S. erichsonii, woody hook-tendril [BH]
6. S. cogens, inner 2° vein pair splitting above base, 3° veins ± parallel, horizontal to long axis [FRD]
7. S. glabra, 2° veins alternate and splitting above base, inner 2° vein pair forming 'leaf within leaf' [FRD]
8. S. mitscherlichii, a) leafy climbing branch with tendril-hooks; b) dense inflorescence of tubular flowers; c) single flower, corolla tube 0.6-0.7 cm long, anthers borne at mouth, style extending slightly out of throat; d) corolla cut lengthwise to reveal long style, short anthers, and hairy throat; e) globose fruits on branch; f) fruit cut open to reveal disc-shaped seeds [NCB]

LOGANIACEAE

1.3. Strychnos medeola Sagot ex Progel

Common name: orari/urari (Ca-GU); wïrari (Tr)

Low-climbing liana or shrub; lightly hairy to hairless. Stems: older stems small, ± 3 cm diameter. **Leaves: lanceolate, small, 3-7 x 1.5-2 cm,** base rounded, apex finely pointed; soft and thin to papery, hairless except along veins, underside dull and with tiny white dots; palmate veins 3, **inner pair splitting 0-0.3 cm from base, 3° veins faint** above and below; petiole 0.1-0.2 cm long, with mixed long and short hairs. **Inflorescences:** terminal or axillary, loose, few-flowered clusters with reddish-brown hairs; calyx lobes narrow and pointed, to 0.3 cm long; corolla tube to 1.3 cm long. **Fruits: berries ± round, 2-2.7 cm diameter,** smooth, shell thin; seeds 1-2, each ± 1.3 x 1 cm. **Use:** A second priority curare species for the Trio in Suriname.

Common; non-flooded old-growth forest, seasonal forest, savanna, secondary forest, often on rocks. Fr. Guiana, Suriname; also Amazonian Brazil (Amapá and Pará).

1.4. Strychnos melinoniana Baill.

Common name: devildoer/white devildoer (GU); dobrudua (Sr); kumarawa (Ca-GU); kwabanero (Ar); mamakure (Tr); oraridan (Ca-GU); paruisy-iton (Par); uraridan (Ar).

Very large woody canopy liana; ± hairless. Stems: branchlets greyish-white; older stems round, large, > 15 cm diameter, **bark light-colored with abundant orange-brown lenticels. Leaves:** shape variable, elliptic to ovate, **medium-sized, 8-15 x 4.5-7 cm,** base obtuse to acute, apex acuminate; papery to leathery, hairless and shiny, drying grey-green; palmate veins 3(5), **inner vein pair alternate, splitting 0-1 cm above base, 3° veins barely raised;** petiole 0.4-1.1 cm long, drying blackish, hairless. **Inflorescences:** terminal, dense, few-flowered clusters, up to 1.5 cm long, with dense short hairs; corolla 0.3-0.4 cm long, white, woolly-haired

within and on lobes. **Fruits: berries ± round to oval, small, to 3 x 2 cm,** with a fine point at apex, shell thin, fleshy; pulp sweet; seeds 2, ca. 0.4 x 1.6 cm. **Uses:** One of the most commonly used and sold species for aphrodisiac and medicinal tonics in the Guianas (van Andel and Ruysschaert 2011). Traditionally used by the Warau of Guyana and Venezuela as a curare arrow poison.

Common; wide ecological range, including non-flooded and flooded forests. Guianas; also adjoining Brazil (Amapá, Pará).

1.5. Strychnos mitscherlichii M.R.Schomb.

Common name: dobrudua (Sr); kumarawa (Ca-GU); kwabanero (Ar); white devildoer (GU).

Large woody canopy liana; ± hairless. Stems: older stems round, large, > 10 cm diameter. **Leaves:** variable, lanceolate to ovate, **small- to medium-sized, 4-22 x 3.5-10.5 cm,** base obtuse to acute, apex rounded to long-pointed; leathery, shiny above and below; palmate veins 3-5, **inner pair splitting 0.2-1.5 cm above base; 3° veins prominent,** petiole 0.3-1.2 cm long. **Inflorescences:** axillary, opposite, narrow clusters to 4 cm long, branchlets black, hairless; corolla tube cream-colored, 0.6-0.7 cm long, throat hairy; stamens at mouth, very short, style extending slightly out of throat. **Fruits: berries ± round, 2-4.5 cm diameter,** rough with yellow lenticels, shell thin; seeds many, each ± 1.8 x 0.6 cm. **Uses:** Aphrodisiac and medicinal tonic. Curare arrow poison (Ak, Are). Reportedly with a 'weak curare-like action' (Krukoff 1972).

Common; non-flooded old-growth forest. Guianas; occurs widely across tropical South America.

1.6. Strychnos tomentosa Benth.

Common name: apotai (Ca-GU); wilali (Way)

Large woody canopy liana; very hairy. Stems: branchlets hairy; older woody stems medium or large, to 10 cm diameter. **Leaves:** elliptic, **medium-sized, 6-10 x 4-5 cm,** base rounded to obtuse, apex acute or obtuse with a narrow point; thin to sub-leathery, older leaves shiny, dark green above, with red-brown, short-curved hairs below; palmate veins 3(5), **inner pair splitting at 0-0.4 cm from base, 3° veins faint;** petiole 0.4 cm long, with dense, short-curved hairs. **Inflorescences:** terminal, loose flower clusters, with dense yellow-brown hairs; calyx lobes ovate to lanceolate, 0.2- 0.3 cm long; corolla tube up to 1.5 cm long, hairy outside, hairs mixed short yellow-brown and long rusty-brown; stamens extending beyond throat. **Fruits: berries round, 6-10 cm diameter, greenish-yellow to almost black,** ± shiny, smooth, shell skin thin and soft, peeling; pulp edible, with pleasant scent and taste; seeds many, each to 2 x 1.5 cm. **Uses:** Important traditional curare arrow poison ingredient in French Guiana (Way, Wpi) and Brazil.

Uncommon; non-flooded old-growth forest, seasonal forest. Guianas; also Amazonian Brazil (Amapá, Belém) and Venezuela.

1.7. Strychnos toxifera R.H.Schomb. ex Lindl.

Common name: curare (GU); devildoer (GU); orari/urari (Ca-many); wilali (Way); wïrari (Tr).

Medium-sized woody canopy liana; very hairy. Stems: older woody stems medium-sized, up to 5 cm diameter. **Leaves:** variable, narrow to oblanceolate, **small- to medium-sized, 6-20 x 3-8 cm;** base subcordate to obtuse, apex often long-acuminate; thin and soft to papery, **with golden-brown soft hairs above and below,** also often 'warty' below; palmate veins 5, **inner pair splitting 0.2-2 cm above base, 3° veins faint to slightly raised;** petiole 0.1-0.6 cm long, with mixed size hairs. **Inflorescences:** terminal, loose clusters of flowers, with long, spreading reddish-brown hairs (hairs > 0.1 cm long); calyx lobes linear-lanceolate, to 0.6 cm long; corolla tube to 1.5 cm long, densely hairy outside [*hirsute*]; stamens barely extending from throat. **Fruits:**

berries round, to 7 cm diameter, blue-green to orange, smooth, shell thick; seeds 10-15, each ± 2.3 cm long. **Uses:** Principal ingredient in traditional curare arrow poison recipes in Guyana (Mac, Wap) and Fr. Guiana (Wpi). Also reported as curare ingredient in Brazil (Tecuna), Colombia (Kofan), Ecuador (Canelo), and Venezuela (Cunipusana).

Uncommon; old-growth forest (flooded, non-flooded, creek) and secondary forest. Guianas; also NW South America to Panama.

9. Strychnos asperula node with horizontal line (Brazil) [FRD]
10. Strychnos froesii node with hairs (Brazil) [FRD]
11. Strychnos froesii, typical Strychnos inflorescence (Brazil) [FRD]
12. Strychnos melinoniana, immature disc-shaped seeds within fruit [BH]

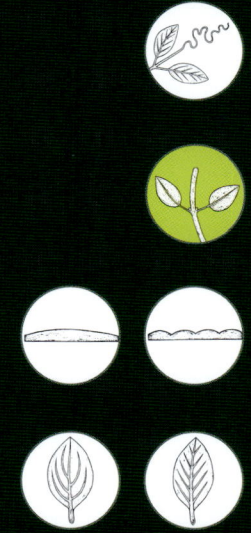

MALPIGHIACEAE

MALPIGHIACEAE

Woody canopy lianas and perennial vines; also erect shrubs and trees. Climbing via twining shoots. Leaves are opposite, simple, and entire. Stipules are present. Flowers are 5-parted, with sepals bearing 8-10 conspicuous oil glands and clawed petals with fringed or toothed margins, mostly yellow, pink, or white; ovary superior with 3 styles. Fruits are mostly of three fused nuts with conspicuous wings or ridges.

General: Vegetation often conspicuously hairy and shiny due to a dense layer of **flattened hairs** [*sericeous or tomentose*] lost with age or persistent. Individual hairs typically 2-armed, **Y- or T-shaped** (rarely star-shaped). Glands common on leaves, petioles and inflorescences. Colored sap and distinctive scent absent.

Stems: Branchlets round, lightly to densely hairy, bearing lenticels; nodes often with a horizontal ridge or line, **stipules mostly tiny and inconspicuous, of diverse locations and forms:** i) discrete, on the stem near the petiole [*interpetiolar*]; ii) 2 fused across the node and leaving a horizontal scar [*interpetiolar*]; iii) discrete, at the petiole base [*intrapetiolar*]; iv) 2 fused and extending up onto the petiole as a sheath [*epipetiolar, Hiraea*]. Also occurring as indistinct bumps (Stigmaphyllon) or absent. **Older stems round, 2-cabled, lobed, or twisted and multi-cabled** and then attaining enormous sizes, up to 35 cm diameter; bark mostly thick and fibrous, silver to reddish-brown, bark often cracked and with raised lenticels, often furrowed; inner bark sometimes orange compound-lobed or divided (multiple cables forming one mega-stem, cables visible individually or immersed). Some species producing **large underground tubers (Stigmaphyllon).**

Leaves: Shape variable, ovate to heart-shaped (Stigmaphyllon), **margins mostly entire, rarely round-toothed** (Stigmaphyllon); veins pinnate, 3° veins net-like [*reticulate*] or parallel; **leaf glands often present,** including: i) **large gland-pairs** - at the leaf base/petiole apex or lower on petiole (nipples, mounds, discs, craters); ii) **small flat glands** – circular/oval glands in lengthwise rows or scattered on the leaf blade, mostly at or near the leaf margin or underside of blade; petioles 1-2 cm long, longer in most Stigmaphyllon, grooved above, often hairy.

Inflorescences: Showy racemes or panicles with conspicuous leaflets [*bracts, bracteoles*] within the branchlets; terminal units [*umbels, cymes*] of

4-6 flowers or otherwise, individual flower stalks usually jointed with a small leaf pair [*bracteoles*] and basal leaf [*bract*] below the joint; flowers 5-parted; **calyx 5, usually in a "4+1" pattern,** with four similar sepals each bearing 1-2 **raised oil glands,** and a fifth unique sepal usually lacking oil glands; **corolla 5, usually in a "4+1" pattern,** with 4 similar lateral petals (alternating with sepals) and one dissimilar, erect 'flag' petal; **petals claw-shaped with fringed or toothed margins, bright yellow, pink (white, lilac);** bisexual; stamens usually 10, similar or dissimilar; ovary superior, 3-chambered [*3-locular*]; **styles 3,** united or spreading, stigmas diverse - rounded, truncate, hooked, sometimes with a leafy appendage.

Fruits: In lianas, fruits are composed **of three fused nuts [*samaras, mericarps*] with wings, winglets, discs, crests or ridges of diverse shapes and sizes.** Wings and other structures are either 'dorsal' (pointing straight up like a fish's dorsal fin) or 'lateral' (angled outwards like a fish's side fins). Intermediate winglets or crests often occur between large wings.

Ecology: Insect pollinated - certain bee species are attracted to conspicuous oil glands on the calyx. The unique fifth petal of the corolla in most species serves as a "flag" to attract and orient insect pollinators. Seeds dispersed via wind or water (animal-dispersed only in some shrubs and trees). Water-dispersed species often have fruits with crests or ridges rather than wings and an internal air chamber that allows seeds to float.

Distribution: Largely neotropics and subtropics: 67 genera/1200 species; Guianas: 17 liana genera/62 liana species; all liana genera/33 species described here.

Use: Several species used in traditional medicines and rituals [*e.g., genital steam baths, menstruation pain, eye medicine, aphrodisiacs, 'bush yaws' (leishmaniasis) treatment - van Andel & Ruysschaert 2011*]. Banisteriopsis caapi Morton is a well-known component of the psychotropic ayahuasca or yagé drink but does not occur in the Guianas.

Notes: Malpighiaceae has been revised considerably in the 21st century and we aim to present the most current taxonomic concepts and names (Anderson & Davis 2007). The most notable changes are the transfer of many species out of Banisteriopsis and Mascagnia and into new or revised genera such as Alicia, Carolus, and Christianella. Families possibly confused with the Malpighiaceae include Apocynaceae (white milky sap), Celastraceae (green branchlets, toothed margins, colored sap), Combretaceae (without stipules, T-shaped hairs or leaf glands), Rubiaceae (with interpetiolar stipules, branchlets swollen below nodes), and Trigoniaceae (with free stipules but without T-shaped hairs or leaf glands).

1. Alicia W.R.Anderson

Woody lianas. Climbing via twining shoots. S American tropics: 2 species; Guianas: 1 species.

1.1. Alicia macrodisca (Triana & Planch) W.R.Anderson

[Syn: Mascagnia macrodisca (Triana & Planch) Nied.]

Shrubby liana; densely hairy [*tomentose, sericeous*]. **Stems:** branchlets round, to 0.5 cm diameter, hairs grey; nodes with tiny, triangular, discrete stipules, on or near petiole; older stem round, with fibrous bark; stem x-s pattern lobed to compound-lobed/divided – with multiple lobed cylinders. **Leaves:** variable, ovate to oblong, 5-17 x 2-9 cm, base acute or obtuse; older leaves hairy below; **blade bearing 2-5**

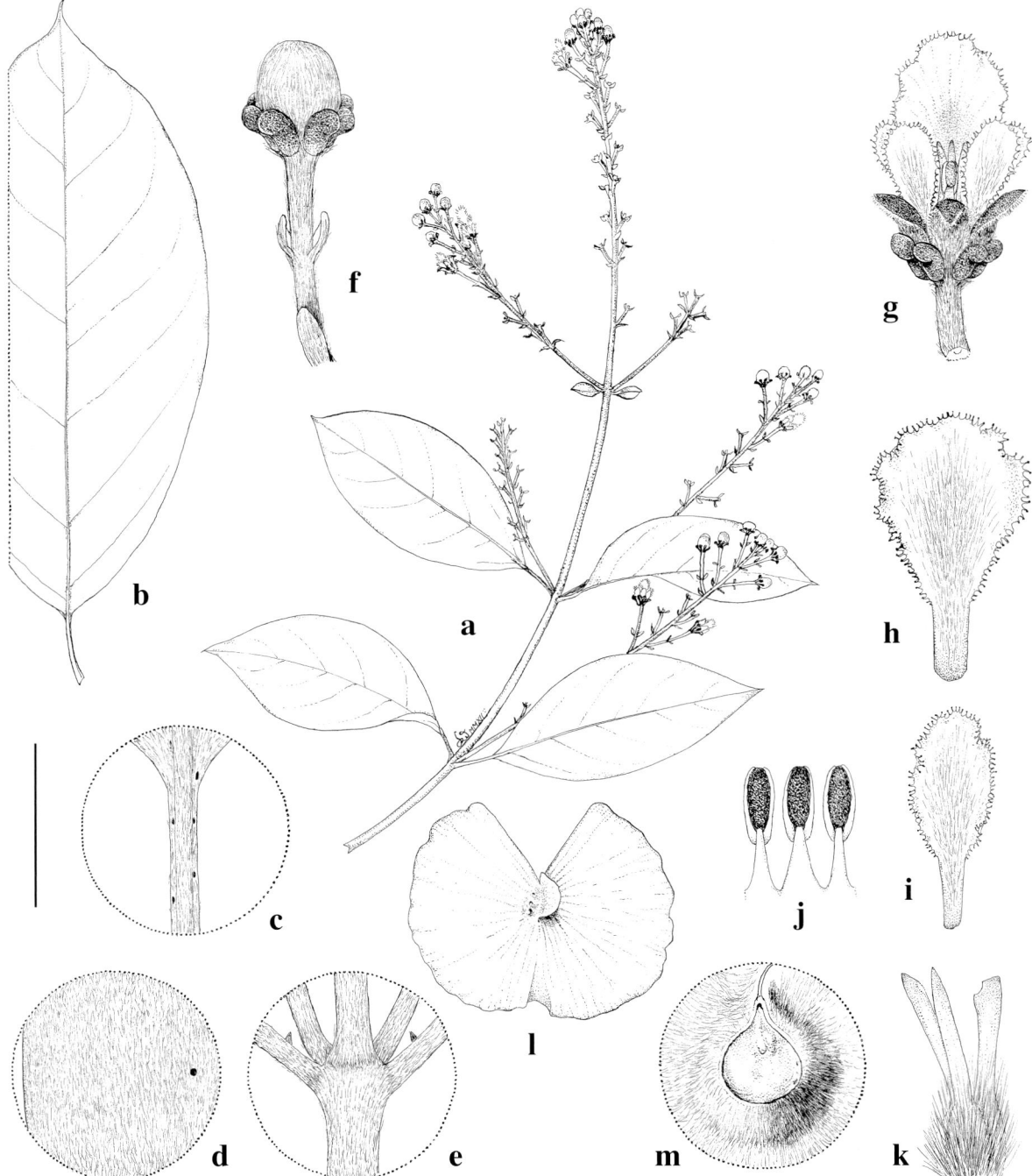

5

1. Stigmaphyllon puberum habit [PT]

2. Close-up of flowers with clawed, fringed petals [PT]

3. Stigmaphyllon convolvulifolium habit [PT]

4. Mascagnia arenicola, robust canopy liana in old-growth forest, Brownsberg Nature Park, Suriname [BH]

5. Alicia macrodisca illustration. a. Flowering branch. b. Large leaf from above. c. Underside view of petiole, small glands. d. Straight hairs and tiny glands on leaf underside. e. Node, tiny triangular stipules. f. Flower bud with persistent oil glands (dark), bracteoles on flower stalk, and basal bract. g. Flower, unique fifth petal uppermost. h. Fifth petal. i. Lateral petal. j. Three stamens. k. Ovary with three styles, one dissimilar. l. Samara, outerside. m. Nut of samara from innerside to show areole. Scale bars: a, b, 4 cm; c, 8 mm; d, 4 mm; e, 8 mm; f, g, 5.7 mm; h, i, 4 mm; j, k, 2.7 mm; l, 4 cm; m, 6.7 mm. [Drawn by Karin Douthit, reproduced with permission of the University of Michigan Herbarium]

basal glands, **1-2 small gland-pairs in two rows on the petiole; and small flat glands in rows on the blade underside;** veins prominent below; petioles thick, grooved, glandular. **Inflorescences:** axillary or terminal, hairy racemes or panicles, flower stalks bearing bracts; flowers 1-1.5 cm diameter; calyx 8-glandular, **sepals concealing petals in bud; petals white, white and pink, pink or lilac,** thin, densely hairy outside, subequal; stamens 10, hairless, similar; ovary superior, styles straight. **Fruits:** highly variable, samaras 4-7.5 cm across, with **large, papery, disc-shaped, lateral wings,** 2-4 cm high x 4-8 cm broad, lateral wing round at base, deeply divided to seed body (nut) at apex, margin entire or wavy; dorsal crest unevenly elliptic or ovate, 0.5-2 cm high x 0.6-1 cm broad, wavy; intermediate winglets absent; seed body subglobose, 0.5-0.6 cm diam.

Uncommon, locally abundant; in secondary forest. Guyana, Suriname; also widespread in Amazonia.

2. Banisteriopsis C.B.Robinson

Woody lianas and climbing shrubs; also erect shrubs or treelets. Climbing via twining shoots.

Stems: As for family; nodes with tiny, triangular, discrete stipules, on or near petiole. Older stems robust, often twisted, stem x-s pattern lobed to compound-lobed/divided – with multiple lobed cylinders.

Leaves: Elliptic to ovate, base of leaf never heart-shaped; hairless or lightly hairy below; **glands often present at leaf base, on petiole and in rows along leaf blade margin.**

Inflorescences: Axillary, panicles or cymes with **4-flowered clusters** [*umbels*]; flower stalk bracts and bracteoles present, without glands; calyx 4 + 1, 8(0)-glandular; corolla 4 + 1, **petals yellow, pink, or white, hairy or hairless;** stamens 10, anthers dissimilar; styles hairy or hairless, spreading; stigmas strictly terminal, sometimes head-like.

Fruits: samaras with a **large dorsal wing, thickened at upper margin,** lower margin thin, in some species reduced to crest; nut smooth, veined, or with 1-many lateral winglets.

Ecology: Bee-pollinated; Seed dispersal via wind or water.

Distribution: Widespread in the neotropics, extending to Argentina and Mexico: ca. 118 species; Guianas: 6 species, 2 species described here.

Notes: Within the Malpighiaceae, this genus is most similar to Bronwenia and Diplopterys.

2.1. Banisteriopsis martiniana (A.Juss.) Cuatrec.

Common name: mabudehi (Ar)

Woody liana or scandent shrub; slightly hairy. **Stems:** branchlets with small brown hairs and lenticels; older stems round, hairless, dark reddish-brown with conspicuous dark lenticels. **Leaves:** ovate to elliptic, 5-13 x 2-7 cm, base obtuse to subcordate, apex obtuse to acuminate, margin flat, not recurved; leathery; hairs few; basilaminar and marginal cup-like or flat glands present; petiole 0.4-2 cm long. **Inflorescences:** to 40 cm long, branches minutely reddish-brown hairy; flower stalks 0.6-1.4 cm long; flowers small, petals to 0.95 cm long, yellow, toothed, upright 'flag' petal with gland-tipped teeth, hairless; styles hairless. **Fruits:** samaras straight, with dorsal wing semi-obovate, slender, to 3.2 x 1.5 cm; nut small and smooth, without ridges, crests, or lateral winglets; hairs irritating.

Uncommon, locally abundant; old-growth non-flooded forest, savanna, mountain xerophytic forest and scrub on sandstone. Guianas; also Venezuela, Brazil, and Trinidad.

2.2. Banisteriopsis muricata (Cav.) Cuatrec

Woody liana or vining shrub; with dense, white appressed hairs [*sericeous or velutinous*]. **Stems:** branchlets flattened, hairy; older stems round, bark grey-brown to dark brown with conspicuous dark lenticels. **Leaves:** ovate, elliptic or rotund, 5-13 x 5-8 cm, base cuneate to cordate, margin flat; hairs few above, often dense and white-shining below; basilaminar gland pairs present (on petiole) or absent; petiole 0.5-1.6 cm long, hairy. **Inflorescences:** rachis white hairy; flower stalks 0.3-1.2 cm long; flowers small, petals to 0.85 cm long, pink, upright 'flag' petal with yellow basal half, fringed with glands at base; ovary hairy, styles slender and diverging; stigmas head-like. **Fruits:** samaras straight, hairy, with dorsal wing 2-3.4 cm long x 1-1.6 cm wide, nut hairy, without ridges, crests, or winglets. **Notes: The most widely-distributed and variable species in the genus.** Guiana Shield specimens are densely white-hairy compared to the sparser golden or silver hairs of specimens from other areas (Bronwen 1982).

Common; old-growth forest and open woodlands. Guianas; also widespread in the neotropics, extending to Argentina and Mexico.

3. Bronwenia W.R.Anderson & C.Davis

Woody lianas or sprawling shrubs. Climbing via twining shoots. Neotropics: 10 species; Guianas: 3 species, one described here.

3.1. Bronwenia wurdackii (B.Gates) W.R.Anderson & C.Davis

[Syn: Banisteriopsis wurdackii B. Gates]

Woody liana; hairy [*sericeous*]. **Stems:** branchlets round, to 0.5 cm diameter, pale to dark brown with pale raised lenticels; nodes with tiny, triangular, discrete stipules near petiole; older stems round, with fibrous bark; stem x-s pattern with a few to many "cables". **Leaves:** variable, 9-22 x 4.5-13 cm, lanceolate to broadly elliptic, base acute to obtuse, apex acuminate, margin ± curled; hairless above, lightly hairy below; **large gland-pairs present at leaf base, small flat glands present along margins below;** 2° veins raised below; petiole grooved, lightly hairy. **Inflorescences:** axillary or terminal, short, dense racemes of 6–45 flowers; bracts and bracteoles without glands, persistent; in bud, sepals not covering petals; flowers small; calyx 4 + 1, 8(0)-glandular, **oil glands attached below free part of sepals;** corolla 4 + 1, **petals yellow, hairless,** 'flag' petal erect, 0.65 cm long, margins short-frilly; stamens 10; styles 3, distinct, ± alike; stigmas terminal, truncate or capitate. **Fruits: dorsal wing of samara thickened on the outer edge** with the veins bending toward the thinner inner edge, 2.4-3.8 cm long x 1-1.5 cm wide, narrowly oblong, with appendage; **nut of samara globose, smooth or bearing a single ridge or winglet** on each side; nut hairless within center. **Notes:** Bronwenia is a member of the very large Stigmaphyllon group. Similar taxa include: Banisteriopsis, Diplopterys, and Stigmaphyllon.

Common. Old-growth forest at low elevations; Guianas; also widespread from N South America to Bolivia and C America.

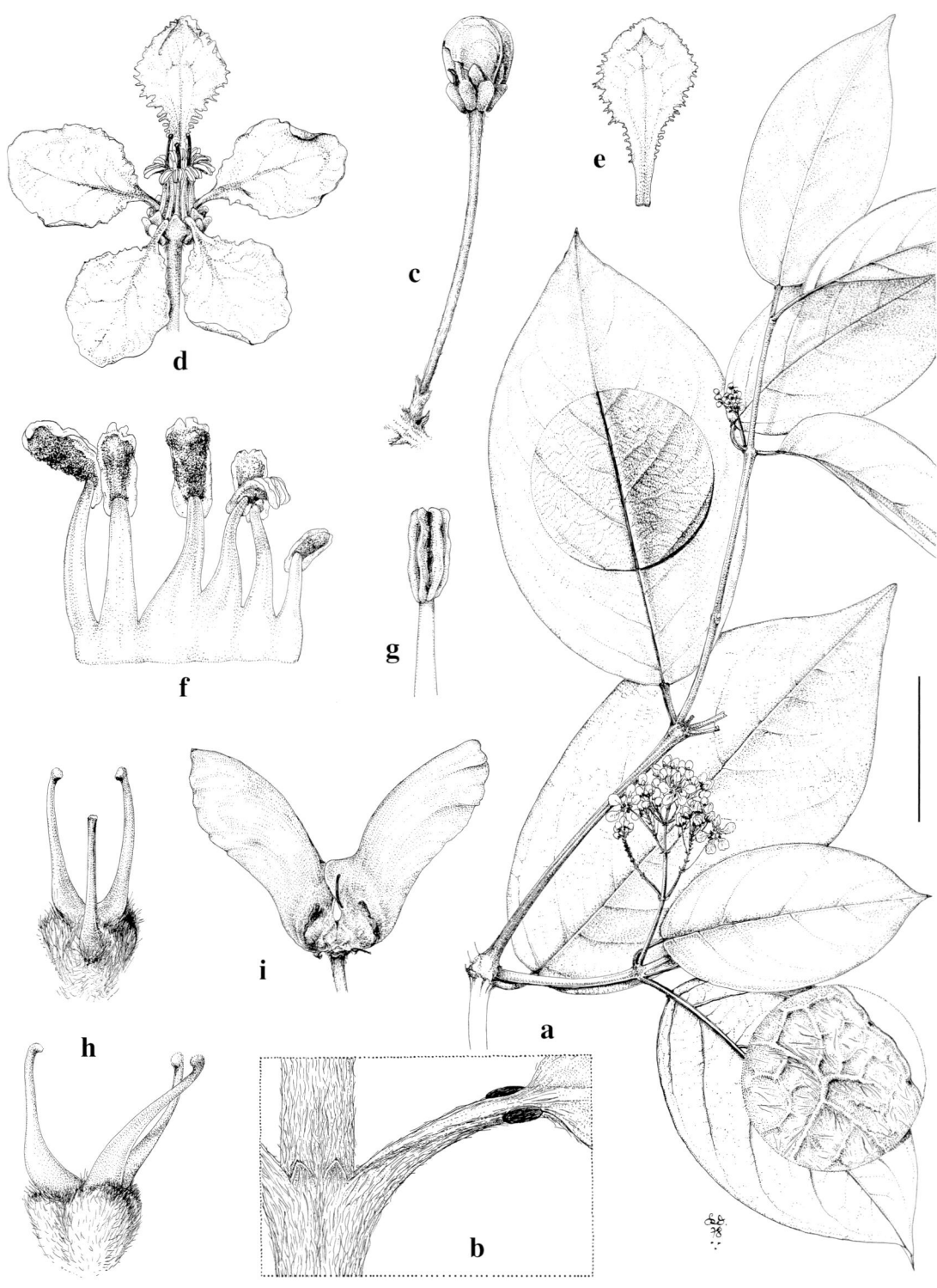

6. Bronwenia wurdackii illustration. a. Flowering branch, with enlargement of upper and lower surfaces. b. A node, showing interpetiolar stipules and large gland pair on petiole. c. Flower bud. d. Flower, 'unique' fifth petal uppermost. e. Fifth petal. f. Stamens. g. Anther. h. Ovary with three styles, one dissimilar. i. Fruit with two samaras developed. Scale bars: a, 4 cm (upper circle 2 cm, lower circle 4 mm); b, 4 mm; c–e, 5.7 mm; f–h, 2 mm; i, 2.7 cm. [Drawn by Karin Douthit, reproduced with permission of the University of Michigan Herbarium]

4. Carolus W.R.Anderson

Woody lianas. Climbing via twining shoots. Mexico, Central America, the Lesser Antilles, and South America: 6 species; Guianas: 1 species.

4.1. Carolus sinemariensis (Aubl.) W.R.Anderson

[Syn: Mascagnia sinemarensis (Aubl.) Griseb.]

Robust woody liana; often hairy [*sericeous, tomentose*] when young. **Stems:** branchlets round, to 0.5 cm diameter; nodes with tiny, triangular, discrete stipules; older stem round, with fibrous bark; stem x-s pattern lobed or compound-lobed/ divided. **Leaves:** variable, **2–4(6) small glands present at petiole apex** (or absent); **small flat glands present along blade margin;** petiole grooved, glandular. **Inflorescences:** axillary or terminal, compact, unbranched or paniculate; flower stalk bracts and bracteoles without glands; sepals not covering petals in bud; flowers small; calyx 4 + 1, 8(0)-glandular, **oil glands attached on receptacle below free part of the sepals; corolla** 4 + 1 or 5, **petals yellow,** densely hairy [*sericeous*]; 'flag' petal erect, toothed or frilly; stamens 10, of different lengths; styles 3, erect or spreading, ± alike, dorsally rounded, truncate, or acute. **Fruits:** samaras most commonly butterfly-shaped. In the Guianas and adjacent Brazil, the **lateral wings are transformed into two long, narrow, swept-back projections (French Guiana) or are reduced to corky outgrowths on an enlarged nut (Guyana and Brazil).** These are most likely adaptations for water dispersal. **Ecology:** Bee-pollinated; Seed dispersal via water (in the Guianas).

Distribution: Uncommon. Old growth forest; seasonally-flooded forest; riparian forest. Fr. Guiana and Guyana; also widespread from S Brazil to S Mexico and the Lesser Antilles.

5. Christianella W.R.Anderson

Woody vines or shrubs in open habitats. Climbing via twining shoots. Widespread in the neotropics, from Brazil to Mexico: 5 species. Guianas: 1 species.

5.1. Christianella surinamensis (Kosterm.) W.R.Anderson

[Syn: Mascagnia surinamensis Kosterm.]

Woody liana or climbing shrub; most surfaces with dense, flattened hairs [*sericeous, tomentose, wooly, velvety*]. **Stems:** branchlets round to flattened, hairy; nodes with tiny, triangular, discrete stipules born on petiole near base; older stem not seen. **Leaves:** size and shape variable; **large to small flat glands present in two rows on petioles and along leaf margin** (or absent). **Inflorescences:** terminal or lateral, single or in panicles, densely hairy; raised or stalked glands present on bracts and bracteoles (or absent); sepals not concealing petals in bud; calyx 4 + 1 with 8 oil glands, **all 5 sepals bearing a row of un-stalked to long-stalked marginal glands;** corolla 4 + 1, **petals yellow, densely hairy outside,** hairless within, 'flag' petal with a longer and thicker claw than others, fringed; 10 stamens dissimilar (of different lengths), apex of styles truncate to short-hooked [*stigma internal*]. **Fruits: samaras disc- to butterfly-shaped,** lateral wings papery, net-veined, with stiff hairs and wavy or toothed margin; dorsal wing extending forward at apex through opening in lateral wing, intermediate winglets mostly lacking.

Uncommon, locally abundant. Non-flooded old-growth forest, shrubby savannas. Guyana & Suriname; also adjoining Amazonian Brazil (Pará, Amapá), Venezuela, Peru and Bolivia.

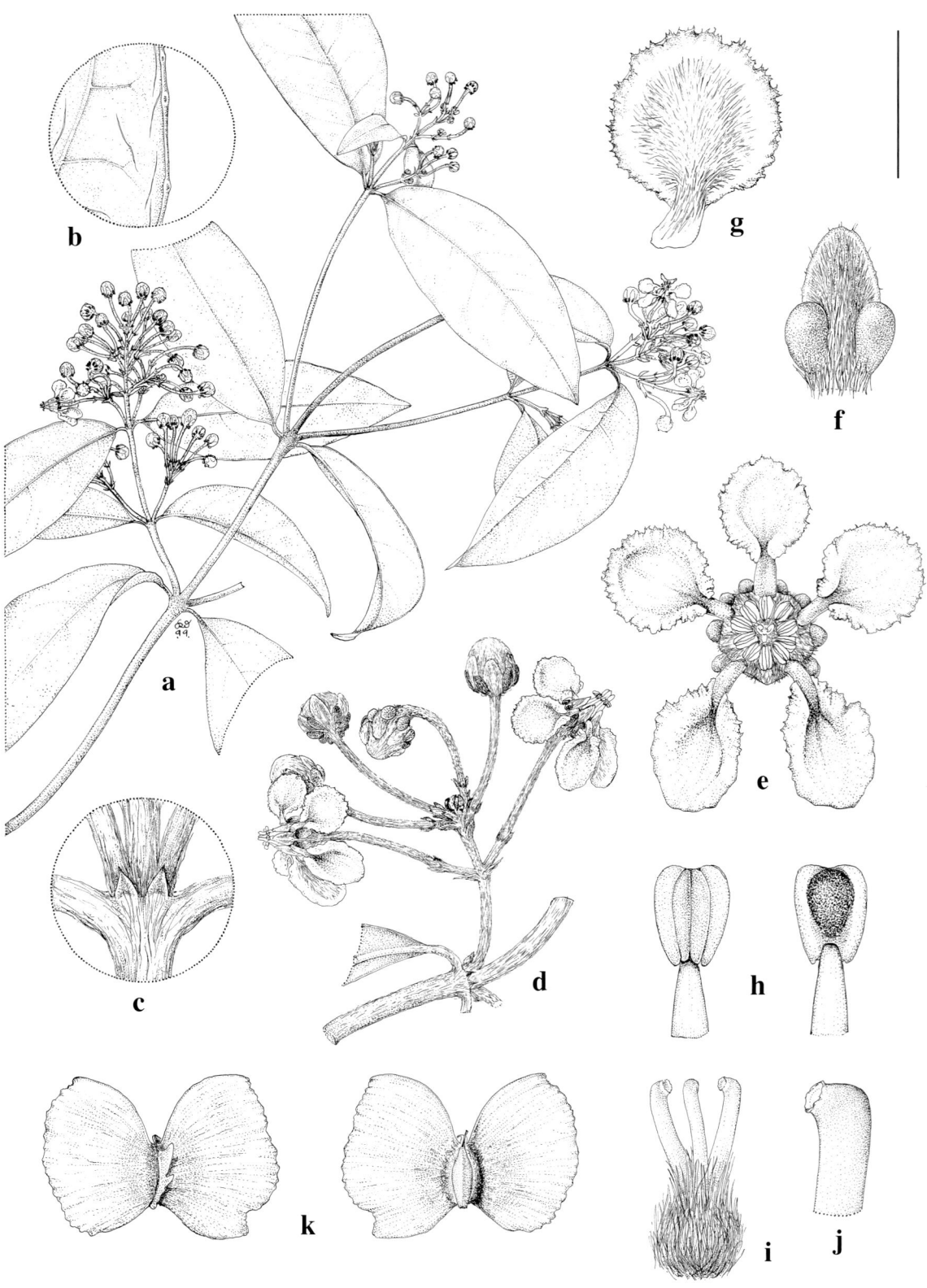

7.   Carolus sinemariensis illustration. a. Flowering branch. b. Leaf margin below. c. Node with stipules. d. Axillary inflorescence. e. Flower, unique fifth petal upright. f. Lateral sepal with oil glands, outer side. g. Lateral petal. h. Anthers. i. Ovary. j. Apex of style (internal stigma). k. Samaras, outer side (left), inner side (right). Scale bars: a, 4 cm; b, c, 4 mm; d, 1.3 cm; e, 7 mm; f, 2.7 mm; g, 4 mm; h, 2 mm; i, 2.7 mm; j, 1.3 mm; k, 2 cm. [Drawn by Karin Douthit, reproduced with permission of the University of Michigan Herbarium]

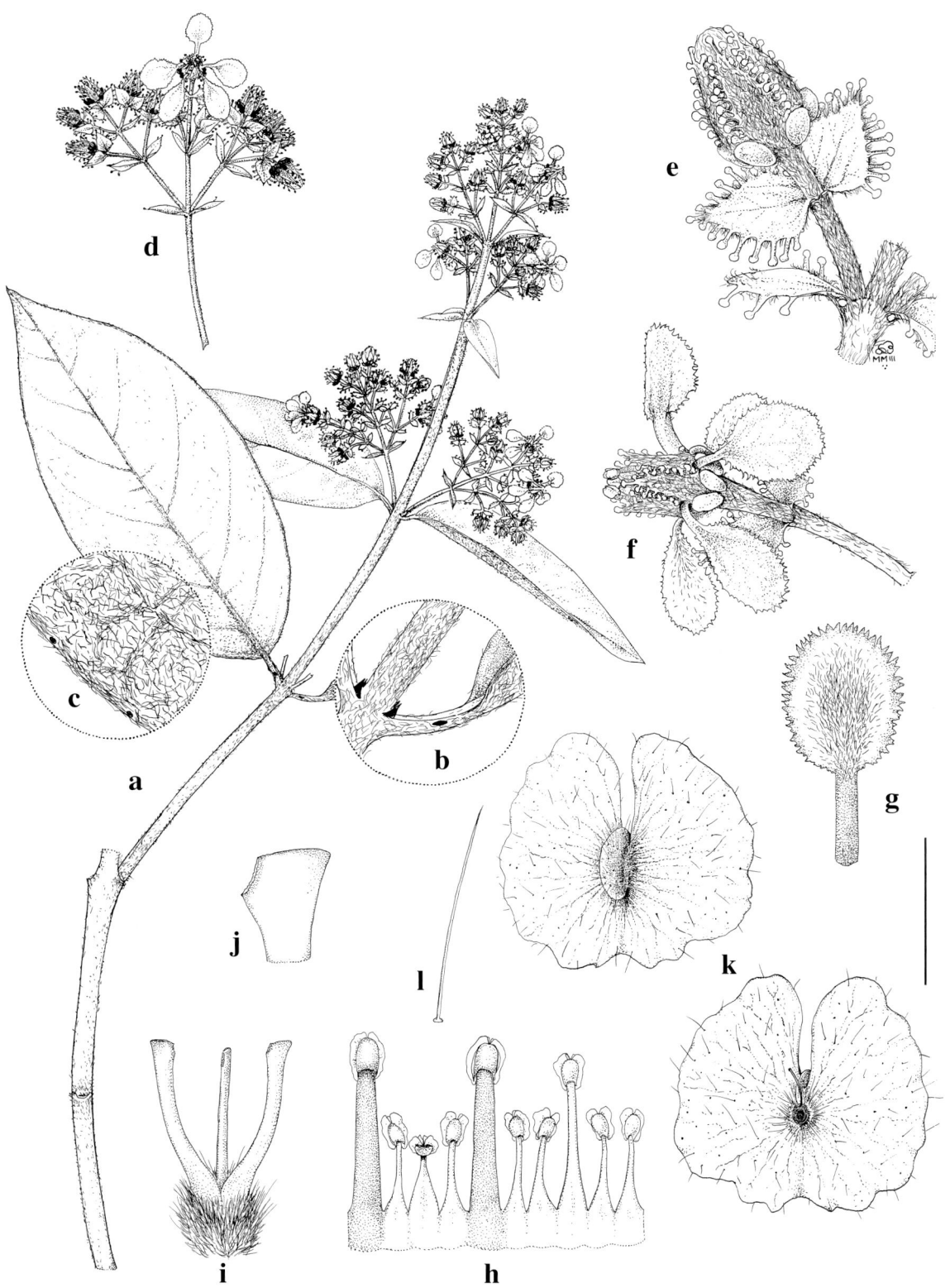

8. *Christianella surinamensis* illustration. a. Flowering branch. b. Node with stipules and gland on petiole. c. Edge of leaf below - marginal glands. d. Portion of inflorescence. e. Flower bud with glandular bracteoles and basal bract. f. Flower, unique fifth petal bent upwards. g. Fifth petal, outer side. h. Stamens of different sizes. i. Ovary with dissimilar styles. j. Apex of style (internal stigma). k. Samara. l. Single hair from samara. Scale bars: a, 4 cm; b, 1 cm; c, 4 mm; d, 2 cm; e, 4 mm; f, 8 mm; g, 4 mm; h, i, 2.7 mm; j, 0.8 mm; k, 2 cm; l, 2 mm. [Drawn by Karin Douthit, reproduced with permission of the University of Michigan Herbarium]

6. Diplopterys A.Juss.

Robust woody lianas or twining shrubs. Climbing via twining shoots.

General: Surfaces often with flattened hairs [*sericeous, tomentose, wooly, velvety*].

Stems: Branchlets round, green, hairy; nodes with small, triangular, discrete stipules. Older stems large, > 20 cm diameter, bark fibrous, stem x-s pattern variable, simple, lobed or divided.

Leaves: Size and shape variable; **large gland-pairs present (or absent) at petiole apex, small flat glands present (or absent) along leaf margins.**

Inflorescences: Axillary 4-flowered umbels; bracts and bracteoles without glands; calyx 4 + 1, 8(0), glands attached below free part of sepals corolla 4 + 1, **petals yellow, long-clawed, densely hairy outside,** hairless within; stamens 10, similar, extending beyond corolla mouth; styles 3, dissimilar, stigmas truncate to head-like.

Fruits: Samaras with **conspicuous dorsal wing, thickened on upper edge,** often with small lateral winglets; dorsal wing sometimes reduced to 3-7 crest-like lateral wings; nut smooth-sided, wrinkled, ribbed.

Ecology: Flowers often produced when plants are leafless. Seed dispersal via wind or water; often occurring along waterways.

Distribution: Widespread in the neotropics: 31 species; Guianas: 3 species, 2 presented here.

Notes: Closely related to Banisteriopsis and Bronwenia.

6.1. Diplopterys cristata (Griseb.) W.R.Anderson & C.Davis
[Syn: Banisteriopsis cristata (Griseb.) Cuatrec.

Woody vine or climbing shrub. Stems: older stems medium-sized, round; bark with dark brown corky ridges. **Leaves:** ovate to almost round, 2.7-9(-12) x 4-8 cm, base cuneate to obtuse; margin flat; glands present at petiole apex, absent on leaf margin; petiole 0.5-1.5 cm long. **Inflorescences:** calyx glands absent; petals hairless; styles hairy. **Fruits:** samaras with dorsal wing 2.5-3.4 x 0.9-1.1 cm; nut bearing 3–7 crestlike lateral wings on each side.

Uncommon; savannah and deciduous forest. Guyana, Suriname; also Venezuela.

6.2. Diplopterys lucida (Rich.) W.R.Anderson & C.Davis
[Syn: Banisteriopsis lucida (L.C. Rich.) Small]

Common name: karinama (Tr)

Robust liana. Stem: older woody stem large, > 20 cm diameter, twisted; bark fibrous; stem x-s pattern lobed to compound-lobed, with multiple distinct cylinders. **Leaves:** elliptic, 6-13 x 3-6 cm, base cuneate, apex short- to long-acuminate, margin ± curled; sparsely hairy below; **large, sunken gland-pairs present on petiole; small flat glands present on leaf margin;** petiole 0.4-1.0 cm long. **Inflorescences:** calyx 8-glandular, densely golden-hairy outside; petals densely hairy [*sericeous*] outside; styles hairy. **Fruits:** dorsal wing well-developed, obovate, 5.5 long x 2 cm wide, with many tiny lateral winglets and ridges, yellow.

Common; old-growth non-flooded, seasonally-flooded, and riparian forest. Guianas; also Amazon basin from Bolivia and E Brazil to Colombia.

---

9. Heteropterys macrostachya with sunken petiole and leaf glands [BH]

10. Heteropterys macrostachya in flower with reddish-brown leaf undersides [BH]

11. Heteropterys macrostachya winged samara [BH]

10

11

7. Excentradenia W.R.Anderson

Woody lianas. Climbing via twining shoots. N South America: 4 species; Guianas: 2 species, 1 presented here. Rarely collected genus.

7.1. Excentradenia propinqua (W.R.Anderson) W.R.Anderson

[Syn: Hiraea propinqua W.R.Anderson]

Woody liana; many surfaces lightly to densely hairy [*sericeous, farinose*]. **Stems:** branchlets round, bark hairless or finely powdery [*farinose*], smooth or with small lenticels; nodes without horizontal ridge or line; stipules small, triangular, discrete, on stem or at petiole base; older stems not seen. **Leaves:** sometimes sub-opposite, elliptic, broadly ovate, or ± orbicular, base sometimes extending onto the petiole [*decurrent*]; leathery to subleathery, rough, hairless or sericeous below; **large gland-pair mounds present on upper petiole; small flat glands often along margins, never scattered on blade;** 1° and 2° veins prominent below, **3° veins conspicuously parallel** (slanted or at 90° to the midvein). **Inflorescences:** axillary, ultimate units 4-flowered umbels; flower stalk bracteoles (1 of each pair) glandular; calyx 4 + 1, 8(0)-glandular; corolla 4 + 1, **yellow,** hairless, 'flag' petal fringed and

with tiny glands; stamens 10, anthers ± uniform; ovary hairy; styles dissimilar. **Fruits:** samaras with **large, papery, disc-shaped, lateral wings,** approximately 5 cm diameter, round below, notched deeply to the seed body at apex; dorsal wing small; intermediate winglets absent; wavy at margin, with very fine, white appressed hairs: dorsal wing 0.4 cm wide, 0.6 cm high, entire or with coarse teeth. **Ecology:** Seed dispersal via wind. **Notes:** Excentradenia is closely related to Hiraea (parallel 3° veins, marginal leaf glands, umbels as terminal flower units, ± separate carpels, and papery disc-like lateral wings). Hiraea is distinguished by elongated stipules that clasp the petiole and terminate as a distinctive "forked tongue" on the upper petiole.

Uncommon (less than 5 collections). Old-growth forest, riparian forest; Guianas only. Excentradenia adenophora collected in the coastal plain of Guyana and in eastern Venezuela.

8. Heteropterys H.B.K.

Woody canopy lianas; also shrubs and trees. Climbing via twining shoots.

Stems: Branchlets round or flattened at nodes, often bearing lenticels; nodes without horizontal ridge, stipules small, triangular, discrete, near petiole base. **Older stems often robust (> 20 cm diameter), composed of a single column and not distinct cables;** bark fibrous and thick, light brown, often bearing lenticels; stem x-s pattern simple or lobed.

Leaves: Variable in size and shape, **base never heart-shaped;** thin to leathery; hairless or persistently hairy below; 3° veins net-veined (not parallel); **large gland-pair present at petiole apex**

(or absent), **small flat glands present along leaf margins** (or absent).

Inflorescences: Axillary or terminal, with compound racemes, corymbs, or umbels; flower stalk bracts and bracteoles with or without glands; **calyx 8-10 glandular (with 4-5 bi-glandular sepals) or glands absent, sepals completely concealing the petals in bud,** reflexed (or straight in H. macro-stachya) at maturity; corolla 4 + 1, with a unique 'flag' petal, **petals yellow, hairless;** stamens 10, uniform; carpels ± fused, styles 3, alike [*stigmas internal, with apex variable*].

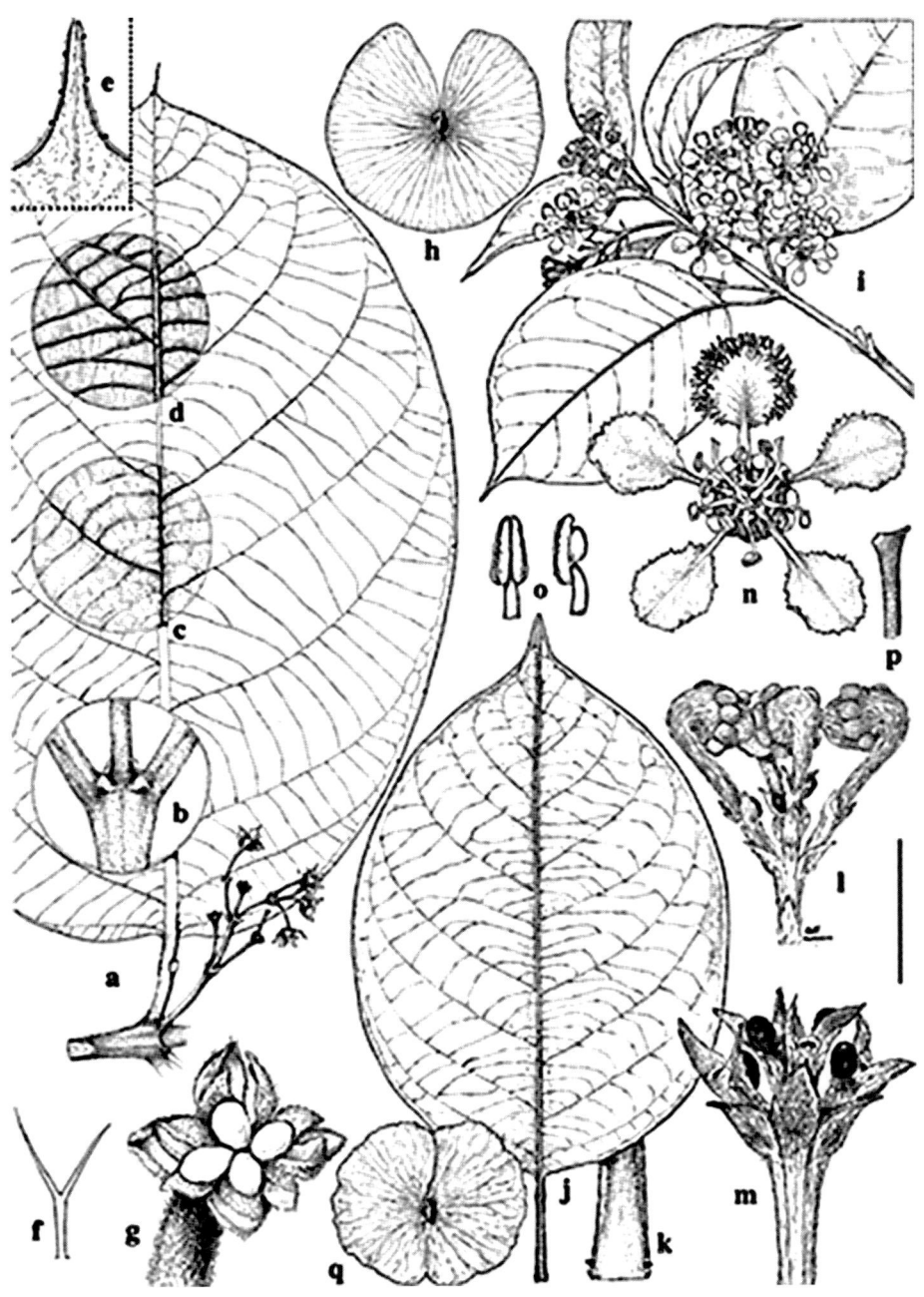

12. Excentradenia. a–h, E. primaeva. a. Leaf and old infructescence. b. Node with stipules. c. Detail of leaf blade underside, minus hairs. d. Detail of leaf blade underside, minus hairs. e. Leaf apex enlarged to show marginal glands. f. Hair from lower surface of leaf blade. g. Bracts and bracteoles of umbel, with unique glands on four bracteoles. h. Samara, outer side. i–p, E. adenophora. i. Flowering branch. j. Detached large leaf, inner side view. k. Base of petiole with stipules. l. Umbel of four circinate flower buds. m. Base of umbel enlarged to show unique glands on four bracteoles. n. Flower, unique fifth petal uppermost (in nature the stamens would be less spreading, probably nearly erect). o. Anthers, outside view (left) and side view (right). p. Distal portion of style. q, E. propinqua. q. Samara, inner side. Scale bar equivalents: a, 4 cm; b, 2 cm; c, d, 4 cm; e, 8 mm; f, 0.4 mm; g, 2.7 mm; h–j, 4 cm; k, l, 8 mm; m, 4 mm; n, 8 mm; o, p, 2 mm; q, 4 cm. [Drawn by Karin Douthit, reproduced with permission of the University of Michigan Herbarium]

Fruits: **Large dorsal wing unique – thickened lower margin, thin upper margin** (sometimes reduced); nut with small crests or smooth.

Ecology: Seed dispersal via wind or water.

Distribution: Widespread in the neotropics; also the West Indies and 1 species in W Africa: ± 25 species; Guianas: 13 species, 4 presented here.

Notes: The fruits are especially important for species identification in this genus.

8.1. Heteropterys leona (Cav.) Exell

[Syn: Heteropterys multiflora (DC.) Hochr.]

Woody liana; shrub or tree; surfaces mostly **reddish-brown hairy** *[sericeous]*. **Stems:** branchlets flattened, hairy; stipules tiny or absent; older stems round, to 10 cm diameter, bark furrowed, pinkish-brown, bearing lenticels. **Leaves:** variable shape, large, to 38 cm x 20 cm, base obtuse or rounded; leathery; large gland-pair absent, **small flat glands and tiny blisters or nipple-like projections** *[papillae]* **present below;** 1° and 2° veins strongly raised below, 3° veins parallel or net-like; petiole to 1.5 cm long. **Inflorescences:** 20-40-flowered hairy racemes, calyx with **oil glands 8(0), sepals reflexed. Fruits:** samaras hairy; dorsal wing semi-circular, thick-leathery, to 3.5 x 2.5 cm, veins fan-like, upper margin ± round, straight or curved; nut to 2 cm diameter. Water dispersed.

Common; especially in riparian forest. Guianas; also N South America, Jamaica, W Africa.

8.2. Heteropterys macradena (DC.) W.R.Anderson

High-climbing woody liana; shrub or treelet; surfaces **mostly lightly hairy. Stems:** branchlets round, grey, with many lenticels. **Leaves:** highly variable, oval or elliptic, 6-15 x 3-8 cm, base cuneate or rounded; conspicuously leathery; **large gland-pair absent, small flat glands present below, in submarginal rows;** 3° veins prominent above and below, net-veined; petiole 0.4-1 cm long. **Inflorescences:** paniculate, to 5 cm long; bracts and bracteoles without glands; calyx with **oil glands 8(0), sepals reflexed. Fruits:** samaras green to pink; dorsal wing obovate or curved, 2.5-4 x 1-1.4 cm, lower edge bent upward ± abruptly beyond nut; nut to 1 cm diameter.

Uncommon; non-flooded or seasonally-flooded riverbank forest and rapids. Guianas; also Amazon basin.

8.3. Heteropterys macrostachya A.Juss.

Common name: wetibaka (Pa)

Liana or climbing shrub; silver or golden-yellow hairy *[sericeous]*. **Stems:** branchlets hairy; older stem bark brown with scattered horizontal lenticels and green sub-bark layer; inner bark orangish-yellow; wood white. **Leaves:** variable, 10-18 x 5-11 cm, base rounded; conspicuously leathery; densely hairy below, **large gland-pair present in upper half of petiole, small flat glands present below, in submarginal rows;** veins slightly raised above and below, 3° veins parallel or net-like; petiole 0.8-1.3 cm long, glandular. **Inflorescences:** narrow panicles or racemes of 4-6-flowered umbels; calyx oil glands 8(0), **sepals erect. Fruits:** samaras hairy; dorsal wing straight or ± curved, 2.5-6 cm x 1.2-3 cm, thick, pointing upwards, basally reddish-brown; nuts to 1 cm diameter, reddish-brown, hairless.

Uncommon; riparian and non-flooded old-growth forest. Guianas; widespread in the neotropics, from Bolivia to S Mexico.

MALPIGHIACEAE

8.4. Heteropterys nervosa A.Juss.

Common name: huriaballi (Ar)

Liana or climbing shrub; reddish-brown hairy [*sericeous*] **or hairless. Stems:** twigs flattened, hairy; older stems with rough, silver-grey bark, deeply furrowed, with tubercles. **Leaves:** variable, 7-11 x 4-6 cm, base obtuse or rounded; leathery; hairless and shiny above and below, **large gland-pairs absent, small flat glands present;** 3° veins net-like; petiole hairless, grooved above, 0.6-2.5 cm long. **Inflorescences:** 1-3 pairs of flowers below terminal 4-6-flowered umbels, flowers 0.5 cm diameter; calyx oil **glands 8(0), sepals reflexed. Fruits:** samara hairy; dorsal wing ± pear-shaped to 4.5 x 1.5 cm, veined, weakly curved, pointing upwards; 0.6 x 0.2 cm; nuts 0.7 x 0.4 cm, ± hairless.

Common; riparian forest, swamps, and savanna forest, also on granite. Guianas; N South America and the West Indies.

9. Hiraea Jacq.

Woody canopy lianas, woody vines or sprawling shrubs. Climbing via twining shoots.

Stems: Branchlets round, often hairy, interpetiolar ridge present; **nodes with conspicuous fused stipules that extend onto the petiole and terminate as a 'forked tongue' or teeth.** Older stems robust, round, bark fibrous and thick, with lenticels; **stem x-s pattern lobed** (not divided with multiple cables).

Leaves: elliptic to ovate, **base never cordate,** hairless to persistently hairy below; tiny marginal glands often present; 3° veins often conspicuously parallel, 90° to midvein; petiole of standard length, often with a raised gland pair above the mid-point.

Inflorescences: axillary, short, 1–several 4-flowered umbels in a condensed cyme (Guianas species); infl. stalk not well developed as in Mascagnia; bracts and bracteoles without glands; calyx 4 + 1, 8-glandular (or 0), sepals reflexed; corolla 4 + 1, **petals yellow, rarely red, hairless;** stamens 10, dissimilar; styles 3, similar [*stigmas internal, apex dorsally rounded to hooked*].

Fruits: Schizocarps 3-(1-2)-parted; **samaras with 2 fan-like or butterfly-shaped lateral wings;** dorsal winglets present and small or absent.

Ecology: Seed dispersal via wind.

Distribution: Widespread in the neotropics, extending to Mexico and the West Indies: 47 species; Guianas: 8 species; 3 species described here.

9.1. Hiraea affinis Miq.

Common name: kawtitei (Sr)

Woody canopy liana. Stems: older stem round, to 9 cm diameter, specimen observed with large stems coiled on the forest floor; bark shiny grey, furrowed, sub-bark reddish, cracked along length, bearing lenticels. **Leaves:** variable in shape, 12-23 x 8-10 cm, base rounded or ± cordate; papery, ± hairless below, 3° veins finely parallel (0.2-0.5 cm apart); **small lamina glands absent below;** petioles 0.4-1.1 cm long; **raised gland-pair present at petiole apex; stipules visible above or below petiole midpoint,** only bases remaining in older leaves. **Inflorescences:** umbels 4-flowered, in vertical rows in axils; petals light yellow, thin; [*style with a short rounded dorsal hook*]. **Fruits:** lateral wings to 5 x 3 cm, thin, greenish-white; dorsal crest small or absent; nut globose, to 0.6 cm diameter, hairy.

Uncommon. Locally abundant; riparian and non-flooded old-growth forest. Guianas; also South American tropics, exc. Colombia.

9.2. Hiraea fagifolia (DC.) A.Juss.

Woody liana. Stems: branchlets round, bark grey or reddish, bearing lenticels; interpetiolar ridge very prominent. **Leaves:** variable, 8-16 x 3-7 cm, base rounded or subcordate; 3° veins finely parallel (0.05-0.15 cm apart); **small marginal glands present near apex;** petioles 0.6-1.5 cm long, **raised gland-pair present at petiole apex, stipules to 0.5 cm long, reaching within**

0.2 cm of petiole apex. Inflorescences: axillary, short, three 4-flowered umbels per cyme; petals yellow, thin [*style hooked*]. **Fruits:** lateral wings to 5 x 2.5 cm, mostly thin, sometimes leathery, strongly veined; dorsal crest short, minutely hairy; nuts globose, 0.4 cm diam., hairy.

Common; non-flooded old-growth forest, riparian forest. Guianas; widespread in the neotropics.

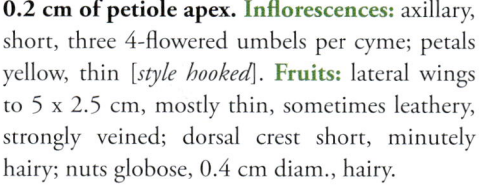

13. Heteropterys cristata, leafy branch with conspicuous white-green paired glands at base of reddish petioles [BH]
14. Heteropterys cristata, pink and white flowers with crested petals [BH]
15. Hiraea fagifolia leaves with 2-pronged stipule on petiole and circular, papery samaras [BH]
16. Hiraea fagifolia woody stem with silver bark and slash [BH]

MALPIGHIACEAE

9.3. Hiraea faginea (Sw.) Nied.

[Syn: H. chrysophylla A. Juss.]

Common name: aláta tátai (Sa); alata titei (Sr); alátu páu (Sa); awari parata (Tr); kurumpïnpïrïkï (Tr); pinya pau (Ma); tiriki (Ca); watra ingi (Sr).

Woody liana or climbing shrub. Leaves: obovate or elliptic, 9-19 x 3-8 cm, rounded or subcordate at base; **densely golden or silvery-hairy below;** 3° veins finely parallel (0.05-0.2 cm apart); **small marginal lamina glands present near apex;** petioles 0.4-1.1 cm long, **raised gland-pair present at petiole apex, stipules to 0.5 cm long, ending above the petiole midpoint. Inflorescences:** axillary, short, finely hairy, with three 4-flowered umbels per cyme; petals yellow, thick, flag petal glandular-margined [*technical: style hooked*]. **Fruits:** small lateral wings to 1.5 x 0.5 cm, mostly lobed and leathery, free at base and apex, sometimes completely divided into 2 (cf. Tetrapterys); dorsal crest almost entire, 0.15-0.5 cm long; nut 0.4-0.6 cm diameter, hairy. **Notes:** Vs. H. fagifolia - dense golden or silver hairs on leaves below, fruit with thick-leathery lobes and occurrence in watery habitats. **Use:** stems and/or leaves in genital steambaths, 'bush yaws' cure, health tonic, aphrodisiac.

Common; riparian forest. Guianas; N Brazil, Venezuela, Central America and the Lesser Antilles.

10. Jubelina A. Juss.

Woody lianas. Climbing via twining shoots. Neotropics: 6 species; Guianas: 2 species, one presented here.

10.1. Jubelina rosea (Miq.) Nied.

[Syn: Diplopterys rosea (Miq.) Nied.]

Common name: bomia titei (Pa); cat-claw liana (EN); kaikui amoi tëpu (Tr); kattepoot liaan (Sr);

Woody canopy liana; densely hairy. **Stems:** branchlets flattened, densely hairy; stipules small, triangular, discrete, near base of petiole; older stems round, robust, in twisted and furrowed single column; stem x-s pattern lobed. **Leaves:** broadly elliptic to round, 10-20 x 5-13 cm, base cuneate or truncate, **never cordate; yellowish-green-hairy [*velutinous*] on both sides;** large gland-pairs absent; **small flat glands in rows of along margins;** 3° veins parallel, 90° **to midvein;** petioles 0.7–2.1 cm long, grooved above. **Inflorescences:** axillary or terminal panicles, terminal units 4- or 6-flowered; infl. stalks bearing 2-4 umbels; bracteoles and bracts pink; calyx 4 + 1, **sepals pink, uni-glandular (4 glands per flower), reflexed;** petals 4 + 1, pink, hairy outside (flag petal hairless); stamens dissimilar; styles dissimilar [*stigmas internal, apex truncate or very short-hooked*]. **Fruits:** samaras with lateral wings disc-shaped, dorsal winglet present; deeply lobed winglets present between dorsal and lateral wings. **Ecology:** Seeds dispersed via wind.

Distribution: Uncommon. Non-flooded old-growth forest. Fr. Guiana, Suriname only.

11. Lophopterys A. Juss.

[Syn: Dolichopterys Kosterm.]

Lianas or climbing shrubs. Climbing via twining shoots.

General: Often conspicuously golden, brown or grey hairy [*sericeous, velutinous*].

Stems: Branchlets round, hairs dark brown, dark blonde or grey; stipules absent or inconspicuous, short-lived; older stem not seen.

Leaves: Obovate-oblong or elliptic (L. splendens), 20-35 cm long, base variable (never cordate); leathery; **dark-blonde hairs present below; leaf blade glands absent, petiole with 1-many small gland-pairs** (absent); 3° veins net-veined and raised above and below; petioles 1.5-6 cm long, thick, grooved.

Inflorescences: terminal or axillary panicles, compound, branchlets hairy; bracts and bracteoles on flower stalks 0.1-0.5 cm long; calyx 4+1, **sepals triangular, uni-glandular (4 glands per flower);** corolla 4+1, petals bright yellow, margins often fringed and glandular; stamens 10, dissimilar; styles spreading, of distinct heights; stigmas large and wider than tall.

Fruits: L. euryptera, L. surinamensis: with one short dorsal, squarish wing and two narrow forward-pointing lateral wings, each 3x longer than wide; L. splendens: with one short dorsal crest and the lateral wings reduced to ridges or absent.

Ecology: Seeds dispersed via wind or water.

Distribution: South America including the Guianas: 7 species; Guianas: 3 species.

Notes: Wind-dispersed Lophopterys have 1 oil gland per sepal and uniquely winged fruits.

11.1. Lophopterys euryptera Sandwith

Woody liana to 35 m long. Leaves: obovate-oblong, large, 16-35 x 10-23 cm, base obovate, truncate or short attenuate, apex truncate; leathery; **hairs densely interwoven and persistent below; petiole 2-6 cm long,** bearing 1-4 pairs of small glands or glands absent. **Fruits:** nut 0.9-1.1 cm diameter, hairy, smooth on sides; lateral wings 4.5–6.2 x 1–1.8 cm, narrowly elliptical and gradually tapering to tip; dorsal wing encircling nut, to 2 cm wide x 2.5 cm high.

Uncommon, rarely collected. Guyana (north and central); eastern Venezuela.

11.2. Lophopterys surinamensis (Kosterm.) Sandw.

[Syn: Dolichopterys surinamensis Kosterm.]

Woody liana. Leaves: obovate-oblong, 15-20 x 8-10 cm, base rounded, apex broadly rounded and deeply notched; leathery; **hairs loosely interwoven and persistent below; petiole 1.5-2 cm long,** bearing 1-3 small gland-pairs (sunken) on upper petiole (or absent). **Fruits:** nut 1-1.4 cm diameter, smooth on sides and between wings; lateral wings 4.8-6 x 0.65-1.1 cm; dorsal wing encircling nut, to 1.5 cm wide x 2.5 cm high.

Uncommon, rarely collected. Suriname only (Gran Lio, Upper Suriname River).

11.3. Lophopterys splendens A.Juss.

Woody liana. Leaves: elliptical to obovate, 16-27 x 5-13 cm, base cuneate, apex acuminate; leathery; **hairs tightly interwoven and persistent below; petiole 1.2-2.2 cm long,** bearing 1-5 small gland-pairs (sunken) in middle third of petiole (or absent). **Fruits: nut large, 1.5-2 cm diameter,** hairy, with parallel veins meeting at a ridge near apex, lateral wings absent; dorsal crest partially encircling nut (1/2-2/3).

Uncommon, rarely collected. French Guiana only (coastal forest to east).

12. Malpighiodes Niedenzu

Woody lianas. Climbing via twining shoots. N South America: 4 species; Guianas: 1 species.

12.1. Malpighiodes guianensis (W.R.Anderson) W.R.Anderson

[Syn: Mascagnia guianensis W.R.Anderson]

Woody liana; densely to lightly hairy [*tomentose*]. **Stems:** branchlets round, nodes with horizontal line; stipules tiny, triangular, discrete, borne on lower half of petiole [*epipetiolar*]; older stem not seen. **Leaves:** elliptic, 6-13 x 3-6 cm, base cuneate to rotund, apex rotund to apiculate, margin often curled; densely hairy above but lost with age, persistently and densely hairy below, hairs stalked; **1 large gland-pair present at petiole mid-point, small flat glands impressed in lower surface in 1-3 rows between midrib and margin;** petiole 0.8-2 cm long, hairy, glandular. **Inflorescences:** terminal or lateral, axes hairy,

ultimate branchlets in umbels of 3-4 flowers [*dichasia*]; flower stalk below joint well-developed, bracteoles large and oval, without glands; calyx 4 + 1, oil-glands 10, **all sepals bi-glandular, long and concealing petals in bud, recurved when flowers open;** corolla 4 + 1, **petals yellow, turning red in age,** hairless; stamens 10, of different lengths; styles 6, straight, dissimilar, stigma terminal. **Fruits: samaras with lateral wings disc-shaped,** each to 3.5 cm x 2.4 cm, round at base, deeply notched at apex, margin entire, wavy or few-toothed; dorsal wing small, 0.4-0.7 cm wide, free from lateral wing at base; intermediate winglets absent or tiny.

Locally common; non-flooded old-growth forest. Guianas.

13. Mascagnia (Bert. ex DC.) Colla

Woody lianas or shrubs. Climbing via twining shoots. Neotropics: 37 species; Guianas: 3 species, one presented here.

Notes: Mascagnia once included many species incorrectly assumed to be closely related due to similar disc-like winged fruits. Mascagnia is currently a more closely-related, smaller genus. Former Mascagnia species in this guide include Alicia macrodisca, Carolus sinemarensis, and Christianella surinamensis.

13.1. Mascagnia arenicola C.E.Anderson

[Misapplied name: Mascagnia sepium]

Robust canopy liana; densely hairy to lightly hairy [*sericeous*]. **Stems:** branchlets round, lenticels large; node with horizontal ridge, stipules small, narrowly triangular or linear, 0.5-0.7 cm long, discrete on stem or rarely fused across node [*interpetiolar*]; **older stems becoming enourmous,** ± round with length-wise furrows, **often**

a twisted column of multiple cables; bark thick and rough; stem x-s pattern lobed, compound-lobed, divided. **Leaves:** elliptic to ovate, 6.5-12.5 x 4-8 cm, base truncate to ± heart-shaped, apex apiculate or acuminate, margin curled; hairless or reddish-hairy below; **large gland-pairs (1-4) at petiole mid-point or apex/leaf base; small flat glands present on leaf underside, never on margin (or absent);** 3° veins net-veined; petiole 0.7-2 cm long, hairy. **Inflorescences:** terminal or axillary, 10-22-flowered racemes to 3 cm long; flowers 1-1.5 cm diameter; bracts/bracteoles without glands; calyx 4+1, oil glands 8(10), sepals ovate, not reflexed with age; corolla 4+1, **petals yellow, pink, white, or lilac;** stamens dissimilar (different heights); styles 3, slender to stout, dissimilar stigmas large. **Fruits:** samaras 3-4 cm diameter, **lateral wings ± disc-shaped,** base round, apex notched or divided into 2

wings; dorsal crest 0.8-1.5 cm long, 0.3-0.55 cm high; nut subglobose, to 0.45 cm high. **Ecology:** Seed dispersal via wind.

Uncommon, locally abundant; secondary forest on white sand, forest edges, swamp, and riparian forest. Guianas; also Amazonian Colombia and Brazil.

17. Malpighioides guianensis, leafy branch [BH]

18. Malpighioides guianensis, pink-winged samaras and soft leaves [BH]

19. Mascagnia divaricata, pink inflorescence [PT]

20. Mascagnia arenicola, young winged samaras [PT]

21. Mascagnia arenicola, robust canopy liana in old-growth forest, Brownsberg Nature Park, Suriname [BH]

14. Mezia Nied.

Woody vines; shrubs or small trees. Climbing via twining shoots. Notable for very large, globe-shaped bracts [*bracteoles*] enclosing flowers in bud. Samaras with lateral wings dominant, ± round, notched on one side, or butterfly-shaped with dorsal wing crest and winglets. South America: 10 species; Guianas: 2 species; 1 species described here.

14.1. Mezia includens (Benth.) Cuatrec.

Woody vines; hairs densely reddish- or brown-hairy [*sericeous*]. Stems: branchlets round, hairy; stipules tiny, triangular, borne on stem near petiole [*interpetiolar*], falling away early; older stem not seen. **Leaves:** elliptic to ovate, 13-20 x 6-10.5 cm, base cuneate to rounded, apex short-acuminate; hairy below, hairs lost with age; **large gland-pair present at leaf base** (or absent), **small flat glands absent;** petiole 1.5-2.7 cm long. **Inflorescences:** axillary or terminal, ultimate units in 4-flowered umbels, rachis with many leafy bracts, flower stalks unjointed, persistent, **globe-shaped bracteoles enclosing flower buds 0.6-1.1 cm long;** calyx 4+1, oil glands 8(10) or fused and 4(5); **sepals very narrow and hairy on underside,** not reflexed with age; corolla 4+1, **petals lemon-yellow, the 'flag' petal often red;** stamens dissimilar; styles 3, stout, dissimilar, stigma large. Samaras 7-10 x 5-9 cm, **lateral wings ± disc-shaped,** 3-4.5 cm wide, base round, apex notched, dorsal wing 0.7-1 cm wide, with a complex of crests and winglets. **Notes:** Mezia species are similar to Jubelina in flower and fruit characters. Jubelina species differ in tiny stipules located on the petiole (epipetiolar), flower stalks with a joint and bracteoles that do not enclose flower buds. Mezia angelica differs from Mezia includens in the smaller lateral wings, 2.5-3.1 cm wide.

Uncommon; non-flooded old-growth forest. Guianas; Amazon basin of S America.

15. Niedenzuella W.R.Anderson

Woody vines; sometimes shrubby. Climbing via twining shoots. Central and South America: 16 species; Guianas: 2 species; one described here.

15.1. Niedenzuella acutifolia (Cav.) W.R.Anderson

[Syn: Tetrapterys acutifolia Cav.]

Woody liana; light to densely hairy [*sericeous*]. **Stem:** branchlets flattened, slender, yellow- or grey-haired when young; stipules tiny, triangular, discrete, borne on the petiole base; older stems round, smooth, dark-violet, nearly hairless; stem x-s pattern lobed. **Leaves:** elliptic to ovate, to 8-13 x 3-5.5 cm, base obtuse or rounded, apex acute or long-acuminate; leathery, glossy dark green above, light green below; **1-2 small gland pairs often present at or above the petiole midpoint,** small flat glands present or absent on leaf margins; petiole 0.4-0.7 cm long, slender and channeled. **Inflorescences:** (sub)terminal panicles, 4-6 cm long racemes; flowers 1 cm diameter; flower stalks below joint very short, with 1 basal bract and 2 bracteoles; calyx 4+1, 8-glandular, hairy outside, **sepals long, strongly recurved, concealing petals in bud;** corolla 4+1, **petals yellow, thin, hairy outside;** stamens 10, (dis)similar; ovary with styles stout, straight or spreading. **Fruits:** samaras with **four lateral wings, ± equal, forming an 'X',** leathery, ovate or oblong, 1-1.25 x 0.4-0.7 cm, dorsal crest ring-like, 0.2-0.4 cm high; with a row of bumpy or wing-like appendages present between the crest and main wings; nuts ovoid, 0.4 cm long, hairy.

Notes: This species was once in the genus Tetrapterys due to its similar, 4-winged samaras. One key identification character is gland pairs near the petiole mid-point. The second species occurring in the Guianas is Niedenzuella poeppigiana (A. Juss.) W.R.Anderson, which was formerly in the Mascagnia genus.

Uncommon; especially in riparian forest. Guianas; also N Brazil.

16. Stigmaphyllon A. Juss.

Woody lianas, climbing shrubs, and herbaceous vines, often low-climbing; rarely erect shrubs. Climbing via twining shoots.

Common name: konkonikasaba titei (Sr); matukru (Tr), matukruimë (Tr)

General: Surfaces hairless to densely hairy [*appressed-sericeous*] and shiny; **some species producing large underground tubers that appear similar to cassava tubers (image no. 26).**

Stems: Branchlets round or flattened, hairy, bearing lenticels; node often with horizontal ridge or line, stipules small, discrete, interpetiolar, sometimes inconspicuous as 2 bumps between petioles. **Older stems round or 2- to multi-cabled, stem x-s pattern lobed, compound or divided.**

Leaves: Variable in size and shape, lanceolate to round, **base sometimes heart-shaped [*cordate*] or ear-shaped [*auriculate*];** margins entire or with small rounded teeth [*crenate*]; often hairy and shiny below; **large gland-pairs often present at petiole apex** (stalked, unstalked, or flat); small flat glands present or absent along margins; **petioles exceptionally long, up to 13 cm,** sometimes regular (S. banisterioides).

Inflorescences: Terminal or axillary, evenly-branched [*dichotomous*], ultimate units corymbs or umbels; flower stalk below joint [*peduncle*] well-developed; calyx 4+1, oil glands 8 or absent; corolla 4+1, **petals yellow, yellow-red, or orange-red,** margins often frilly or toothed; stamens of different lengths; **styles 3, each bearing a leafy appendage** (exc. S. banisterioides).

Fruits: Samaras producing a large dorsal wing, papery and elongate, 3-7 cm long, **upper margin thick, lower margin thin** (sometimes reduced or absent in water-dispersed species); nut ovoid, surface bumpy and veined, with 1 to many winglets or crests, or smooth.

Ecology: Commonly found in open, disturbed areas, secondary forest, riparian forest, coastal lowlands; Seeds dispersed via wind or water.

Distribution: Widespread in the neotropics: 92 species; Guianas: 5 species, 3 presented here.

16.1. Stigmaphyllon bannisterioides (L.) C.E.Anderson

Woody liana or herbaceous vine; small shrub; lightly hairy. **Leaves:** variable in shape, 4-13 x 1.2-5.5 cm, base attenuate or truncate; **margin entire; lightly hairy below;** large gland-pairs present (flat), small flat glands absent; **petioles short,** to 2 cm long. **Inflorescences:** axillary, 4-flowered umbels; petals yellow with red, hairless; **stigmas lacking leafy appendages. Fruits:** dorsal wing highly reduced, 0.5-0.6 cm long; nut enlarged, to 1.1 cm diameter, subglobose, with lateral, warty ridges. **Notes:** Seeds water-dispersed (hence reduced dorsal wing).

Common; brackish swamps, coastal ridges, river deltas. Guianas; widespread in the American tropics; also known in western Africa.

16.2. Stigmaphyllon convolvulifolium (Cav.) A.Juss.

Common name: akuri i'kererï (Ca); fungu-ansi titei (Sa); foengoe-mira titei (Sr); konkoni kasaba-titei (Sr)

High-climbing liana; lightly hairy. **Stems:** branchlets hairy, stipules tiny and discrete or absent; **older stem robust, to 10 cm diameter,** bark reddish-brown, soft, with scattered lenticels, inner bark orange-brown, wood whitish-yellow. **Leaves: broadly ovate to heart-shaped,** 5-15 x 4.5-11.5 cm; **margin entire; lightly hairy below,** large gland-pair present (flat), small flat glands present at margins; veins red-colored above and below; petioles 2-10 cm, hairy, fused across the nodes. **Inflorescences:** axillary and terminal, dense racemes; flower stalk above joint 0.3-1 cm long; **stigmas with a leafy appendage. Fruits:** dorsal wing nearly erect, 3-5 x 1.2-2 cm, red, upper margin straight, base with a blunt tooth; nut to 0.3 cm diameter, hairy, bearing radial crests apically enlarged into winglets.

Common; riparian areas, secondary forest, old sand ridges. Guianas; also NE Brazil and Venezuela.

16.3. Stigmaphyllon puberum (Rich.) A.Juss.

Common name: fungu ansi-titei (Sa)

Subwoody to woody liana; often densely hairy. **Stems:** node with conspicuous horizontal ridge, stipules inconspicuous. **Leaves:** shape variable, 8-20 x 3-13 cm, base rarely heart-shaped, apex acuminate; **margin entire; hairy below;** large gland-pair present (flat), small flat glands present at margins; **petiole to 7.2 cm long, densely hairy. Inflorescences:** axillary, branchlets densely hairy, 8-15-flowered umbels; flower stalks above joint 2.5-7.5 cm long; **stigmas with a leafy appendage. Fruits:** dorsal wing 4-7 x 1.5 cm or highly reduced, 3 x 2 cm, dark red; nut to 1 x 0.7 cm, smooth or with 1-5 ribs. **Notes:**

Seeds water- and (possibly) wind-dispersed.

Common; old-growth forest, especially riparian forests of the coastal plain. Guianas; Amazonia to Central America and the West Indies.

16.4. Stigmaphyllon sinuatum (DC.) A.Juss.

Common name: akuri i'kererï (Ca); djangáfútu tatai (Sa); fungu ans' tatái (Sa); karia (Ar); konkonikasaba titei (Sr); kusai tatái (Sa); koratiwiwiri (Sr); matu kasaba (Sa); matukru (Tr).

Softly woody liana or herbaceous vine; often hairy; producing large underground tubers. **Leaves:** round, kidney-shaped, or heart-shaped in the Guianas, large, 6-21 x 4.5-20 cm, **margin with rounded teeth [crenate]; densely silver-sericeous below;** large gland-pair present (flat), small flat glands present along margins; petiole to 13 cm long, hairy, fused across the nodes. **Inflorescences:** axillary and terminal, 15-40-flowered dense racemes, flower stalks above joint 0.3-1 cm long; **stigmas with a leafy appendage. Fruits:** dorsal wing 3.5-6 x 1-1.8 cm, dark red, upper margin straight, thick, base with a blunt tooth; nut 0.3-0.45 cm in diameter, smooth or with 1-3 lateral winglets, spurs or crests. **Notes:** A highly variable species. **Use:** The large tubers are inedible but cited by Maroon informants in Suriname for medicinal treatment of eye pain, 'bosyaws' sores (leishmaniasis), stomach pain, and menstruation problems. Additional uses for leaves include treatment of rheumatism, trembling limbs, paralysis, genital steambaths, and malaria (van Andel & Ruysschaert 2011, van Andel 2000).

Common; secondary forest on white sand, savanna forest, and along old-growth non-flooded forest margins. Guianas; also N Amazon.

17. Tetrapterys Cav.

Lianas and climbing shrubs; erect shrubs. Climbing via twining apical stems.

General: Surfaces hairless to densely hairy [*flattened-sericeous*] and shiny; some species producing large, woody underground rhizomes or tubers.

Stems: Branchlets round or flattened, hairy or hairless, bark fibrous, reddish-brown, brown or silver, with abundant lenticels; node usually with an interpetiolar ridge or line, **stipules either small and discrete on the stem, or pairs fused across the node** (cf. Rubiaceae) and then leaving a single broad scar; sometimes borne at the petiole base. Older stems round, bark thick and fibrous, sub-bark orange, stem x-s pattern simple or lobed, without multiple cables.

Leaves: Variable in size and shape, narrowly elliptic to obovate, **base never cordate; margins entire;** papery to leathery; **large gland-pairs rare, small flat glands common in lengthwise rows between midvein and margin,** sometimes close to margin; 3° veins strongly net-veined or parallel; petiole length short to medium.

Inflorescences: Terminal or axillary, flowers borne in umbels, corymbs, or pseudoracemes; flower stalk with 2 leafy bracteoles, with or without glands; calyx 4+1, oil glands 8-10 or absent; corolla 4+1, **petals yellow (pink), margins often frilly or toothed;** stamens 10, ± similar (in length); styles 3, similar, not bearing leafy appendages, hooked or truncate.

Fruits: Samaras with dominant lateral wings, divided into 4 equal or unequal segments, forming an "X" shape; dorsal wings and intermediate winglets or crests sometimes present; all wings strongly reduced in some species; nut subglobose.

Distribution: Widespread in the New World tropics, including most of South America to Mexico and the West Indies: 70 species; Guianas: 12 species, 4 described here.

17.1. Tetrapterys crispa A.Juss.

Woody liana; hairy [*sericeous*]. **Stems:** branchlets round, hollow, white or yellowish-brown hairy; stipules fused across the node, leaving a 0.25-0.55 cm wide scar. **Leaves:** variable, 13-20 x 6-10.5 cm; large gland-pairs rarely present, **small flat glands usually present in lengthwise rows; 3° veins parallel;** petioles 1.2-2.7 cm long. **Inflorescences:** axillary panicles, 4-flowered umbels; flower stalks below joint to 0.75 cm long, with 2 leafy bracteoles; petals yellow; stigmas simple. **Fruits:** upper wings leathery, narrowly obovate, 2.2-3.5 x 1.3 cm wide, lower wings elliptic, 1.1-2 x 0.5 cm, winglets sometimes present; dorsal crest 1 cm long; nuts ovoid, 0.5 cm long, smooth (without projections) aside from wings.

Uncommon; especially riparian forest overhanging river. Guianas; also N South America.

17.2. Tetrapterys discolor (G. Mey.) DC.

Common name: matukruimë (Tr)

Woody liana; hairy when young. **Stems: stipules fused across node,** triangular, leaving scar 0.2-0.25 cm wide. **Leaves:** elliptic, 9-14 x 4-6.5 cm, base rounded or cuneate; leathery; large gland-pairs absent at base, **small flat glands present in lengthwise rows between midrib and margin;** petiole 0.7-1.2 cm long. **Inflorescences:** terminal panicles, ultimate units 4-flowered umbels, flower stalks below joint 0.25-0.6 cm long, with 2 oval, purplish-brown bracteoles; petals yellow, hairless; stigmas simple. **Fruits:** lateral wings leathery, obovate, upper pair 1.2-2.5 x 0.6 cm, 3x larger than lower pair; dorsal winglet 0.8 x 0.4 cm; several intermediate winglets present; nuts globose, 0.5 cm diam.

Common; riparian forest and along roadsides. Guianas; Bolivia to Guatemala and the West Indies.

17.3. Tetrapterys fimbripetala A. Juss.

Woody liana or shrub to 2.5 m; golden-hairy [*sericeous*]. **Stems:** branchlets round, persistently hairy; **stipules tiny, discrete. Leaves:** elliptic or obovate, 4.5-15 x 2-6 cm wide, base cuneate, often curled; leathery; large gland pairs absent, **small flat glands in lengthwise rows (6-20)** between midrib and margin; petiole 0.6-1.1 cm long. **Inflorescences:** axillary, pseudo-racemes 4-16 flowered; flower stalks below joint to 0.9 cm long, with 2 oval bracteoles bearing 2-6 small glands each; **calyx 10-glandular;** petals yellow, occasionally tinged with red, margins fringed, hairless. **Fruits:** lateral wings papery, ± equal, narrowly-ovate, 1.2-2.3 x 0.5-0.8 cm; dorsal crest uneven-4-sided, 0.6 x 0.5 cm, dorsal tip protruding; nut round, 0.2-0.3 cm diam., radial-veined.

Common; especially in open areas along streams and shrub-savannas. Guianas; also S Venezuela and adjacent Brazil.

17.4. Tetrapterys mucronata Cav.

Woody liana or shrub; lightly hairy to hairless. **Stems:** branchlets round; **stipules tiny, discrete. Leaves:** ovate or elliptic, 6-15 x 3-8 cm; base cuneate or rounded, apex acuminate to obtuse; hairless above, lightly hairy below; **large gland-pair present at base, small flat glands present in lengthwise rows between midrib and margin;** petiole 0.5-1.5 cm long. **Inflorescences:** axillary or terminal, panicles of 4-6-flowered umbels; flower stalk below joint 0.6-1 cm long, slender, with 2 small bracteoles without glands; **calyx usually lacking glands (rarely 8-glandular);** petals yellow, sometimes red-flecked, hairless. **Fruits:** lateral wings leathery, 0.5-2.2 cm long, upper pair 2-3x longer then lower pair; dorsal crest 1 cm long x 0.25 cm high, basally thorn-shaped, margin wavy; intermediate winglets thorn-shaped on each side; nut subglobose, 0.4 cm diam.

Common; roadsides, especially riverine and swamp forest. Guianas; also Brazil and Bolivia N to Panama and the Lesser Antilles.

17.5. Tetrapterys styloptera A. Juss.

Woody liana; rarely a shrub. **Stems:** apical branchlets hairy; **stipules tiny and discrete, or absent. Leaves:** ovate or rarely elliptic, 2.5-7 x 5-15 cm; base obtuse or rounded; hairless above, lightly hairy below, **large gland-pair present at base, small flat glands present in lengthwise rows between midrib and margin;** petiole 0.4-0.9 cm long. **Inflorescences:** axillary or terminal, racemes to 10 cm long, straight, hairy; flower stalk below joint to 0.6 cm long with 2 small, oval bracteoles each bearing 1-2 glands; **calyx 10-glandular,** ± hairy, recurved; petals yellow, with tiny marginal teeth. **Fruits:** lateral wings ± equal, oblong or shovel-shaped, 0.7-1.5 x 0.35 cm, hairless; dorsal crest semi-circular, to 0.5 cm long, 0.2-0.3 cm high, apex protruding between upper wing segments; intermediate crests several; nut 0.3 cm diameter. **Notes:** Leaves, bracteoles, and floral characters highly variable.

Common; especially riparian and secondary forest, white sand savanna, granite and sandstone mountains. Guianas; widespread in tropical S America.

22. S. convolvulifolium flower and winged samaras [PT]

23. S. puberum, close-up of lobed stem cross-section [BH]

24. Stigmaphyllon sinuatum, cross-section of 2-cabled woody stem [BH]

25. Stigmaphyllon puberum, lobed stem cross-section [BH]

26. Stigmaphyllon sinuatum, huge, edible (but not tasty) tuber [PT]

27. Tetrapterys crispa, close-up of winged fruits [BH]

28. Tetrapterys crispa, flowering branch with conspicuous oil glands [BH]

29. Tetrapterys crispa, fruiting branch and samaras with four robust lateral wings [BH]

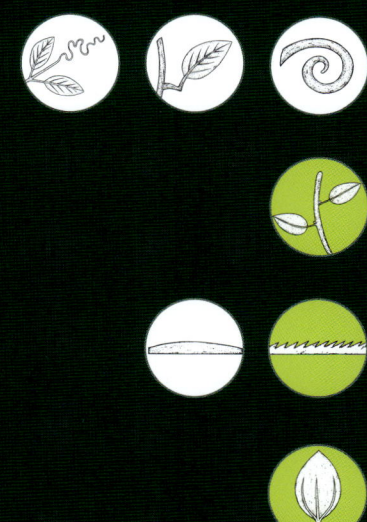

MALVACEAE

MALVACEAE

Rarely lianas or vines; mostly trees, herbs and shrubs. Cosmopolitan; Neotropics: 129 genera/ 2000 species; Guianas: 1 genus with climbers.

1. Byttneria Loefl.

Slender, spiny lianas or vines; also erect herbs, shrubs or treelets. Climbing via twining stems, lateral branchlets, and flexed petioles. Leaves are alternate or whorled, simple, 3-veined from the base with toothed or entire margins. Stipules are present. Flowers are radial, 5-parted, and small, with a fused calyx, free petals, a stamen tube, and a superior ovary with 5 styles. Fruits are round, conspicuously spiny capsules.

General: Weedy habit; **recurved "cat claw" spines; hairs star-shaped [*stellate*]; rough-surfaced** [*asperous*]; sap a clear, sticky gel.

Stems: Branchlets herbaceous or lightly woody, **round or 5-ridged, often conspicuously spiny;** bark tough, peeling in strips or strings; center hollow or filled with white soft pith; nodes smooth and without horizontal ridge or line; stipules linear, inconspicuous or lost. Older stems similar to branchlets, to 2 cm diameter producing clear, sticky gel from central chamber; stem x-s pattern simple or lobed.

Leaves: Elliptic to heart-shaped; thin; margins **sharp-toothed, round-toothed, wavy** or entire; **palmate-3-5-veined at base pinnate-veined above** towards tip, main veins strongly raised below, 3° veins often parallel and 90° to main

palmate veins; **long, sunken glands present on palmate veins near the leaf base;** spines present; **petiole short or very long, often flexed, swollen at apex** [*pulvinus*].

Inflorescences: Axillary, opposite to leaves, or terminal, cymose; **flowers radial, small, 5-parted; calyx cup-shaped to tubular at base,** sepals fused < ½ of their length, often white and conspicuous; **petals free,** purplish-black, purplish–red, or white, cupped or hooded and **bearing unique strap-like 'appendages';** stamens 10, united into a short tube (5 fertile, 5 infertile); ovary superior [*5-locular*], style and stigma 1.

Fruits: Round, spiny capsules, walls collapsing to release seeds; seeds 1.

Ecology: Many Byttneria species are "ant plants"

with ants occupying hollow, enlarged stems and feeding at extra-floral nectaries. Seed dispersal via animal.

Distribution: Pantropical; ca. 136 species, 81 in South America; Guianas: 9 species. 3 species described here.

Notes: In the past, this genus was a member of the Sterculiaceae family, including Theobroma (cacao) and Herrenia, or considered to represent its own family, the Byttneriaceae. Today, all of the "Malvalean" families – Sterculiaceae, Malvaceae, Bombacaceae, and Tiliaceae – are combined into one family, the Malvaceae. Within this mega-family, Byttneria is placed within its own sub-family, the Byttnerioideae.

1.1. Byttneria cordifolia Sagot

Common name: pitïkitïkïkai (Tr)

Slender liana, clambering shrub or vine; finely hairy and asperous. Stems: branchlets round or 5-sided, with dense short spines, finely hairy. **Leaves:** broadly to narrowly heart-shaped, to 12 x 7 cm, **margins toothed;** thin, bright green, hairless; sunken glands present on veins at leaf base; petiole up to 7 cm long. **Inflorescences:** sepals wine-colored within, green to tannish-pink outside; corolla "appendages" dark-red. **Fruits:** capsules to 2 cm diameter, greenish-white, densely soft-spiny.

Common; open or disturbed areas, forest edges, riparian zones. Fr. Guiana, Suriname; also C & E Amazonia.

1.2. Byttneria divaricata Benth.

Slender weedy liana, clambering shrub or vine; stellate hairs present. Stems: branchlets round, spiny, densely hairy; older stems ± 5-sided, to 1.5 cm diameter; bark reddish-brown with sparse spines and lenticels; hollow within. **Leaves:** ovate or ovate-oblong, 6-8 x 2-4.5 cm,

base rounded or heart-shaped, **margins toothed;** thin, hairy above [*hirsute*] and below [*stellate*]; sunken glands present on veins at leaf base; petiole short, < 1 cm long. **Inflorescences:** sepals wine-colored within, green to purplish-black outside; corolla "appendages" purple with red tips. **Fruits:** capsules to 2.5 cm diameter, spiny, with 1 cm long fleshy spines or tubercles.

Common; open or disturbed areas, forest edges, riparian zones. Guianas; widespread from S Brazil and Bolivia to N South America.

1.3. Byttneria scabra L.

Common name: wayamaka andïkïrï (Ca)

Slender liana, vine, or clambering shrub; hairless and asperous. Stems: branchlets round or 5-sided, with dense short spines. **Leaves: very narrow and long,** to 7-12 x 1-1.5 cm, apex with fine point, **margins entire;** leathery; sunken glands present on veins at leaf base; petiole 1 cm long. **Inflorescences:** axillary, sepals white, corolla "appendages" purple with darker purple tips. **Fruits:** capsules to 1 cm diameter, 5-lobed, slightly flattened, spiny.

Uncommon; forest patch edges in savanna; Guianas; widespread in tropical South America and the West Indies.

1. Byttneria benensis, flowering shoot [BH]
2. B. cordifolia, leaves and spiny fruit [OG]
3. B. cordifolia, flowers with petal-like sepals and dark red corolla 'appendages' (Costa Rica species) [OG]
4. B. aculeata, spiny young branchlet, palmate-veined leaf base (Costa Rica species) [RA]
5. B. aculeata, hollow stem [RA]
6. B. cordifolia, palmate venation and thickened petiole apex [FRD]
7. B. uaupensis, woody stem (Guyana species, photo only) [BH]
8. B. uaupensis, stem cross-section with "lobed" pattern [BH]

MALVACEAE

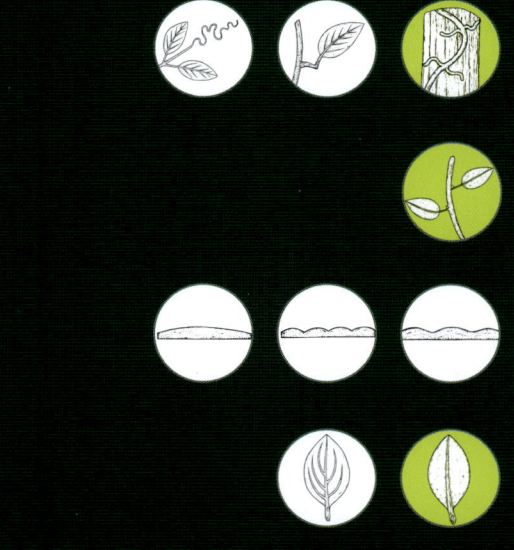

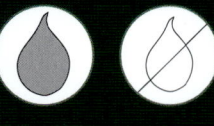

MARCGRAVIACEAE

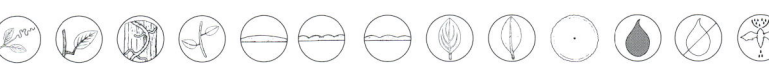

MARCGRAVIACEAE

Lianas, hemi-epiphytes, epiphytic shrubs; also erect shrubs. Climbing via twining shoots, angled branchlets, or roots borne on the stem. Leaves are simple and alternate. Stipules are absent. Inflorescences with dull to brightly-colored nectaries (bracts) of diverse forms. Flowers are 4-5 parted, with a superior ovary and a single 5-lobed style. Fruits are many-seeded berries or capsules.

General: Vegetation fleshy or leathery and hairless, **torn leaves with visible fibers [sclereids] appearing as fine, broken hairs.** Marcgravia species with strikingly different juvenile and adult growth forms.

Stems: Branchlets round or strongly flattened, bearing raised lenticels; nodes smooth, stipules absent. Older stems round to strongly flattened, slender and sub-woody (Marcgravia) to robustly woody (Norantea); stem x-s pattern simple.

Leaves: Terminal leaves often rolled, pointed and enclosing terminal buds. Leaf attachment is two-ranked in Marcgravia, spiral in other genera, margins entire, wavy or blunt-toothed [*crenate*]; leathery to fleshy; **glands present at leaf base and/or along leaf margins;** veins invisible or visible below; petioles short.

Inflorescences: Terminal, umbel-like (Marcgravia and Margraviastrum) **or spike-like** (Norantea and Souroubea), **erect or hanging; with conspicuous green or red-orange nectaries;** flowers bisexual, radial; flower stalk with 2 'bracteoles', often resembling sepals; sepals 4-5; petals 4-5, fused at base (in Marcgravia forming a unique, ovoid "calyptra" cap); stamens 3-numerous, free or basally fused; ovary superior, 2-20-chambered [locular], style simple, crowned by a 5-lobed stigma.

Fruits: Berries or capsules; seeds numerous and embedded in fleshy, often sticky pulp.

Distribution: Widespread in the New World tropics and subtropics: 7 genera/130 species; Guianas: 5 genera; 4 described here. [*Sarcoptera excluded (S. tepuiensis (de Roon) Bedell occurs in the Pakaraima Mtns. of Guyana).*]

1. Marcgravia L.

Slender lianas or hemi-epiphytes; with strikingly different juvenile and adult growth forms.

Stems: Juvenile stems conspicuously angular or flattened, grey and bearing many lenticels, creeping with stem-borne, adhesive roots. Older branches spreading, creeping, or hanging, with two-ranked leaves. Older woody stems round or bumpy, 3-4 cm diameter, bark often silvery-grey and bearing many wart-like lenticels, sometimes with horizontal ridges; stem x-s pattern simple.

Leaves: Juvenile plant leaves small, thin, rounded, and unstalked, **often overlapping along the stem. Adult plant leaves** large, 6-17 x 1.5-5 cm, elongate, leathery-succulent, margins entire or wavy, **not overlapping along stem;** basal glands and marginal black dots present; veins often obscure above, prominent or obscure below.

Inflorescences: Umbel-like, outer whorl of (fertile) flowers, each hidden by a cap-like 'calyptra', **inner whorl of bag-shaped nectar-bracts;** sepals 4, persistent; petals 4, fused to form the calyptra; stamens 6-many; ovary 4-20-locular.

Fruits: Leathery capsules, splitting from the base; seeds many and small.

Ecology: Pollination via bats (and likely birds); seed dispersal via animal (through bird gut).

Distribution: Widespread in the neotropics; 65 species; Guianas: 6 species, 2 species described here.

1.1. Marcgravia coriacea Vahl

Common name: aruakabo (Ar); kaikui amu (Way); katyusi anyarari, kaikusi amararï (Ca); kunepepe (Tr)

Lianas or climbing shrubs. Leaves: elliptic to oblong-lanceolate, 6-16 x 1.5-4.5 cm, base ± rounded, apex acuminate or cuspidate; dull green above and below; **2° veins inconspicuous below;** petiole to 0.3-0.6 cm long. **Inflorescences: infl. stalk [*peduncle*] short, nectaries 4-6 per inflorescence, large, each 3.5-4.5 cm long,** on 0.6-1.5 cm long stalks; flowers 10-26 per umbel, flower stalks 2-3 cm long, thick; stamens 20-35. **Fruits:** ellipsoid to globose capsules, to 1.3 x 1.2 cm. **Use:** Bark and leaves used to treat centipede stings (Trio - common name kunepepe means "centipede"), sap used for sore eyes in NW Guyana; curare ingredient (Trio).

Common; riparian forest. Guianas; also Venezuela and Brazil.

1.2. Marcgravia pedunculosa Triana & Planch.

[Syn: M. parviflora L.C. Rich. ex Wittm.)

Common name: aruakabo (Ar); boedi-diabatere (Sr); kaikui wenaru or kaikui tepu (Tr)

Lianas or climbing shrubs. Leaves: elliptic- to oblanceolate-oblong, 6-17 x 3.5-7.5 cm, base acute, apex cuspidate, often with a fine tip; shiny above and below; basal glands sometimes absent; 2° veins prominent below; petiole 0.4-0.8 cm long. **Inflorescences: infl. stalk [*peduncle*] long, ca. 20 cm; nectaries 5-8 per inflorescence, short, each 0.5-0.6 cm long,** on 0.3-0.7 cm long stalks; flowers 20-30 per umbel, flower stalks 1.5-2.5 cm long, slender; calyptra unevenly conical-ovoid, brownish-yellow; stamens 12-15. **Fruits:** globose capsules, 0.6-0.8 cm diameter.

Uncommon; non-flooded old-growth forest and savanna woodland, especially along river and creek margins. Guianas only.

1. Norantea guianensis, erect flowering spikes with showy red-or-ange nectaries (nectar-bracts) [PT]

2. Marcgravia coriacea overhanging creek with juvenile and adult shoots [PT]

3. Mature Marcgravia sp. stem with large lenticels [FRD]

4. Marcgravia sp., inflorescence [PT]

5. Marcgravia sp., inflorescence, outer ring of flowers in bud, with green nectar-bracts (modified infertile flowers) in center [FRD]

6. Marcgravia sp. , inflorescence [PT]

7. Marcgravia pedunculosa, dried older stem with aerial rootlets [BH]

2. Marcgraviastrum (Wittm. ex Szyszyl.) de Roon & S. Dressler

Lianas or scandent shrubs. Neotropics: 15 species; Guianas: 1 species.

2.1. Marcgraviastrum pendulum (Lanj. & Heerdt) Bedell

[Syn: Norantea pendula Lanj. & Heerdt]

Liana or hemi-epiphytic shrub; juvenile and adult stems similar (vs. strikingly different in Marcgravia). **Leaves: spirally arranged,** oblong, 9-15 x 3-6 cm, base rounded or cuneate, apex acute, obtuse, or with an extended fine tip, margin entire, ± curled; leathery; **2° veins inconspicuous below,** ascending sharply from midvein; **leaf glands in a linear row of 3-7 per side below,** submarginal to 0.7 cm from the margin; petiole 0.3-0.6 cm long, thick. **Inflorescences: umbel-like; bag-like nectaries present,** each 1.5-3 x 0.4-0.8 cm, greenish-yellow to -red, **attached directly to flower stalks at 1-1.5 cm from base;** flowers 6-15 per umbel, all fertile, flowers stalks 4-7 cm long, slender or stout; flowers radial, 5-parted, 0.5 cm diameter; sepals broadly ovate; petals pinkish-green, obovate-oblong, 1.2 cm long, free; stamens 50-65; calyptra cap absent. **Fruits:** globose berry, ca. 2.5 cm diameter; seeds immersed in sticky pulp, kidney-shaped, black, shiny. **Ecology:** Pollination likely by bats and birds; Seed dispersal via animal (through bird gut).

Uncommon; montane forest, high-elevation savanna, and rock outcrops. Guianas; also Brazil and Venezuela.

3. Norantea Aubl.

Lianas or hemi-epiphytic climbing shrubs. Tropical South America, Lesser Antilles, Costa Rica: 2 species; Guianas: 1 species.

3.1. Norantea guianensis Aubl.

Common name: alalalabu (Sa); atarakare (Tr); karakara (Ar); konopo yorokorï (Ca); rafrutere, ravetere (Sr).

Large liana or hemi-epiphytic shrub; juvenile and adult stems similar (vs. strikingly different in Marcgravia). **General:** entirely hairless. **Stems: round, robustly woody, observed to 12 cm diameter,** bearing long aerial (adventitious) roots; bark silvery-grey or dark brown, fibrous and deeply furrowed, inner bark thick, red to orangish-brown, pith pinkish-white; **releasing copious drinkable water when cut. Leaves: spirally arranged,** obovate or obovate-elliptic, 8-18 x 3.5-8 cm, base acute to attenuate, apex obtuse to rounded, margin entire; leathery, glossy dark green above, dull below; 1° and 2° veins inconspicuous or slightly prominent beneath; **2 basal glands present,** marginal glands absent; petiole 1-2 cm long. **Inflorescences: terminal, erect racemes,** axes thick, 0.25-1 m long, flowers 100-350 per raceme; flowers radial, 0.5 cm diameter; pedicels 0.3-0.6 cm (in fruit 1-1.5 cm long); **long-stemmed nectaries borne midway on petiole,** 2.5-4 cm long, bright red or orange-red; sepals broadly ovate, to 0.2 cm long; petals ovate to ovate-oblong, 0.3-0.6 cm long, pinkish-white, red, or purple; stamens 20-40, fused with corolla; ovary 3-6-locular. **Fruits:** globose berries, 0.5 - 1 cm diameter, leathery, fairly rough; seeds several, subround, ca. 0.3-0.4 x 0.1 cm, almost black, shiny. **Ecology:** Pollination by perching birds; seed dispersal via animal (through bird gut). **Uses:** Trio Indians use the stem water against colds, coughing, and voice loss. An infusion of the bark is used to control

MARCGRAVIACEAE

menstrual bleeding and against internal worms.

Common; non-flooded old-growth forest, riparian forest, higher elevation savanna woodland, granite mountains; Guianas; also N South America, Lesser Antilles, Costa Rica.

4. Souroubea Aubl.

Lianas or hemi-epiphytic climbing shrubs. Widespread in the neotropics; 19 species; Guianas: 1 species.

4.1. Souroubea guianensis Aubl. subsp. guianensis

Common name: diabatitei (Pa); diabetere (Sr); karakara (Ar); konopo yorokorï (Ca); kwerimuru (Ar)

Large liana or hemi-epiphytic shrub; juvenile and adult stems similar (vs. strikingly different in Marcgravia). **General:** succulent, entirely hairless. **Stems:** round or slightly flattened, lenticels present, bearing long aerial roots. **Leaves: spirally arranged,** highly variable in shape, narrowly elliptic to ovate or obovate, often oblong, 7-15 x 3-7 cm, base cuneate or rounded, apex acute or rounded, often with a sharp point; leathery, thick; upper surface darker green, rough-surfaced; venation immersed above and below; **basal glands absent, small lamina glands present below, scattered or in linear rows, 3-12 per side;** petiole 0.4-0.8 cm long, furrowed above. **Inflorescences: terminal, erect racemes, 10-25 cm long, with showy red nectar-bracts;** flowers ca. 1 cm diameter; flower stalks robust, 1.5-2 cm long; **nectaries 2-3 cm long, 3-parted, with a middle spur 1-1.2 cm long;** 2 bracteoles resembling sepals; sepals almost round, 0.3 cm long; petals fleshy, 0.8-1.1 cm long, yellow to orange; stamens 5, borne upon sepals; ovary superior [*3-5-locular*]. **Fruits:** Oval to round, leathery berries, ca. 2 cm diameter, light brown, wrinkled, obscurely 5-ribbed, almost woody; seeds few, curved, 0.3-0.5 long. **Ecology:** Pollination by insects; seed dispersal via animal (gut of birds). **Use:** Sap from the bark is used to treat cuts and wounds.

Common; rocky areas, riparian forest near rapids, granite mountains. Guianas; also Brazil, Colombia, and Venezuela.

8. Marcgraviastrum gigantophyllum, climbing habit with flowering and fruiting umbels (Ecuador species) [AK]

9. M. gigantophyllum, fruiting umbel with berries, seeds in orange pulp (Ecuador species) [AK]

10. M. gigantophyllum, berry spit to reveal seeds in orange pulp (Ecuador species) [AK]

11. Norantea guianensis, with erect spikes and showy red-orange nectaries (nectar-bracts) [BH]

12. N. guianensis, older stem with dense, fibrous brown bark and simple cross-section anatomy [BH]

13. N. guianensis berries [PT]

14/15. N. guianensis, close-up of flowers [OG]

16. N. guianensis, stem with aerial/adventitious roots [BH]

17. Souroubea guianensis, close-up of flowers with red nectaries and yellow flowers [PT]

18. S. guianensis, habit with flowering branches [BH]

19. S. guianensis, stems with aerial/adventitious roots [BH]

20. S. guianensis, leaves with youngest leaf forming protective tube around apical bud [BH]

21. S. guianensis, inflorescence with red nectaries and yellow flowers [BH]

MELASTOMATACEAE

1
2
3
4
5

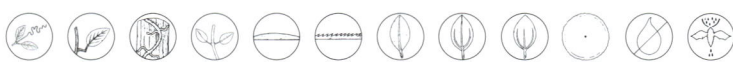

MELASTOMATACEAE

Occasional woody lianas, climbing shrubs, hemi-epiphytes, and subwoody vines; mostly erect herbs, shrubs, or trees. Climbing via twining shoots, clambering branches, or aerial roots produced at stem nodes. Leaves are opposite, simple, and palmately-veined with ladder-like cross veins raised or faint. Stipules are absent. Flowers are showy, radial, 4-6-parted, with a persistent calyx, free white or pink petals, and showy stamens; the ovary superior or inferior with one style. Fruits are dehiscent capsules or indehiscent berries.

General: Hairs simple or forked when present (in climbing taxa covered below); glands absent; distinctive sap, scent or taste absent.

Stems: Shoots round to flattened, green and smooth or hairy, often with raised lenticels; nodes often with a swelling or ridge, stipules and glands absent. Older stems sub-woody to densely woody, round or irregular and bumpy, small- to medium-diameter; bark light grey to dark brown, often thin, lightly to densely covered with raised lenticels; stem x-s pattern simple.

Leaves: Broadly ovate (in climbers); margins entire, toothed or hair-lined [*ciliate*]; texture leathery to thin; mostly hairless, sometimes densely hairy; **3-7-palmately-veined,** diverging from leaf base or just above base, re-uniting at the apex, **with conspicuous ladder-like cross-veins,** veins often faint on upper surface; petioles short or long.

Inflorescences: Axillary or terminal, flowers solitary or in cymes, often showy; **flowers bisexual, radial, 5-parted** (Adelobotrys) or **6-parted** (Totobea); hypanthium present - a cup bearing sepals, petals, and stamens; calyx persistent and often toothed; petals free, white, pink or purple, often contorted; stamens 2x the number of petals, uniform or dissimilar [*dimorphic*]; **ovary superior (Adelobotrys) or ± inferior (Topobea),** 2-5-chambered; stigma 1.

Fruits: Capsules with a toothed calyx (Adelobotrys) or dry berries (Topobea); seeds numerous, tiny, ovoid to linear.

Ecology: Seed dispersal via animal (Topobea, passing through bird gut) or self-dispersal, often via explosive release (Adelobotrys).

Distribution: Global tropics and subtropics: > 150 genera/5000 species; Guianas: 41 genera/300 species, 3 genera with ± woody climbers.

Besides the Adelobotrys and Topobea species described below, Leandra procumbens Ule is a rare clambering shrub with 4-parted pink flowers. It was collected once in the Guianas, on the slopes of Mt. Roraima.

Notes: In Melastomataceae, stamens usually have elbow-like 'appendages' in the stalks [filaments] that are a good character for family recognition. However, woody climbers in the Guianas generally lack such appendages.

1. Adelobotrys DC.

Lianas, shrubby hemi-epiphytes, or vines; also shrubs. Climbing via twining shoots, clambering branches and aerial roots.

General: **Hairs either large and simple or fine and 2-forked.**

Stems: Branchlets round to flattened. Older stems round or bumpy, up to 4 cm diameter; bark grey to silver, smooth, living bark greenish-brown or dark yellow; wood white.

Leaves: Leaf pairs at nodes equal or conspicuously unequal in size; mostly ovate or round, **margins usually finely to obscurely toothed and ciliate** (hairs extending from the margin like eyelashes); densely hairy on veins and petioles or hairless; main veins 5-7, from the base or above the base, all veins prominent and visible below, main veins and cross-veins obscure or visible above.

Inflorescences: Terminal or axillary panicles, **flowers 5-parted;** hypanthium round, sometimes ribbed; calyx conspicuously lobed and often toothed; corolla pink or white petals obovate and rounded; **stamens 10, straight, without appendages;** ovary superior, 5-locular in climbers,

Fruits: 5-locular capsules; seeds numerous and narrow-spear-shaped.

Distribution: Tropical S America to S Mexico, Jamaica: 25 species; Guianas: 6 species. 4 described here.

1.1. Adelobotrys adscendens (Swartz) Triana

Woody liana; hairs forked only. Leaves: ovate to ± oblong, **large, 8-16 x 5-12 cm,** base rounded to ± cordate; sub-leathery, **hairless;** main veins 5-7, strongly prominent below, diverging from the base; cross-veins ± immersed; petiole 2.5-5.5 cm long. **Inflorescences:** terminal, many-flowered panicles, 9-20 cm long, lightly hairy; calyx to 0.45 cm long; **petals white or pink,** 0.8-1 cm long; stamens dissimilar. **Fruits:** capsules ellipsoid, 1 x 0.6 cm, with 10 ribs; seeds winged above.

Uncommon, locally abundant; old-growth forest, creek forest, and mountain areas. Guianas; from the Amazon basin to S. Mexico and Jamaica.

1.2. Adelobotrys ciliata (Naudin) Triana

Woody liana; hairs forked and simple. Leaves: oblong-ovate, **large, 10-15 x 6-8 cm,** base obtuse; sub-leathery, **hairless;** main veins 5, diverging from the base, cross-veins ± immersed; petiole 2-5.5 cm long. **Inflorescences:** terminal, many-flowered panicles, 4-12 cm long, lightly hairy; calyx 0.15-0.18 cm long; **petals pink,** 1.2-1.5 cm long; stamens dissimilar. **Fruits:** unseen.

Uncommon; lowland old-growth forest and submontane areas. French Guiana, Suriname; also SE Venezuela.

1.3. Adelobotrys monticola Gleason

Woody liana or hemi-epiphyte; hairs forked only. Leaves: broadly elliptic, **small, 3.5-7.5 x 2.5-5 cm,** base obtuse to rounded; sub-leathery, **hairless or sparsely hairy;** main veins 5, diverging from the base, cross-veins slightly raised; petiole 0.3-1.5 cm long. **Inflorescences:** terminal, panicles 3-7 cm long, few-flowered; calyx 0.15 cm long and obscurely lobed; **petals white,** 0.8-0.9 cm long; stamens dissimilar. Fruits: unseen.

Uncommon; higher elevation old-growth forest to 1000 m. Suriname only; also S. Venezuela and adjoining Brazil.

1.4. Adelobotrys scandens (Aubl.) DC.

Woody liana; hairs forked and simple.
Leaves: elliptic-ovate to elliptic, **small, 3.5-6 x 2-4.5 cm,** base obtuse to rounded; sub-leathery, densely hairy - **reddish-brown or purplish hairs on veins below and at leaf base;** main veins 5, diverging from above base, cross-veins ± immersed; petiole 1-2 cm long, densely hairy. **Inflorescences:** terminal, compact panicles, 1-2 cm long; bracts leafy; calyx ca. 0.14 cm long; **petals white,** to 1.3 cm long; stamens dissimilar. **Fruits:** unseen.

Uncommon; old-growth forest; French Guiana, Suriname; widespread in northern South America.

2. Topobea Aubl.

Occasional climbing or epiphytic shrubs; usually erect shrubs or trees; often hairy. Climbing via clambering branches and aerial roots. Neotropics, from Amazon to S Mexico: ca. 50 species; Guianas: 1 species.

2.1. Topobea parasitica Aubl.

Liana-like hemi-epiphyte. Stems: branchlets round, thick, with petiole scars, bark brown or grey, smooth, soft, younger growth reddish-brown hairy [*strigose*] or powdery. **Leaves:** clustered at branch ends; round, ovate-oblong or elliptic, 12-20 x 7-12 cm, base rounded or broadly obtuse, apex acuminate, **margin ± entire;** hairless and green above, main veins 5-7, all veins immersed but visible above, raised below, often dark pinkish-purple and hairy; petiole 2-8 cm long. **Inflorescences:** axillary, clustered in 4-6-flowered fascicles in upper branchlets; hypanthium round, enclosed by 2 pairs of long bracts, 0.8-1.5 cm long; **flowers 6-parted;** calyx ca. 3.5 cm long, usually persistent; **petals dark pink or magenta;** large, broadly ovate, 1.5-1.7 cm long, stamens 12, uniform; ovary 6-chambered, ± inferior. **Fruits:** dry, globose berries, 1-1.5 cm diameter, yellow or red at maturity; crowned by calyx tube, with 6 small lobes at apex; seeds very small, numerous.
Ecology: Seeds dispersed via animal (through bird gut).

Uncommon: old-growth forest, on granite and laterite soils. French Guiana, Suriname; also E Brazil.

1. Topobea parasitica, flowering branches [MR©]
2. Adelobotrys adscendens, flowering shoot (Costa Rica) [RA]
3. Adelobotrys scandens in fruit [FRD]
4. Topobea parasitica with purple leaf undersides [RA]
5. A. adscendens, woody stem with lenticels [FRD]

MENISPERMACEAE

MENISPERMACEAE

Woody lianas and vines; rarely shrubs or treelets; plants male or female. Climbing via twining terminal shoots, tendril-like lateral shoots, and wiry, flexed petioles. Stipules are absent. Leaves are alternate, simple, entire, mostly palmate-veined. Inflorescences many branched or long and narrow, with small 3(4)-parted, radial flowers. Ovary is superior. Fruits are ovoid drupes.

General: Vegetation mostly **shiny green and smooth (hairless),** hairs simple when present; plant parts **bitter tasting** due to alkaloids, 'poison' scent noted on occasion; very rarely with colored sap (white-yellow) or onion scent.

Stems: Branchlets green; nodes smooth, without interpetiolar ridge or line, stipules and tendrils absent. Older stems robust, round or strongly flattened (e.g, Bauhinia, Machaerium); bark hard and fibrous with furrows, or soft and corky, often light brownish-yellow; **stem x-s pattern concentric, mostly off-center** (alternating color bands) or with **spoke-like radial rays** (Borismene, Odontocarya); wood usually yellow.

Leaves: Shapes highly variable, margins entire (rarely wavy or few-toothed); leathery or subleathery, less commonly thin; **leaves 3-7 palmate-veined, palmate/pinnate-veined** (palmate at base, with alternate or opposite 2° lateral veins above base), **or fully pinnate-veined** (Telitoxicum), 3° and finer venation ladder-like [*scalariform*] or net-veined; **petiole (leaf stalk)** short to long, rigid, **apex commonly flexed or twisted and with a cylindrical swelling** [*1st pulvinus*]**,** base simple or with a half-globe-like swelling attached to stem [*2nd pulvinus*], **rarely twisted or flexed only at base** (Borismene, Disciphania, Odontocarya).

Inflorescences: Axillary, sub-axillary, or borne on old wood [*cauliflorous*]. Inflorescences unisexual, **male flowers often in shortly-branched panicles, female flowers often in spikes or long racemes,** sometimes with leafy bracts; **flowers tiny or small, radial, usually 3-parted** (4-parted in Cissampelos), up to 0.6 cm long, greenish-white, yellow, pink, orange, or brown; sepals 6-12, petals 6-12(0), free or fused into small inner ring; male flowers with 3-12 stamens, free or fused; female flowers with superior ovary, carpels 3-6.

Fruits: Ovoid, single-seeded drupes [*drupe-like monocarps*], a very few or dozens per inflorescence, 1-3 per flower (6-15 in Sciadotenia); bright red, orange, yellow, tan, or purplish-black at maturity, often fuzzy, with a leathery skin [*exocarp*], fibrous or fleshy inner layer [*mesocarp*] and a bony, woody or papery inner wall [*endocarp*]. Internal characters of the seed and seed food storage [*endosperm*] are important. The embryo within the seed can be straight, U-shaped, or J-shaped. The endosperm around the embryo can be wrinkled [*ruminate*], smooth or absent.

Ecology: Dioecious (separate male and female plants); pollination via insects; seed dispersal via animal (monkeys, bats, rodents).

Distribution: Pan-tropical and subtropical: 70 genera/400-500 species; neotropics: 17 genera/> 180 species; Guianas: 13 genera/37 species, all genera and 17 species described here.

Uses: An alkaloid-rich family, common source of poisons and medicines. For centuries, Menisper-maceae alkaloids were popular ingredients of curare arrow poison, known for muscle relaxant properties (Abuta, Chondrodendron, Curarea, Telitoxicum) (Bisset 2002).

Notes: A diverse assortment of liana species are vegetatively similar to Menispermaceae. Ampelozizyphus (Rhamnaceae) – leaf stalk thick (unswollen, unflexed); Aristolochia (Aristolochiaceae) – leaf stalk not wiry, flexed at base, wintergreen scent; Cucurbitaceae – spiral tendrils at leaf axils; Dioscorea (Dioscoreaceae) – stem with swollen nodes, petiole flexed at base; Euphorbiaceae – large gland-pairs on leaves, white latex; Smilax (Smilacaceae) – tendril pair on petiole; and Sparattanthelium (Hernandiaceae) – 3-palmate-veined from below leaf base, wintergreen scent.

1. Abuta Aubl

Lianas; occasional shrubs or small trees. Climbing via twining shoots and angled petioles.

General: Mostly green and hairless, occasionally densely, softly hairy.

Stems: Branchlets often slightly zig-zag, **wiry and very tough.** Older stems round or flattened, to 20 cm diameter or more, bark yellowish-brown, hard and fibrous with furrows, lenticels present, wood often yellow; stem x-s pattern off-center **concentric** - often visible in smaller branchlets as well as in older stems.

Leaves: Elliptic to ovate, margins entire or wavy; usually leathery, hairless or grey-haired below; **3-5 palmate-veined from base** (or above base), **with a few pinnate veins toward apex,** 3° veins ladder-like or net-veined; **petiole apex usually with a flexed, cylindrical swelling [pulvinus],** base sometimes swollen.

Inflorescences: Axillary or borne slightly above leaf axil; male flowers in compound panicles, female flowers in loose, simple racemes; sepals in 2 or 3 cycles of 3; **petals absent or very small; stamens 6.**

Fruits: Ovoid drupes, **hard, bright yellow to orange** when ripe [*seed U-shaped, endosperm wrinkled*].

Ecology: Seed dispersal via animal (gut of large-bodied monkeys).

Distribution: Tropical and subtropical forests from SE Brazil to Mexico: 10-30 species; Guianas: 9 species, 4 described here.

Notes: Abuta includes more species than most genera in the family, but many species remain poorly documented.

1.1. Abuta imene (Mart.) Eichler

Woody canopy liana; mostly hairless. Stems: older stem flattened, bark fibrous, yellow-orange, densely covered in lenticels. **Leaves:** elliptic to ovate, 10-16 x 5-9 cm; leathery; **3-veined from base,** 2 additional weak veins diverging from midrib above base, 3° veins ladder-like; **petiole**

2-4 cm long, apex with a small or cylindrical swelling, base with a small swelling. **Inflorescences: axes minutely grey-white hairy,** male inflorescences simple, 4-12 cm long, flowers yellow; female inflorescences 1.5-4.5 cm long; sepals grey-white hairy. **Fruits:** drupes 1-2, 1.8-2.5 x 1-1.6 cm, hairless and yellow.

Rare; lowland old-growth forest, mountain savanna forest, gallery forest. Guianas; widespread in Amazon Basin.

1.2. Abuta obovata Diels

Common name: wirapa ijokë (Tr)

Woody canopy liana; mostly hairless. Stems: yellow sap sometimes present in wood slash. **Leaves:** ± ovate, 4-8 x 3-5 cm, margin entire and curled [*revolute*]; leathery, surface arched upwards [*bullate*]; **strongly 3-veined from base or above base,** a few pinnate veins towards apex, 3° veins ladder-like; **petiole short, 1-1.5 cm, apex flexed, with a cylindrical swelling, base swollen. Inflorescences:** male inflorescences narrow panicles, solitary or paired, 5-10 cm long, branchlets mostly 3-flowered. **Fruits:** drupes 1-2, 1.8-2.3 x 1.1-1.3 cm, minutely grey-hairy when young, yellow when mature, base uneven.

Rare; xerophytic forest and riparian forest zones. Guianas, also N. Amazonian Brazil and Venezuela.

1.3. Abuta rufescens Aubl.

Common names: búndu tatái (Sa); tamokumpë marakë (Tr).

High-climbing liana; densely and softly hairy, hairs white to reddish-brown. **Stems:** branchlets hairy; older stems flattened to round, >10 cm diameter. **Leaves:** broad, ovate to round, 10-28 x 8.5-22 cm, **margins entire to wavy** (rarely few-toothed); leathery, **densely hairy below; 5-palmate-veined at base,** 1-2 pairs of pinnate veins towards apex, 3° veins ladder-like; **petiole**

long, 4-9(-13) cm, apex flexed, with a cylindrical swelling. **Inflorescences: axes hairy;** male inflorescences 10-20 cm long; female inflorescences 3-10 cm long, flowers tiny, hairy, yellowish-green. **Fruits:** drupes 1-2, 2 x 3 cm, orange and fuzzy; endosperm wrinkled [*ruminate*]. **Use:** Cited as a curare ingredient (Bisset 2002).

Rare; non-flooded old-growth and mountain savanna forest. Guianas; also widespread in NW Amazonia.

1.4. Abuta sandwithiana Krukoff & Barneby

Low-climbing liana; also an erect shrub or treelet; **mostly hairless. Leaves:** narrowly to broadly elliptic, 8-19 x 4-8 cm, margins entire; leathery, surface arched upwards [*bullate*]; **3-veined from base,** lateral vein pair running close to margin, pinnate-veined towards apex; **petiole 2-3 cm long, apex flexed, with a cylindrical swelling. Inflorescences:** male inflorescences unknown; female inflorescences in racemes, to 2 cm long. **Fruits:** in small clusters on stem, drupes ellipsoid, 2-2.2 x 1.2 cm, yellow and fuzzy.

Rare; non-flooded and seasonally-flooded old-growth, riparian forests. Guianas; also Amazonian Brazil & Bolivia.

1. Abuta sp. with tendril-like, coiled lateral shoot for climbing [BH]
2. Typical Menisperm leaf stalks (petioles) — apex swollen and bent, base swollen (A. grandiflora) [OG]
3. Less common leaf stalk variation — twisted at base, unswollen at apex; immature flower stalk in leaf axil (Disciphania ernstii), Costa Rica [RA]
4. Typical Menisperm leaves — simple, alternate, palmately-veined (Abuta grandiflora) [OG]
5. Single-seeded, hairy fruits (drupes, Abuta rufescens) [FRD]
6. Single-seeded, smooth, orange fruits (drupes, Abuta grandiflora) [OG]
7. Off-center concentric stem, cross-section pattern (Abuta rufescens) [BH]
8. Bark of robust liana stem (Abuta rufescens) [BH]
9. Concentric stem cross-section pattern (Abuta condollei) [BH]
10. Bark of robust liana stem (Abuta condollei) [BH]
11. Abuta imene leaf [FRD]
12. Abuta obovata leaf [FRD]
13. Abuta sandwithiana leaf [FRD]
14. Abuta rufescens leaf [FRD]

2. Anomospermum Miers

Woody lianas. Climbing via twining shoots and angled petioles. Pantropical: 24 species; Neotropics: 6 species; French Guiana, Guyana: 3 species; one species described here.

2.1. Anomospermum chloranthum Diels

Woody canopy liana, rarely a shrub or tree; **mostly hairless. Stems:** branchlets wiry, finely furrowed or ridged; older stems round, large; **stem x-s pattern concentric — slash usually with blue-purple and white bands. Leaves:** ovate to elliptic, relatively small, 6-12.5 x 4-9.5 cm, strongly 3-veined from base, outermost lateral vein pair weak, pinnate-veined toward apex, 3° veins net-veined; petiole **2-6 cm long, apex flexed, with a narrow cylindrical swelling, base shortly thickened. Inflorescences:** male inflorescences axillary or borne on old stems, to 18 cm long; male flower 6-parted, orange; carpels 3. **Fruits:** drupes ovoid, 1-2-clustered, each ± 3 cm long, **orange and leathery** [*embryo J-shaped, endosperm unwrinkled*]. **Notes:** Sterile Orthomene schomburgkii and Anomospermum chloranthum have a similar appearance.

Uncommon; old-growth forest. French Guiana; also greater Amazonia.

15. Anomospermum solimoesa [FRD]
16. A. solimoesa (Brazil) [FRD]

3. Borismene Barneby

Monospecific genus. Climbing via twining shoots and angled petioles.

3.1. Borismene japurensis (Mart.) Barneby

Liana; mostly hairless. Stems: woody stems and branchlets with lengthwise furrows or grooves; older stems round, to 5 cm diameter; bark yellowish, light brown or grey, thick, soft, vertical-furrowed or -cracked; **stem x-s pattern with conspicuous radial rays, concentric circles absent. Leaves:** narrowly elliptic to ovate, 8-25 x 4-14 cm, margin often wavy; **3-5-palmate-veined from above base,** outermost vein pair weak and marginal, venation pinnate towards apex, with many diverging 2° veins; **petiole 3-18 cm long,** wiry, apex usually unthickened, **base often twisted or flexed. Inflorescences:** male and female inflorescences axillary, less than 15 cm long, in male flowers sepals 9; petals 6; stamens 6, free; female flowers yellowish-green, carpels 3. **Fruits: drupes bright red-orange,** 1-2 per flower, nearly egg-shaped, 1.9-2.9 cm long [*embryo straight, endosperm smooth*]. **Notes:** This species might be confused with Odontocarya species. Noted as a "strangler" in Venezuela (Steyermark coll. 93197).

Uncommon; non-flooded old-growth forest. Suriname; widespread in South American tropics.

MENISPERMACEAE

4. Caryomene Barneby & Krukoff

Large lianas. Climbing via twining shoots and angled petioles. Neotropics: 5-6 species; Guianas: 1 species.

4.1. Caryomene foveolata Barneby & Krukoff

Woody canopy liana; mostly hairless. Stems: stem x-s pattern not observed. **Leaves:** ovate with broadly rounded apex, **small 'bumps' [*papillae*] below; 3-5-palmate veined from leaf base,** 2° veins steeply ascending; 3° veins net-veined; petiole > 2 cm long, slender, slightly swollen at both ends. **Inflorescences:** flowers unknown. **Fruits:**

drupes 3 per flower, **yellow, very large and hard (woody), to 7 x 6 cm,** unevenly egg-shaped [*embryo U-shaped, endosperm wrinkled*]. **Notes:** For comparison, Abuta species lack tiny bumps on leaves below, have mostly ladder-like 3° cross-veins, and smaller, softer drupes.

Distribution: Very rare; old-growth lowland forest. Suriname; also SE Venezuela, Amazonian Brazil and Peru.

5. Cissampelos L.

Slender weedy lianas or herbaceous vines. Climbing via twining shoots and angled petioles; often hairy. Pantropical: 22 species; Neotropics: 10 species; Guianas: 4 species, 1 slender liana or vine described here (C. pareira is a herbaceous vine). Well-collected group due to common occurrence in secondary forests and other disturbed areas.

5.1. Cissampelos andromorpha DC.

Common names: aritaritamë (Tr); mibi tokoro (Ar); napewa (Way)

Liana or herbaceous twiner; hairless to softly hairy. Stems: cylindrical or flattened, slender (up to 0.5 cm diameter), bearing many lenticels; stem x-s pattern concentric. **Leaves: heart-or triangle-shaped,** 3-15 x 3-14.5 cm, with fine point at apex [*mucronate*]; venation palmate to peltate; **petiole slender, strongly twisted at apex** and ± thickened at the base. **Inflorescences:** rachis 20-26 cm long, male infl. often conspicuous leafy, lacking bracts, **male flowers 4-parted. Fruits: drupes single, berry-like,** soft and fleshy, asymmetrical, flat, 0.6 x 0.5 x 0.25 cm, becoming red, softly hairy; [*embryo straight*]. **Ecology:** Leafless during flowering. Seed dispersal via animal (bird).

Distribution: Common; lowland old-growth

and montane forest, secondary vegetation, sandy ridges, savanna edges. Guianas; ranging from subtropical South America to Central America.

17. Cissampelos andromorpha, heart-shaped leaves [FRD]

MENISPERMACEAE

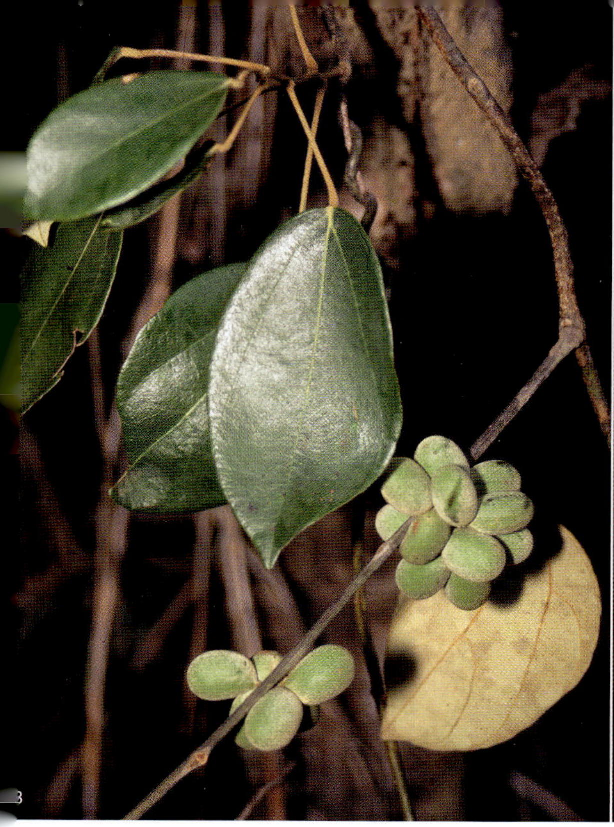

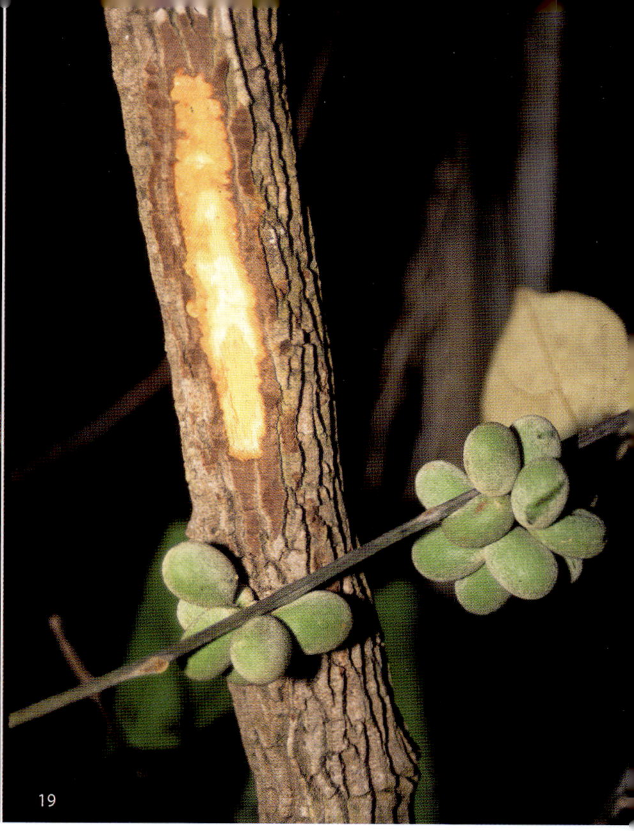

6. Curarea Barneby & Krukoff

Robust lianas; climbing via twining stems and angled petioles; Neotropics: 4 species; Guianas: 1 species.

6.1. Curarea candicans (Rich. ex DC.) Barneby & Krukoff

Common names: andïkïrï (Ca); granny backbone (white) (GU); teteabo (Ar); wanëkë (Tr)

Woody liana; with grey-white matted (flattened) hairs. **Stems:** branchlets wiry, usually shortly-hairy; **older stems flat, ribbon-like,** often large, to 10 cm wide x 2 cm thick; bark light brown to greyish, soft, sub-bark chocolate brown; wood yellow; stem x-s pattern **strongly (off-center) concentric. Leaves:** ovate-elliptic, 6-17 cm x 3-10 cm, **dark green above, conspicuously white or grey-matted below; 3-5 palmate-veined from well above leaf base,** 3° veins ladder-like, widely-spaced, ultimate venation net-veined; petiole long (to 8 cm), **apex flexed, with a cylindrical swelling,** at 90° angle from stem. **Inflorescences:** axillary or on older stems; male inflorescences to 15 cm long, axes hairy, male flowers small, greenish-white, sepals 6, petals 3-6, hairless; stamens 6; female inflorescences poorly known; carpels 3. **Fruits:** drupes 3, ovoid, 1.8-2.2 x 1.6-1.8 cm, surface fuzzy [*embryo U-shaped, endosperm smooth*]. **Use:** The name Curarea refers to the use of some species in curare (arrow poison) preparation. The bark is used by coastlanders in Guyana to make a bitter tea or tonic for fevers, sores and impotence (van Andel 2000). The Trio of southern Suriname also use this species for medicinal purposes.

Uncommon; non-flooded old-growth forest, lower or mid-level canopy. Guianas; also Venezuela and Brazil (Pará).

MENISPERMACEAE

7. Disciphania

Slender lianas and vines with corky bark. Climbing via twining shoots and twisted petiole bases. Leaves highly variable, unlobed to palmately-lobed (palmately-compound outside Guianas), narrow to ovate, 3-7-palmate venation. Neotropics, extending into warm-temperate Mexico, Brazil and Paraguay: ca. 25 species; Guianas: 3 species; 1 species described here.

7.1. Disciphania unilateralis Barneby

Common name: alasiku (Way)

Slender liana or vine. Stems: branchlets slender, conspicuously twisted, striate; older stems < 1 cm diameter, bearing lenticels; stem x-s pattern not observed. **Leaves:** narrowly elliptic, 13 x 4 cm, base rounded or acute, apex acuminate with an extended tip; thin; **3-palmate veined from base,** pinnate-veined towards apex, 3° veins net-veined; **petiole base twisted and slightly thickened,** apex usually not flexed or swollen. **Inflorescences:** axillary, above axils or borne on older wood, spikes or racemes, male and female flowers unstalked. Male flowers: sepals 6; petals 3-6, often of inner and outer whorls, fused to form a fleshy disc, pink; stamens 3. Female flowers; same as male flowers but carpels 3. **Fruits:** 1-2 drupes per flower, oblong-ellipsoid, 1.5 x 1 cm, chocolate brown at maturity, fleshy, seed coat 8-ribbed lengthwise and the ribs commonly winged or fringed [*embryo straight, endosperm smooth*]. **Notes:** *D. moriorum* Barneby has heart-shaped, rough-surfaced leaves; flowers with light green outer petals and yellow inner petals. *D. lobatum* has leaves wider than long, with 3 fine-pointed lobes (similar to a bat wing and similar to some Passiflora species); flowers pinkish-red throughout.

MENISPERMACEAE

8. Elephantomene Barneby & Krukoff

Monotypic genus. Climbing via twining stems and petioles.

8.1. Elephantomene eburnea Barneby & Krukoff

High climbing liana; finely hairy. Stems: branchlets round or flat-sided, finely vertical-grooved, whitish to brown hairy; lenticels absent; older stems irregularly grooved, very large, to 25 cm diameter, bark thick, corky, whitish; **stem x-s pattern off-center concentric. Leaves:** broad, ovate to almost round, large, 11-24 x 9-23 cm;

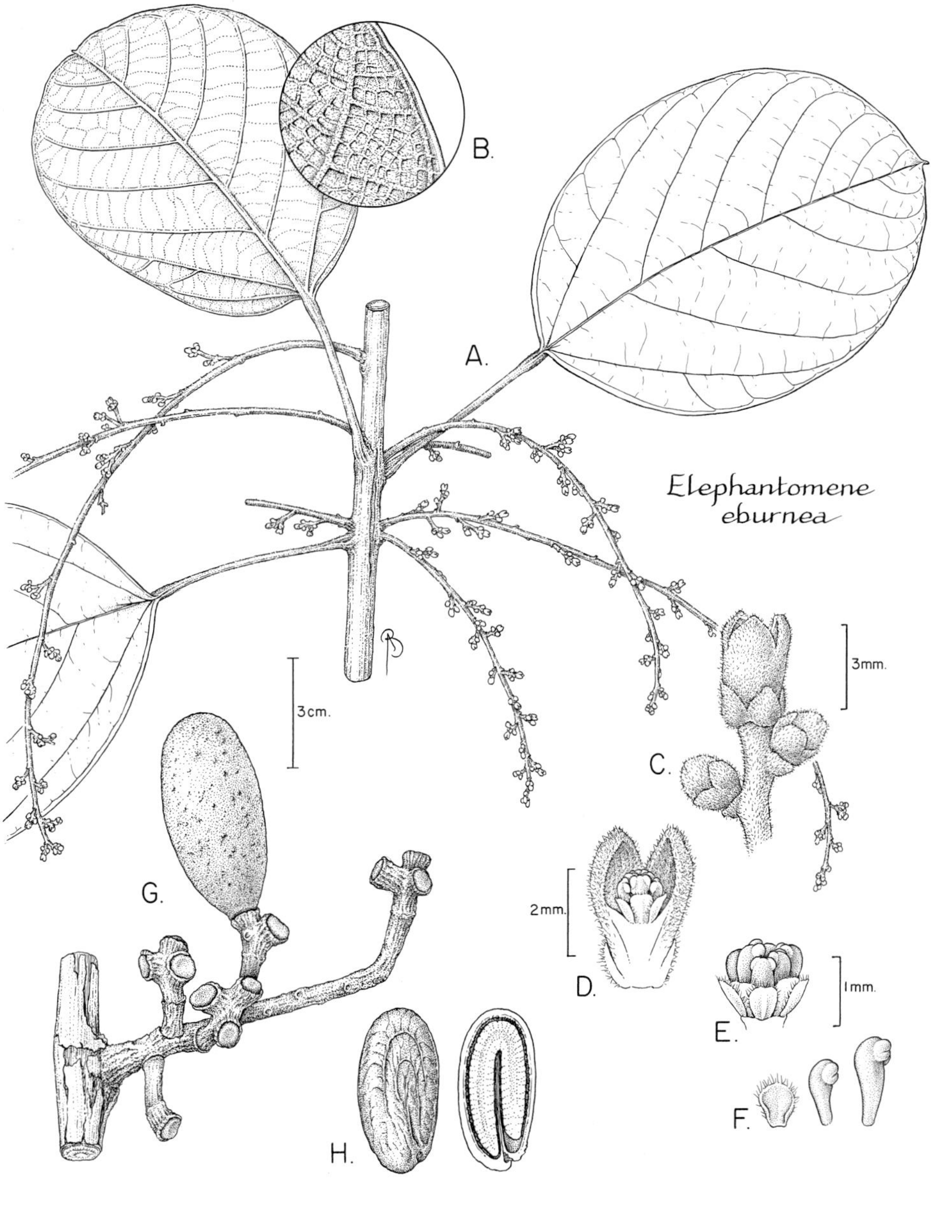

Elephantomene eburnea

leathery, surface arched upwards, smooth and shiny above with fine, matted white or brown hairs below; **3-5-palmate veined from leafbase** — outermost pair weak and close to margin, pinnate-veined towards apex, 3° veins ladder-like; **petiole 6.5-8 cm long, strongly to slightly thickened and twisted at apex. Inflorescences:** axillary or just above axil; male inflorescences narrow panicles, 11-23 cm long, flowers brown-ish-cream to light orange, calyx fused into tube; petals 6; stamens 6; female flowers unknown. **Fruits:** drupes 3, on 3-sided elongate receptacle, oval-shaped, 5-6 x 2.7-3 cm, with greyish, green, or brown fuzzy [*endosperm wrinkled*].

Very rare; non-flooded old-growth forest. Fr. Guiana, Suriname; also W Amazonian Brazil and Ecuador.

9. Hyperbaena Miers ex Benth.

Woody canopy lianas; shrubs and trees. Climbing via apical twining shoots, lateral branchlets or angled petioles. Neotropics: 22 species; Guianas: 1 species.

9.1. Hyperbaena domingensis (DC.) Benth.

Large canopy liana; rarely a shrub or tree; **mostly hairless. Stems:** branchlets smooth, striate older stems round to 5-sided, deeply furrowed, > 25 cm diameter, bark soft, flaky, reddish-brown, easily-peeled; **stem x-s pattern off-center-concentric. Leaves:** elliptic to narrow-ovate, size variable, 4.5-20 x 2.5-12 cm, sub-leathery, margin entire to wavy; **weakly 3-palmately-veined from the leaf base, appearing mostly pinnate,** 3° veins net-veined; petiole 1.5-5 cm long, wiry and thin, apex flexed and slightly swollen. **Inflorescences:** axillary, to 20 cm long, axes white-golden hairy; male and female flowers greenish- to yellowish-brown. **Fruits:** drupes 1-3, egg-shaped, 1.5 x 1.1 cm, hairless, at maturity reddish-purple to almost black [*seed J-shaped, endosperm absent*]. **Notes:** Similar to Orthomene schomburgkii and Anomospermum species. O. schomburgkii has oval, small leaves with weak lateral veins, wrinkled endosperm and straight embryo. Anomospermum has oval, small leaves with strong lateral veins, wrinkled endosperm and J-shaped embryo.

Rare; riparian and floodplain forests. Guianas. Widely distributed, neotropics and West Indies.

26

10. Odontocarya Miers

Medium-sized canopy lianas. Climbing via twining shoots and twisted petiole bases.

General: Lightly hairy to hairless.

Stems: Branchlets conspicuously twisted, finely-vertical grooved. Older stems round, bark grey to reddish-brown, corky, smooth and papery, often peeling, bearing lenticels; **stem x-s pattern with conspicuous radial rays.**

Leaves: Broadly heart-shaped, 3-lobed or ovate; thin; **3-7-palmate veined from base,** 3° veins ladder-like or net-veined; **petiole twisted and thickened at base,** apex usually not flexed or swollen.

Inflorescences: Axillary or stem-borne racemes, sometimes many-branched, up to 60 cm long; male and female flowers stalked, sepals and petals 6, stamens 3-6; carpels 3.

Fruits: 1-2 drupes per flower, yellow to red, smooth or warty, toothed at the apex or at both ends [*endosperm smooth, seed uncurved*].

Distribution: Neotropics to S Brazil and Argentina: 35 species; Guianas: 2 species.

Notes: This genus might be confused with Borismene japurensis.

10.1. Odontocarya tamoides (DC.) Miers

Slender, weak liana. Leaves: heart-shaped or arrowhead-shaped, 5-8 x 4.5-8 cm, apex with a tiny extended point [*mucronate*]; thin, hairy or hairless; 3-7-palmate-veined, often with small pockets or hairs in leaf axils below [*domatia*]; petiole 3.5-7.5 cm long, slightly flexed at base. **Inflorescences:** axillary, simple racemes, 7-12 cm long. **Fruits:** oval drupes, 1-1.2 x 0.6-0.8 cm, reddish-orange, surface wrinkled.

Rare; riparian and creek forest, coastlines, disturbed areas, to 200 m. Guyana, Suriname; widespread in the neotropics.

10.2. Odontocarya wullschlaegelii (Eichler) Barneby

Robust liana. Stem: to 15 cm diameter. **Leaves: ovate-elliptic,** 15-30 x 7-19 cm, thin, shining above, light-grey hairy or hairless below; 3-palmate-veined. **Inflorescences:** male inflorescence borne on stem, panicles to 60 cm long and 45 cm wide. **Fruits:** oval drupes, to 1.5 x 1.3 cm, yellow, with 1 obscure lengthwise rib on one side.

Rare; lowland old-growth forest, ridge and montane forest, secondary forest. Guianas; also Amazonian Brazil and Peru.

27

20. Disciphania lobata, male inflorescence [OG]

21. Disciphania lobata leaf [FRD]

22. Disciphania ernstii, fruiting stalk (Costa Rica plant) [RA]

23. Disciphania ernstii , male flowering plant [RA]

24. Disciphania ernstii, male flowers close-up [RA]

25. Elephantome eburnea. a. Stem with leaves and male inflorescence, b. Veins on underside of leaf, c. Male branchlet on inflorescence, d. Male flower, one sepal removed, e. Male flower, all sepals removed, f. Petal and two stamens, g. Female fruiting stalk, h. Seed coat removed, inner wall [*endocarp*] to left, x-s to right [BA]

26. Hyperbaena dominiguensis, twisted woody stem with leaves [BH]

27. Odontocarya wullschlaegelii, leaf [FRD]

MENISPERMACEAE

11. Orthomene Barneby & Krukoff

Woody lianas. Climbing via tendril-like shoots and flexed petioles. South American tropics to Panama; Guianas: 3 species; 1 described here.

11.1. Orthomene schomburgkii (Miers) Barneby & Krukoff

Common names: ituri-ishi-lokodo (Ar); kanari wiwiri (Sr); tamakaremu (Ca); wirapa ijokë (Tr)

Slender lianas. Stems: branchlets round, smooth, tough and wiry; older stems round, 10-15 cm diameter; bark yellow, fibrous, smooth, with vertical and/or horizontal lines, bearing lenticels; stem x-s pattern **concentric with white and yellow bands. Leaves:** narrowly elliptic, 5.5-10 x 1.7-3.6 cm, apex with an extended point [*mucronate*]; leathery and shiny, **3-palmate-veined from base, the outer vein pair faint and marginal,** otherwise pinnate; 2° and 3° veins obscure above, barely raised below; **petiole 0.6-1.5 cm long, apex flexed with cylindrical swelling. Inflorescences:** axillary, male and female flowers solitary or few, dull yellowish-green. **Fruits:** drupes 1-3, long, ovoid-cylindrical, 2.5-4 x 1.2-2.5 cm, bright yellow-orange, hairless, with faint length-wise ribs [*endosperm wrinkled, embryo straight*].

Very common; non-flooded lowland old-growth forest, along rivers and creeks. Guianas; also widespread in tropical South America to Panama.

28. Orthomene schomburgkii, flowering shoot [AP]

29. O. schomburgkii, male flower [AP]

30. O. schomburgkii, fruit cut lengthwise [AP]

12. Sciadotenia Miers

Slender to robust lianas; shrubs. Climbing via twining shoots and angled petioles. South American tropics to Panama: 18 species; Guianas: 4 species; 1 described here.

12.1. Sciadotenia cayennensis Benth.

Common names: wanëkë (Tr)

Slender liana; hairless to fine-haired. **Stems:** Branchlets round, thin, twisted, with short, soft hairs; older stems round, to 1 cm diameter, woody, with vertical grooves; bark thin, yellowish-brown, with few, large lenticels; stem x-s pattern not observed. **Leaves:** Broadly-elliptic to ovate, 5-13 x 2-5.5 cm, apex pointed, margins entire; thin, hairless and shiny above, softly hairy to hairless below; **3-5-palmate-veined from just above base,** sometimes with a faint vein pair at base, without pinnate veins towards apex, **3° veins ladder-like,** veins conspicuous above and below; **petiole thin, wiry, to 4 cm long, apex flexed with slender swelling. Inflorescences:** Male inflorescences axillary, 4-5 cm long, flowers "cone-like" , in small clusters, sepals many (9-18), petals 6, stamens 6; **female flowers on long stalks (to 20 cm) produced at base of plant, below leaves. Fruits:** Drupes 6-15 per flower, round, each ca. 1.1 x 1.2 cm, orange to dark purple, fuzzy [*embryo U-shaped, endosperm wrinkled*]. **Ecology:** Seed dispersal via animal (bats).

Use: Trio Indians use water from stem for earaches (Plotkin 1987). **Notes:** Sciadotenia species can be confused with Curarea candicans. C. candicans is distinguished by the white-haired leaf undersides, broader leaves, and less conspicuous ladder-like 3° venation. In addition, Sciadotenia species produce many more fruits per flower (up to 15) and have wrinkled endosperm (vs. 1-3 fruits per flower and smooth endosperm in Curarea).

Common; riparian forest, non-flooded old-growth lowland forest, savanna edges, and bauxite mountains; Guianas; also Brazil (state of Pará).

13. Telitoxicum Moldenke

Woody canopy lianas; cut parts sometimes with onion-scent. Climbing via twining shoots. Neotropics: 6 species; Guianas: 3 species, 1 described here.

13.1. Telitoxicum inopinatum Moldenke

Liana; mostly hairless. Stems: branchlets with raised socket-like leaf scars; older stems round or flat, robust, furrowed; bark greyish, lenticels large, raised and scattered; stem x-s pattern concentric. **Leaves:** elliptic, 8-13 x 3.5-5.5 cm, with abruptly short tip, surface arched upwards [*bullate*]; leathery, hairless and shiny; mostly **pinnate-veined (3-palmate veined),** 3° veins ladder-like or net-veined; petiole slender, 2.5-4.5 cm long, **apex unflexed, swollen, base with socket-like swelling. Inflorescences:** axillary or borne on old wood, racemes 13-24 cm long, 1-2 cm wide, with soft, long hairs; male and female flowers 6-parted. **Fruits:** drupes 1-2, oval-curved, to 3 x 2.2 cm; leathery [*seed U-shaped, endosperm wrinkled*]. **Ecology:** Bat-dispersed. **Notes:** The generic name is derived from the Latin terms 'telum', meaning weapon, and 'toxicum', meaning poison — referring to the use of this species in the preparation of arrow poisons (Krukoff & Moldenke 1938).

Rare; savannas, lowland old-growth forest, riverine forest, bush islands. Guianas; adjoining Venezuela.

31. Telitoxicum inopinatum woody stem, concentric x-s, leafy shoot [BH]

NYCTAGINACEAE

1. Pisonia aculeata [TR]
2. Pisonia macranthocarpa habit [DJ]
3. P. macranthocarpa, subopposite leaves, leaf undersides [DJ]
4. P. macranthocarpa habit [DJ]
5. P. macranthocarpa, axillary spines [DJ]
6. P. macranthocarpa, leaf dark green above, veins immersed [DJ]
7. P. macranthocarpa, leaf light green below, veins immersed [DJ]

NYCTAGINACEAE

Rarely lianas or clambering shrubs; often herbs, shrubs, or trees. Mostly in (sub)tropics of the New World: ca. 31 genera/400 species; Guianas: 6 genera/14 species; 1 genus including a woody climber.

1. Pisonia L.

Rarely lianas or clambering shrubs; mostly trees and shrubs; often armed, sometimes with glandular hairs. Climbing via rigid, horizontal branchlets and thick, straight to recurved spines in leaf axils. Stipules are absent. Leaves are opposite to sub-opposite, simple, margins entire. Flowers small, radial; calyx corolla-like; petals absent; stamens 5, tubular at base; ovary superior, 1-chambered. **Fruits small, drupe-like,** seeds 1 per fruit.

Distribution: Tropical and subtropical regions: 40 species; Guianas: 1 species occurring in Guyana.

1.1. Pisonia macranthocarpa (Donn.Sm.) Donn.Sm.

Scrambling shrub or liana with many arch-like branches; climbing to > 6 m. **General:** finely hairy, lost with age. **Stems:** branchlets round, bark smooth, dark green with white lenticels; **nodes often swollen and with axillary spines,** straight to slightly recurved, 0.3-1.1 cm long; older stems round, bark grey or reddish and covered with lenticels, to 5 cm diameter; stem x-s pattern simple. **Leaves:** (sub)opposite (looking whorled), entire, leathery to thin, oval-elliptic, 11 x 6 cm; veins immersed; petiole to 3 cm long.

Inflorescences: at apex of short shoots on 1-5 cm long stalk; flowers unisexual – male flowers greenish-yellow, bell-shaped, 0.2-0.4 x 0.2 cm, with short hairs; female flowers tube-shaped, 0.15-0.3 cm long. **Fruits:** capsule on elongated stalk with 5 rows of glands, protected by persistent calyx, dry, leathery, green, 1.6-2.0 x 0.8-0.9 cm, 5-angled/ribbed with dense hairs; seeds 1. **Ecology:** Seed dispersal via animal (birds and monkeys, hitchhiking or through gut). **Distribution:** Occurring in thickets on wooded, elevated 'islands' in savanna vegetation. Guyana; also Central and South America.

OLACACEAE

OLACACEAE

Rarely lianas, climbing shrubs, or hemi-epiphytes; usually understory shrubs, treelets, or large trees. Pantropical ; ± 28 genera/180 species; Guianas: 10 genera/17 species, 1-2 genera including liana species.

1. Heisteria Jacq.

Rarely lianas; usually understory shrubs or treelets. Climbing via twining stems and petioles. Leaves are alternate, simple, with entire or wavy margins; venation pinnate or 3-palmate. Stipules are absent. Inflorescences are axillary clusters of small, white, radial, 5-parted flowers (in climbers); ovary superior, stigma 3. Fruits are 1-seeded drupes. Largely neotropical, ± 36 species; 1 liana species in the Guianas.

1.1. Heisteria scandens Ducke

High-climbing liana. General: mostly hairless, cut parts reported with scant white sap (not observed). **Stems: branchlets conspicuously green and smooth,** growing forward zig-zag; older woody stems round, bearing lenticels; bark grey to reddish-brown with horizontal cracks, sub-bark layer green, inner bark thin and yellowish-white, wood white; stem x-s pattern simple. **Leaves:** ovate-oblong, 6-12 x 2-4.5 cm, base acute, apex acuminate; sub-leathery, **pinnately-veined,** 2° vein pairs 5-6, widely-spaced, **sometimes with a basal 2° vein pair (= palmate venation);** veins prominent below, 3° veins mostly obscure; **petiole curved upwards and thickest where it joins the leaf blade. Inflorescences:** axillary, in 10-20-flowered bundles; flower stalks slender, 0.6-0.8 cm long; calyx cupular, 5-lobed, greenish-white; **petals 5,** white, < 0.25 cm long; stamens 10 (5 on petals, 5 on sepals); ovary 3-chambered, style 1, **stigma with 3 lobes. Fruits: drupes bright red or orange, elongate-ovoid, 1.3 x 0.9 cm, subtended by a frilly, persistent, skirt-like green calyx,** to 1.5 cm diameter, seeds 1. **Ecology:** Seed dispersal via animal (ingested by birds and monkeys). **Notes:** Heisteria scandens bears a highly distinctive, unmistakable fruit. Without flowers or fruits, many Olacaceae have few notable characters to

OLACACEAE

aid identification. Among lianas, sterile Hysteria scandens might be mistaken for some Menispermaceae due to the sub-leathery leaves, thickened petiole, and the absence of tendrils. Other alternate/simple leaf taxa for comparison include Ampelozizyphus, Convolvulaceae, Corynostylis, Guatteria, Icacinaceae, Pisonia, Plukentia, Polygalaceae, Seguieria, and Solanaceae.

Cathedra acuminata is an Olacaceae tree that was described as a woody liana in the Reserva Ducke near Manaus, Brazil (Ribeiro et al. 1999). Branchlets green and smooth. Stem with orange or reddish sub-bark. Leaves lanceolate to oblong, 6-15 cm long; petioles twisted, not recurved or thicker towards apex. Flowers white, 6-parted. Drupe 2 cm diameter, yellowish-orange, almost fully surrounded by persistent nectar disc. French Guiana, Guyana, expected in Suriname; also widespread across Amazon basin. This species has not yet been documented as a liana in the Guianas.

Uncommon, non-flooded old-growth forest. Guianas; widely distributed in the neotropics, from Bolivia and Brazil to Nicaragua.

1   Heisteria scandens, hanging branches with fruits [BH]
2   Heisteria scandens, fruits with skirt-like persistent calyces [PT]

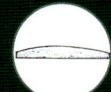

PASSIFLORACEAE

PASSIFLORACEAE

Mostly lianas, climbing shrubs and vines; rarely herbs, shrubs or trees. Climbing via axillary tendrils. Stipules are present. Leaves are alternate, simple (rarely compound), of diverse shapes and textures, with entire or toothed margins, palmate or pinnate venation, and conspicuous glands. Inflorescences are axillary and few flowered; flowers are radial, 5(4)-parted, showy and complex; stamens 5(8-10); ovary superior, 1-chambered, stigmas 3(4). Fruits are ovoid berries or angled capsules with few to many seeds.

Distribution: Pantropical and subtropical: 16 genera/ ± 700 species; Neotropics: 4 genera/ ± 450 species; Guianas: 4 genera/ ± 75 species.

Notes: The Passiflora genus, with 5-parted flowers, dominates the family. In the Guianas, three additional poorly-known genera occur, all of them with 4-parted flowers. These include Ancistrothyrsus (1-2 sp.), Dilkea (2 sp.) and Mitostemma (2 sp.). This chapter focuses first upon Passiflora, and ends with a brief overview of the 4-petal genera.

1. Passiflora L.

Occasional robust woody canopy lianas, more often sub-woody lianas, basally-woody vines, or entirely herbaceous vines. Climbing via single tendrils at leaf axils or inflorescence branchlets.

Common name: kunaparu (Tr); marakusya (Pa, Sa); markusa (Sr).

General: Surfaces often **hairless,** hairs simple and mostly reddish-brown; **nectar glands** very common; cut parts often with the fragrant **scent of passionflower fruit.**

Stems: Branchlets round to 3-5 angled in cross-section, rarely winged, smooth or with lengthwise grooves and ridges; nodes without interpetiolar ridge or line, **stipules usually small** (bristle-like, linear, triangular) and short-lived, rarely large and leafy. **Older woody stems round, 3-5-angled, or deeply lobed,** rarely > 5 cm diameter; **bark softly woody and flaky,** black to reddish-brown, **slash often with a green sub-bark layer; stem x-s pattern simple or lobed** (most commonly, lobes internal and invisible without cutting stem, also sometimes entire stem visibly lobed without cutting stem).

Leaves: Spirally arranged along the stem. **In woody lianas, usually simple and unlobed to shallow-lobed, with 3 main (palmate) veins, sometimes entirely pinnate,** margins entire, wavy, toothed or gland/spine-tipped; veins deeply immersed above and raised below; **nectar glands (1-2 knobs, flaps or scars) present on the petiole or leaf blade.**

Inflorescences: Few flowered on short axillary shoots; **flowers showy, complex, radial, 5-parted, white, green, red-pink, to blue-purple;** receptacle bearing 3 basal bracts, 5 petal-like sepal lobes (broader than petals), 5 petals (often long, narrow), and 1-many rings of filaments or membranes [*corona*]; male and female parts often borne on a central stalk [*androgynophore*]; stamens fused below ovary stalk, free and 5 above; ovary superior, ovoid or 3-sided, 1-chambered [*3 carpellate*], styles 3, free, stigmas large.

Fruits: Globose to ovoid berries, blue-black to orange, soft to leathery, fragrant; seeds numerous, ± flattened, surface rough, often embedded in an edible, sweet to sour pulp.

Ecology: Insect-pollinated (bees, butterflies, moths, and flies); seed dispersal via animal (birds or monkeys).

Distribution: Widespread in low- to high-elevation old-growth forests; vines common in forest gaps, edges, secondary forest, and other disturbed areas. Ten of the woodier species in the Guianas described below.

Use: The ripe fruits of many species are edible and some are important local and regional market products as fruits and juices (P. edulis, P. foetida). Passiflora species are often cultivated as ornamentals due to their unique, beautiful flowers. The leaves, bark, and flowers of various species are used as medicines - sometimes to kill intestinal worms or induce sleep. Vegetation and im-

mature fruits may be toxic [*cyanogenic glycosides*].

Notes: Tendril position and branching distinguish Passiflora from other woody climbers with spiral-coiled tendrils. Passiflora tendrils emerge directly from the leaf axil and are unbranched. Cucurbitaceae tendrils emerge 90° from the petiole and branch evenly 2-7 times. Vitaceae (Cissus) tendrils emerge 180° (opposite) to the leaf axil, branching evenly 1-2 times. Smilacaceae (Smilax) have paired, unbranched tendrils arising from petiole sheaths.

1.1. Passiflora acuminata DC.

Woody liana; hairless; glands on petiole and leaf margins. **Stems:** branchlets round to angular, nodes with tiny linear stipules, 0.4 cm long, and tough tendrils; older woody stems round, robust, bark reddish-brown and cracked; stem x-s pattern lobed. **Leaves:** lanceolate to oblong-lanceolate 7-14 x 2-5 cm, base rounded, apex acute, **margins entire;** sub-leathery to thick-leathery, not waxy below; petiole slender, 1 cm long, with **2-4 knob-like glands in lower half. Inflorescences:** flowers bowl-shaped at base, 5-6 cm wide, **perianth red;** bracts oblong, to 4 x 1.5 cm; sepal lobes narrow, greenish-white outside, bright red within, petal lobes entirely red; corona 5-ringed, fleshy, red, innermost ring toothed. **Fruits:** ovoid berries with 3 vertical grooves, green with white-yellow spots.

Uncommon. Old-growth lowland forest. Guianas; also N Brazil.

1.2. Passiflora amoena L.K.Escobar

Woody liana; mostly hairless; glands on petiole only (similar to P. fuschsiiflora). **Stems:** branchlets round, nodes with small, bristle-like stipules and tough, persistent tendrils; older woody stems round, bark light-brown and cracked; stem x-s pattern lobed. **Leaves:** elliptic, 9.5-13 × 4–7.5 cm, base obtuse to rounded, apex obtuse, abruptly acuminate, **margins entire;** sub-leath-

2
3
4
5
6

7

8

9

10

11

12

13

14

ery to thick-leathery, shiny above, blue-grey to white waxy [*glaucous*] below; petiole thick, 1.5-5 cm long, with **2 knob-like glands at apex. Inflorescences:** 1-many flowered per stalk; flowers long-tubular; basal bracts triangular and small; **perianth lobes bright pink and fleshy,** flat to recurved; corona fleshy, 3-5-ringed, filaments flat-triangular, not-thread-like, with bright-yellow tips (see images 5 and 6). **Fruits:** ellipsoidal berries, 6-9 x 3.5-5 cm, leathery, reddish-green with pink.

Rare. Lowland old-growth forest. Guianas only (observed in southern forests near Brazilian border).

1.3. Passiflora auriculata Kunth

Common name: sosoporo (Ca); wilde markusa (SD)

Woody, subwoody, or herbaceous climber; hairless or with fine white hairs (not reddish-brown); glands on petiole and throughout leaf blade. **Stems:** branchlets angled, nodes with small thread-like stipules and tough, persistent tendrils; older woody stems 3-5 angled to deeply lobed; stem x-s pattern lobed. **Leaves:** lanceolate to ovate, 5-15 x 2-10 cm, base subcordate or rounded, apex acuminate, margin wavy; subleathery; **3-palmately veined, two main 2° veins ending in marginal spines; flat glands in vertical rows in-between main veins and along leaf margin; petiole** 0.5-2.0 cm long, **with earlike [*auriculate*] gland pairs near the base. Inflorescences:** flowers 2-per-node; bowl-shaped, 2-2.5 cm wide; basal bracts bristle-like, 0.2 cm long; sepals narrow, greenish-yellow; **petals linear, 0.5-0.7 cm long, white;** corona 2-ringed, outer ring thread-like, red to purple at base, yellowish-green above. **Fruits:** globose berries, 1-1.5 cm diameter, maturing black; leathery.

Common. Lowland old-growth forest, secondary forest, and savanna scrub. Guianas; also Amazonia to C America.

1.4. Passiflora candida (Poepp. & Endl.) Mast.

Woody liana; hairs reddish-brown; glands on petiole and leaf margin. **Stems:** branchlets round to angled, or deeply furrowed; nodes with tiny stipules and tough, persistent tendrils; older stems round or lightly furrowed, bark reddish-brown; stem x-s pattern lobed. **Leaves:** broadly ovate or ovate-oblong, 8-18(-50) × 7–14 cm, base rounded, apex shortly acuminate, **margin obscurely toothed** with small sub-marginal glands; leathery, lightly hairy below [*pilose*]; **petiole** short and robust, **with 2 large, knob-like glands at apex. Inflorescences:** flower broadly funnel-shaped, 3-4 cm long, perianth white; sepals and petals not spreading; corona 4-ringed, mostly white with a yellow and reddish-brown spotted outer ring. **Fruits:** berry ellipsoidal, 4-6 x 2-3 cm; skin thick-leathery, green.

Uncommon. Lowland old-growth forest. Guianas; also Venezuelan and Brazilian Amazon.

1.5. Passiflora coccinea Aubl.

Common name: araha enapiru (Tr); marudi oura (Ar); snekimarkusa (Sr)

Liana or subwoody vine; hairless or hairs white-yellow to red-brown; glands on petiole and leaf margin. **Stems:** branchlets round to angled, nodes with linear stipules, 0.4–0.6 cm long, entire or with glandular teeth; tendrils tough and persistent; older stems round to deeply lobed; bark grey; stem x-s pattern lobed. **Leaves:** ovate-oblong, 6–14 × 3–7 cm, base subcordate, apex acute, **margins with round- to sharp-teeth and small glands;** thin, densely hairy below [*tomentose*]; petiole to 3.5 cm long, **with 2 large, knob-like glands near base. Inflorescences:** Flower stalks 1 per node, stout, to 8 cm long, hairy; flowers bowl-shaped, 7-9 cm wide; bracts ovate, to 6 x 3.5 cm, red-orange, hairy [*pilose*]; **perianth bright red to red-orange, sepal and petal lobes narrow;** corona 1-ringed, white to light pink in lower 1/2, deep red to purple above.

Fruits: subglobose to ovoid berries, 4.5-6.5 x 3.5-5.5 cm, orange or yellow, green-striped, hairy; ovary stalk red.

Common; old-growth forest, riparian forest, savanna forest, open and disturbed areas. Guianas; also widespread in Amazonia.

1.6. Passiflora costata Mast.

Common name: markusa (Sr); marakusya (Pa)

Woody liana or climbing shrub; lightly hairy; nectar glands on petiole only. **Leaves:** obovate to orbicular, 5-25 x 4-16 cm, base rounded, apex rounded, **margins entire;** leathery, hairy below; **veins pinnate,** with 12-15 parallel 2° vein pairs; petiole 2-3.5 cm long, **with 2 glands or flat scars at apex. Inflorescences:** flower stalks 2 per node, each 3-4 cm long and 1-flowered; flowers 6–7 cm wide, finely hairy outside, fragrant; basal bracts linear and small; **perianth white;** sepals oblong-lanceolate; petals oblong-spatulate; corona ± 6-ringed, base orange, reddish in mid-section. **Fruits:** ellipsoid to ovoid berries, to 5.5 x 4 cm, blue-green with white spots.

Common; seasonally-flooded old-growth forest, along rivers and on rocky islands, especially near rapids. Guianas; also widespread in Amazonia.

1.7. Passiflora fuchsiiflora Hemsl.

Common name: horotoballi (Ar); markusa (Sr)

Large woody liana; hairless; glands on petiole only; similar to P. amoena. **Stems:** branchlets round, nodes with linear stipules, 0.7-0.9 cm long, falling early; tendrils stout; older stems round, bark grey; stem x-s pattern lobed. **Leaves:** broadly ovate or suborbicular, 10-12 x 8-10 cm, base truncate, apex obtuse or emarginate, **margins entire;** thick-leathery, shiny above, bluish-green and waxy below [*glaucous*]; **veins completely pinnate,** with 5-7 curving 2° vein pairs; **petiole to 8 cm long, with 2 glands or**

flat scars at apex. **Inflorescences:** often leafless during flowering, flowers in crowded bundles or short racemes to 15 cm long, borne on old wood or trunk, 10-30-flowered; flowers tubular, tiny bracts scattered on infl. stalk; **perianth orange-yellow to pink;** sepals lanceolate, to 2 cm long, thin; petals similar but slightly shorter; corona ± 3-ringed, orange-red with filaments widening in upper half. **Fruits:** ovoid to spindle-shaped berries, to 6 cm long, pink to red, with 3 thick styles; ovary stalk to 5.5 cm long.

Rare; old-growth forest. Guianas; also adjoining areas of Venezuela.

1.8. Passiflora garckei Mast.

Common name: markusa (Sr); takusi wokuru (Ca)

Woody liana, subwoody or herbaceous vine; hairless; glands on petioles, leaf margins and stipules. **Stems:** branchlets round to angled, **nodes with very large, leathery stipules, obovate or kidney-shaped, 3-5 cm long, bearing glands;** older stems round to subangular; stem x-s pattern lobed. **Leaves: 3-lobed to middle,** 7-15 x 10-25 cm, base ± peltate, truncate, or rounded, **lobes ovate to oblong, margins entire or toothed,** flat glands present; leathery, blue-grey to white waxy below [*glaucous*]; **palmate-5-7-veined,** veins pink below; **petiole** 3–10 cm long, **with 4-6 scattered glands. Inflorescences:** infl. stalks 1.5-6 cm long and 1-flowered; flowers bowl-shaped, 7-9 cm wide; bracts narrowly ovate, to 1.3 cm long, located 1-1.5 cm below flower; **perianth blue-purple;** sepals oblong, to 4 cm long, petals oblong, to 3.5 cm long, corona 2-3-ringed, thread- or hair-like, outer rings violet-blue with white or yellow tips, inner rings white; styles persistent, 1 cm long. **Fruits:** ellipsoid berries, 5.5 x 3.5 cm, green; skin white, spongy; ovary stalk to 3.5 cm long.

Uncommon; gaps and borders in lowland old-growth and riparian forest. Guianas; Venezuela. Notes: Sometimes naturalized or cultivated due to edible fruits.

1.9. Passiflora glandulosa Cav.

Common name: ampuku titei (Au); araha enapiro (Tr); bimititokon (Ar); kuleto (Ar); losaboto kalawiru (Ca); marakusya (Pa); ndjuka makodja (Sa); pomme liane (Sr); roode markusa (SD); yawahü merekuya (Ar)

Woody liana, or subwoody vine; mostly hairless; glands present on petioles and leaf margins. **Stems:** branchlets round to angled, purplish, slender; **nodes with bristly or linear stipules;** tendrils tough and persistent; older stems round to deeply lobed; bark grey; stem x-s pattern lobed. **Leaves:** oblong-ovate to -lanceolate, 6-15 x 4-10 cm, base slightly cordate to acute, apex acute to rounded, with a 'drip-tip', margins entire, slightly recurved; with small submarginal glands below; petiole 1–2.5 cm long, **with paired glands below mid-point. Inflorescences:** racemes 3-8 cm long with 1-2 flowers; flowers tubular, 1.5-2.5 cm long (tube only), **perianth red or scarlet;** bracts narrowly ovate or triangular, 0.5-1 cm long; sepal lobes narrow to oblong, 3–5 cm long, with a few glands; petals shorter than sepals, becoming recurved; corona 2-ringed, the outer ring with white or pink awl-shaped filaments to 1 cm long. **Fruits:** ovoid berries, 5-6 x 2.5-3 cm, green to deep red, apex ± acute; ovary stalk to 5 cm long. **Notes:** Sometimes leafless during flowering; seed dispersal via animal (esp. monkeys).

Very common; old-growth lowland forest, bush islands, secondary forest. Guianas; also Brazil and Venezuela.

PASSIFLORACEAE

1.10. Passiflora nitida Kunth

Subwoody liana or vine; hairless. **Stems:** branchlets angled (x-s), **stipules awl-shaped, 0.4-0.6 cm long, bearing glands;** older stems round to angular; stem x-s pattern round to lobed. **Leaves: ovate to broadly ovate,** 12-17 x 7-13 cm, base rounded, apex mucronate, margins toothed/wavy; (sub)leathery; **pinnate-veined; petiole** 0.5-6 cm long, **with 2 raised glands at apex. Inflorescences:** axillary, infl. stalks 3-7 cm long, 1-flowered; flowers 9-11 cm wide; basal bracts green, ovate, to 3.5 cm long, glandular; **perianth white,** often reflexed; sepals oblong, to 5 cm long; petals narrower; corona outer ring hair-like, base maroon-white-banded, upper 3/4 violet-blue-white-banded, inner rings white; androgynophore white, 2-3 cm long, styles 0.7 cm long. **Fruits:** berries ovoid, 5-7 cm diameter, inner wall to 1 cm thick. **Ecology:** Seed dispersal via animal (monkeys).

Common; secondary forest, riparian forest. Guyana; also Amazon Basin.

2. Additional Passifloraceae climbers: 3 rare genera with 4-parted flowers

Ancistrothyrsus tessmannii Harms (FG, GU); Dilkea acuminata Mast. (FG); Dilkea retusa Mast. (SU); Mitostemma glaziovii Mast. (GU); Mitostemma jenmanii Mast. (GU)

Robustly woody lianas and climbing shrubs. **Climbing via tendrils, but these are often lost with age.** Ancistrothyrsus tendrils occur only upon inflorescence branches with a single terminal thickened hook. Dilkea tendrils occur in leaf axils and inflorescence branches, spiral-coiled with three tiny terminal claws. Mitostemma tendrils occur in leaf axils and are simple.

Stems: Branchlets round, nodes smooth with thread-like stipules, small and short-lived. Older woody stems round, robustly woody; **stem x-s pattern simple (not lobed as in many Passiflora).**

Leaves: **Simple, unlobed, margins entire;** leathery, blades hairy with reddish glandular scales (Ansistrothyrsus) or mostly hairless (Dilkea, Mitostemma); glands absent; petiole thick at base.

Inflorescences: Short racemes or clusters, often borne on woody stems [*cauliflorous*]; floral tube and calyx absent or small, **flowers 4-parted, white (red); corona 1-3-ringed;** stamens 4-8(10), free or slightly fused at base; **ovary superior, either stalked, united at base, splitting to 4 above midpoint** (Ancistrothyrsus, Dilkea) **or unstalked, with 4 styles straight from base** (Mitostemma).

Fruits: Globose to ovoid berries with a fleshy, sweet aril (Dilkea, Mitostemma) or a 4-angled capsule (Ancistrothyrsus).

Distribution: Rare; Guianas collections mostly in far southern lowland forests; many species also in N Amazonia.

Notes: Dilkea saplings may be confused with the common treelet, Clavija (Theophrastaceae).

1. Passiflora nitida, flower with green basal bracts, white perianth, and a purple-blue-white-magenta-banded corona [BH]

2. Passiflora acuminata, with 'lobed' stem cross-section (x-s) pattern common in Passiflora [FRD]

3. P. acuminata, a diagonal stem slash and the 'lobed' stem x-s pattern [FRD]

4. P. acuminata, tendril borne in leaf axil [FRD]

5. Passiflora amoena, a woody liana in old secondary forest [BH]

6. P. amoena, pink perianth, yellow corona [BH]

7. Passiflora auriculata, white perianth with orange corona [PT]

8. P. auriculata, globose black berries [FRD]

9. P. auriculata, paired ear-like glands at petiole base [FRD]

10. Passiflora coccinea, flowers in an open area, with red basal bracts [BH]

11. P. coccinea, leaf base with palmate venation [FRD]

12. P. coccinea, leaf with gland-tipped marginal teeth [FRD]

13. P. coccinea node, with stipule, globose petiole glands, and axillary tendril [FRD]

14. P. coccinea, typical secondary forest habitat [PT]

15. Passiflora glandulosa, light red perianth, white corona [PT]

16. P. glandulosa, lobed x-s pattern [BH]

17. P. glandulosa, immature elongate fruit [PT]

18. Passiflora nitida, flower with a dark blue 'corona' [NH]

19. P. nitida, sectioned immature fruit with many seeds [BH]

20. P. nitida, flower with part of 'corona' removed to show the androgynophore (fused male and female reproductive organs) [BH]

PHYTOLACCACEAE

PHYTOLACCACEAE

Woody lianas or clambering shrubs; also trees, succulent and woody shrubs and herbs. Climbing via twining shoots, paired spines, or lateral branchlets. Stipules are present as paired spines or inconspicuous. Leaves are alternate, simple, fleshy and pinnate-veined. Inflorescences are loose racemes or panicles (in climbers), with 4-5 parted radial flowers and a superior ovary. Fruits are fleshy drupes, berries or winged samaras.

General: Armed or unarmed; vegetative surfaces green, fleshy, smooth and hairless; hairs simple or many-forked when present; colorful sap and nectar glands absent.

Stems: Branchlets round or angled, green and smooth, with lengthwise furrows, often hollow or soft; **paired stipular spines present** (Seguieria) **or short-lived, tiny stipules** (Trichostigma). Older stems round, 3-15 cm diameter, green and smooth, turning brown with age; **stem x-s pattern simple or with very fine concentric circles (Seguieria).**

Leaves: Elliptic, medium-sized, margins entire (Seguieria) or wavy/scalloped (Trichostigma); **mostly green, ± fleshy, smooth and hairless, drying olive or black;** leaf glands absent; **petioles sometimes long and thin** (Trichostigma) (Gentry 1993).

Inflorescences: Axillary or terminal, many-flowered, panicles of 5-parted flowers (Seguieria) or loose racemes of 4-parted flowers (Trichostigma), bracts often present; **flowers small, radial, with one whorl of free sepals** (petals absent), greenish white to pink or purple; stamens 8 to many; ovary superior, 1-chambered in climbers.

Fruits: Winged samara (Seguieria) **or fleshy drupe** (Trichostigma), sepals and stamens often persistent; seeds 1.

Ecology: Seed dispersal via animal (birds) or wind.

Distribution: Mostly neotropical: 13 genera/60 species; Guianas: 6 genera, 2 genera with woody to subwoody climbers.

Use: Some shrub and weedy herb species are used in traditional medicine.

Notes: Securidaca (Polygalaceae) and Machaerium (Leguminosae) liana species have winged samaras similar to Seguieria. Machaerium has axillary spines (modified stipules) on stems and tendril-like shoots. Securidaca and Machaerium both have "Leguminosae-Papilionoideae" pea-like flower forms, which are distinctly different from the open, radial flowers of Seguieria.

1. Seguieria Loefl.

Woody lianas or clambering shrubs; occasional trees; usually armed.

General: Armed; vegetative surfaces green, fleshy, smooth and hairless (inflorescence sometimes hairy).

Stems: Branchlets and older stems with **robust paired spines or thorns** on raised base above petiole, spines straight or recurved, often spreading, small to > 1 cm long.

Leaves: Highly variable in shape and size, **margins entire;** petiole 0.3-0.5 cm long.

Inflorescences: Terminal or axillary, many-flowered panicles, often with leafy, spear-shaped to triangular bracts; **flowers 5-parted, sepals white to yellow-green,** 0.45-0.75 cm long, reflexed in fruit; **stamens 15-65;** ovary 1-chambered, with 1 style, stigma with tiny bumps [*papillose*].

Fruits: **Dry, winged samara,** 4-5 cm long, often with winglets near base of fruit; seed 1, seed coat black or reddish-brown.

Ecology: Seed dispersal via wind.

Distribution: Widespread in the American tropics and subtropics, from Argentina to Trinidad: 6 species; Guianas: 2 species (Rohwer 1982).

1.1. Seguieria americana L.
[Syn: S. aculeata L.]

Liana or clambering shrub; also treelet. **Stems:** spines straight or recurved, up to 2 cm long. **Leaves:** variable in shape, to 15 x 7 cm; papery to sub-leathery. **Inflorescences:** panicles to 40 cm long, many flowered (< 100), light to densely hairy; bracts leaflike; sepals to 0.75 cm long, stamens 0.5-0.6 cm long. **Fruits:** winged samara to 5 cm long, turning greenish or yellowish upon drying (never black); seed coat reddish-brown.

Common; partially-open areas, forest and river margins. Guianas; widespread in the neotropics.

1.2. Seguieria macrophylla Benth.

Woody liana or clambering shrub. Stems: older branches hollow; spines recurved, up to 1.2 cm long. **Leaves:** mostly elliptic, to 18 x 8.5 cm; often leathery. **Inflorescences:** panicles to 50 cm long, often > 100-flowered, light to densely hairy; bracts small; sepals small, to 0.45 cm long; stamens < 0.5 cm long. **Fruits:** winged samara to 4 cm long, turning dark to black upon drying; seed coat black.

Uncommon; partially-open areas, forest and river margins, old-growth forest. Guianas; widespread in tropical America from Bolivia to Panama and Trinidad.

2. Trichostigma A.Rich.

Woody lianas or clambering shrubs; occasional trees; unarmed. Climbing via lateral branchlets.

Distribution: Neotropics: 3 species; Guianas: 1 species.

2.1. Trichostigma octandra (L.) H.Walter

Liana or clambering shrub; also erect shrub or treelet. **Stems: lateral branchlets hanging downwards,** green, round, bearing many lenticels; nodes unarmed, without paired spines, stipules present but tiny, short-lived; older stem round, 5-15 cm diameter, reaching 5-30 m length; stem x-s with scattered vascular bundles visible. **Leaves:** elliptic or oblong, 4-12 × 2-5 cm, base acute or obtuse, apex acute or acuminate, **margins wavy or scalloped/round-toothed** [*crenulate*]; dark green and smooth above, light green with 3° veins inconspicuous below; petioles to 3 cm long, swollen at the base. **Inflorescences:** axillary, loose, many-flowered racemes 4-12(-15) cm long, **flowers 4-parted,** radial, **sepals white, small** (0.3-0.5 cm long); **stamens 8-16;** ovary ovoid, with many threadlike branch-es, 1-chambered. **Fruits:** a **fleshy, ovoid drupe,** red or purple, 0.5-0.9 cm diameter, with old, dry sepals and stamens remaining attached; seed 1. **Ecology:** Seed dispersal via animal (bird). Flowering and fruiting from April to October. **Use:** Cultivated as a decorative plant or bower, as a craft fiber, and for medicine.

Common (locally abundant); non-flooded old-growth forest, secondary forest, open areas. Guianas; widespread in the neotropics and the Caribbean.

---

1. Trichostigma octandra, with alternate, simple leaves [RH]
2. Securidaca americana, leafy shoot with green, smooth stems [BH]
3. T. octandra, with racemes of white flowers [RH]
4. S. americana, with paired axillary spines [BH]
5. T. octandra, 4-parted flowers with white sepals and petals absent [RH]

PHYTOLACCACEAE

POLYGALACEAE

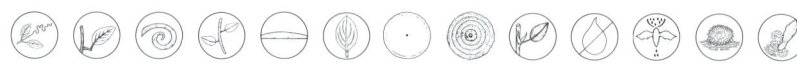

POLYGALACEAE

Woody lianas, climbing shrubs, and herbaceous vines; also annual herbs and small trees. Climbing via twining shoots, clasping branches, and spines. Leaves are alternate and simple with entire margins. Stipules are present, often modified as glands or spines. Flowers are legume-like or half-tubular; ovary superior. Fruit forms are diverse, including dry capsules, fleshy drupes, and winged samaras.

General: Mostly hairless, sometimes densely hairy, hairs simple when present; cut parts without distinctive sap, color, scent or taste.

Stems: Branchlets often green, flexible, smooth and hairless, round to irregularly angled or flattened, sometimes with tiny, scattered spines (Moutabea); **nodes often with raised mound bearing 2 circular glands or spines** (= modified stipules) **where the leaf stalk [*petiole*] joins the stem;** stipules sometimes unmodified and inconspicuous at leaf axils. Older woody stems round, with persistent raised mounds and/or spines at nodes, bark smooth to rough and fibrous, **stem x-s pattern simple or off-center concentric,** center sometimes hollow.

Leaves: Shape and size highly variable; **leaf glands common, petioles short.** In Moutabea, leaf blades are thick-leathery to fleshy with invisible or faint veins. In the remaining genera, leaf blades are leathery to thin with all veins visible and 3° veins net-veined [*reticulate*].

Inflorescences: Terminal or axillary, many-flowered racemes or panicles, or solitary; flowers bisexual, small- to medium-sized, 5-parted; stamens 8, free or clustered; ovary superior, 2-5-chambered [2-5-locular], style 1, stigmas 1 or 2. Two flower forms: i) **legume-flower-like** [*zygomorphic*] with 2 sepal-wings, a boat-shaped lower petal, and 2 lateral petals (Bredemeyera, Securidaca); or ii) **half-tubular** with free or fused sepals and petals (Barnhartia and Moutabea).

Fruits: Capsules (Bredemeyera), drupes (Barnhartia, Moutabea), or winged samaras (Securidaca).

Ecology: Due to diverse flower forms and colors, the effective pollinators are diverse – including bees, beetles, butterflies, moths, birds and bats. Similarly, a wide diversity of fruit forms reflects diverse seed dispersal agents — including animals, wind, and water.

Distribution: Cosmopolitan (primarily annual herbs in temperate regions): ca. 21 genera and 1000 species; Guianas: 4 genera including woody climbers.

Use: Medicine; cordage; water source.

Notes: Polygalaceae climbers may be confused with Maripa (Convolvulaceae), Seguieria (Phytolaccaceae) and many species in the Menispermaceae. These taxa often have a similar vegetative appearance and a 'concentric' (concentric circles) woody stem x-s pattern. The leaf glands and glands/spines at stem nodes are important in identifying Polygalaceae species (Seguieria also has paired spines at nodes). Polygalaceae with legume-like flowers can be distinguished from Fabaceae climbers by their simple leaves. To identify Polygalaceae to the species level, leaves and fruits are often more useful than flowers.

1. Barnhartia Gleason
Monotypic genus.

1.1. Barnhartia floribunda Gleason

Common name: abuyamibia (Ar)

Slender woody liana or a weakly climbing shrub. Climbing via twining shoots — clasping branches and spines absent. **Stems:** branchlets round to flattened, bark smooth, hairless and reddish-green, lenticels few; **nodes without a raised mound bearing glands or spines, stipules inconspicuous;** older woody stems flattened, less than 1 cm diameter, smooth, lenticels few; stem x-s pattern simple. **Leaves:** elliptic-oblong or lanceolate, 9-14 x 2.4-5 cm, base cuneate, apex acute to acuminate; **subleathery,** hairless, **with 2 pairs of circular glands, one at the leaf base and another at the petiole apex; 2° veins often with minor, parallel inter-second-** ary veins; petiole 0.5-1 cm long, lightly hairy; stipules and paired glands/spines at petiole base absent. **Inflorescences:** axillary or terminal, in compound racemes, 2-8 cm long; leafy bracts persistent, paired basal glands absent; flowers partially tubular, small, 0.6-0.8 cm long, off-white to yellowish; sepals free; petals linear, with one free, others paired; stamens in 3 bundles; ovary 2-chambered, stigma 1. **Fruits:** drupe-like (similar to Moutabea), globose, 0.7-1 cm diameter, hairless, yellow to orange at maturity, 2-seeded; seeds ± ellipsoid, to 0.7 x 0.5 cm. **Ecology:** Seed dispersal via animal (ingested by monkeys). **Notes:** This species can be identified from leaf gland and vein characters alone.

Uncommon; non-flooded old-growth forest and savannah forest. Guianas; also Amazonian Brazil.

2. Bredemeyera Willd.

Woody lianas and weakly-climbing shrubs; also treelets. Climbing via twining shoots, lateral branches and petioles — clasping branches and spines absent.

General: hairless or hairy with simple hairs.

Stems: Branchlets round, bark smooth, hairless or short-haired, leaf scars conspicuous, lenticels absent; **nodes without a raised mound bearing glands or spines, stipules absent.**

Older woody stems round, smooth, lenticels absent; **stem x-s pattern concentric.**

Leaves: Size and shape highly variable, elliptic to orbicular; **leathery or subleathery; circular glands absent;** main 2° veins ± straight and parallel, often with minor, parallel inter-secondary veins.

Inflorescences: Terminal or axillary panicles, axes whitish-yellow, with persistent leafy bracts, paired basal glands absent; **flowers pea-like and very small,** 0.2-0.6 cm long, greenish-white to yellowish-white; sepals 5 with 2 wings, falling away in fruit; petals 3, keel-shaped lower petal without a crest, tri-lobed; stamens in 3-bundles; ovary 2-chambered, stigma bi-lobed.

Ecology: Seed dispersal via wind.

Fruits: **Diagnostic flattened, spoon-shaped [*spathulate*] capsules** with an apical notch; seeds 2, each with a silky hairtuft longer than the seed, aril short or absent.

Distribution: Neotropics, West Indies, Australia: 60 species known; Guianas: 1-2 species.

Use: medicine — tonics, stimulants, and aphrodisiacs.

2.1. Bredemeyera densiflora A.W.Benn.

Common name: mibi (Ar)

Woody liana, sprawling shrub; also treelet. **Stems:** branchlets round, with whitish-yellow hairs [*villose or tomentose*]. **Leaves:** ovate, 5-7 x 2.3-3 cm, base rounded, apex attenuate; strongly leathery, hairless and shiny above and below; petiole 0.5-1 cm long. **Inflorescences:** terminal, dense panicles, 7-15 cm long, flowers 0.2-0.3 cm long; corolla greenish-white, mostly hairless, ±equal to calyx "wings" in length. **Fruits:** capsule ca. 1 x 0.4 cm, bright green, shiny, black when dry.

Common; savannah, shrub savanna, and savanna forest on white sand. Guyana, Suriname; also N Amazonian Brazil.

2.2. Bredemeyera lucida (Benth.) Klotzsch ex Hassk.

Liana or sprawling shrub. **Stems:** branchlets round, bark soft and flaky, with orangish-white hairs [*tomentose*] or hairless. **Leaves:** elliptic-ovate, 6.7-14 x 3-6.5 cm, base rounded, apex obtuse to acuminate; leathery, shiny above, minutely hairy below; petiole 0.5-1 cm long. **Inflorescences:** axillary, large, open panicles, 2-30 cm long; flowers 0.3-0.4 cm long; corolla greenish-white, hairless, ± equal to calyx "wings" in length. **Fruits:** capsule 1-1.3 x 0.4 cm, greenish-tan, shiny, black when dry. **Notes:** B. Hoffman coll. no. 5330, Central Suriname Nature Reserve.

Uncommon; non-flooded old-growth, savanna and riverine forests. Guyana, Suriname; also N Amazonian Brazil and Mato Grosso, north to Mexico and Trinidad.

3. Moutabea Aubl.

Woody canopy lianas and sprawling shrubs; also erect shrubs. Climbing via twining shoots and lateral branches — clasping branches and nodal spines absent.

General: Hairless or hairy with simple hairs.

Stems: Branchlets round, bark furrowed, with small spines, lenticels few; **nodes with raised mound bearing paired circular glands** (modified stipules). Older woody stems round to flattened, twisted, with small scattered spines and furrows; **stem x-s pattern simple or off-center concentric.**

Leaves: Thick-leathery, succulent, hairless or lightly hairy, **2° and 3° venation immersed, invisible or faint.**

Inflorescences: Axillary, short racemes with persistent leafy bracts/bracteoles, with or without circular glands at the base; **semi-tubular,** calyx tubular, with 5 ± equal lobes, densely hairy; **corolla with 5 off-white petal lobes,** shorter than calyx tube, 1 petal free, others in 2 pairs; 8 stamens in 2 groups; ovary 4-5-chambered, style straight, stigma head-like.

Fruits: Leathery globose berries to 5 cm diameter, yellow-orange at maturity, pulp edible and sweet; seeds 2-5, ovoid, black.

Ecology: Seed dispersed via animal (ingested by spider monkeys).

Distribution: Widespread in South American tropics: 8 species; Guianas: 2-3 species.

Use: Older woody stems with abundant drinkable water.

3.1. Moutabea aculeata (Ruiz & Pav.) Poepp. & End

Shrubby, high-climbing liana. Stems: older woody stems observed to 5 cm diameter, with vertical furrows and horizontal ridges; **bark chocolate brown, thin and easily shattered,** inner bark reddish-brown, wood yellow-white, **stem x-s pattern simple, not concentric. Leaves:** oblanceolate, to 17 x 4.5 cm, base cuneate to truncate, apex acute; petiole dark, 0.3-0.5 cm long. **Inflorescences:** axillary, slender racemes 1.5-2.5 cm long, axes hairless, **bracts without 2 basal glands;** flowers 1.5-2 cm long, corolla white and hairless. **Fruit:** Not seen.

Uncommon; non-flooded old-growth forest. Guianas; also Amazonian S America to Costa Rica.

3.2. Moutabea guianensis Aubl.

Common name: aymoutabou (Ca); grão-de-macaco (BP); kuturi (Tr); lianappel (NL)

Woody canopy liana or sprawling shrub. Stems: older woody stems observed to 15 cm diameter; **bark grey- to yellowish-brown, fibrous,** living bark and wood yellow; **stem x-s pattern with alternating yellow and white rings. Leaves:** oblong to oblanceolate, 4-10 x 1.5-4 cm, base cuneate to acute, apex variable; petiole 0.3-0.7 cm long. **Inflorescences:** axillary, short racemes, 1-3 cm long, lightly hairy; **bracts with 2 basal glands;** flowers 1.5-2.5 cm long; calyx yellow; corolla off-white and hairless. **Fruits:** globose berries to 5 cm diameter, maturing yellow to orange.

Common; rain, savannah, and creek forests. Guianas; Brazil.

---

1. Securidaca diversifolia, inflorescence close-up with zygomorphic, legume-like flowers [PT]
2. Bredemeyera densiflora, inflorescence [PT]
3. B. densiflora, leaves and inflorescences [PT]
4. Moutabea guianensis, growing as a shrub with edible, globose berries [BH]
5. Moutabea sp. with white, irregularly tubular flowers [FRD]
6. Moutabea guianensis, ripe globose berry [BH]
7. M. guianensis, woody stem x-s with narrow, off-center concentric rings [BH]

4. Securidaca L.

Woody lianas and climbing shrubs; also treelets. Climbing via twining shoots, lateral branches, clasping branches and nodal spines.

General: Hairless or lightly hairy with simple hairs.

Stems: Branchlets round to flattened, often looping; bark smooth, olive-green, usually with vertical lines or ridges, lenticels absent; **nodes with raised mound bearing 2 circular glands or spines (modified stipules), stipules sometimes unmodified and inconspicuous.** Older woody stems round to flattened, often furrowed and twisted; bark smooth or rough and bearing lenticels; **stem x-s pattern usually off-center concentric.**

Leaves: Elliptic to oblong or ovate, subleathery to thin; circular glands absent on leaf blade; midvein vein immersed above, 2° and 3° venation visible; petioles short.

Inflorescences: Axillary or terminal, often in many-flowered, showy panicles or with solitary flowers; **leaves often present within the inflorescence rachis; flowers legume-like;** bracts and bracteoles with paired glands present or absent; calyx 5 with 2 sepal-wings, falling away in fruit; **corolla magenta, purple, or white, often bi-colored,** petals 3, **keel-shaped petal with or without a crest;** 8 stamens in 2 groups, forming a sheath; ovary 1-chambered, style short and curved, stigma terminal.

Fruits: Winged samaras, primary wing broad, seldom short, often with a smaller secondary wing; seed 1.

Ecology: Seeds dispersed via wind or water.

Distribution: Pantropical genus, primarily neotropics: 80 species; Guianas: 17 species; 5 presented here.

Uses: The bark of some species is used for cordage.

Notes: Compare to other taxa with twisted, clasping branchlets and shoots – Convolvulaceae (Maripa), Euphorbiaceae (Omphalea), Leguminosae (Machaerium) and many Celastraceae (Salacia, Tontalea).

4.1. Securidaca coriacea Bonpl.

Woody liana; hairy. **Stems:** branchlets flattened, young growth densely hairy; **nodes bearing 2 circular glands. Leaves:** ovate to broadly ovate, 3-6 x 3-5 cm, base truncate or cordate, apex obtuse, margins curling; subleathery, softly hairy below; petiole 0.1-0.4 cm long, densely hairy. **Inflorescences:** terminal panicles, hairy, subunits 4-15 cm long; bracts lanceolate; infl. leaves to 2.7 cm long; wings and petals rose to violet, **keel with apical crest. Fruits:** samaras 4-4.5 cm long, 1° wing to 4 cm long, 2° wing to 1.4 cm long. **Notes:** S. coriacea is differentiated from S. diversifolia by denser and softer hairs on leaves and branches, and a longer secondary wing.

Uncommon; old-growth forest. Guyana, Suriname; also widespread in tropical S America.

---

8.   Moutabea sp. woody stem [FRD]

9.   Moutabea guianensis immature fruits [PT]

10. Securidaca diversifolia liana habit with showy inflorescences [PT]

4.2. Securidaca diversifolia (L.) S.F. Blake

Common name: sukrudjani (Sr)

Woody liana or sprawling shrub; hairy. **Stems:** branchlets round, softly-hairy, green; **nodes bearing 2 circular glands. Leaves:** highly variable, ovate to oblong, 3-9 x 1.5-3.5 cm, base truncate, apex obtuse, margins curling; leathery, shortly-hairy [*pilose*] below; petiole 0.2-0.4 cm long, densely hairy. **Inflorescences:** terminal, axes orange, densely hairy [*strigulose*], sub-units 4-12 cm long; bracts narrowly ovate; infl. leaves broadly ovate to 3 cm long; wings and petals purple or magenta, **keel with apical crest. Fruits:** samaras 3.5-5 cm long, 1° wing to 4 cm long, 2° wing to 0.4 cm long.

Common; ridge and riverine forests, disturbed areas. Guianas; widespread, from Peru to Mexico and the Lesser Antilles.

4.3. Securidaca paniculata L.C. Rich.

Common name: kurumuri (Tr) ; matu lémíki (Sa)

Woody liana or sprawling shrub. **Stems:** branchlets bamboo-like, smooth with vertical lines, minor scratches reveal green subbark layer; **nodes bearing paired spines, 1-2 cm long** (nipple-like projections when young). Older woody stems round to oval, to 10 cm diameter, nodes bearing spines or spine bases. **Leaves:** oblong-elliptic, 6-14 x 2.5-7 cm, base cordate or truncate, apex with long, narrow tip, margins flat; subleathery, hairless and shiny above, mostly hairless below; petioles 0.3-0.7 cm long, lightly hairy. **Inflorescences:** axillary (terminal) panicles, axes densely hairy, sub-units 2.5-6 cm long; bracts ovate; flowers 0.6-0.8 cm long, wings pink; corolla pink or purple, **keel with apical crest. Fruits:** samaras 2.5-3 cm long, green; 1° wing to 2 cm long, 2° wing to 1 cm long or absent.

Common; creek and river banks, forest edges. Guianas; also N South American tropics to Peru.

4.4. Securidaca pubescens DC.

Common name: kwata-ede-sula (Sr); matu lémíki (Sa)

Woody liana or sprawling shrub; treelet; hairy. **Stems:** branchlets dark brownish-green, densely hairy; **nodes bearing 2 circular glands. Leaves:** ovate to broadly ovate, 4-9 x 2.5-6 cm, base acute to obtuse, apex acute to acuminate; subleathery, greyish-green, shiny above, hairy below; venation straight, not prominent above; petioles 0.5-0.8 cm long, densely hairy. **Inflorescences:** axillary and terminal, panicles many-flowered, axes pinkish; flowers 1-1.2 cm long, wings dark-purple, maturing light-purple; petals magenta, **keel with apical crest. Fruits:** samaras 4.5 cm long (only young ones seen), green to pinkish; 1° wing thin, ± rhomboid, to 4 cm long, darkly veined, 2° wing very small.

Uncommon; non-flooded and floodplain old-growth forest, riverbank forest. Guianas; also N South America.

4.5. Securidaca uniflora Oort

Woody liana or sprawling shrub; hairless. **Stems:** older stems and branchlets flattened, slender; **nodes with small lanceolate stipules, small and falling early, not reduced to glands. Leaves:** elliptic to ovate, 3-8 x 2-4.5 cm, base rounded, apex acute, margins entire; subleathery, greyish-green and shiny above, dull green below; petiole 0.2-0.3 cm long, conspicuously wrinkled. **Inflorescences:** axillary or terminal, flowers 1(2) on short shoots; flowers 0.5-0.7 cm long, fragrant; wings white, **keel very broad, without a conspicuous apical crest. Fruits: samaras, very large, to 7 cm long,** brown, hairless; 1° wing thin, to 5.5 cm long, 2° wing orbicular, very short.

Uncommon; rock savannas and edges of non-flooded old-growth forest on bauxite. Guianas; also Brazil (Amazonas).

POLYGONACEAE

POLYGONACEAE

Woody lianas, climbing shrubs, and vines; also herbs, shrubs, or trees. Cosmopolitan: 31 genera/1500 species; Guianas: 5 genera/32 species, one non-native vine (Antigonon) and one genus of native woody climbers.

1. Coccoloba P.Browne

Mostly woody lianas or climbing shrubs; also a few erect shrubs or trees. Climbing via twining shoots. Leaves are alternate, simple, and leathery with entire or wavy margins and pinnate venation. Stipules are present as a tubular sheath. Inflorescences are long spikes or racemes of tiny white flowers; ovary superior. Fruits pea-sized, ovoid or angled, single-seeded.

Common Name: apukú tatái (Sa); bradilifi (Sr); góni lopu (Sa), pakia tatái (Sa); tamo (Tr)

General: Hairless to finely-hairy [*puberulous*]; **cut parts often with conspicuous scents (e.g., musky, onion-like).**

Stems: Branchlets solid or hollow, with a jointed appearance; bark smooth and bearing many lenticels, often dark green; **nodes with a closed stipule sheath** [*ochrea*] encircling apical bud in new growth, persisting as an open, papery tube and **leaving a circular scar on the stem;** ochrea fine-pointed [*acuminate*] or with a flat apex [*truncate*]; **older stems often 2-cabled** (resembling a double-barrel shotgun); bark hard and fibrous, not peeling; wood very hard; stem x-s pattern mostly simple, rarely with concentric rings in cabled stems.

Leaves: Often of different sizes from top to bottom on a stem; **blade often uneven at base, thin to leathery, dark green;** glands usually absent; venation pinnate raised, 3° veins often parallel; **petioles curved,** inserted below base of stipule sheath, short- to medium-length.

Inflorescences: Terminal or axillary, **spikes or narrow racemes of tiny white or whitish-green flowers** (0.2-0.4 cm long), axes single or in bundles [*fascicles*] **and with a tiny tubular sheath** [*ochreole*] **at base;** flowers 5-parted, unisexual; male flowers — 8 stamens fused into a ring; female flowers — ovary superior [*1-locular, 3-carpellate*], styles 3, free or united.

Fruits: Round or ovoid achenes, green to pink or red, hard, 2-3-4 angled and often pointed, surrounded by expanded, persistent parts of the flower; seeds 1.

Ecology: Seed dispersal via animal (through the gut of birds, monkeys, or fish).

Distribution: A New World group with approximately 400 species; Guianas: 19 species, 14 woody climbers, 5 species described here.

Use: Medicine for eye, ears, ant bites and abdominal complaints, and against lice; stems and leaves used in Maroon herbal baths to please forest spirits; the fruits of some species are edible, although not popular as food.

Notes: Stipule sheath [*ochrea*] characters and slash scent are useful for family recognition and distinguishing between different Coccoloba species.

1.1. Coccoloba ascendens Duss ex Lindau

Liana or climbing shrub; without strong scent. **Stems: ochrea split, long-acuminate, 1-2.5 cm long;** older stems round, to 7 cm diameter, bark whitish-grey, with furrows and lenticels, sub-bark dark red. **Leaves:** narrowly elliptic (ovate), long, 9-27 x 5-16 cm, base rounded to nearly heart-shaped [*sub-cordate*], apex rounded to obtuse; thick-leathery; petiole 0.5-7 cm long. **Inflorescences:** racemes 5-20 cm long, axes angled; **ochreolae sheath bell-shaped, 0.1-0.2 cm long;** flowers solitary or paired, dark pink with green tip. **Fruits:** ovoid, to 1.7 x 1.6 cm, greenish-red to blue, tapering toward tip, juicy; achene 3-angled.

Uncommon; non-flooded old-growth forest on laterite, riparian forest, open areas. Guianas; also widespread, from Brazil (Mato Grosso, Rondonia) to Costa Rica and the Caribbean.

1.2. Coccoloba excelsa Benth.

[Syn: Coccoloba parimensis Benth.]

Liana or climbing shrub; also shrub or treelet; **with rank or onion scent. Stems: ochrea long-acuminate, 1-6 cm long, brown-hairy;** older stems round or flattened, **2-cabled,** bark grey to brownish-black, with furrows and lenticels. **Leaves:** elliptic to obovate, 9-25 x 3.5-20 cm, base rounded or cordate, apex ± acuminate; leathery to thin, waxy above, hairy below; petiole 1.5-4 cm long. **Inflorescences:** racemes 15-18 cm long; **ochreolae narrow-tubular, 0.2-0.3 cm long;** flowers paired, greenish-white. **Fruits:** ovoid, to 0.9 x 0.7 cm, green to wine-red to black, base tapering, lobes ca. 0.15 cm long; achene lens-shaped to ovoid.

Common; coastal savanna, shrubby savanna, riparian zones, forest edges and older secondary forest. Guianas; widespread, from Brazil (Mato Grosso, Acre, Para) to Panama.

1.3. Coccoloba gymnorrhachis Sandw.

Liana or climbing shrub; erect shrub or treelet; scent unknown. **Stems: ochrea split, unevenly truncate, < 1 cm long;** older stems flattened, **2-cabled,** bark gray with brown lenticels. **Leaves:** elliptic to obovate, 6-14 x 4-9 cm, base truncate or sub-cordate, apex long-acuminate; papery to leathery; petiole 0.7-2.0 cm long. **Inflorescences:** terminal on lateral branches, racemes 14-23 (30) cm long, loosely flowered; **ochreolae tubular, 0.2-0.3 cm long;** flowers paired. **Fruits:** sub-ovoid, to 1.5 x 1.3 cm, dark green, base obtuse, apex 4-5 obscurely furrowed.

Uncommon; old-growth forest, creek forest, savanna; erect shrub in savanna woodland. Guianas only.

1.4. Coccoloba lucidula Benth.

High-climbing liana; rank or onion scent present. Stems: branchlets with reddish-brown hairs and lenticels, **ochrea truncate, < 1 cm long, thin;** older stems angled, furrowed, and irregularly branched. **Leaves:** oblong to obovate, 6 -10 x 2.5-5 cm; thin to leathery. **Inflorescences:** racemes to 6 cm long. **Fruits:** globose to nearly pear-shaped [*pyriform*], ca. 1.2 cm diameter, green to red to purple-black, finely ribbed.

Uncommon; riparian zones and shrubby savanna. Guianas; also from Brazil (Mato Grosso, Bahia, Rondonia) to Venezuela.

1.5. Coccoloba marginata Benth.

Liana, climbing shrub, or treelet; rank or onion scent present. Stems: branchlets with reddish-brown hairs and lenticels, **ochrea unevenly truncate, < 1 cm long;** older stems round or oval, not 2-cabled, bark brown or black, lenticellate. **Leaves:** variable, 8-17 cm long, base rounded to subcordate, apex short-acuminate; leathery; 3° veins prominent above and below; petioles 0.5-2.5 cm long, **Inflorescences:** racemes 8-15 cm long, **ochreolae 0.1-0.15 cm long, thin, 2-lobed;** flowers solitary. **Fruits:** ovoid, to 0.9 x 0.6 cm, green to reddish-brown, lobes to 0.2 cm long.

Common; old-growth forest on ridges, riparian zones, secondary forest, and savanna. Guianas; also widespread from S Brazil to Amazonian Peru, Colombia and Venezuela.

1. Coccoloba sp., with 2-cabled, woody stem, tough leaves and green, smooth branchlets [BH]
2. Coccoloba sp., node with ochrea sheath [BH]
3. C. marginata stem slash [BH]
4. Coccoloba sp. in fruit [PT]

POLYGONACEAE

RHAMNACEAE

1-5

RHAMNACEAE

Woody canopy lianas or shrubby climbers; mostly erect shrubs and trees. Climbing via twining shoots or tendrils. Leaves are alternate and simple. The two genera with climbers in this family - Ampelozizyphus and Gouania – are so different in appearance that few people would guess they are related. Ampelozizyphus amazonicus is a large, woody, twining forest liana with entire, palmate-veined leaves. Gouania species are shrubby climbers of open areas with often toothed, pinnate-veined leaves and unique tendrils. Stipules are present in both but short-lived. Inflorescences are compact clusters in Ampelozizyphus and mostly long spikes in Gouania; both bear small, greenish-white, 4-5-parted flowers with free, clawed petals that partly enclose the stamens; stamens 4-5; ovary superior to inferior [hypanthium], 3-chambered with a 3-lobed style. Fruits are ovoid drupes in Ampelozizyphus and 3-winged, propeller-like capsules in Gouania.

Distribution: Widespread in the tropics, subtropics, and southern temperate zones: 52 genera/ 900-1000 species; Guianas: 3 genera total, 2 genera with woody climbers.

1. Ampelozizyphus Ducke

1.1. Ampelozizyphus amazonicus Ducke

Common name: bergimanbebe (Sr); weweimë (Tr)

Robust woody liana, in loops on forest floor or climbing high into canopy. Climbing via twining stems. **General:** few notable characters – hairless, sometimes with **camphor or spearmint scent in cut parts. Stems:** branchlets angular, ± reddish-brown hairy, bark smooth with abundant lenticels; nodes nondescript, with **tiny, short-**lived stipules; older stems round - furrows and grooves absent, observed to 5 cm diameter, bark thick-fibrous, inner layer dark red, wood white; stem x-s pattern simple. **Leaves: oblong-elliptic and large,** 15-30 x 7-15 cm, base rounded to obtuse, apex shortly acuminate, **margins entire;** thick and leathery, occasionally rough-surfaced [*asperous*]; conspicuously **3(-5)-palmate-veined from near base with ladder-like cross-veins,** upper leaf with pinnate 2° veins forming loops, 3° veins net-veined [*reticulate*]; **leaf stalk [*peti-***

ole] **with a groove or canal on upper side,** often recurved, 1-2.5 cm long. **Inflorescences:** axillary, compact clusters [*cymes*], stalks 2-4 cm long, reddish-brown-hairy; flowers greenish-yellow. **Fruits:** round or obovate, fleshy drupes, 1.3-2 cm diameter, 3-chambered. **Ecology:** Insect-pollinated. Seed dispersal via animal (thru monkeys or bird gut). **Distribution:** Rare, locally abundant; old-growth forest, scrubby forest on white sand, southern forests. Guianas; also N South America. **Use:** The bark and wood is used as a traditional medicine by the Trios for sev-

eral medicinal purposes. **Notes:** Climbers with leathery, simple-alternate, palmate-veined leaves and no tendrils occur in relatively few families. Compare to Aristolochiaceae (Aristolochia), Hernandiaceae (Sparattanthelium) and various Menispermaceae. Two petiole characters are useful in identification: a topside groove or channel and an evenly thick stalk - not flexed or swollen at the ends as in the Menispermaceae.

2. Gouania Jacq.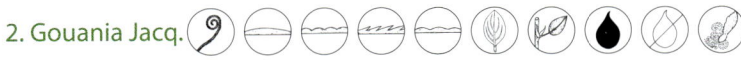

Lianas and shrubby climbers. Climbing via unique tendrils (modified branchlets).

General: Conspicuously to lightly hairy; **sometimes with mint scent (G. blanchetiana) or watery red sap (G. velutina).**

Stems: Branchlets 6-8 ridged or furrowed, often hairy (blond to reddish-brown), wood soft; nodes with **paired, leafy stipules,** short-lived; **tendril coiled in one plane like a butterfly tongue or watchspring, borne on branchlets in terminal leaf axils.** Older stems round to strongly grooved or fluted, mostly slender but sometimes to 5 cm diameter, bark often corky; stem x-s pattern simple.

Leaves: Narrowly-elliptic to broadly-ovate, thin, **margins often toothed (large),** also round-toothed, wavy or ± entire; **cone-shaped glands sometimes present at leaf base;** veins mostly pinnate, sometimes weakly 3-nerved at base, **2° veins close together** and curving sharply upwards into teeth, **teeth often glandular,** 3° veins often parallel and angled [*oblique*] to 2° veins; petioles canal-shaped.

Inflorescences: Axillary and terminal, spikes with numerous tiny, white, 5-parted flowers; ovary inferior, sunken in fringed nectar disc.

Fruits: Three-parted propeller-like capsules, small, 0.4-1.5 cm diameter, crowned by persis-

tent calyx, splitting into three 2-winged parts [*mericarps*]; seeds 1 per segment, flattened, convex.

Ecology: Insect-pollinated; seed dispersal via wind.

Distribution: Flooded and non-flooded old-growth forest and secondary forest, often in open areas. Widespread in the tropics and subtropics: 35 species; Guianas: 8 species, 3 common species described here.

Notes: The unique tendrils and lack of spines allow for easy distinction of this genus from other species with similar leaves and shrubby habit such as Celtis iguanae (Cannabaceae) and Byttneria spp. (Malvaceae).

2.1. Gouania blanchetiana Miq.

Common name: djumi-toto-tongo (Sr); ërukërukë (Tr); misa (Wa)

Liana; lightly hairy. **Stems:** nodes with **stipules long and narrow,** ca. 0.5 cm; older stems round, 1.5 cm diameter, bark light gray, smooth with fine length-wise cracks; **slash with mint scent,** wood pinkish-white. **Leaves:** ovate-elliptic, to

7.5 x 3 cm, base sometimes uneven, apex acute to long-acuminate, margin toothed to ± entire; blade hairless except in vein axils below; petiole 0.6-1.4 cm long. **Inflorescences:** spikes 7-25 cm long, flowers in clusters of 3-4. **Fruits:** capsules to 0.8 x 1.1 cm, hairless; seed brown, flattened, 0.3-0.4 cm long.

Common; non-flooded old-growth forest, roadsides, secondary forest, especially on bauxite. Guianas; also Venezuela and Brazil, to Panama.

2.2. Gouania polygama (Jacq.) Urb.

Shrubby climber; densely hairy, tan to reddish-brown [*tomentose*]. **Stems:** nodes with **stipules broadly triangular,** to 0.6 cm long. **Leaves:** elliptic to ovate, to 14 x 8 cm, base rounded or truncate, apex acuminate, margin coarsely-toothed to round-toothed. **Inflorescences:** spikes 16-25 cm long, flowers densely clustered. **Fruits:** capsules to 0.8 x 1.3 cm, hairless or lightly hairy.

Uncommon; riparian forest, secondary forest and open places. Guyana, Suriname, possibly French Guiana; widely distributed in the neotropics.

2.3. Gouania velutina Reissek

Liana; densely hairy, reddish-brown [*tomentose*]. **Stems:** older stems round, **slash sometimes with a little watery red sap. Leaves:** broadly ovate, to 9 x 6 cm, base rounded or subcordate, apex acuminate, margin sharp- to round-toothed. **Inflorescences:** spikes short, 5-12 cm long, flowers in clusters of 8-10. **Fruits:** capsules ca. 0.4 cm diameter, hairy. **Use:** Leaf extract for itching and fever (Trios)

Uncommon; riparian forest. Guyana, Suriname; also Amazonian Brazil, Ecuador, Venezuela.

1. Ampelozizyphus amazonicus habit, leaves palmate-veined with ladder-like cross-veins [BH]
2. A. amazonicus, robust woody stem [BH]
3. A. amazonicus, older stem in cross-section [FRD]
4. A. amazonicus fleshy fruit [FRD]
5. Gouania blanchetiana habit [OG]
6. G. lupuloides with tendril (Costa Rica) [RA]
7. G. polygama, stipules (Costa Rica) [RA]
8. G. blanchetiana, flowers close-up [OG]

RUBIACEAE

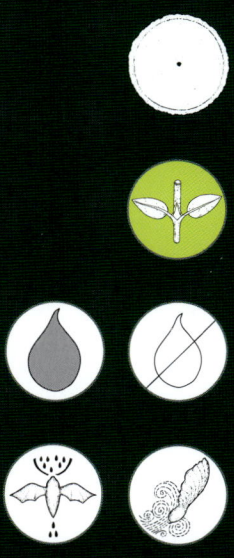

RUBIACEAE

A very large and diverse family with occasional lianas, vines, scrambling shrubs or hemi-epiphytes; predominantly herbs, shrubs, or trees. Climbing via branchlets, twining shoots, axillary hooks, spines, or aerial climbing roots. Leaves are opposite or whorled, simple, and entire. Interpetiolar stipules are present. Flowers are usually tubular, 4-5-parted, with a nectar disk and inferior ovary. Fruits are diverse, including berries, drupes and capsules with winged seeds.

General: Surfaces hairless to densely hairy, **hairs simple when present;** armed or unarmed; pure water sometimes present; colored sap, conspicuous glands, and conspicuous scent absent.

Stems: Climbing branchlets flattened, mostly smooth and hairless, sometimes armed with spines (Guettarda, Randia, Uncaria), **commonly swollen below nodes; interpetiolar stipules present,** narrow or expanded across nodes, erect or reflexed, free or fused, leaving a horizontal scar upon falling. Older stems variable in shape: cylindrical, fluted, 4-sided or flattened, slender to medium in diameter; bark variable: smooth and papery to thick and fibrous; **stem x-s pattern simple.**

Leaves: Variable in shape and size, margins entire; texture often glossy, rarely hairy; raised or flat leaf glands absent; **venation pinnate,** 2° veins widely spaced and steeply ascending, usually without marginal loops, **3° veins usually net-veined,** sometimes parallel and angled [*oblique*] towards midvein (Guettarda, Malanea, Sabicea) or almost invisible (Hillia, Schradera); petiole variable in length, without glands.

Inflorescences: Highly variable: cymes, racemes, panicles or heads, **often subtended by showy bracts;** flowers usually bisexual, radial [*actinomorphic*], 4-5-parted; calyx cup-shaped or reduced, persistent in fruit; **corolla usually tubular and 4-5 lobed;** nectar disc usually present, attached below the ovary [*hypogenous*]; stamens attached to corolla [*epipetalous*] and alternate with lobes; **ovary inferior,** usually 2 (1-many)-chambered [*locular*]; stigma 1 or 2-lobed.

Fruits: Variable: many-seeded berries, few-seeded drupes or capsules with many winged seeds.

Ecology: Pollination via bird, bat or insect; seed dispersal via animal or wind.

Distribution: Cosmopolitan, 500-700 genera/ ± 10,000 species; Guianas: > 90 genera/500 species; with 13 genera/48 species of (sub)woody lianas, vines and liana-like hemi-epiphytes; 8 genera & 15 species with diverse climbing growth forms described here.

Use: Diverse traditional uses including food and medicine (see genera for more detail).

1. Guettarda L.

Rarely lianas; mostly shrubs, treelets or trees; sometimes armed. Climbing via twining shoots and spines. Mostly neotropical, center of diversity in the Antilles: 60-80 species; Guianas: 10 species, with 1 woody climber.

1.1. Guettarda aromatica Poepp.

Common names: popokaimaka (Sr)

Liana; sometimes shrub or treelet; hairy and armed. Stems: branchlets with flat, long, silky hairs [*sericeous*], hairs lost with age; slightly re-curved **spines present in leaf axils,** 0.5-1.3 cm long, **stipules narrowly triangular, ca. 0.5 cm long,** persistent or short-lived; older stems nearly cylindrical, robustly woody; bark thin, reddish-brown with lenticels; stem x-s pattern simple.. **Leaves:** (Ovate-)elliptic, 10(18) x 6.5 cm; **densely golden silky-hairy above and below** in young leafy shoots; 3° veins parallel, slanted to midvein; petiole conspicuously long (2-3 cm).

Inflorescences: axillary, flat-topped many-flowered heads [*corymbs*] with 5-8 cm long stalks, axes shallow-furrowed and densely hairy [*pilose*]; flowers tubular, 5-6-parted; calyx short; corolla to 2.2 cm long, **tube straight, white or yellow,** densely silky-hairy outside; stamens attached at mouth of tube; ovary 2-9-chambered. **Fruits:** ovoid drupes, less than 2 cm diameter, purplish-black, covered with short, spreading hairs [*velutinous*], outer layer of fruit [*exocarp*] fleshy, crowned by persistent calyx; seeds 3, unwinged.

Uncommon; seasonally-flooded old-growth forest along rivers. Guianas; also Brazil.

2. Hillia Jacq.

Liana-like hemi-epiphytes. Climbing via adventitious roots.

General: **Surfaces hairless, green, fleshy; unarmed.**

Stems: Branchlets stout, remotely 4-sided, succulent; **stipules large, 1.5 to 6 cm long,** strap-shaped to broadly spear-shaped, short-lived. Older stems slender, 1-3 cm diameter, bark grey; stem x-s pattern simple, radial rays inconspicuous.

Leaves: Elliptic to broadly ovate; leathery to sub-leathery; 2° veins ± obscure, pinnate to almost palmate, 3° veins obscure; leaf pockets [*domatia*] with hair tufts in leaf axils below.

Inflorescences: Terminal, solitary or 2-3-flowered (not in heads as in Schradera); **flowers large and 6-parted,** stalked or unstalked; calyx cylindrical, free or ± fused; **corolla funnel- or trumpet-shaped, white, green, or yellowish-green,** tube 2.8-11 cm long; stamens 4-7, attached within the corolla mouth, not protruding from mouth; ovary 2-chambered.

Fruits: **Oblong, round capsules, > 5 cm long, crowned by persistent calyx lobes** (due to inferior ovary); seeds numerous, cigar-shaped, each with an apical hair tuft (coma).

Ecology: Pollination by bats is expected because plants flower at night. Seed dispersal via wind.

Distribution: Widespread in the neotropics and West Indies; 24 species known; 2-3 in Suriname.

Notes: **Hillia parasitica Jacq.** is an uncommon species notable for an extra-long corolla tube, to 11 cm.

2.1. Hillia illustris (Vell.) K. Schum.

Common name: kumpuruni (Tr)

Hemi-epiphytic, sub-woody climber. Stems: stipules 2.5-5 x 0.6-1.4 cm. Leaves: elliptic to ovate, 9.5-16 x 2.5-8.5 cm, base cuneate, apex acuminate; 2° veins pinnate. **Inflorescences:** terminal, flowers 1-3; calyx tube tiny, enlarged in fruit, truncate or lobed; **corolla funnelform, 4.5-6.5 cm long,** lobes ovate, erect. **Fruits:** capsules 7-12.5 x 1.1-1.5 cm, green to grayish-brown, with obscure ribs; seeds small, hairtuft 1.5-2.2 cm long.

Common; creek or riverine forest, on rocky islands. Guianas; also Brazil.

2.2. Hillia ulei K. Krause

Hemi-epiphytic, sub-woody climber (much smaller than Hillia illustris). **Stems: stipules to 1.5-0.8 cm. Leaves:** elliptic to broadly ovate, 2-6.5 x 1.3-4 cm; base rounded to cuneate, apex acute to acuminate; **2° veins subpalmate. Inflorescences:** terminal, flowers 1; calyx 1-1.2 cm long, deeply lobed; **corolla broadly funnelform, short, 2.8-4 cm long,** lobes short and broad, erect. **Fruits:** capsules 6-10 cm long, green; seeds small, hairtuft 1-2.4 cm long.

Rare; creek or riverine forest. Guyana, Suriname; widespread across the Amazon Basin, especially the Andean foothills.

3

3. Malanea Aubl.

Usually woody lianas; also shrubs and treelets. Climbing via twining shoots and lateral branchlets, sometimes strangling other plants when shrubby.

General: Surfaces softly hairy to hairless; **unarmed.**

Stems: Branchlets nearly cylindrical, **flattened at nodes,** often slightly hairy; **stipules large and conspicuous, broadly ovate to strap-shaped** [*ligulate*]. Older stems cylindrical, fluted or flattened, bark light-colored, moderately thick, ± smooth, bearing many lenticels.

Leaves: Usually broadly ovate, margin slightly curled [*revolute*]; ± leathery, dark green and shiny above, **pale or silvery-green hairy or hairless below;** 3° veins net-veined or parallel/slanted.

Inflorescences: Many-branched pairs of panicles at leaf axils, often densely hairy; bracts and bracteoles small; **flowers 4-parted, small;** calyx 4-toothed; corolla rotate or shortly funnel-shaped densely soft hairy within; stamens 4, attached at corolla mouth, protruding [*exserted*]; ovary 2-chambered.

Fruits: Drupes, ovoid or narrowly oblong, 1-2-chambered; crowned by persistent calyx; inner wall [*endocarp*] bony; seeds 1, round, seeds without wings or hair tuft.

Ecology: Seed dispersal via animal.

Distribution: Neotropics: 10-30 species; Guianas: 11 species, 3 described here.

3.1. Malanea glabra A.Rich.

Liana or scrambling shrub. Stems: branchlets sparsely hairy or hairless, smooth; nodes with **stipules ligulate, very large, 1.1-2.5 cm long,** mostly hairless outside. **Leaves:** broadly ovate, 6.5-14.5 x 3.5-9 cm; silky-hairy below; 2° veins prominent below, 3° veins nearly parallel; petiole 1 cm long. **Inflorescences:** panicles 2.5-7 cm long; calyx < 0.1 cm long; **corolla funnel-form, pale yellow,** 0.4-0.7 cm long, densely hairy outside. **Fruits:** drupes 0.8-1 x 0.3-0.5 cm, purplish-black, hairless, smooth.

Common; savanna woodland, savanna, and riverine forest. Guianas; also Venezuela, N. Brazil, Trinidad and the Antilles.

3.2. Malanea sarmentosa Aubl.

Liana or scrambling shrub. Stems: branchlets densely reddish-brown hairy; nodes with stipules oblong-triangular, 0.8-1.2 cm long, densely reddish-brown hairy outside. **Leaves:** broadly ovate, 6.5-14.5 x 3.5-9 cm; leathery, surface raised above the veins [*bullate*], densely-hairy below; 2° veins prominent below, 3° veins nearly parallel; petiole 1-2 cm long. **Inflorescences:** panicles 5-14 cm long, densely reddish-brown hairy [hirsute]; calyx 0.8 cm long; **corolla nearly radial,** tube 0.4-0.6 cm long, sparsely hairy outside, soft hairy within. **Fruits:** drupes, 0.6-0.8 x 0.4 cm.

Uncommon; creek and savanna forest. Guianas; also Trinidad, Venezuelan Guayana and N. Brazil.

3.3. Malanea tafelbergensis Steyerm.

Liana or scrambling shrub. Stems: branchlets hairless, smooth; nodes with stipules oblong-triangular, ca. 0.3 cm long. Leaves: broadly ovate, 5-8 x 3-4.5 cm; leathery, hairless above and below; 2° veins prominent below, 3° veins net-veined or obscure; petiole <1 cm long. **Inflorescences:** panicles short and many-branched, ca. 3 cm long, densely reddish-brown hairy [*hirsute*]. **Fruits:** drupes, 0.8 x 0.3 cm, ribbed.

Rare; Guyana, Suriname, known only from high elevation 'tepui' mountains.

4. Manettia Mutis ex L.

Slender subwoody lianas and vines. Climbing via twining shoots.

General: Surfaces softly hairy to hairless; **unarmed.**

Stems: Slender, cylindrical to 4-sided, hairy, twisted, < 0.5 cm diameter; nodes with **stipules broad and short, often fused to the petiole and persistent.**

Leaves: Small, variable in shape; thin to ± leathery; 2° veins curving upwards close to margins, without marginal loops, 3° veins net-veined, obscure.

Inflorescences: Cymes at leaf axils, in clusters or umbels; calyx persistent; **corolla tubular or funnel-shaped, usually 4-parted, small,** white, red, blue, or lavender, tube short or long; stamens 4, attached at corolla mouth and extending out; ovary 2-chambered.

Fruits: Small ovoid capsules, smooth, weakly 2-furrowed, **crowned by a 4-lobed, persistent calyx,** splitting into 2 from the apex; seeds 2-parted, small, flat, winged.

Ecology: Seed dispersal via wind.

Distribution: Widespread in tropical America: ca. 80 species; Guianas: 3 species, 2 described here.

4.1. Manettia alba (Aubl.) Wernham

Common names: wangsiri wetë (Ca)

Subwoody liana or vine; minutely hairy. Stems: nodes with **triangular stipules, small, 0.1 cm long. Leaves:** elliptic-oblong-lanceolate, 3.5-7 x 1.5-3.5 cm, apex with a fine point, margins curled; **± leathery, bullate, hairy below;** petiole 0.5-1.2 cm long. **Inflorescences:** axillary, usually paired at nodes, **many-flowered (12-24),** infl stalk 1-2 cm long; calyx lobes < 0.35 cm long; **corolla flat and circular [*rotate*],** 0.5 cm long, **white, tube shorter than the lobes. Fruits:** capsules ovoid, to 4.5 x 3 cm.

Uncommon; old-growth forest, often higher elevations. Guianas; Venezuela, NE Brazil (Guiana Shield Highlands).

4.2. Manettia reclinata L.

Subwoody liana or vine; hairless. **Stems:** nodes with **ovate stipules, small, 0.15 cm long. Leaves:** lanceolate to ovate, to 5 x 2.6 cm, apex with a fine point, margins curled; **thin and soft, not leathery;** petiole 0.4-1 cm long. **Inflorescences:** axillary, **usually single flowered,** infl. stalk 0.1-1(3) cm long; calyx lobes 0.35-1 cm long; **corolla platter-shaped [*hypocrateriform*],** 1.6-3 cm long, **red, tube 1 cm long. Fruits:** capsules ovoid, to 0.9 x 0.7 cm.

Uncommon; riverine forest. Guianas; also Brazil, Venezuela, and Colombia.

5. Randia L.

Some climbing shrubs; mostly erect shrubs or small trees. Climbing via stiff lateral branchlets (short shoots) and spines. Widespread genus, from the southern United States to Chile and Argentina: ca. 100 species; Guianas: 6 species; 1 woody climber.

5.1. Randia armata (Swartz) DC.

Common name: djumbi-maka (Sa); përë pana (Tr); spikri-maka (Sr).

Climbing shrub or erect treelet to 11 m; **armed and hairless. Stems:** branchlets cylindrical to flattened, **with erect, persistent, woody spines in groups of 2-4 at nodes; stipules long-triangular, to 0.6 cm, leaving an obvious interpetiolar scar;** older stems ± cylindrical, to 5 cm diameter, bark reddish-brown to grey, with scattered, conspicuous lenticels. **Leaves:** in clusters at branch ends, elliptic to nearly lanceolate, to 15 x 6 cm, papery; petiole 0.3-0.5 cm long; 2° and 3° veins obscure, net-veined. **Inflorescences:** axillary or terminal, male flowers clustered terminally, female flowers solitary on short shoots (flowering short shoots with 1-2 pairs of sharp, axillary spines); flowers 5-parted; calyx 0.3 cm long, enlarged in fruit; **corolla tube white, maturing yellow, 2.9-4.5 cm long;** stamens 5; ovary 2-chambered. **Fruits: globose, leathery berries,** 1.5-2.5(-3) cm diameter, green to orange-yellow, crowned by persistent calyx; seeds in 2 rows, flattened, black. **Ecology:** Plants dioecious, individual plants with male and female flowers. Seed dispersal via animal.

Distribution: Common; secondary and riverine forest, rock savannas. Guianas; widespread in the American tropics and subtropics.

6. Sabicea Aubl.

Woody to subwoody lianas and vines; also herbs and shrubs. Climbing via twining shoots.

Common name: aripëmë (Tr), awarimë pïtëkënë (Tr); kibriwiwiri (Sr); zauzanpatu (Sa)

General: **Surfaces typically densely hairy, hairs cobwebby [*arachnoid*] or stiff and short [*hirsute*], white, yellowish-white or red; unarmed.**

Stems: Branchlets hairy with reddish-brown bark; nodes with **stipules broadly ovate to triangular,** entire or 2-3-lobed, often persistent, interpetiolar line obscure. Older stems cylindrical, 2-5 cm diameter, bark often thick and fibrous, lenticels not prominent; stem x-s pattern simple.

Leaves: Elliptic to ovate, medium-sized, mostly thin and soft; **often dark green above and densely-hairy below.**

Inflorescences: Axillary, globose clusters, few-flowered, without an infl. stalk; flowers small, calyx persistent, lobes 3-6; **corolla tubular, white, 4-6-lobed,** stamens 4-6, inserted within tube; ovary 3-5-locular.

Fruits: Berries globose to ellipsoid, 0.5-1.3 cm diameter, fleshy, red to black, 2-5-chambered; seeds two-parted, egg-shaped or angular, small.

Ecology: Seeds dispersed via animal (through bird gut).

Distribution: Common in open areas, especially riverbanks and secondary vegetation. Pantropical genus: 120-135 species; Guianas: 11 species, 3 described here.

6.1. Sabicea cinerea Aubl.

Small liana. Stems: branchlets with flattened, short hairs; **stipules ovate to round,** 2-3-lobed, 0.6-0.9 cm long, densely hairy outside. **Leaves:** 5-12 cm x 2-5.5 cm, **sparsely hairy above, densely hairy below;** petiole 0.5-1 cm long. **Inflorescences:** 3-7-flowered; calyx tube 1.3-1.6 cm long, densely soft hairy outside, calyx lobes lanceolate, to 1 cm long; corolla tubular, 1.8-2.1 cm long, yellowish-white, mostly densely spreading-hairy outside. **Fruits:** berry red at maturity.

Uncommon; riverbanks and secondary vegetation. French Guiana, Suriname; also NE Venezuela and Puerto Rico.

6

6.2. Sabicea glabrescens Benth.

Low-climbing liana or shrub. Stems: branchlets with soft [*velutinous*] or stiff [*strigose*] flattened hairs; **stipules broadly ovate,** reflexed, to 1 cm long, hairless. **Leaves:** 7-13 x 3-5 cm, dark green and **hairless above, sparsely hairy below;** petiole 0.5-1.5 cm long. **Inflorescences:** 5-6-flowered, unstalked; calyx lobes larger than tube, calyx tube 0.12-0.13 cm long, hairless outside, corolla platter-shaped [*hypocrateriform*], 1.3-1.4 cm long, densely hairy outside. **Fruits:** berry white to light blue, reddish-brown hairy, 0.9-1.3 x 0.8 cm.

Very common; riverbanks and secondary vegetation. Guianas; also Venezuela and N Brazil.

6.3. Sabicea oblongifolia (Miq.) Steyerm.

Small liana or vine. Stems: branchlets with coarse, stiff hairs [*hirsute*]; **stipules triangular,** obtuse to acute at apex, erect or flat, 0.4-0.9 cm long, densely hairy. **Leaves:** 7-15 x 3-8.5 cm, **with soft, long hairs [*pilose*] above and felty short hairs below;** petiole 0.4-1.5 cm long. **Inflorescences:** 5-12-flowered, longer than wide, densely hairy; calyx tube densely tomentose outside, corolla tubular, 1.1-1.3 cm long, white. **Fruits:** berry reddish-purple, enclosed by green calyx.

Locally common; riverbanks and secondary forest. Guianas, also NE Venezuela and Trinidad.

1. Uncaria guianensis in flower [BH]
2. Hillia illustris, liana-like hemi-epiphyte with large tubular flowers and inferior ovary [PT]
3. Malanea sp., leaf upper surface (Costa Rica) [RA]
4. Guettarda aromatica, woody stem with paired robust spines at nodes [BH]
5. Randia armata, paired, opposed spines on young shoots
6. Sabicea cinerea, robust leaves and white tubular flower [PT]

7. Schradera Vahl

Lianas and liana-like hemi-epiphytes. Climbing via twining shoots and/or adventitious roots. Neotropical; ca. 15 species; Guianas: 4 species, 1 described here.

7.1. Schradera polycephala DC.

Woody liana or hemi-epiphyte; mostly hairless and unarmed. Stems: branchlets strongly 4-sided; stipules large, to 1.6 x 1 cm, rounded, erect or reflexed, basally fused; older stems 4-sided, ca. 1.5 cm diameter; bark grey or brown, smooth, succulent; slash **with white wood and lengthwise purple-brown lines;** stem x-s pattern simple, but radial rays apparent. **Leaves:** obovate, 5-12 x 1.5-4 cm, apex rounded, margin slightly curled; tough and fleshy; veins invisible above, impressed and fine below, **3° veins obscure;** petiole 0.5-2 cm long. **Inflorescences:** terminal, long-stalked heads subtended by two large bracts; flower heads 8-13-flowered; flowers tubular, 5-10-parted; calyx 0.35-5 cm long, hairless; **corolla platter-shaped, white, 5-6-lobed, 1-2 cm long,** hairless outside. **Fruits:** berries, 2-3-chambered; seeds 2-parted and flat. **Notes:** Other liana-like hemi-epiphytic climbers with opposite, fleshy leaves include Clusia (Clusiaceae), Drymonia (Gesneriaceae), Hillia (Rubiaceae), and Schlegelia (Bignoniaceae). Clusia in particular has similar leaves but is easily distinguished by brightly colored latex, lack of stipules and tree-like habit.

Uncommon. Guianas; Brazilian Amazon and Guayanan Venezuela.

8. Uncaria Schreb.

Woody canopy lianas. Climbing via thorn-hooks.

General: Surfaces mostly hairless or short-hairy; armed; abundant water present occasionally in larger stems.

Stems: Branchlets sharply 4-sided in cross-section, glossy, green, hairless; **nodes with paired hooks or spines (thorn-hooks); interpetiolar stipules ovate-triangular to triangular,** leaving conspicuous scar. Older stems developing to become ± cylindrical with age, > 5 cm diameter; bark brown to reddish-brown, thick and corky, with lengthwise platelets; stem x-s pattern simple.

Leaves: elliptic to ovate, ± leathery, hairless or short-hairy below; 2° veins yellow, not looping at margins, 3° veins visible to obscure, net-veined.

Inflorescences: Axillary and terminal, with **globose heads of tubular flowers, stalked (U. guianensis) or unstalked (U. tomentosa);** flowers 5-parted; calyx lobes triangular; corolla platter-shaped, white turning orange-yellow, hairy outside.

Fruits: Globose heads of cigar-shaped capsules, 2-valved, dehiscent; outer layer becoming free from inner layer; seeds small, each 2-winged, apical wing small, basal wing minute and split.

Ecology: Seeds dispersed via wind.

Distribution: Pantropical, center of diversity in Old World tropics: 35 species; Guianas: 2 species.

Use: Uncaria species are important traditional medicines in the Guianas for a wide variety of ailments. They sometimes contain abundant pure drinking water in large stems, and the bark is a globally-recognized herbal immune system booster sold as "cat's claw".

8.1. Uncaria guianensis (Aubl.) J.F.Gmel.

Common name: ampumaka (Pa); buribada (Ar), bái bái lopu (Sa); pijana oroi (Tr); popokai-nangra (Sr),

High-climbing woody liana; mostly hairless. **Stems:** stipules oblong; **1-2 spines at nodes very strongly re-curved like a cat's claw, hairless. Leaves:** elliptic, to 9 x 4.5 cm, thin to nearly leathery, **hairless above and below; petiole 0.5 cm long. Inflorescences:** axillary and terminal, **globose heads solitary; flowers stalked,** hairy outside [*villous*]; **calyx 0.35-0.4 cm long; corolla tubular, to 2.5 cm long. Fruits:** globose heads, capsules cigar-shaped, **1.8-2.5 cm long, stalked;** seeds to 0.65 cm long (incl. wings).

Common; old-growth forest, forest edges in disturbed areas. Guianas; widespread across Amazon region.

8.2. Uncaria tomentosa (Willd. ex Schult.) DC.

Common name: pijana oroi (Tr)

High-climbing liana; mostly hairy [*tomentose*]. **Stems:** stipules broadly ovate-triangular, persistent; **1-2 spines at nodes slightly recurved, hairy. Leaves:** elliptic, to 11 x 6 cm; thin to nearly leathery, hairless above, **grey or brown-hairy below, petiole 1 cm long. Inflorescences:** terminal, **globose heads solitary or paniculate; flowers unstalked,** hairy outside; **calyx tiny (< 0.1 cm long); corolla tubular, 0.5-0.7 cm long. Fruits:** globose heads, capsules cigar-shaped, **0.6-1.7 cm long, unstalked.**

Uncommon. Guianas; also N South America and Central America.

7.   Schradera sp. inflorescence [TB]

8.   Schradera polycephala, fused interpetiolar stipules [FRD]

9.   Uncaria guianensis, robust woody stem [BH]

SAPINDACEAE

1 2 3 4 5

SAPINDACEAE

Commonly woody lianas, rarely herbaceous vines; also shrubs and trees. Climbing via forked tendrils. Stipules present in climbing species. Leaves are alternate and compound. Flowers are mostly bilateral-symmetric in climbers [*zygomorphic*], also radial or irregular, white or yellowish, 4-5-parted, nectary disk present, ovary superior. Fruits are either fleshy capsules that split open to reveal seeds with arils, or dry, indehiscent, winged nuts [*winged samaras*].

General: Vegetation hairless to densely hairy, hairs simple when present; cut parts **with or without white, watery sap or white, resinous sap,** both often sparse when present; **tendrils of two forms:** i) borne on leafy, sterile branchlets and splitting into two spiral (3D) tendril arms; or ii) borne on inflorescence stalks and splitting into two (2D) rolled-up arms – like a butterfly tongue or 'watch spring'.

Stems: Branchlets round or sharply 3-6-sided (in cross-section), sometimes winged; nodes with stipules, small or large, sometimes clasping the stem. Older stems cylindrical, sharply-angled (3-6-9-sided), lobed, or flattened; bark often with vertical ridges or furrows; stem x-s pattern **simple, lobed or compound** (with a large central column [*stele*] and multiple (3-10) smaller, satellite columns); wood often soft, pith sometimes hollow.

Leaves: 3-foliolate (Thinouia, Urvillea), **odd-pinnate,** 5-13 leaflets (Paullinea, Serjania), or **pal-mately-compound** with ultimate units of 3-leaflets (Paullinea, Serjania), terminal leaflet usu. larger; **margins toothed, wavy,** or entire; 3° veins often parallel; **petioles and rachis often winged,** never with a cylindrical swelling [*pulvinus*] at base; raised and flat leaf glands absent.

Inflorescences: Terminal, axillary, or borne on stem [*cauliflorous*], in dense spikes and racemes, simple or compound, branchlets often angled or winged; **flowers small to tiny, white or white-yellow,** bilateral symmetric (Paullinia, Serjania), radial (Thinouia) or irregular (Urvillea), **4-5-parted,** calyx free or ± fused, overlapping [*imbricate*]; **petals free and with inner "appendages" of various shapes;** nectar disc present; stamens ± 8, free, similar or dissimilar (in height); ovary superior, (2)3(5)-chambered; style 1 or 3.

Fruits: Fleshy to woody, 3-valved, dehiscent **capsules** (Paullinia) or dry, **3-winged fruits** [*mericarps*] (Serjania, Thinouia, and Urvillea) that split off into three 1-seeded, winged samaras; seeds 1 per chamber or samara, seed coat black and shiny, thin, shell-like or bony, with a juicy, edible aril (Paullinia).

Ecology: All flowers appearing bisexual, but sometimes only male or female part fertile. Insect-pollinated (small bees); seed dispersal via wind, water or animal (through gut).

Distribution: Widespread in the tropics and subtropics: ± 140 genera/2,000 species; Guianas: 22 genera/114 species, 4 genera with woody climbers (Cardiospermum species are herbaceous, annual vines), ± 46 (sub)woody species known.

Use: Binding and weaving material, edible aril, medicines (diabetes, intestinal complaints, pain), spiritual baths, (caffeine-rich) stimulants, fish and arrow poisons.

Notes: Paullinia and Serjania climbers are easily confused in vegetative and flowering condition. In general, Serjania species have more leaflets than Paullinia species. Of the winged genera, Serjania fruits have wings expanded at the base (near stalk) and seeds fused at the apex, Thinouia fruits are opposite with apical wings and basal seeds, Urvillea fruits have seeds fused in the midsection and wings at both ends. (Gentry 1993, Acevedo-Rodríguez 2012).

Sterile characteristics of the Leguminosae that can distinguish them from the Sapindaceae: petiole with swollen cylindrical pulvinus; leaf glands; presence of red sap (absence of white sap); stem x-s pattern sometimes concentric; spines and prickles; "green-bean" or "cucumber" scents; absence of toothed leaf margins.

1. Paullinia L.

Mostly woody lianas, sometimes climbing shrubs or herbaceous vines; climbing via forked tendrils.

Common names: botermelkenkaas (SD); feififinga (Au, Sr); kamasuri (Ca); kutupu (Way); kutupuran (Ca); pakale (Wpi); taaukasi (Wpi); tukukuimë (Tr); vijfvingerblad (SD); yesikushi (Ar)

General: See family description.

Stems: See family description.

Leaves: Odd-pinnate, mostly 5-foliolate, sometimes 11-13-foliolate or bipinnate; **margins toothed, wavy,** or entire with a few remote apical teeth, teeth often gland-tipped; **petioles and leaf rachis often winged.**

Inflorescences: Solitary spikes, racemes or large terminal panicles; flower tiny, white to greenish-yellow, bilateral; calyx 4-5-parted; corolla with **4 distinct petals, appendages hood-shaped with a fleshy, yellowish apex; nectar disc 2-4 glandular on one side;** stamens 8, unequal in height, usually hairy; ovary 3-chambered with one ovule.

Fruits: Thick-walled capsules, usually red (green or yellow), valves often winged or ribbed; **seeds** (1-)3, globose, **mostly enclosed within a fleshy coat and with a conspicuous white aril.**

Ecology: Fruits split open but seeds remain attached as a display to attract dispersers [*septifragal*]; Seed dispersed via animal (through gut of birds and monkeys).

Distribution: Neotropics only (except Paullinia pinnata, which also occurs in Africa and Madagascar): > 200 species; Guianas: 36 species, 8 described here.

1.1. Paullinia alata (Ruiz & Pav.) G.Don.

High-climbing liana, usually many-branched from base; **white sap scarce or absent,** lightly hairy. **Stems:** branchlets sharply 3-angled and 3-furrowed, **stipules awl-shaped** [*subulate*], ca. 0.3 cm long; older stem **x-s pattern compound** (1 central + 3 peripheral steles). **Leaves: 5-foliolate,** petiole and rachis conspicuously **winged;** leaflets elongate-elliptic, to 14.5 x 9 cm, base attenuate to obtuse, apex obtusely pointed [*acuminate*], **remotely toothed;** sub-leathery, shiny above, lightly hairy below. **Inflorescences:** cauliflorous, with dense clusters of subglobose flowers; corolla to 0.4 cm long. **Fruits: capsules pear-shaped** [*pyriform*], to 1.5 x 1 cm, **unwinged,** ± 3-lobed, pink to red, pointed; seeds 3, 0.7 x 0.6 cm, black, surrounded to 2/3 by white aril.

Uncommon; non-flooded old-growth forest edges and riparian forest. Fr. Guiana and Suriname; N South America to Panama, south to Ecuador and Brazil.

1.2. Paullinia capreolata (Aubl.) Radlk.

Woody liana, 1.5-25 m long; **white sap abundant; reddish-brown hairy** [*tomentose*]. **Stems:** branchlets round, with fine lines or grooves [*striate*] and lenticels; stipules inconspicuous, short-lived; older stems to 15 cm diameter at base, round, x-s pattern simple. **Leaves: 5-foliolate, unwinged;** leaflets papery, elliptic to nearly ovate, 6-15-(27) × 2-4-(12.5) cm, margins entire or remotely toothed in upper 1/3. **Inflorescences:** panicles large, stout, often hairy [*tomentellous*]; flowers in clusters on ca. 1 cm long infl. stalks; calyx 4; petals white, 0.3 cm long. **Fruits: capsules sub-globose,** to 1.7 x 2 cm, **unwinged,** with 6 obscure ribs or grooves, yellow to orange, hairy; seeds 1, brown, 1.2-1.5 cm wide, aril absent.

Rare; non-flooded and flooded lowland forest, especially riparian forest. Guianas; also widespread in South America and present in Costa Rica.

SAPINDACEAE

1.3. Paullinia fuscescens Kunth

Woody liana, 3-7 m long; **white sap scarce;** lightly hairy. **Stems:** branchlets round, bearing lenticels, lightly hairy [*pilose*]; stipules triangular, small, 0.1 cm long; older stems round, x-s pattern simple. **Leaves: 1-palmately-compound, with ultimate units of 3-leaflets,** rachis narrowly winged, largest leaves 10 x 4.5 cm; leaflets elliptic to inversely lanceolate, ± papery, **finely-toothed. Inflorescences:** axillary, crowded at branch tips, densely hairy, flowers white; corolla to 0.45 cm long. **Fruits: remotely 3-sided, subglobose to broadly obovate capsules,** to 1 x 1.5 cm, **often winged,** red; seeds 1-3, aril white.

Uncommon; in coastal thickets, savannas and disturbed vegetation. Guianas; widespread from N South America to Central America and Mexico.

1.4. Paullinia ingifolia Rich. ex Juss.

Woody liana, 5-10 m long, 2 m tall shrub when young; **white sap scarce;** mostly hairless. **Stems:** branchlets round with fine lines or grooves [*striate*] and lenticels; **stipules large, to 2-4 cm long, clasping the stem, leaving scar to 0.5 cm wide;** older stems round, x-s pattern simple. **Leaves: 7-(9)-pinnately compound,** rachis mostly winged; leaflets hairless, 4.5-20 × 2.5-7.6 cm, margins remotely toothed. **Inflorescences:** solitary, densely flowered, rusty hairy; calyx greenish white with 4 sepals, petals white, ca. 5 mm long. **Fruits: capsules sub-globose, unwinged,** with (3)6-9-ridges, yellow, woody, 1-seeded, aril white.

Common; lowland old-growth forests; Guianas; widespread in the neotropics, from Bolivia to Costa Rica.

1.5. Paullinia pinnata L.

Common names: botermelkenkaas (SD); feifi-finga (Au, Sr); jesikoesji (Ar); kamasoeli (Ca); kutupu (Way); liana carre (FG); tukukuimë (Tr); vijfvingerblad (SD)

Woody liana, 5-15 m long, usually many-branched from base; **white sap scarce;** mostly hairless. **Stems: branchlets 3-angled;** stipules awl-shaped [*subulate*], 0.5 cm long; older stems usually bluntly 3-5 angled, **x-s pattern compound** (1 central + 3-7 peripheral steles). **Leaves: 5-foliolate,** petiole and rachis broadly winged; leaflets leathery, elliptic to oblong, to 10 x 4 cm, margin with sharp or rounded-teeth. **Inflorescences:** axillary, single racemes, to 10 cm long, with 2 basal tendrils; flowers tightly clustered, unstalked; calyx lobes 5; corolla white, ca. 0.4 cm long. **Fruits: capsules pear-shaped to oblong,** to 3 x 1-1.4 cm, **unwinged,** green to pinkish-red, valves thick, woody, hairless inside; seeds usually 1, ellipsoid, blue-purple to black, white aril. **Use:** Traditionally used as a fish poison by some indigenous groups in the Guianas and Amazon basin.

Very common; especially in young coastal belt, swamps, riverbanks, disturbed sites. Guianas; widespread in the neotropics and present in tropical Africa.

1.6. Paullinia plagioptera Radlk.

Woody liana, 5-7 m long; **white sap absent;** mostly hairless. **Stems:** branchlets cylindrical, with dense white lenticels; stipules small, pointed, ca. 0.2 cm long, short-lived; older stems round to remote-angled, stem x-s pattern simple. **Leaves: 5-foliolate,** petiole and rachis narrowly winged, hairy; leaflets elliptic to oblong, 4-9 x 2-3.2 cm, base obtuse to rounded, unequal except terminal leaflet, margin coarsely toothed; nearly leathery. **Inflorescences:** axillary, panicles to 12 cm long; calyx lobes 4, corolla small, white. **Fruits: capsules sub-globose** 1.5 x 1.2 cm, red, **with 3 wings reaching halfway to apex;** hairless inside; seeds 1, globose, to 1 cm diameter, more than half surrounded by white aril.

Rare; upland rain and riverine forest; endemic to the Guianas.

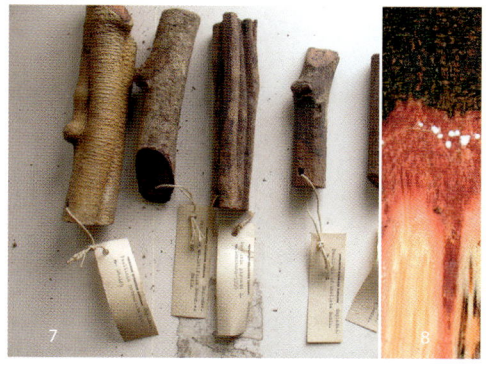

globose to 2.1 x 1.9 cm, **unwinged, 6-ribbed (3 obvious, 3 faint),** light-green speckled, woolly-haired [*lanate*] inside; seeds 1-3, subglobose, aril whitish.

Uncommon; flooded and non-flooded old-growth forest; Fr. Guiana, Suriname; widespread in Amazon basin.

1.8. Paullinia stellata Radlk.

Woody liana, 2-10 m long; **white sap sparse; hairs golden-orange to brown** [*tomentose*]. **Stems:** branchlets 3-angular, grooved, and hairy when young; **stipules deeply dissected with ovate outline;** 0.3-0.4 cm long, persistent; older stem x-s pattern simple. **Leaves: 5-foliolate, rachis unwinged;** leaflets elliptic or oblong, 7-14.5 × 3.5-7 cm, papery, minutely hairy below. **Inflorescences:** stalks hairy, calyx lobes 5, corolla 0.45-0.5 cm long. **Fruits: capsules subglobose, to 6-angular, unwinged,** densely hairy; seeds 1, dark brown, aril present.

Uncommon; old-growth forest; Fr. Guiana, Suriname; also Venezuela.

1.7. Paullinia sphaerocarpa Rich. ex Juss.

Woody liana, to 10 m long; **white sap present; hairs reddish-brown. Stems:** branchlets round to oval, hairy when young; stipules small, short-lived; older stems round to remotely angled, bark grey with vertical lines of small white lenticels, stem x-s pattern simple. **Leaves: 5-foliolate,** rachis slightly winged, to 12 x 5 cm; leaflets elliptic or oblong, base rounded, apex acuminate, coarsely toothed; ± papery, hairless except midrib above. **Inflorescences:** axillary, panicles slender, hairy when young, to 30 cm long; calyx lobes 4; corolla very small, white. **Fruits: capsules sub-**

2. Serjania Mill.

Mostly lianas, sometimes climbing shrubs; climbing via forked tendrils.

Common names: kashiri (Ar); neku (Sr); seybi-finga wiwiri (Sr); tukukuimë; tuhkukuimë (Tr); tuhpokuimë (Tr).

General: See family description; cut parts often with **resinous white sap as well as watery white sap.**

Stems: See family description; stipules small, short-lived; pith sometimes hollow.

Leaves: Mostly palmately-compound, with ultimate units of 3-leaflets, sometimes 5-7 odd-pinnate; **margins entire, wavy or (fine) toothed;** petioles and rachis often winged.

Inflorescences: Axillary, terminal (never cauliflorous), in racemes or panicles; **flowers small, white to greenish-yellow, bilateral;** calyx 4-5-parted; corolla with **4 distinct petals, appendages hood-shaped** with a fleshy, yellowish apex; nectar disc 2-4 glandular, at one side; stamens 8, unequal in height; ovary 3-chambered with one ovule.

Fruits: Dry, indehiscent fruits [*schizocarp*], samaras with wings expanded at the base (near stalk) and seeds fused at the apex. Seeds 1 per samara, with small arils.

Distribution: Tropical and subtropical America: 230 species; Guianas: 8 species; 4 described here.

2.1. Serjania grandifolia Sagot

Common name: kutupu (Wpi)

Woody liana, to 10 m long or more; **white sap abundant;** mostly hairless. **Stems:** branchlets oval or 5-6-sided; older stems oval to 5-6-sided, ± 3 cm diameter; **stem x-s pattern simple;** pith usually hollow. **Leaves: bi-pinnate;** petiole and rachis furrowed, unwinged; leaflets elliptic to oblong, 5-15 x 3-7 cm, **margins usually entire,** papery. Inflorescences: erect panicles to 35 cm long; calyx 5; corolla white, 0.3 cm long, anthers pinkish. **Fruits: samaras very large, 4-6 cm long,** wings reddish, net-veined; seed stone-like.

Uncommon; non-flooded forests. French Guiana, Suriname; widespread in the South American tropics.

2.2. Serjania membranacea Splitg.

Liana to 20 m long, often subwoody and creeping; **white sap sometimes present;** mostly hairless. **Stems:** branchlets 5-sided; **stem x-s pattern simple;** pith usually hollow. **Leaves: bi-pinnate;** petiole and rachis furrowed, unwinged; leaflets 2.5-9 × 1.5-4.7 cm, **margins finely and regularly toothed,** teeth ending in glands. **Inflorescences:** racemes 7-15 cm long, dense, with short hairs; calyx 5; corolla white, 0.4 cm long, anthers yellow. **Fruits:** samaras 2-2.5 cm long, wings thin, net-veined.

Common; open and disturbed areas, non-flooded forest and riparian forest. Guianas: widespread in the neotropics from Brazil and Peru to Guatemala.

2.3. Serjania paucidentata DC.

Common names: aboho (Ar); drikanti (Sr); hebechiabo (Ar); kashiri (Ar); kotupuru (Ca); kutupu (Ca); old man's bark (GU); pahaglaime (Way); taitetulea (Wpi).

Liana, to 10 m long; **white sap sometimes present;** mostly hairless. **Stems:** branchlets grooved when young; older stems 3-6-sided, ribbed; **stem x-s pattern compound,** with one central and 10 peripheral columns; pith red. **Leaves: bi-pinnate;** rachis (slightly) winged; leaflets ± elliptic, 3.5-16 x 1.5-6 cm, base tapering to narrow point [*attenuate*], **margins entire or with a few apical teeth,** curled under [*revolute*]. **Inflorescences:** axillary, panicles to 26 cm long, reddish-brown hairy when young; calyx 5; petals white, ca. 0.5 cm long, soft hairy. **Fruits:** samaras 2-2.4 cm long, pale yellowish-green to pinkish-red, wings thin, net-veined.

Common; riparian, savanna bush islands and secondary forest, shrubby vegetation. Guianas: widespread in the neotropics from Brazil and Peru to Mexico.

2.4. Serjania pyramidata Radlk.

Liana, growing high in canopy; **white sap abundant;** hairs reddish-brown to brown. **Stems:** branchlets hairy when young; older stems round, to 10 cm diameter, with 8-10 vertical stripes; **stem x-s pattern compound,** with one central and 10 peripheral columns. **Leaves: bi-pinnate;** leaflets ± spear-shaped, to 12 x 6 cm, **margins fine and regularly toothed** [*serrate*]; sparsely hairy with purplish-red veins below. **Inflorescences:** terminal and axillary, panicles 15-35 cm long, axes hairy when young; calyx 5, corolla white, ca. 0.25 cm long. **Fruits:** samaras oval-shaped, ca. 2.5 cm long, hairless, maturing bright pink- to purplish-red.

Uncommon, locally abundant; old-growth forest and along rivers. Guianas; widespread in the neotropics, from Brazil to Mexico.

1. Paullinia pinnata, with forked 'butterfly-tongue" tendrils [PT]

2. P. pinnata inflorescence close-up [PT]

3. P. pinnata, pear-shaped capsules [PT]

4. P. pinnata, close-up of zygomorphic flowers [*bilateral symmetric*] [OG]

5. P. pinnata, capsules opened to show shiny black seeds and white arils [PT]

6. P. pinnata, habit [OG]

7. Paullinia species wood specimens at the Suriname National Herbarium [BH]

8. Paullinia capreolata, stem slash with white sap visible [FRD]

9. Serjania grandifollia winged samaras (mericarps), with seeds at apex. [OG]

10. S. grandifollia, close-up of samaras [OG]

11. Serjania paucidentata, leaves palmately-compound (3-leaflet units, 9 leaflets total) [OG]

12. S. paucidentata, inflorescence [OG]

13. Compound stem x-s pattern in Sapindaceae (Thinouea myriandra) [BH]

14. S. paucidentata, close-up of flowers with white petals and yellow hooded petal "appendages" [OG]

15. Thinouia myriandra, robust leafy stem with "butterfly tongue" forked tendrils [BH]

16. T. myriandra, winged samaras (mericarps), seeds at base [BH]

17. Urvillea ulmacea, inflorescence close-up [OG]

18. U. ulmacea, flowering branch [OG]

3. Thinouia Triana & Planch.

Lianas, scandent shrubs or trees. Climbing via forked, coiled tendrils. Neotropics: 12 species; Guianas: 1 species.

3.1. Thinouia myriantha Triana & Planch.

Common names: meudetah (Tr)

Liana, to 10 m long; **white resinous sap usually absent;** mostly hairless. **Stems:** branchlets round to oval, bark brown with small white lenticels; stipules inconspicuous, short-lived; older stems to 12 cm diameter, with deep lengthwise grooves, bark reddish-brown, flaky, with horizontal stripes; **stem x-s pattern compound,** with one central column and 5-10 smaller peripheral columns. **Leaves: always 3-foliolate, rachis unwinged;** leaflets elliptic to broadly ovate, 6.5-14 x 3.5-7 cm; **margins mostly entire with 2-3** remote teeth at apex; papery; 2° veins wide-spaced, curving, not forming loops; 3° veins ± parallel, all veins immersed above, prominent below. **Inflorescences:** axillary or terminal; flowers radial, calyx cup-shaped, hairy, corolla ca. 0.7-1 mm long, greenish, appendages longer than petals. **Fruits:** Dry, indehiscent fruits [*schizocarp*], samaras with wings expanded at the apex and seeds fused at the base (near fruit stalk). Wing 4-5.5 cm long, veined; seeds 1, aril absent.

Uncommon; open, disturbed vegetation and old-growth forest along rivers. Guianas; widespread from N South America to Mexico.

4. Urvillea Kunth

Woody to subwoody lianas, herbaceous vines when young. Climbing via forked, coiled tendrils; Neotropics: 15 species; Guianas: 1 species.

4.1. Urvillea ulmacea Kunth

Subwoody liana, 3-7 m long; **white resinous sap usually absent;** hairless or lightly hairy. **Stems:** branchlets round or 3-5 lobed, bark brown with small white lenticels; stipules inconspicuous, short-lived; older stems with deep lengthwise grooves, **stem x-s pattern compound,** with one central column and 5-10 smaller peripheral columns. **Leaves: always 3-foliolate, rachis unwinged;** leaflets elliptic to deltoid, terminal leaflet to 10 x 5 cm, **margins toothed with large blunt teeth on sides and apex;** 2° veins slightly curving, not forming loops, ending at teeth. **Inflorescences:** axillary, 3-12 cm long, short hairy; flowers asymmetric, corolla to 0.4 cm long, white with an apically yellow appendage; nectar disc 4-lobed. **Fruits:** Dry, indehiscent fruits [*schizocarp*], samaras with seeds fused in the mid-section and oval wings present at both ends. Wings 4-5.5 cm long, veined, turning from green to straw-colored; seeds 1, aril white.

Uncommon; open or disturbed areas, old-growth forest and river margins. Guianas; widespread from northern Argentina to Mexico.

SCHLEGELIACEAE

SCHLEGELIACEAE

Hemi-epiphytic climbing shrubs and lianas; sometimes small trees. Neotropics and the Caribbean: 5 genera/43 species. Only one climbing genus present in the Guianas.

1. Schlegelia Miq.

Hemi-epiphytic climbing shrubs and lianas. Climbing via adventitious roots. Leaves are opposite, simple and entire. Stipules are absent. Flowers are tubular, with a persistent calyx, 5-lobed corolla, 4-stamens and superior ovary. Fruits are many-seeded berries.

General: Surfaces mostly **hairless,** often slightly fleshy, hairs simple when present; **robust lenticels present** on fertile and infertile branches; slash without notable color, sap or scent.

Stem: Branchlets round, bark soft, greyish-brown with white lenticels; nodes without an interpetiolar ridge or gland fields, prophylls (axillary buds) present, small, pointed or round: **adventitious roots present; tendrils absent.** Older stem similar to branchlets, bearing large, raised lenticels; stem x-s pattern simple.

Leaves: Elliptic to ovate, **strongly leathery or fleshy,** hairless and glossy; veins immersed above, prominent below, **glands present at midrib base;** petiole robust, curved.

Inflorescences: Terminal or axillary, often borne directly on stem, long racemes or short panicles;

calyx cup-shaped or two-lobed, persisting in fruit; **corolla tube-shaped with reflexed or flat lobes,** 1.6-4 cm long, often purple-pink or pure white, rarely red, with conspicuous glandular scales within; stamens 4; nectar disk present, ovary inferior.

Fruits: **Round berry,** 1-3 cm diameter, partly **covered by persistent calyx at the base,** fruit wall dry and thin; seeds numerous, tiny, embedded in pulp.

Ecology: Seed dispersal via animal (especially birds through gut).

Distribution: Widespread in the Neotropics, reaching Mexico and the Antilles: 14 species; Guianas: 6 species, 2 of the Guianas-wide species described here.

Notes: This genus was included in the Bignoniaceae for many years and has recently been placed in its own family, the Schlegeliaceae (APG II, 2003; Olmstead et al., 2009). Schlegelia was unique as the only regional climber of the Bignoniaceae family with simple, opposite leaves and no tendrils. Sterile Schlegelia might be confused with other opposite-leaved hemi-epiphytes in the Guianas, including Clusia (Clusiaceae), Drymonia (Gesneriaceae), Hillia (Rubiaceae), and Schradera (Rubiaceae).

1.1. Schlegelia paraensis Ducke

Liana or hemi-epiphytic shrub. Stems: prophylls narrow, pointed and 2-lobed. Leaves: elliptic, 8-14.5 x 4.5-8 cm; fleshy to leathery. **Inflorescences:** axillary or terminal, short with 3 flowers clustered on 1 cm long stalk; **calyx 2-lobed to cupular; corolla tube white, lobes pink with white margins, 3.5-4 cm long. Fruits:** nearly round, to 2 cm diameter, purple-brown with fine dots; apex with short, blunt prickle.

Rare; non-flooded old-growth and riparian forest edges; Guianas.

1.2. Schlegelia violacea Griseb.

Common names: bulta-takubia (Ar); diaba-titei (Sr); kupayaran (Ca)

Lianas or hemi-epiphytic shrub. Stems: prophylls flat-rectangular. Leaves: narrowly elliptic to ovate, 7-19 x 3-12, curling at sides; strongly leathery. **Inflorescences:** terminal, long and narrow, to 20 cm long, many-flowered; **calyx cupular,** irregularly lobed, purplish-pink, 0.6-0.9 cm long; **corolla tube light pink to purplish-pink, 1.2-1.5 cm long. Fruits:** elliptic to round, to 1.4 x 1.3 cm, dark purplish-green, finely wrinkled, persistent calyx covering lower 2/3; apex with a short, sharp prickle.

Locally common; secondary forest, seasonally-flooded old-growth forest, lower montane forest, 50-800 m. Guianas; N Amazonian Brazil, E Venezuela.

---

1. Schlegelia violacea, a hemi-epiphytic liana growing downwards from a host tree [BH]

2. S. violacea, on forest edge with large, showy inflorescence and robust leaves [PT]

3. S. parviflora, leaf underside, with robust, curved, swollen petioles and flat leaf glands (Costa Rica) [RA]

4. S. violaceae, inflorescence close-up, with cupular calyx, and tubular, 5-lobed corolla [PT]

5. S. parviflora, leafy branchlet with raised lenticels, axillary bud 'prophylls', and glands at leaf bases (Costa Rica) [RA]

6. S. parviflora, infructescence of pink berries, each berry partially covered by an expanded, persistent sepal (Costa Rica) [RA]

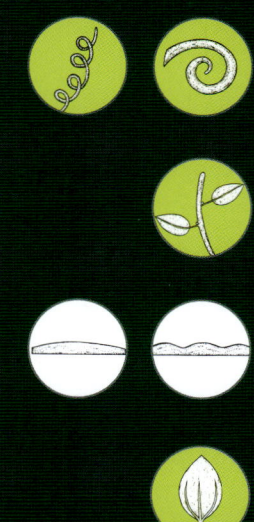

SMILACACEAE

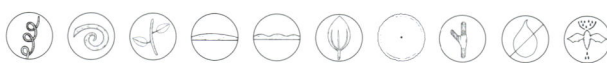

SMILACACEAE

Slender subwoody lianas, climbing shrubs, or vines; monocot. Cosmopolitan, including temperate and tropical regions. Only one genus in the Americas.

1. Smilax L.

Small lianas, scrambling shrubs, or vines; with green prickly or spiny stems and large rhizomes or tuberous roots. Climbing via paired tendrils. Stipules absent. Leaves are alternate, simple, entire, palmately-veined, often with a winged, stipule-like sheath on the petiole. Individual plants are either male or female [dioecious], with inflorescences of small 3-parted male or female flowers. Fruits are brightly colored, round, fleshy berries.

General: **Most surfaces smooth and green;** sometimes rough-surfaced [*asperous*] or with tiny, black recurved bumps [*tubercles*]; notable scent or sap absent.

Stems: Branchlets and older stems **subwoody and armed** - bearing recurved hooks, thick spines or needle-like prickles, **1-2 persistent papery flaps or sheaths present at branch/stem bases.** Older stems x-s pattern simple (typical for monocot).

Leaves: Diverse shapes, narrow to broad or heart-shaped, base extending slightly down [*decurrent*] onto the leaf stalk [*petiole*] apex, margin entire, sometimes prickly along edges; leathery to papery, gland-dots sometimes present; **veins 3-5-(9)-palmate, finer veins net-veined** [*reticulate*]; **leaf stalk thick and flexed or jointed** [*geniculate*], **usually with a stipule-like winged "sheath" bearing two tendrils,** or the tendrils borne halfway on the leaf stalk with no visible sheath, tendrils spiral-coiled or straight.

Inflorescences: Axillary or terminal on short shoots, **flowers in umbels or compound umbels,** "receptacle" (raised area bearing flower stalks) curved or symmetrical and globe-shaped; flowers **small, regular** [*radial*], **white to greenish-white,** 3-parted, unisexual — male flowers with 6 free stamens; female flowers with 1-3 stigmas.

Fruits: Round berries, maturing orange, red, or black-purple; seeds 1-6, round, smooth.

Ecology: Climber of the forest understory and open places, common along trails, sometimes weedy; insect-pollinated; seed dispersal via animal (birds, bats).

Distribution: Cosmopolitan, more diverse in the tropics: ca. 300 species; Neotropics: 100 species; Guianas: 12 species, four common and woody species described here. Most Guiana Shield species rarely collected.

Uses: Rhizomes, tubers, and bark are commonly used as medicines (e.g. genitourinary, venereal diseases) and aphrodisiacs in the Guianas; stems are sometimes used in cordage and basketry; berries have been used to poison fish.

Notes: The genus most likely to be confused with Smilax is Dioscorea (Dioscoreaceae) – small, prickly vines with broad, palmate-veined leaves and tubers. Dioscorea is distinguished by ladder-like 3° veins, tendrils absent, petiole swollen at base and apex, flowers in spikes (vs. umbels in Smilax) and 3-winged capsules (vs. round berries in Smilax). Also compare with Aristolochiaceae, Hernandiaceae, Menispermaceae and Rhamnaceae.

1.1. Smilax domingensis Willd.

Common name: busi-nyamsi (Sr); sepé atano (Ca)

Slender subwoody liana, trailing or climbing. **Stems:** flattened, smooth (not asperous), **hooked prickles** present or absent; black tubercles absent; **basal scales 2. Leaves:** variable, mostly ovate, 6-15(-25) x 3-8.5 cm, base truncate, rounded or lobed, margins flat; leathery, main palmate veins 5; **petiole to 1 cm long, sheath to 0.6 cm long. Inflorescences:** axillary umbels, flower stalks 0.5-0.9 cm long, shorter than petiole sheath; **receptables globe-shaped,** male umbels ± 25-flowered, female umbels ± 65-flowered. Fruits: berries to 1 cm diameter, purplish-black.

Common; open savanna, rock savanna, and granitic outcrops. Guianas; also Venezuela.

1.2. Smilax oblongata Sw.

[Syn: Smilax cumanensis Humb. & Bonpl. ex Willd.; Smilax surinamensis Miq.]

Common name: agbagomaka (Sa); agwagomaka (Au); aguwagomaka (Bo); akikiwa (Ca); busi nyamsi (Sr); cockshun (GU); dorokwaro-balli (Ar); ikhana (Ar); maka-titei (Sr); nyamsimaka (Sr); sarsaparilla (EN); sipatamu (Ca); sipatu (Tr).

Slender subwoody liana. Stems: flattened to round, **unarmed** or (rarely) with stout prickles, slightly asperous, black tubercles absent; **basal scales 1. Leaves:** variable, 8-14 (-17) x 4-9 cm, base flat or lobed, apex fine-pointed, margins flat; papery to sub-leathery, gland-dotted below; main palmate veins 5-7; **petiole 0.4-1 cm long, sheath to 0.6 cm long. Inflorescences:** axillary, flower stalks 0.1-0.3 cm long, shorter than adjoining petioles, male receptacles globe-shaped, 15-60 flowered, female receptacles curved, 8-25-flowered. **Fruits:** berries to 1.2 cm diameter, red to bluish-black; seed 1, 0.6 x 0.3 cm.

Common; on riverbanks, sand- and shell-ridges in coastal plain, secondary vegetation, non-flooded old-growth forest. Guianas; N South America to Costa Rica and Tobago.

1.3. Smilax santaremensis A.DC.

Slender subwoody liana. Stems: flattened to round, smooth (not asperous), **base with hooked prickles,** black tubercles absent; **basal scales 2. Leaves:** ovate or narrowly ovate, 9-13 x 3.5-6 cm, base flat or lobed, gradually narrowing to petiole, margins flat; papery, gland-dotted above; main palmate veins 3-5; **petiole 0.3-0.6 cm long, jointed near the middle; sheath to 0.5 cm long. Inflorescences:** axillary, male and female umbels with receptacles curved, 8-15 flowered, flower stalks 1.7-2.5 cm long. **Fruits:** berries to 0.6 cm diameter, red.

Common; along forest edges in savanna areas. Guianas; also Brazil.

1.4. Smilax schomburgkiana Kunth

Common name: agbagomaka-tatai (Sa); bagou (FG); barakut (Pal); cockshun (GU); dorokwaro-pimpla (GU); dorokwaro-balli (Ar); hameluna-balli (GU); liane bagotte (FG); maka-titei (Sr); nyamsimaka (Sr); nyamsi-titei (Sr); pletauw (Way); salsepareille (FG); sipatai (Tr); sipatamu (Ca); sipa-tu (Tr); spotai (Ak)

Subwoody liana, with large woody tuber. **Stems:** flattened, slightly asperous, often armed with a few large **straight or curved spines, with many black tubercles; basal scales 2. Leaves:** extremely variable, 12-30(48) x 5-16(24) cm, smaller on fertile branchlets, base truncate, rounded or lobed, margin flat; papery to leathery, hairless, gland-dotted below; main palmate veins 3-7; **petiole 0.5-4 cm long, sheath 0.5-3.5 cm long. Inflorescences:** axillary or terminal, male inflorescences in panicles of umbels, 5-25 cm long, female flowers in simple umbels; for both sexes, receptacles curved, flower stalks 0.5-6 cm long, flowers up to 30 (rarely males > 50) per umbel. **Fruits:** berries 0.8-1.2 cm diameter, orange-red to black; seeds 1-2 (3). **Use:** One of the more widely used medicinal and aphrodisiac plants in the Guianas. The roots contain the hormones testosterone and progesterone.

Common; riverbanks, non-flooded old-growth forest, dry forest, cultivated areas. Guianas; Venezuela, NE Brazil.

1. Smilax poeppigii, umbels of orange fruits [FG]
2. Smilax schomburgkiana, with green, subwoody, armed stem [PT]
3. S. schomburgkiana, with green, subwoody, armed stem [PT]
4. S. schomburgkiana, base of plant with tuber [BH]
5. Smilax sp., young shoot with prickles and reddish petiole sheath [BH]
6. Smilax sp., tendril pair borne on petiole sheath (immature leaf) [BH]
7. Smilax sp., leaves with palmate veins and flexed, thickened petioles [BH]
8. Smilax sp., male flowers [BH]

SOLANACEAE

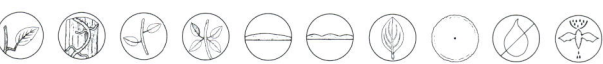

SOLANACEAE

Sub-woody lianas (Lysianthes, Solanum), hemi-epiphytes (Markea, Solandra), climbing shrubs, or vines; mostly herbs, shrubs, or small trees. Climbing via twining petioles, leaning branches and clasping aerial roots. Leaves are alternate to spiral-clustered, simple, with entire margins in the climbers. Stipules are absent. Inflorescences are often showy, with 5-parted, long-tubular, short-tubular or bell-shaped flowers, a persistent calyx, and superior or partially superior ovary. Fruits fleshy or leathery berries with many seeds, sometimes like small tomatoes.

General: Armed with prickles (Solanum) or unarmed; **often with a unpleasant "rank" tomato-leaf-like scent;** smooth and hairless to densely hairy and appearing "dusty", **hairs star-shaped** [*stellate*] or 2-armed (Y-shaped or T-shaped).

Stems: Branchlets growing forward in a 'zig-zag' pattern or curving; round, oval, or unevenly flattened; ± fleshy and bearing lenticels, often greyish-green; nodes with stipules absent. Older stems round to flattened, small- to medium-diameter, **bark softly woody, green sub-bark with white wood;** stem x-s pattern simple.

Leaves: Often spiral-clustered at branch ends (hemi-epiphytes) or main leaves with a small, opposite 'companion' leaf at nodes (lianas); margins entire, wavy or with a few rough teeth; thick-leathery to papery; veins pinnate, curving steeply, often looping, 3° veins net-like [*reticulate*]; clear gland dots often present; petiole twining in liana species.

Inflorescences: Terminal, axillary, or inserted above the axil; flowers in cymes or solitary, often hanging downwards, rachis often with leafy bracts or bracteoles; flowers 5-parted; calyx lobed or entire; corolla long-tubular, short-tubular or wide-bell-shaped, lobed, sometimes with finger-like appendages (Lysianthes); stamens 5, inserted on petals, equal or unequal, anthers sometimes pressed together to form a yellow cone (Lysianthes, Solanum); ovary superior to half-inferior [*2-locular*], style 1, stigma 1-2-lobed.

Fruits: Fleshy to leathery berries in climbers, otherwise capsules or drupes; seeds 1-many.

Ecology: Pollination by bees, butterflies and bats. In Lysianthes and Solanum, anthers release pollen only through tiny apical pores [*poricidal dehiscence*]. Visiting bees must vibrate their bodies at a certain frequency to drive pollen out of the anthers. Hemi-epiphytes live upon other plants but are not necessarily parasites. They produce several types of roots at nodes - one is primarily for climbing and holding on to host plants and the other is extended to the forest floor to obtain nutrients and water. Seed dispersal via animal (ingested by birds, monkeys).

Distribution: Cosmopolitan, mostly neotropics: ± 80 genera/3000 species; Guianas: 4 genera with ± woody climbers (13 species).

Notes: The leaves and stems of some Convolvulaceae, Icacinaceae, and Solanaceae lianas have a similar look. Woody Convolvulaceae climbers are distinguished by long petioles and off-center concentric rings in cross-section. Icacinaceae has petioles twisted and canal-like, 2-cabled mature stems, and cut stems oxidizing orange.

Use: Foods, medicines, hallucinogens.

1. Lycianthes (Dun.) Hassl.

Lianas; also herbs and shrubs. Climbing mostly via twining petioles. Pantropical, center of diversity in the neotropics: 200 species; Guianas: 4 species, 1 described here.

1.1. Lycianthes pauciflora (Vahl) Bitter

Common name: bos tomaat (SD)

Liana or climbing shrub; unarmed, hairy in parts [*stellate, tomentose*]. **Stems:** branchlets round or flattened, zigzag, often swollen at nodes, bark smooth, with conspicuous raised lenticels. **Leaves: mature leaves with a small, opposite 'companion' leaf; elliptic,** 7.5-15 x 4-7.5 cm, base rounded or slanted, apex attenuate; thin, ± hairless above, densely hairy below; petiole 0.8-1.3 cm long. **Inflorescences:** axillary, solitary or 2-4-flowered; **calyx with (5)10 finger-like projections,** each 0.2-0.3 cm long; **corolla white,** **widely bell-shaped,** ± 1.5 cm long, not showy. **Fruits:** round to triangular **tomato-like berries,** 1.5-3 cm diameter, with a pale-green to red skin; pulp orange, calyx projections persistent in fruit; seeds round or angled, ca. 0.3 x 0.3 x 0.05 cm. **Ecology:** Bee-pollinated. Seed dispersal via animal (birds). **Notes:** Flowers similar to Solanum but distinguished by calyx with ten finger-like projections.

Uncommon; open areas in tropical lowland and savanna forest, especially along paths. Guianas; also widespread in the neotropics, from Paraguay to Panama.

2. Markea L.C. Rich.

Slender liana-like hemi-epiphytes; or shrubs. Climbing via clambering stems and clasping aerial roots.

General: "Rank" scent present in cut shoots and leaves; hairless to lightly hairy.

Stems: Branchlets oval, slender and smooth, often with large, raised, white lenticels. Older stems oval, to ca. 2 cm diameter, smooth, bark whitish-grey to blackish-brown, softly woody, thin, with a large pith.

Leaves: **Widely-spaced on lower branchlets, clustered towards branch ends,** mature leaves without small, opposite 'companion' leaves; margins entire, sometimes with marginal hairs [*ciliate*]; leathery, succulent, or thin, **often with flat, clear glands.**

Inflorescences: Terminal on twigs, solitary or few-flowered, often hanging below the leaves; **flowers showy to inconspicuous, red, orange,** yellow, or white; 5-parted; calyx 5-lobed, hairless, 0.5-3.5 cm long, fleshy and globose or tubular; corolla tubular to bell-shaped, 0.5 to 10 cm long; stamens 5; ovary superior to half-inferior.

Fruits: Cone- or egg-shaped [obovate] berries, leathery to fleshy.

Ecology: Some Markea species have night-opening flowers and are likely pollinated by bats, others are hummingbird- or bee-pollinated; most species likely involved in co-evolutionary interactions with ants.

Distribution: Neotropics: 20 species; Guianas: 4 species.

Notes: Markea is not well documented, and considerable taxonomic revision is expected.

2.1. Markea coccinea L.C. Rich.

Common name: businenge pogasi (Au)

Slender liana-like hemi-epiphyte growing in shrubs and trees. Leaves: oblanceolate, 12-15 x 5-5.5 cm, base acute, apex apiculate, margin entire, without marginal hairs; sub-leathery; petiole 0.4-0.7 cm long. Inflorescences: flowers solitary or few, infl. stalk slender, to 30 cm long; calyx green with purple veins, veins inconspicuous, lobes lanceolate, divided nearly to base, to 3 cm long; corolla narrowly tubular, bright reddish-orange, to 5 cm long. Fruits: oblong berries, to 1.6 x 0.5 cm, surrounded by a persistent calyx. Ecology: likely hummingbird-pollinated.

Common; old-growth forest, swamps, savanna forest, especially along creeks and rivers. Guianas; also central Amazonia.

2.2. Markea formicarum Dammer

Slender liana-like hemi-epiphyte growing in shrubs and trees. Leaves: broadly elliptic, large, 11-17 x 3.5-8 cm, margin with conspicuous

hairs [ciliate]; thin (not leathery). Inflorescences: flowers solitary or paired; calyx green, veins inconspicuous, deeply 5-lobed, to 3 cm long; corolla funnel-shaped, cream-colored with 2-lipped purplish lobes, 6-7.5 cm long. Fruits: ovoid berries, cream-colored.

Uncommon; old-growth forest, swamps, savanna forest, especially along creeks and rivers. Fr. Guiana, Suriname; widespread in the Amazon Basin, expected in Guyana.

2.3. Markea longiflora Miers

Woody vine growing in shrubs and trees. Leaves: ovate, 10-15 x 5-8 cm; margin entire, without marginal hairs; sub-leathery. Inflorescences: flowers in racemes or clusters; calyx with main vein network conspicuous, dark purplish-brown, deeply 5-lobed, to 3 cm long; corolla funnel-shaped, cream-colored with net-veins, 6.5-7.5 cm long. Fruits: ovoid berries, cream-colored.

Common; old-growth forest, swamps, especially along creeks and rivers. Guianas; also central Amazonia.

2.4. Markea sessiliflora Ducke

Liana-like hemi-epiphyte, growing in shrubs and trees. Leaves: in clusters of 2-3, elliptic to obovate, large, 10-17 x 5-9 cm, base truncate to subcordate, margin entire, without marginal hairs; leathery. Inflorescences: solitary or few, infl. stalk to 4 cm long, thick, corky; calyx light green, veins inconspicuous, deeply 5-lobed, to 2.5 cm long; corolla tubular, light yellow or creamy white, < 7 cm long. Fruits: ovoid berries, ca. 1.7 x 1 cm, yellow, surrounded by calyx.

Uncommon; old-growth forest. Guianas; N Amazon Basin.

3. Solandra L.

High climbing, liana-like hemi-epiphytes with huge tubular flowers. Climbing via clambering stems and aerial clasping roots. Small neotropical genus ranging from SE Brazil to Mexico and the West Indies: 9-10 species. Guianas: 2 species.

3.1. Solandra longiflora Tussac

Common name: kumpuruni (Tr)

Liana-like hemi-epiphyte; mostly hairless. **Stems:** branchlets ovate to angled; bark grey, smooth, soft, and furrowed. **Leaves: clustered at branch ends,** mature leaves without small, opposite 'companion' leaves; elliptic to obovate, to 13 x 5 cm, often asymmetrical, base cuneate to acute, apex shortly acuminate, entire; **sub-leathery; petiole long, to 6 cm long. Inflorescences:** terminal, flowers 1-several per inflorescence, flowers large (to 40 cm long), fragrant and showy; flower stalks short and stout; calyx tubular, 6.5-11 cm long, irregularly lobed, hairless; **corolla funnel-shaped, 23-36 x 5-7.5 cm, yellow, with 10 purplish nectar lines inside the tube,** 5-lobed, lobes overlapping in bud; stamens 5, subequal, anthers 0.6-1 cm long; ovary 4-chambered, half-inferior, stigma 2-lobed. **Fruits:** leathery conical berry, calyx persisting and splitting in fruit; seeds compressed, disc- or kidney-shaped, 0.4-0.65 cm long. **Use:** Genus contains alkaloids such as atropine, noratropine, hyoscyamine, and tropine and is used by some indigenous groups in Mexico as a hallucinogen (Bernardello & Hunziker 1987, Knab 1977). **Notes:** Solandra paraensis Ducke is a rare Guiana Shield species known from French Guiana, Colombia and N and NE Brazil.

Uncommon; old-growth forest, sometimes an ornamental. Guianas; also Colombia, Ecuador, Venezuela, Ecuador, and the West Indies.

4. Solanum L.

Rarely lianas or climbing shrubs, usually erect herbs, shrubs, or treelets. Climbing mostly via twining petioles.

General: Unarmed (S. pensile) or armed with prickles (S. coriaceum, S. rubiginosum); typical "rank" scent present; hairs commonly branched or stellate.

Stems: Branchlets sparsely to densely hairy, often greyish-green. Older stems round or oval, sub-woody, to 2 cm diameter.

Leaves: Not clustered at branch ends, main leaves with a small, opposite 'companion' leaf at nodes; leaf bases often unequal, margins entire.

Inflorescences: Axillary or terminal, 2-to-many-flowered cymes; flowers 5-parted; calyx cupular, persistent and often growing larger after flowering, without finger-like projections; corolla lobes spreading, white, yellow, or purple; stamens 5, usually equal, anthers forming a yellow cone.

Fruits: Tomato-like globose berries; seeds numerous, flat, small.

Ecology: Bee-pollinated. Seed dispersal via animal (birds).

Use: Foods, medicines, hallucinogens.

Common in disturbed areas or secondary forests. Widespread in temperate and tropical regions: > 1250 species; Guianas: 3 sub-woody climbers.

SOLANACEAE

4.1. Solanum coriaceum Dunal

High climbing liana; armed with short, reflexed "cat claw" prickles, hairless to densely hairy [*tomentose*]. **Leaves:** oblong, 6-18 x 3-9.5 cm, base unequal; sub-leathery; prickly below on midvein; petiole 0.5-1 cm long, prickly. **Inflorescences: axillary racemes,** 8-15 cm long; calyx irregularly 5-lobed, lobes rounded; corolla deeply lobed, bluish- to reddish-purple, hairy. **Fruits:** berries up to 2.2 cm diameter, maturing yellowish-green, hairless; seeds ca. 0.25 cm long and wide.

Common; forest margins and riverbanks. Guianas; also E Brazilian Amazon.

4.2. Solanum pensile Sendtn.

Liana or climbing shrub; unarmed, hairless or lightly hairy. Leaves: ovate, 7-14 x 4-8 cm, base obtuse to subcordate; petiole 3-4 cm long, twisted or coiled. **Inflorescences: terminal panicle with hanging, recurved branchlets;** calyx shallow-lobed; corolla deeply cut, lobes spear-shaped to linear, purplish-blue outside, yellow inside. **Fruits:** berries, to 1.4 cm diameter, maturing dull blue-purple.

Common; riparian forest, secondary forest, open areas. Guianas; also Central and NW South America to Nicaragua.

4.3. Solanum rubiginosum Vahl

Common name: bofrumaka (Sr); sobolero (Ar)

High climbing liana; armed with short, reflexed "cat claw" prickles, hairless to densely hairy [*tomentose*]. **Leaves:** elliptic, 7-14 x 3.5-7 cm, base rounded, apex attenuate; sub-leathery, dark green with sparse stellate hairs and unarmed above, petiole 1-3.5 cm long, prickly. **Inflorescences: terminal, many branched,** to 12 cm long, reddish-brown hairy; calyx deeply-cut, lobes triangular; corolla deeply cut, lobes spear-

shaped, bluish- to reddish-purple, to 2 cm diameter. **Fruits:** berries to 1.5 cm diameter, densely hairy; seeds ca. 0.35 cm long and flattened.

Uncommon; forest margins and along riverbanks. Guianas; also NE Brazil, lower Amazon River basin.

1. Markea longiflora in flower [OG]
2. M. longiflora, robust leaves with inflorescence [OG]
3. Markea coccinea in flower [OG]
4. Solanum subinerme, typical Solanum-type flower (this species not in the Guianas) [FRD]
5. Lycianthes pauciflora, fruiting branch, berry with finger-like calyx lobes [AP]
6. M. coccinea, close-up, corolla lobes, stamens extending out of throat [OG]
7. L. pauciflora, mature berries [DC]
8. L. pauciflora, tiny 'companion leaf' visible [AP]
9. M. coccinea, leafy shoot in lower canopy [OG]
10. Solandra longiflora, yellow-white corolla [SZ]
11. Solanum pensile, habit [BH]

SOLANACEAE

TRIGONIACEAE

TRIGONIACEAE

Mostly woody lianas and climbing or scrambling shrubs; sometimes erect shrubs, treelets, and trees. Pantropical: 3 genera, 40 species. One genus with woody climbers in the neotropics.

1. Trigonia Aubl.

Robust woody lianas and weakly climbing shrubs. Climbing via twining stems and opposite branchlets. Leaves are opposite and simple with entire margins, and often hairy or white below. Stipules are present. Flowers are legume-like, with stamens united at base and a superior ovary. Fruits are 3-parted dry capsules with hairy seeds.

General: **Often conspicuously and densely hairy; hairs simple, variable—soft and felty, long-flattened [*villose*], short-flattened [*strigose*], woolly, or spider-web-like;** conspicuous sap, glands and scent absent.

Stems: Branchlets round, nodes with a swollen ridge and **four interpetiolar stipules or scars, triangular to narrow and pointed (awl-shaped).** Younger woody stem bark silver, brown, to dark-red, densely covered with raised lenticels and bumps [*tubercles*]. Older stems round to oval, often twisted and fluted, sometimes > 10 cm diameter; **outer bark thick, soft, yellowish-white to light-brown,** bearing lenticels; inner bark pink/white; stem x-s pattern simple or slightly lobed.

Leaves: Leathery to thin, smooth, not rough-to-touch, **often dark green above and conspicuously white, lighter green and/or hairy below;** 2° veins widely-spaced, curving or ± straight, not reaching margin, sometimes looping, raised below, 3° veins net-like [*reticulate*] to ± parallel; petiole tiny or short, canal-shaped.

Inflorescences: Narrow racemes or panicles, axillary or terminal; rachis commonly with leafy bracts and bracteoles; **flowers legume-like** (two lateral 'wing' petals, upright 'banner' petal, and two underside 'keel' petals), whitish-yellow to pink, small, sweet-smelling; stamens 5-12, fused at base; ovary superior, usually 3-chambered, style 1, stigma 1.

Fruits: **Dry, 3-parted capsules,** elongate to ovoid, splitting lengthwise, hairless to densely hairy; seeds small, round, with hairs of 2 distinct lengths.

Ecology: Insect-pollinated (bees). Seed dispersal via water (hairy seeds) or wind.

Distribution: Wide habitat range, more common along rivers and in disturbed areas; Neotropics: 24 species; Guianas: 8 species, mostly uncommon, 5 species described below.

Notes: The genus name combines two Greek words, trias (three) and gonos (angles), in reference to the 3-angled fruit. Trigonia species are most likely to be confused with Malpighiaceae (complex hairs, glands) and Rubiaceae, esp. Sabicea (persistent interpetiolar stipules).

1.1. Trigonia hypoleuca Griseb.

Climbing or erect shrub; hairs strigose. **Stems:** branchlets round, smooth; **nodes with stipules awl-shaped,** 0.2-0.3 cm long. **Leaves:** oblong-elliptic to obovate, 8-18 x 4-10 cm, base cuneate to obtuse; papery to sub-leathery, ± hairless above, **white and with air chambers below;** petiole 5-12 cm long, wrinkled, ± hairless. **Inflorescences:** panicles to 30 cm long, branches sometimes contorted; bracts linear, 0.1-0.25 cm long, densely hairy; **flowers 1-4 clustered, reddish-pink;** sepals ovate to oblong, strigose to woolly; central 'banner' petal 0.55-0.85 cm long. **Fruits:** leathery dry capsules, 5-7 x 2.5-4.0 cm, **1-chambered (other species 3-chambered);** hairless, veins raised; seeds 0.6.-0.8 cm long, elliptic, flattened, sticky, covered with stiff long hairs.

Uncommon; seasonally-flooded forest, riverbanks (water-dispersed). Guianas; also NE Brazil.

1.2. Trigonia microcarpa Sagot ex Warm.

[Syn: Trigonia laevis Aubl]

High-climbing liana or scandent shrub; hairs strigose. **Stems: nodes with stipules awl-shaped,** hairy, short, 0.15-0.2 cm long; older stems round, to 10 cm diameter; outer bark brown-grey, with black lenticels and tubercles, inner bark pink, wood whitish-yellow, pith red. **Leaves:** broadly elliptic to oblong-elliptic, 4-9 x 2.2-4.5 cm, base oblique to obtuse; papery to ± leathery, mostly hairless; petiole 0.4-0.8 cm long, often hairy. **Inflorescences:** panicles to 30 cm long; **flowers 3-6-clustered, usually white;** bracts 0.1-0.3 cm long, awl-shaped, hairy; sepals deltoid to oblong, 0.2-0.3 cm long; central 'banner' petal short, 0.35-0.5 cm long. **Fruits: capsules small, 0.6-1 cm diameter, 3-parted;** outer wall thin, bright green, maturing purplish-brown and hairless; seeds 1-3, subglobose, hairs to 1 cm long.

Uncommon; non-flooded old-growth forest, secondary forest. Guianas; also Brazil, Bolivia, Peru, W Venezuela.

1.3. Trigonia nivea Cambess.

High-climbing liana or scandent shrub; hairs mostly wooly. **Stems:** branchlets round, wooly when young; **nodes with stipules triangular,** 0.6-1.0 cm long. **Leaves:** elliptic to oblong-elliptic (ovate), 5-13 x 1-4.5 cm, base obtuse; sub-leathery, mostly hairless above, green below; petiole 0.2-0.8 cm long. **Inflorescences:** 5-10 cm long, dense or open; **flowers 1-4-clustered, white with yellow;** bracts 0.2-0.6 cm long, triangular to awl-shaped, hairy; sepals ovate to oblong, hairy; central 'banner' petal 4.5-6.5 x 3 cm; stamens 10-11. **Fruits: capsules cylindrical, 3-parted,** 3-7 cm x 1-1.5 cm, surface velvety orange-brown, wrinkled; seed hairs ca. 1 cm long.

Uncommon; interior and coastal old-growth forests. Guyana & Suriname (new record for Suriname; B. Hoffman collection no. 5704); also from Venezuela to S Brazil and Paraguay.

1.4. Trigonia subcymosa Benth.

Common name: awari imopi tëkënë, kurumpïn-pïrïkï (Tr)

Liana or scandent shrub; hairs strigose to wooly. **Stems:** branchlets round; **nodes with stipules ± crescent-shaped to** 0.5 x 0.2 cm. **Leaves:** elliptic to oblong, 3-8 x 2-4 cm, base cuneate to obtuse, apex obtuse to fine-pointed; papery, mostly hairless above, **yellow-to white-wooly below;** petiole 0.3-0.4 cm long. **Inflorescences:** terminal, pyramidal panicles, to 15 cm long; **flowers 2-4-clustered, white;** bracts 0.2-0.3 cm long, ovate, hairy; sepals ovate to oblong, hairy; central 'banner' petal 3-3.5 x 2 cm; stamens 8. **Fruits:** Unknown.

Uncommon: Guyana & Suriname (new record for Suriname; B. Hoffman collection no. 6353).

1.5. Trigonia villosa Aubl.

Common name: awari imopi tëkënë (Tr); kurumpïnpïrïkï (Tr)

High-climbing liana or scandent shrub; hairs strigose and villose. **Stems:** branchlets round, golden-brown strigose; **nodes with stipules leathery, awl-shaped,** 0.3-0.6 cm long. **Leaves:** broadly elliptic to obovate, unequal, 5-14 cm x 2-8.5 cm, base cuneate to obtuse; sub-leathery, green below; petiole 0.3-1 cm long. **Inflorescences:** to 25 cm long, **flowers 1-3-clustered, white;** bracts linear or ovate, 0.2-0.3 cm long; sepals ovate or oblong, hairy; standard petal 0.5-0.7 cm long; stamens 10-11. **Fruits:** oblong, 4.5-11 cm long (large), reddish-brown to yellowish-brown velvety; seeds 20 per chamber, with hairs to 2 cm long.

Uncommon; non-flooded old-growth forest. Guianas; also Brazil.

1. Trigonia subcymosa habit, growing in open area [BH]
2. T. subcymosa branchlet with stipules and wooly-white leaf underside [BH]
3. T. subcymosa, close-up of node and stipules [BH]
4. Trigonia hypoleuca, inflorescence branches and fruits (water dispersed) [BH]
5. Trigonia microcarpa, flowering branch (+hidden spider) [BH]
6. T. microcarpa fruit close-up [BH]
7. T. microcarpa, woody stem cross-section, slightly lobed [BH]
8. T. microcarpa, woody stem bark with dense lenticels [BH]
9. T. microcarpa, hairy seeds [BH]

VERBENACEAE

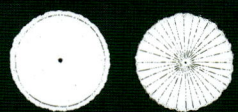

VERBENACEAE

A family of occasional lianas and climbing shrubby species, more commonly aromatic herbs (Verbena), erect shrubs (Lantana), and small trees . Mostly temperate and tropical America, ± 35 genera/1,200 species. Only one liana genus in the Guianas.

1. Petrea Loefl.

Lianas or shrubs. Climbing via twining stems, opposite branching, and petiole bases. Leaves are opposite, simple, and entire. Stipules are absent. Inflorescences are open, showy racemes of tubular, 5-lobed, blue-purple or yellow flowers; stamens 4; ovary superior, stigma 1. Fruits are tiny drupes enclosed by a persistent winged calyx at maturity.

General: Vegetation strongly to lightly rough-surfaced, sand-paper-like [*asperous*]; plants hairless to hairy [*hirsute, pilose*].

Stems: Branchlets mostly oval, rarely 4-sided (P. sulfurea), with lines or ridges along stem; **lenticels abundant; nodes often with conspicuously pointed axillary buds and clusters of scales.** Older stems round to oval, observed to 3 cm diameter, with slanted lines/marks at nodes; **bark soft, often light-colored, wood white and soft,** slash with sugarcane scent (P. bracteata); stem x-s pattern simple or radial, growth rings present.

Leaves: Ovate-elliptic, to 30 cm long, **leathery to flexible and tough** [*cartilaginous*], blade upper surface often raised in-between veins [*bullate*], **with or without tiny pits below;** venation always raised below, enlarged in 2° veins widely spaced and curved, looping at margins; 3° veins net-veined [*reticulate*]; petioles thin to thick, 0.3-3 cm long.

Inflorescences: Axillary and/or terminal, open, hanging racemes to 50 cm long; flowers with conspicuous stalks and basal bracts; **calyx persistent, bell-shaped with a crown** [*corona*] **and 5 papery, petal-like lobes,** reaching 2.5 cm long at maturity, violet-blue, pale green (P. sulfurea) or white; **corolla tubular, trumpet-shaped,** 5-lobed, blue to blue-purple (darker than the calyx, often with white center) or yellow (P. sulfurea); bisexual; stamens 4, attached at corolla midpoint, equal in length; ovary superior, 2-locular, style 1, stigma head-like [*capitate*].

VERBENACEAE

Fruits: Samara-like, with 2 drupes enclosed by the persistent, winged calyx. Even after petals have dried up, the persistent calyx looks like an open flower.

Ecology: Seeds dispersed via wind, possibly water. In Guianan forests, winged Petrea calyces and corollas are commonly seen floating slowly down from the canopy to the soil. An entire sequence of flower and fruiting stages can often be observed on one plant.

Distribution: Tropical S America: 35 species or varieties; Guianas: 5 species, 4 included here.

Uses: Ornamental; medicinal; aphrodisiac; cultural.

Note: Other liana taxa with opposite, simple, rough-surfaced leaves: Asteraceae (Mikania, Tilesia); Celastraceae (Prionostemma aspera).

1.1. Petrea bracteata Steud.

Common names: hachiballi kurero (Ar); hajariballi saléroe (Ar); kapai awara (Tr); parapo (Ca); schuurpapier (SD).

Liana climbing in the lower canopy; rough-surfaced. Stems: branchlets round; older stems oval, twisted, bark white, soft, with scattered lenticels, slash with sugarcane scent, wood white; stem x-s pattern radial. Leaves: elliptic to narrowly elliptic, 10-26 x 3-9 cm, base narrowly cuneate to rounded; mature leaves leathery, bullate, without pits and enlarged veins below; veins sunken above; petiole blackish, 0.3-3 cm long. Inflorescences: terminal or axillary, racemes solitary or in threes, 20-50 cm long, branches purplish-blue, with tiny hairs [pilose]; calyx 1-1.5 cm long, purplish-blue, wings narrowly elliptic to narrowly obovate, 1.5-2 cm long; corolla tube 2-2.5 cm long, dark blue to violet, hairy outside. Fruits: fruiting calyx blue, tube 2.5-3 cm long, wings 1.5-2.5 x 0.4-1.8 cm, fruiting flower stalk to 3.5 cm long; drupes obovoid, 0.6 cm long. Uses: com-

mon ornamental shrub; sandpaper; ritual plant (Sa, Tr); male aphrodisiac (Tr).

Common; non-flooded old-growth forest, riparian forest, and secondary forest along roads; low and high elevations. Guianas; tropical S America.

1.2. Petrea macrostachya Benth.

Common name: sandpaper vine (GU)

Liana climbing into the lower canopy; rough-surfaced. Stem: older stem 3 cm diameter, bark white. Leaves: ± elliptic, 6-23 x 3-14 cm, base acute to rounded; mature leaves leathery, bullate, with pits and enlarged veins below; petiole 0.5-2.0 cm long. Inflorescences: terminal or axillary, racemes long, ± 50 cm, pendant; calyx tube 2-3 cm long at flowering, purplish-blue, wings narrowly elliptic or narrowly obovate, 1.5-2 cm long; corolla tube 1.5-2 cm long, purplish-blue, hairy within. Fruits: fruiting calyx blue, large, tube 3-5 cm long, wings 2-4 x 0.5-2 cm; drupes oblong-ellipsoid, 0.8 cm long, hairless.

Uncommon; Guianas only.

1.3. Petrea sulfurea Jans.-Jac.

Liana; hairy with short glandular hairs; rough-surfaced. Stems: branchlets sharply 4-sided with wings up to 0.2 cm wide. Leaves: Elliptic, 5-13 x 2.5-6 cm, base acute to broafly acute; mature leaves leathery, flat (not bullate); without pits and enlarged veins below; veins slightly raised above; petiole 0.5-1 cm long. Inflorescences: terminal only, racemes up to 15 cm long, infl. stalk 1-2 cm long; flwr. stalks 0.1-0.2 cm long, calyx pale green, tube 2.5 cm long at flowering, wings strongly unequal: 2 narrow (2 cm long), 3 broad-ovate (0.6 cm long); corolla tube yellow, to 3 cm long, with glandular hairs outside, hairless within, lobes rounded and ± 0.5 cm long. Fruits: unknown.

Rarely collected; French Guiana only.

1.4. Petrea volubilis L.

Common name: bonbon gasa, (Sa); kapai awara (Tr); liana gris (FG)

Liana; surfaces only slightly rough. Leaves: elliptic, 5-22 x 3-11 cm, base acute to cuneate; **mature leaves sub-leathery, flat (not bullate), without pits and enlarged veins below;** veins slightly raised above; petiole 0.5-2.0 cm long. **Inflorescences:** terminal or axillary, racemes long, ± 50 cm, hanging; calyx purplish-blue, tube 2.5-3 cm long at flowering, wings 1.5-2.5 cm long; corolla tube purplish-blue, 1.5-2 cm long. **Fruits:** Fruiting calyx violet-blue, 2.5-3 cm long, wings 1.5-2.5 x 0.5-1 cm; drupes oblong-ellipsoid, 0.6 cm long, hairless.

Uncommon; Guianas; widespread in the neotropics from S. Brazil to S Mexico.

1. Petrea macrostachya, older woody stem with flowering shoot; flowers have fallen, leaving persistent flower-like purple calyces (plural of calyx) and tiny fruits [BH]
2. Petrea bracteata, woody stem with slanted lines [BH]
3. P. bracteata, leaf underside venation [BH]
4. P. bracteata, woody stem x-s with spoke-like radial rays [BH]
5. P. bracteata, axillary buds [BH]
6. Petrea volubilis, flowering shoot with corolla and persistent calyx visible [PT]
7. Petrea bracteata, mature woody stem with white bark and conspicuous lenticels [BH]
8. Cross-section of older woody stem of Petrea spp. with spoke-like radial rays [FRD]
9. Petrea bracteata, node with pointed buds and scale clusters in leaf axils [BH]
10. P. bracteata, leaf underside venation close-up [FRD]

VIOLACEAE

1. Corynostylis arborea habit [BH]
2. C. arborea, flowers with nectar spur, capsules green, round, 3-seamed [BH]
3. C. arborea woody stems, twisted and furrowed [BH]

VIOLACEAE

Occasional lianas; mostly herbs, shrubs, and small trees. Cosmopolitan: 17 genera/900 species (mostly neotropical). Guianas: 9 genera/35 species, only one genus including a liana species.

1. Corynostylis Mart.

Occasional lianas; also shrubs and trees. Climbing via twining stems. Leaves are alternate, simple and entire to round-toothed [*crenate*]. Stipules are present but inconspicuous. Flowers are zygomorphic, white, with a nectar spur, 5 free stamens and superior ovary. Fruits are woody, 3-parted capsules. Neotropics, from S America to S Mexico: 4 species; Guianas: 1 species.

1.1. Corynostylis arborea (L.) S.F. Blake

Common name: yapepuku (Ca)

Low-climbing liana; shrub. **General:** mostly hairless, with few distinguishing characteristics. **Stems:** branchlets round, smooth, dark brown, with lenticels; nodes with inconspicuous or short-lived stipules; older stems round, 1-3 cm diameter, twisted, with lengthwise furrows, bark softly woody or corky, yellowish-white; stem x-s pattern simple. **Leaves:** elliptic, 6-11 x 1.5-5 cm, base broadly cuneate, apex shortly acuminate, margin round-toothed to sub-entire; leathery; glands absent; 2° veins widely spaced, 3° veins ± parallel and 90° to midvein, **petiole distinctively curved upwards. Inflorescences:** axil-lary or terminal, in racemes 3.5-9 cm long, or solitary; perianth 5-parted, sepals free; corolla large, white, spurred [*zygomorphic*], 2-3 cm long; stamens 5, free; ovary superior, 1-chambered [*3-carpellate*], style and stigma 1. **Fruits:** woody, subglobose capsules with 3 lengthwise ribs or furrows, 5.5-6 x 4-7 cm, green, dehiscent; seeds many, rough-surfaced, unevenly 4-sided, 3 x 1.6 x 0.4 cm. **Ecology:** water or animal-dispersed. **Notes:** The often round-toothed leaf margins, upward curved petioles and conspicuous flowers help in distinguishing this unique member of the violet family.

Uncommon, locally abundant; riparian vegetation. Guianas; widespread in the neotropics, from Bolivia to Mexico and Cuba.

VIOLACEAE

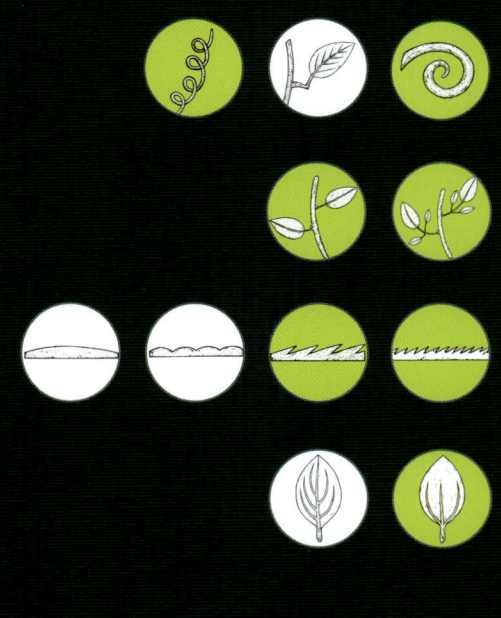

VITACEAE

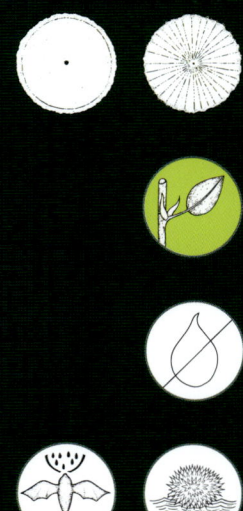

VITACEAE

Lianas, clambering shrubs, and herbaceous vines. Cosmopolitan, tropical and temperate; 13 genera/700-800 species; Neotropics: 6 genera; Guianas: 1 native genus.

1. Cissus L.

Softly woody lianas, clambering shrubs and herbaceous vines. Climbing via tendrils and angled branchlets. Leaves are alternate, simple or with 3-leaflets, often with toothed or spiny margins. Stipules are present. Inflorescences are opposite to leaves, long-stalked, with flat-topped arrays of tiny, red to yellowish-green flowers. Fruits are smooth, green to black berries, ± 1 cm diameter.

General: Overall subwoody/fleshy appearance; vegetation shiny green and with few-to-many white or grey hairs; colored sap and conspicuous scent or taste absent; small, stalked "pearl bodies" (or food bodies) present on young vegetation of most species.

Stems: Branchlets growing forward in a zig-zag pattern; **stem round, oval or 4-sided and winged,** sometimes spiny, reddish-brown to yellowish-green; nodes swollen and jointed, **tendrils opposite to leaves,** simple or branched, growing straight at first, coiling upon contact with a climbing surface; **stipules present** – leafy and wide, sometimes enclosing the entire node (C. haematantha) or small, and falling away early. Older stems round or 4-sided, observed to 6 cm diameter; bark thin, flaky or furrowed; slash with sub-bark green layer, wood soft; **stem x-s pattern simple or radial** (spoke-like radial rays most visible upon drying).

Leaves: Simple and palmate-veined (C. verticellata) **or compound with three pinnate-veined leaflets** (C. erosa, C. spinosa) - sometimes with both leaf types on the same plant; **margins entire, lobed, toothed or spiny;** fleshy to papery; glands rarely present.

Inflorescences: Many-flowered, flat-topped displays [*corymbs*], **usually opposite to leaves** (as with tendrils); flowers bisexual, tiny, **4-parted,** red to greenish-yellow; sepals mostly fused; petals free; stamens opposite petals; ovary superior, 2-chambered, style 1.

Fruits: Berries globose to ovoid, often maturing black, 0.8-1.2 cm diameter; seeds 1(-4), ± round, pointed at the base.

Ecology: Tendrils developing fully before the opposing leaf. Insect-pollinated (butterflies); seed dispersal via animal (bird or fish gut) or water. Often flowering abundantly regardless of season. Pearl bodies (or food bodies) are produced in abundance on young stems and leaves to attract ant species and thereby gain protection against herbivores (Paiva et al. 2009).

Distribution: Common, low climbing lianas in forest understory and on vegetation along river banks. Pantropical and temperate; ca. 350 species, 11 species known in the Guianas: 3 presented here.

Uses: Some species used medicinally (to treat wounds and colds); larger berries may be edible, but in many species the fruits have an unpleasant taste due to 'raphide crystals ' (Lombardi 2009).

Notes: The leaf-opposed tendril distinguishes Vitaceae from sub-woody lianas in Cucurbitaceae (90° angle from petiole), Passifloraceae (axillary tendril), Sapindaceae (forked tendril from inflorescence branch), and Smilacaceae (paired tendrils from petiole sheath).

1.1. Cissus erosa L.C. Rich.

Common name: (rode) bun-ati-mama (Sr); kikwe nyehnyeh (Ak); lebiki-faja (Pa); napëkoimë (Tr); soitongo (Sa); todofinga (Sr); wilde napie (SN)

Subwoody liana; hairless to lightly hairy. **Stems: branchlets sharply 4-sided and narrowly winged,** lenticels present. **Leaves: 3-foliolate,** petiole absent or very short; leaflets lanceolate to obovate, 5-15 x 2.5-6 cm, margins with tiny teeth. **Inflorescences:** flat-top display to 8 cm long, axes **bright red. Fruits:** berries ± round, 0.6-1.2 cm diameter, red to black, hairless.

Common; along rivers, secondary forest, savanna forest, open wet savanna. Guianas; also widespread from South America to S Mexico and the West Indies.

1.2. Cissus spinosa Cambess.

Common name: busi-watramon (Sr)

Sub-woody liana. Stems: branchlets ± 4-sided, with conical spines, unwinged. **Leaves: 3-foliolate,** petiole 1-3.5 cm long; leaflets rhomboid to ovate, 5-10 x 2.5-4.5 cm, base cuneate; margin tiny-toothed to lobed, sparsely hairy above, **densely grey-hairy [*tomentose*] below. Inflorescences:** flat-top display to 13 cm long, axes dark red; flowers with burgundy red petals and yellow nectar disc. **Fruits:** berries ± round, ca. 0.8 cm diameter.

Fairly common in coastal swamps. Guianas; also Brazil.

1.3. Cissus verticillata (L.) Nicholson & Jarvis
[Syn: Cissus sicyoides L.]

Common name: baakakifaia (Pa); (witte) bunatimama (Sr); napëkoimë (Tr); patona (Ca); paramaru (Ca)

Liana or creeper; hairless to lightly hairy. **Stems: branchlets variable** - round to conspicuously 4-winged, without spines; older stems to 6 cm diameter; bark reddish-brown, lenticels present. **Leaves: simple, 3-palmate-veined from base,** ovate to subtriangular, 7-11 x 3-7 cm, base broadly cuneate to cordate; margin fine-toothed. **Inflorescences:** flat-top display to 6 cm long, axes yellow-green, flowers with yellow-white petals and nectar disc. **Fruits:** berries ± round, ca. 1 cm diameter.

Very common; especially along rivers. Guianas; also widespread from the USA to Argentina.

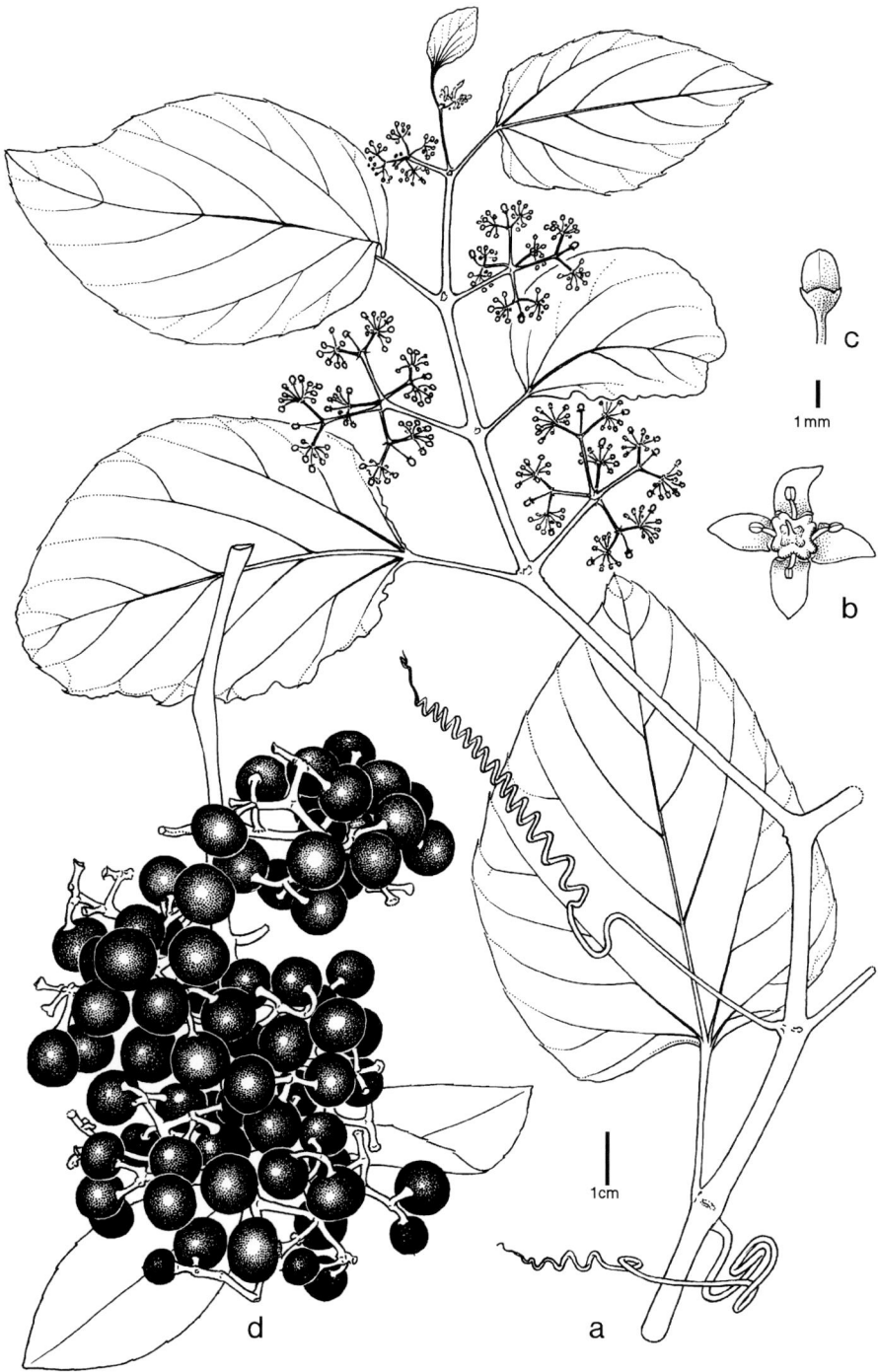

8

1. Cissus spinosa with 3-foliolate leaves and red inflorescences [OG]

2. Cissus haematantha (photo only), large, 3-foliolate leaves [PT]

3. Cissus verticellata, robust older stem [BH]

4. C. verticellata, inflorescence with yellow-green axes [BH]

5. Cissus erosa, inflorescence with red axes [PT]

6. Cissus spinosa, flower and immature fruits [OG]

7. Cissus erosa, square stem and mature fruits [PT]

8. Cissus verticellata. a) branchlet with tendrils and inflorescences opposite to simple leaves;
   b) open, 4-parted flower; c) flower in bud; d) dense display of mature, black fruits [NCB]

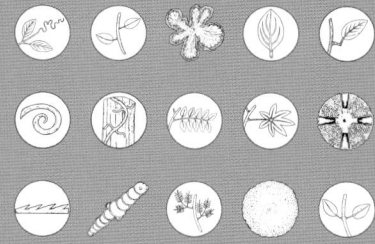

PLANT CHECKLIST

Comprehensive Checklist of Climbing Plant Species, Growth Habits and Territory in the Guianas

[L: woody liana (X-small, stem < 2 cm diam., XX-medium, stem 2-10 cm diam., XXX-large, stem > 10 cm diameter); SC: climbing or clambering shrub; LSW: sub-woody liana; LV: small woody liana, perennial vine (< 0.5 cm diam.); V: herbaceous vine; HEV: vine-like, herbaceous hemi-epiphyte; HEL: liana-like, woody hemi-epiphyte; HET: tree-like hemi-epiphyte; S: shrub; T: tree. Occurrence Data: GU: Guyana; SU: Suriname; FG: French Guiana]. **Only species names in BOLD below are described in the text.**

FAMILY	SPECIES	L	SC	LSW	LV	V	HEV	HEL	HET	S	T	GU	SU	FG
Acanthaceae	Anisacanthus secundus Leonard	X			X							X	X	X
	Mendoncia aspera Ruiz & Pav.	X			X							X	X	X
	Mendoncia bivalvis (L.f.) Merr.	X			X							X	X	X
	Mendoncia crenata Lindau	X			X									X
	Mendoncia glabra Poepp. & Endl.	X			X							X		X
	Mendoncia hoffmannseggiana Nees	X			X							X	X	X
	Mendoncia squamuligera Nees	X			X							X	X	X
	Polylychnis radicans (Nees) Wassh.				X							X		X
	Ruellia inflata Rich.		X							X		X	X
Amaranthaceae	Chamissoa altissima (Jacq.) Kunth	X			X	X						X	X	
Annonaceae	**Annona haematantha Miq.**	X	X									X	X	X
	Guatteria scandens Ducke	XX										X	X	X
Apocynaceae	**Allamanda blanchetii A.DC.**	X	X							X				X
	Allamanda cathartica L.	X	X							X		X	X	X
	Allamanda schottii Pohl	X	X							X				X
	Allamanda setulosa Miq.	X	X							X		X	X	
	Barjonia erecta (Vell.) K.Schum.					X							X	
	Blepharodon glaucescens (Decne.) Fontella					X						X		
	Blepharodon grandiflorum Benth.					X						X		
	Blepharodon pictum (Vahl) W.D.Stevens					X						X	X	X
	Blepharodon tillettii Morillo					X						X		
	Blepharodon ulei Schltr.					X						X		
	Condylocarpon amazonicum (Markgr.) Ducke	XX										X	X	
	Condylocarpon guyanense Desf.	XX										X		X
	Condylocarpon intermedium Müll.Arg.	XX										X	X	X
	Condylocarpon myrtifolium (Miq.) Müll.Arg.	XX										X	X	
	Condylocarpon pubiflorum Müll.Arg.	XX												X
	Cynanchum blandum (Decne.) Sundell					X						X	X	X
	Cynanchum goertsianum Morillo					X							X	X
	Cynanchum jenmanii Morillo					X						X		
	Cynanchum prevostii Morillo					X								X

FAMILY	SPECIES	L	SC	LSW	LV	V	HEV	HEL	HET	S	T	GU	SU	FG
	Fischeria stellata (Vell.) E.Fourn.					X						X		
	Forsteronia acouci (Aubl.) A.DC.	XX										X	X	X
	Forsteronia adenobasis Müll.Arg.	XX										X		X
	Forsteronia diospyrifolia Müll.Arg.	XX										X		
	Forsteronia duckei Markgr.	XX										X		
	Forsteronia gracilis (Benth.) Müll.Arg.	XX										X	X	
	Forsteronia guyanensis Müll.Arg.	XX										X	X	X
	Forsteronia obtusiloba Müll.Arg.	XX										X	X	
	Forsteronia paraensis B.F.Hansen	XX										X		
	Forsteronia schomburgkii A.DC.	XX										X		
	Forsteronia umbellata (Aubl.) Woodson	XX												X
	Gonolobus tobagensis Urb.					X						X		
	Macroditassa carolina Morillo					X						X		
	Macroditassa laurifolia (Decne.) Fontella					X						X		
	Macropharynx spectabilis (Stadelm.) Woodson	XX										X	X	
	Macropharynx strigillosa Woodson											X		
	Macroscepis hirsuta (Vahl) Schltr.				X	X						X		
	Malouetia flavescens (Willd. ex Roem. & Schult.) Müll.Arg.		X		X					X	X	X	X	X
	Malouetia gracilis (Benth.) A.DC.		X		X					X	X	X		
	Malouetia pubescens Markgr.		X		X					X	X	X		
	Malouetia tamaquarina (Aubl.) A.DC.	X	X							X	X	X	X	X
	Mandevilla benthamii (A.DC.) K.Schum.					X						X		
	Mandevilla hirsuta (Rich.) K.Schum.				X	X						X	X	X
	Mandevilla holstii Morillo				X	X						X		
	Mandevilla lasiocarpa (A.DC.) Malme					X								
	Mandevilla leptophylla (A.DC.) K.Schum.				X	X						X		
	Mandevilla rugellosa (Rich.) L.Allorge				X	X						X	X	X
	Mandevilla rugosa (Benth.) Woodson				X	X						X		
	Mandevilla scaberula N.E.Br.				X	X						X		X
	Mandevilla scabra (Hoffmanns. ex Roem. & Schult.) K.Schum.				X	X						X	X	X
	Mandevilla spruceana (Müll.Arg.) K.Schum.				X	X						X		
	Mandevilla steyermarkii Woodson					X								
	Mandevilla subcarnosa (Benth.) Woodson				X	X						X		
	Mandevilla subspicata (Vahl) Markgr.				X	X						X	X	X
	Mandevilla surinamensis (Pulle) Woodson				X	X						X	X	X
	Mandevilla symphitocarpa (G. Mey.) Woodson				X	X						X	X	X

PLANT CHECKLIST

FAMILY	SPECIES	L	SC	LSW	LV	V	HEV	HEL	HET	S	T	GU	SU	FG
	Mandevilla tenuifolia (J.C.Mikan) Woodson				X	X							X	
	Mandevilla trianae Woodson				X	X						X		
	Mandevilla vanheurckii (Müll.Arg.) Benth. ex B.D.Jacks.				X	X						X		
	Marsdenia altissima (Jacq.) Dugand	X			X							X	X	X
	Marsdenia gillespieae Morillo				X							X		
	Marsdenia macrophylla (Humb. & Bonpl. ex Schult.) E.Fourn.				X							X	X	
	Marsdenia rubrofusca E.Fourn.				X							X	X	X
	Marsdenia weddellii (E. Fourn.) Malme				X								X	
	Matelea badilloi Morillo				X	X						X		X
	Matelea bolivarensis Morillo				X	X						X		
	Matelea cayennensis Morillo				X	X								X
	Matelea cremersii Morillo				X	X								X
	Matelea delascioi Morillo				X	X						X		
	Matelea denticulata (Vahl) Fontella & E.A.Schwarz				X	X						X	X	X
	Matelea fournieri Morillo				X	X							X	
	Matelea funkiana Morillo				X	X						X		
	Matelea glaziovii (E. Fourn.) Morillo				X	X							X	
	Matelea gracieae Morillo				X	X								X
	Matelea grenandii Morillo				X	X								X
	Matelea herbacea Woodson				X	X						X		
	Matelea hoffmanii Morillo				X	X						X		
	Matelea jansen-jacobsiae Morillo				X	X							X	
	Matelea lourteigiae Morillo				X	X						X	X	X
	Matelea maritima (Jacq.) Woodson				X	X						X		
	Matelea oldemanii Morillo				X	X								X
	Matelea palustris Aubl.	X	X		X	X				X		X	X	X
	Matelea planiflora (Jacq.) Dugand				X	X						X		
	Matelea riparia Morillo				X	X						X		
	Matelea sastrei Morillo				X	X							X	X
	Matelea squiresii (Rusby) Morillo				X	X						X	X	
	Matelea stenopetala Sandwith				X	X						X		X
	Matelea surinamensis Morillo				X	X							X	
	Mesechites trifidus (Jacq.) Müll.Arg.	X			X	X						X	X	X
	Metalepis albiflora Urb.				X	X								X
	Metalepis prevostiae Morillo				X	X								X
	Metastelma hirtellum (Oliv.) Liede					X						X		

| --- | --- | --- | --- | --- | --- | --- | --- | --- | --- | --- | --- | --- | --- | --- |
	Nephradenia linearis Benth. ex E.Fourn.				X	X						X	X	X
	Odontadenia geminata (Roem. & Schult.) Müll.Arg.	XXX										X	X	X
	Odontadenia kochii Pilg.	XXX										X	X	X
	Odontadenia macrantha (Roem. & Schult.) Markgr.	XXX										X	X	X
	Odontadenia markgrafiana J.F.Morales	XXX												
	Odontadenia nitida (Vahl) Müll.Arg.	XXX										X	X	X
	Odontadenia perrottetii (A.DC.) Woodson	XXX										X	X	X
	Odontadenia puncticulosa (Rich.) Pulle	XXX										X	X	X
	Odontadenia verrucosa (Willd. ex Roem. & Schult.) K.Schum. ex Markgr.	XXX										X	X	X
	Oxypetalum capitatum Mart.				X	X						X	X	
	Pacouria guianensis Aubl.	XXX										X	X	X
	Prestonia annularis (L.f.) G.Don	X			X	X						X	X	X
	Prestonia cayennensis (A.DC.) Pichon	X			X	X						X		X
	Prestonia coalita (Vell.) Woodson	X			X	X						X		X
	Prestonia guianensis Gleason	X			X	X						X		
	Prestonia ipomaeifolia A.DC.	X			X	X								X
	Prestonia megagros (Vell.) Woodson	X			X								X	
	Prestonia quinquangularis (Jacq.) Spreng.	X			X							X	X	
	Prestonia surinamensis Müll.Arg.	X			X							X	X	X
	Prestonia tomentosa R.Br.	X			X	X						X		
	Rhabdadenia biflora (Jacq.) Müll.Arg.				X	X						X	X	X
	Rhabdadenia madida (Vell.) Miers				X	X						X	X	X
	Sarcostemma clausum (Jacq.) Schult.				X	X						X	X	X
	Secondatia densiflora A.DC.	X			X	X						X	X	X
	Tassadia berteroana (Spreng.) W.D.Stevens				X	X						X	X	
	Tassadia decaisneana Miq.				X	X						X	X	
	Tassadia guianensis Decne.				X	X						X	X	X
	Tassadia leptobotrys Decne.				X	X							X	
	Tassadia obovata Decne.				X	X						X	X	X
	Tassadia propinqua Decne.				X	X						X	X	
	Tassadia trailiana (Benth.) Fontella				X	X						X		
	Telminostelma foetidum (Cav.) Fontella & E.A.Schwarz					X						X		
	Vincetoxicum parviflorum Decne.					X						X		
Araceae	Anthurium digitatum (Jacq.) Schott					X								X
	Anthurium eminens Schott					X						X	X	X
	Anthurium expansum Gleason					X						X	X	X

FAMILY	SPECIES	L	SC	LSW	LV	V	HEV	HEL	HET	S	T	GU	SU	FG
	Anthurium moonenii Croat						X							X
	Anthurium pentaphyllum G.Don						X					X	X	X
	Anthurium sinuatum Benth. ex Schott						X					X	X	X
	Heteropsis flexuosa (Kunth) G.S.Bunting						X					X	X	X
	Heteropsis melinonii (Engl.) A.M.E.Jonker & Jonker						X					X	X	X
	Heteropsis oblongifolia Kunth						X					X		
	Heteropsis spruceana Schott						X					X		
	Heteropsis steyermarkii G.S.Bunting						X							X
	Heteropsis tenuispadix G.S.Bunting						X					X	X	X
	Monstera adansonii Schott						X					X	X	X
	Monstera barrieri Croat & Moonen & Poncy						X					X	X	X
	Monstera dubia (Kunth in Humb., Bonpl. & Kunth) Engl. & K.Krause						X					X	X	X
	Monstera obliqua Miq.						X					X	X	X
	Monstera spruceana (Schott) Engl. in Mart.						X					X	X	X
	Philodendron acutatum Schott						X					X	X	X
	Philodendron asplundii Croat & M.L.C.Soares						X							X
	Philodendron ayantepuiense Bunting						X					X		
	Philodendron billietiae Croat						X					X		X
	Philodendron bipennifolium Schott						X					X	X	X
	Philodendron camposportoanum G.M.Barroso						X					X	X	X
	Philodendron carinatum E.G.Gonçalves						X					X	X	X
	Philodendron colombianum R.Schultes						X					X	X	X
	Philodendron cremersii Croat & Grayum						X							X
	Philodendron deflexum Poepp. ex Schott						X					X	X	X
	Philodendron dioscoreoides Gleason						X					X	X	
	Philodendron duckei Croat & Grayum						X					X	X	X
	Philodendron grandifolium (Jacq.) Schott						X					X	X	X
	Philodendron guianense Croat & Grayum						X					X	X	X
	Philodendron guttiferum Kunth						X							X
	Philodendron hederaceum (Jacq.) Schott						X						X	X
	Philodendron houlletianum Engl.						X							X
	Philodendron hylaeae G.S.Bunting						X					X	X	X
	Philodendron insigne Schott						X					X	X	X
	Philodendron jacquinii Schott						X							X
	Philodendron jonkerianum Croat						X						X	
	Philodendron krauseanum Steyermark						X					X	X	X
	Philodendron linnaei Kunth						X					X	X	X

FAMILY	SPECIES	L	SC	LSW	LV	V	HEV	HEL	HET	S	T	GU	SU	FG
	Philodendron macropodum K.Krause						X							X
	Philodendron mawarinumae Bunting						X						X	X
	Philodendron megalophyllum Schott						X					X	X	X
	Philodendron melinonii Brongn. ex Regel						X					X	X	X
	Philodendron moonenii Croat						X							X
	Philodendron muricatum Willd. ex Schott						X					X	X	
	Philodendron myrmecophyllum Engl.						X						X	
	Philodendron ornatum Schott						X					X	X	X
	Philodendron pedatum (Hook.) Kunth						X					X	X	X
	Philodendron placidum Schott						X					X	X	X
	Philodendron platypodium Gleason						X					X	X	X
	Philodendron potaroense Croat						X					X		
	Philodendron riedelianum Schott						X							X
	Philodendron roraimae K.Kr.						X							X
	Philodendron rudgeanum Schott						X					X	X	X
	Philodendron scottmorianaum Croat & J.Moonen						X							X
	Philodendron simsii (Hook.) G.Don in Sweet						X					X		X
	Philodendron solimoesense A.C.Sm.						X					X	X	X
	Philodendron sphalerum Schott						X					X	X	X
	Philodendron splitgerberi Schott						X						X	X
	Philodendron squamiferum Poepp. in Poepp. & Endl.						X						X	X
	Philodendron surinamense (Miq.) Engl. in Mart.						X					X	X	X
	Philodendron ushanum Croat & Moonen						X							X
	Philodendron wittianum Engl.						X					X		X
	Philodendron wullschlaegelii Schott						X							X
	Rhodospatha brachypoda G.S.Bunting						X					X	X	X
	Rhodospatha latifolia Poepp. in Poepp. & Endl.						X					X		X
	Rhodospatha oblongata Poepp.						X					X	X	X
	Rhodospatha venosa Gleason						X					X	X	X
	Stenospermation ammiticum G.S.Bunting						X					X		
	Stenospermation maguirei G.S.Bunting						X					X	X	
	Stenospermation multiovulatum (Engl.) N.E.Br.						X					X	X	X
	Stenospermation spruceanum Schott						X					X	X	
	Stenospermation steyermarkii G.S.Bunting						X					X	X	X
	Syngonium angustifolium Schott						X							X
	Syngonium podophyllum Schott						X					X	X	X

FAMILY	SPECIES	L	SC	LSW	LV	V	HEV	HEL	HET	S	T	GU	SU	FG
	Syngonium yurimaguense Engl.						X							X
Araliaceae	Oreopanax capitatus (Jacq.) Decne. & Planch.		X						X	X	X		X	X
	Schefflera decaphylla (Seem.) Harms		X						X	X	X	X	X	X
Arecaceae	**Desmoncus orthacanthos Mart.**			X								X	X	X
	Desmoncus phoenicocarpus Barb.Rodr.			X								X	X	X
	Desmoncus polyacanthos Mart.			X								X	X	X
Aristolochiaceae	**Aristolochia bukuti Poncy**	XX											X	X
	Aristolochia consimilis Mast.			X	X							X		X
	Aristolochia cornuta Mast.			X	X							X		
	Aristolochia cremersii Poncy			X	X									X
	Aristolochia daemoninoxia Mast.			X	X							X		
	Aristolochia didyma S.Moore			X	X									X
	Aristolochia disticha Mast.			X	X									X
	Aristolochia flava Poncy			X	X									X
	Aristolochia guianensis Poncy					X							X	X
	Aristolochia hians Willd.	X		X	X							X		
	Aristolochia iquitensis O.C.Schmidt	X		X	X							X	X	X
	Aristolochia kanukuensis Feuillet	X		X	X							X		X
	Aristolochia leprieurii Duch.					X						X		X
	Aristolochia melgueiroi Barringer & F.Guán-chez	X		X	X									X
	Aristolochia mossii S.Moore					X								X
	Aristolochia pannosoides Hoehne	X		X	X							X	X	X
	Aristolochia paramaribensis Duch.					X						X	X	
	Aristolochia rugosa Lam.	X		X	X							X		X
	Aristolochia stahelii O.C.Schmidt	X		X	X								X	X
	Aristolochia surinamensis Willd.			X		X						X	X	
	Aristolochia trilobata L.					X						X	X	X
	Aristolochia weddellii Duch.	X		X	X							X	X	X
Asteraceae	**Calea solidaginea Kunth**	X	X	X								X	X	X
	Lepidaploa canescens (Kunth) Cass.		X									X		
	Mikania banisteriae DC.			X	X	X						X	X	X
	Mikania boomii Pruski			X	X	X						X		
	Mikania congesta DC.				X	X						X	X	X
	Mikania cordifolia (L.f.) Willd.				X	X						X	X	X
	Mikania duidensis B.L.Rob.				X	X						X		
	Mikania gleasonii B.L.Rob.			X	X	X					X	X	X	X
	Mikania guaco Humb. & Bonpl.			X	X	X					X	X	X	X
	Mikania houstoniana (L.) B.L.Rob.			X	X	X					X	X	X	X

FAMILY	SPECIES	L	SC	LSW	LV	V	HEV	HEL	HET	S	T	GU	SU	FG
	Mikania lindleyana DC.					X						X	X	
	Mikania mazaruniensis W.C.Holmes & McDaniel			X	X	X						X		
	Mikania micrantha Kunth			X	X	X						X	X	X
	Mikania microptera DC.					X						X	X	X
	Mikania nigropunctulata Hieron.			X	X	X						X	X	
	Mikania pannosa Baker			X	X	X						X		
	Mikania parviflora (Aubl.) H.Karst.			X	X	X				X		X	X	X
	Mikania psilostachya DC.			X	X	X				X		X	X	X
	Mikania rondonensis V.M.Badillo					X						X		
	Mikania sprucei Baker			X	X	X						X		
	Mikania trinitaria DC.			X	X	X				X		X	X	X
	Mikania vitifolia DC.			X	X	X						X	X	
	Pentacalia freemanii (Britton & Greenm.) Cuatrec.		X		X					X		X		
	Piptocarpha leprosa (Less.) Baker		X	X	X					X		X		
	Piptocarpha opaca (Benth.) Baker		X	X	X					X		X		
	Piptocarpha poeppigiana (DC.) Baker		X	X	X					X		X		
	Piptocarpha polycephala Baker		X	X	X					X		X		
	Piptocarpha triflora (Aubl.) Benn. ex Baker		X	X	X					X		X	X	X
	Pseudogynoxys chenopodioides (Kunth) Cabrera					X						X	X	
	Stifftia cayennensis H.Rob. & B.Kahn	X												X
	Tilesia baccata (L.) Pruski		X	X						X		X	X	X
	Wulffia rubens Alexander		X	X								X		
Bignoniaceae	Adenocalymma album (Aubl.) L.G.Lohmann	XX											X	X
	Adenocalymma bracteosum (DC.) L.G.Lohmann	XX										X	X	X
	Adenocalymma flaviflorum (Miq.) L.G.Lohmann	XX										X	X	X
	Adenocalymma heterophyllum Kraenzl.	XX										X		
	Adenocalymma impressum (Rusby) Sandwith	XX											X	X
	Adenocalymma inundatum Mart. ex DC.	XX										X	X	X
	Adenocalymma moringifolium (DC.) L.G.Lohmann	XX										X	X	X
	Adenocalymma prancei A.H.Gentry	XX												X
	Adenocalymma purpurascens Rusby	XX										X		
	Adenocalymma racemosum (A.H.Gentry) L.G.Lohmann	XX										X	X	X
	Adenocalymma saulense A.H.Gentry	XX												X
	Adenocalymma schomburgkii (DC.) L.G.Lohmann	XX										X	X	X

FAMILY	SPECIES	L	SC	LSW	LV	V	HEV	HEL	HET	S	T	GU	SU	FG
	Adenocalymma subincanum Huber	XX												X
	Adenocalymma tanaeciicarpum (A.H.Gentry) L.G.Lohmann	XX										X	X	X
	Adenocalymma validum (K.Schum.) L.G.Lohmann	XX												
	Amphilophium cremersii (A.H.Gentry) L.G.Lohmann	XX											X	X
	Amphilophium crucigerum (L.) L.G.Lohmann	XX										X	X	X
	Amphilophium cuneifolium (DC.) L.G.Lohmann	XX												X
	Amphilophium elongatum (Vahl) L.G.Lohmann	XX										X	X	X
	Amphilophium granulosum (Klotzsch) L.G.Lohmann	XX										X		X
	Amphilophium magnoliifolium (Kunth) L.G.Lohmann	XX										X	X	X
	Amphilophium obovatum (Sandwith) L.G.Lohmann	XX										X		
	Amphilophium paniculatum (L.) Kunth	XX										X	X	X
	Amphilophium parkeri (DC.) L.G.Lohmann	XX										X		
	Amphilophium porphyrotrichum (Sandwith) L.G.Lohmann	XX										X		
	Amphilophium pulverulentum (Sandwith) L.G.Lohmann	XX										X	X	X
	Anemopaegma brevipes S.Moore	XX											X	
	Anemopaegma chamberlaynii (Sims) Bureau & K.Schum.	XX										X		
	Anemopaegma chrysoleucum (Kunth) Sandwith	XX										X	X	X
	Anemopaegma foetidum Bureau & K.Schum.	XX										X	X	
	Anemopaegma granvillei A.H.Gentry	XX												X
	Anemopaegma ionanthum A.H.Gentry	XX										X		X
	Anemopaegma jucundum Bureau & K.Schum.	XX										X		
	Anemopaegma karstenii Bureau & K.Schum.	XX										X		
	Anemopaegma longidens Mart. ex DC.	XX												X
	Anemopaegma oligoneuron (Sprague & Sandwith) A.H.Gentry	XX										X	X	X
	Anemopaegma paraense Bureau & K.Schum.	XX										X	X	X
	Anemopaegma parkeri Sprague	XX										X	X	X
	Anemopaegma robustum Bureau & K.Schum.	XX										X		X
	Adenocalymma validum (K.Schum) L.G.Lohmann	XX										X	X	X
	Bignonia aequinoctialis L.	XX										X	X	X
	Bignonia binata Thunb.	XX										X	X	X
	Bignonia hyacinthina (Standl.) L.G.Lohmann	XX										X		
	Bignonia lilacina (A.H.Gentry) L.G.Lohmann	XX										X	X	X

FAMILY	SPECIES	L	SC	LSW	LV	V	HEV	HEL	HET	S	T	GU	SU	FG
	Bignonia microcalyx G.Mey	XX										X	X	
	Bignonia nocturna (Barb.Rodr.) L.G.Lohmann	XX											X	X
	Bignonia prieurei DC.	XX										X	X	X
	Bignonia sciuripabula (K.Schum.) L.G.Lohmann	XX										X		
	Bignonia sordida (Bureau & K.Schum.) L.G.Lohmann	XX										X	X	X
	Callichlamys latifolia (Rich.) K.Schum.	XX										X	X	X
	Cuspidaria inaequalis (DC. ex Splitg.) L.G.Lohmann	XX										X	X	X
	Cuspidaria subincana A.H.Gentry	XX										X	X	
	Dolichandra steyermarkii (Sandwith) L.G.Lohmann	X			X							X	X	X
	Dolichandra uncata (Andrews) L.G.Lohmann	X			X							X	X	X
	Dolichandra unguis-cati (L.) L.G.Lohmann	X			X							X	X	X
	Fridericia candicans (Rich.) L.G.Lohmann	XX										X	X	X
	Fridericia chica (Bonpl.) L.G.Lohmann	XX											X	X
	Fridericia cinerea (Bureau ex K.Schum.) L.G.Lohmann	XX										X		X
	Fridericia cinnamomea (DC.) L.G.Lohmann	XX											X	X
	Fridericia dichotoma (Jacq.) L.G.Lohmann	XX										X	X	X
	Fridericia egensis (Poepp. ex Bureau & K.Schum.) L.G.Lohmann	XX										X		
	Fridericia fanshawei (Sandwith) L.G.Lohmann	XX										X	X	X
	Fridericia florida (DC.) L.G.Lohmann	XX										X	X	X
	Fridericia grosourdyana (Baill.) L.G.Lohmann	XX										X		X
	Fridericia japurensis (DC.) L.G.Lohmann	XX										X	X	X
	Fridericia mollis (Vahl) L.G.Lohmann	XX										X	X	X
	Fridericia nigrescens (Sandwith) L.G.Lohmann	XX										X	X	X
	Fridericia oligantha (Bureau & K.Schum.) L.G.Lohmann	XX										X	X	X
	Fridericia patellifera (Schltdl.) L.G.Lohmann	XX										X	X	X
	Fridericia pilulifera (Rich.) L.G.Lohmann	XX										X	X	X
	Fridericia pubescens (L.) L.G.Lohmann	XX										X	X	
	Fridericia spicata (Bureau & K.Schum.) L.G.Lohmann	XX												X
	Fridericia trailii (Sprague) L.G.Lohmann	XX										X	X	X
	Fridericia triplinervia (Mart. ex DC.) L.G.Lohmann	XX												X
	Fridericia conjugata (Vell.) L.G.Lohmann	XX										X		X
	Lundia corymbifera (Vahl) Sandwith	XX										X		X
	Lundia densiflora DC.	XX										X	X	

FAMILY	SPECIES	L	SC	LSW	LV	V	HEV	HEL	HET	S	T	GU	SU	FG
	Lundia triphylla (L.) L.G.Lohmann	XX										X	X	X
	Mansoa alliacea (Lam.) A.H.Gentry	XX										X	X	X
	Mansoa onohualcoides A.H.Gentry	XX											X	
	Mansoa standleyi (Steyerm.) A.H.Gentry	XX										X	X	X
	Mansoa verrucifera (Schltdl.) A.H.Gentry	XX										X		
	Martinella iquitoensis A.Samp.	XX										X	X	X
	Martinella obovata (Kunth) Bureau & K.Schum.	XX										X	X	X
	Pachyptera kerere (Aubl.) Sandwith	XX										X	X	X
	Pleonotoma albiflora (Salzm. ex DC.) A.H.Gentry	XX										X	X	X
	Pleonotoma clematis (Kunth) Miers	XX										X	X	X
	Pleonotoma dendrotricha Sandwith	XX												X
	Pleonotoma echitidea Sprague & Sandwith	XX										X	X	
	Pleonotoma variabilis (Jacq.) Miers	XX											X	
	Pyrostegia venusta (Ker Gawl.) Miers	XX										X	X	
	Stizophyllum inaequilaterum Bureau & K.Schum.	XX										X	X	X
	Stizophyllum perforatum (Cham.) Miers	XX										X		
	Stizophyllum riparium (Kunth) Sandwith	XX										X	X	X
	Tanaecium bilabiatum (Sprague) L.G.Lohmann	X										X	X	X
	Tanaecium jaroba Sw.	X										X		X
	Tanaecium pyramidatum (Rich.) L.G.Lohmann	X										X	X	X
	Tanaecium revillae (A.H.Gentry) L.G.Lohmann	X										X	X	
	Tanaecium tetragonolobum (Jacq.) L.G.Lohmann	X										X	X	
	Tynanthus polyanthus (Bureau ex Baill.) Sandwith	X										X		
	Tynanthus pubescens A.H.Gentry	X										X	X	X
	Tynanthus sastrei A.H.Gentry	X											X	X
	Xylophragma seemannianum (Kuntze) Sandwith	X										X	X	
	Tournefortia bicolor Sw.	X	X									X	X	X
	Tournefortia candidula (Miers) I.M.Johnst.	X	X									X		
	Tournefortia cuspidata Kunth	X	X									X	X	X
	Tournefortia hirsutissima L.	X	X									X		
	Tournefortia maculata Jacq.	X	X									X	X	X
	Tournefortia melanochaeta DC.	X	X									X	X	X
	Tournefortia paniculata Cham.	X	X									X	X	X
	Tournefortia ulei Vaupel	X	X									X	X	X

FAMILY	SPECIES	L	SC	LSW	LV	V	HEV	HEL	HET	S	T	GU	SU	FG
	Varronia polycephala Lam	X	X							X		X	X	X
	Varronia schomburgkii (DC.) Borhidi	X	X							X		X	X	X
Cactaceae	Epiphyllum phyllanthus (L.) Haw.						X					X	X	X
	Hylocereus lemairei (Hook.) Britton & Rose						X					X	X	
	Pereskia aculeata Mill.						X					X	X	X
Calophyllaceae	Clusiella axillaris (Engl.) Cuatrec.	X										X		
	Clusiella elegans Planch. & Triana	X										X		
Cannabaceae	**Celtis iguanaea (Jacq.) Sarg.**	X	X	X						X		X	X	
	Celtis schippii Standl.	X	X	X						X		X		
Celastraceae	**Anthodon decussatum Ruiz & Pav.**	XXX										X		X
	Cheiloclinium anomalum Miers	XXX										X	X	
	Cheiloclinium belizense (Standl.) A.C.Sm.	XXX										X	X	X
	Cheiloclinium brevipetiolatum Lombardi	XXX										X	X	X
	Cheiloclinium cognatum (Miers) A.C.Sm.	XXX	X							X	X	X	X	X
	Cheiloclinium diffusiflorum (Miers) A.C.Sm.	XXX										X	X	X
	Cheiloclinium hippocrateoides (Peyr.) A.C.Sm.	XXX										X	X	X
	Cuervea kappleriana (Miq.) A.C.Sm.	XXX										X	X	X
	Elachyptera floribunda (Benth.) A.C.Sm.	XXX										X	X	X
	Hippocratea volubilis L.	XXX										X	X	X
	Hylenaea comosa (Sw.) Miers	XXX										X	X	X
	Hylenaea praecelsa (Miers) A.C.Sm.	XXX										X		
	Hylenaea unguiculata Mennega	XXX											X	
	Peritassa dulcis (Benth.) Miers	XXX										X		
	Peritassa glabra (A.C.Sm.) Lombardi	XXX										X	X	X
	Peritassa huanucana (Loes.) A.C.Sm.	XXX										X	X	X
	Peritassa laevigata (Hoffmanns. ex Link) A.C.Sm.	XXX										X	X	X
	Peritassa nectandrifolia (A.C.Sm.) Lombardi	XXX										X	X	X
	Peritassa pruinosa (Seem.) A.C.Sm.	XXX										X		
	Prionostemma aspera (Lam.) Miers	XXX								X		X	X	X
	Pristimera caudata Mennega	XXX											X	
	Pristimera celastroides (Kunth) A.C.Sm.	XXX										X		
	Pristimera nervosa (Miers) A.C.Sm.	XXX										X	X	X
	Pristimera tenuiflora (Mart. ex Peyr.) A.C.Sm.	XXX											X	X
	Pristimera verrucosa (Kunth) Miers	XXX										X		
	Salacia amplectens A.C.Sm.	XXX										X	X	X
	Salacia cordata (Miers) Mennega	XXX										X	X	
	Salacia elliptica (Mart.) G.Don	XXX										X	X	X
	Salacia impressifolia (Miers) A.C.Sm.	XXX										X	X	X

FAMILY	SPECIES	L	SC	LSW	LV	V	HEV	HEL	HET	S	T	GU	SU	FG
	Salacia insignis A.C.Sm.	XXX									X		X	X
	Salacia juruana Loes.	XXX										X	X	X
	Salacia kanukuensis A.C.Sm.	XXX										X	X	
	Salacia macrantha A.C.Sm.	XXX										X		
	Salacia miqueliana Loes.	XXX											X	X
	Salacia multiflora (Lam.) DC.	XXX										X	X	X
	Tontelea attenuata Miers	XXX										X		X
	Tontelea cylindrocarpa (A.C.Sm.) A.C.Sm.	XXX											X	X
	Tontelea emarginata A.C.Sm.	XXX										X		X
	Tontelea laxiflora (Benth.) A.C.Sm.	XXX										X	X	X
	Tontelea mauritioides (A.C.Sm.) A.C.Sm.	XXX										X	X	X
	Tontelea passiflora (Vell.) Lombardi	XXX										X	X	X
Clusiaceae	Clusia amazonica Planch. & Triana						X	X	X	X				
	Clusia brachystyla Maguire						X	X	X	X				
	Clusia cardonae Maguire						X	X	X	X				
	Clusia cerroana Steyerm.						X	X	X	X	X			
	Clusia chiribiquetensis Maguire						X	X	X	X				
	Clusia columnaris Engl.						X	X	X	X	X			
	Clusia comans (Mart.) Pipoly						X	X	X					X
	Clusia crassifolia Planch. & Triana						X	X	X	X				
	Clusia cuneata Benth.						X	X	X	X				X
	Clusia duidae Gleason						X	X	X	X				
	Clusia flavida (Benth.) Pipoly						X	X	X	X				X
	Clusia flaviflora Engl.						X	X	X		X			
	Clusia fockeana Miq.						X	X	X	X	X		X	
	Clusia gaudichaudii Cambess.						X	X	X	X	X			
	Clusia grandiflora Splitg.						X	X	X	X		X		
	Clusia guayanae Pipoly						X	X	X	X				
	Clusia hammeliana Pipoly						X	X	X	X				
	Clusia huberi Pipoly						X	X	X	X				
	Clusia kanukuana Maguire						X	X	X	X				
	Clusia leprantha Mart.						X	X	X	X	X		X	
	Clusia macropoda Klotzsch ex Engl.						X	X	X	X	X		X	
	Clusia maguireana Pipoly						X	X	X	X				
	Clusia melchiori Gleason						X	X	X	X	X		X	
	Clusia microstemon Planch. & Triana						X	X	X	X	X			
	Clusia minor L.						X	X	X	X	X		X	
	Clusia mutica Maguire						X	X	X	X	X		X	

FAMILY	SPECIES	L	SC	LSW	LV	V	HEV	HEL	HET	S	T	GU	SU	FG
	Clusia myriandra (Benth.) Planch. & Triana								X	X	X	X		
	Clusia nemorosa G.Mey.								X	X	X	X	X	X
	Clusia obovata (Spruce ex Planch. & Triana) Pipoly								X	X	X	X	X	X
	Clusia octandra (Poepp.) Pipoly								X	X	X	X		X
	Clusia palmicida Rich. ex Planch. & Triana								X	X	X	X	X	X
	Clusia panapanari (Aubl.) Choisy								X	X	X	X	X	X
	Clusia parvifolia Maguire								X	X	X	X		
	Clusia platystigma Eyma								X	X	X		X	X
	Clusia pusilla Steyerm.								X	X	X	X	X	
	Clusia renggerioides Planch. & Triana								X	X	X	X	X	
	Clusia robusta Eyma								X	X	X	X	X	
	Clusia rosea Jacq.								X	X	X	X	X	X
	Clusia savannarum Maguire								X	X	X	X		
	Clusia schomburgkiana (Planch. & Triana) Benth. ex Engl.								X	X	X	X		
	Clusia schomburgkii Vesque								X	X	X	X		
	Clusia scrobiculata Benoist								X	X	X	X	X	X
	Clusia sipapoana (Maguire) Pipoly								X	X	X	X		
	Clusia stylosa Maguire								X	X	X		X	
	Clusia tabulamontana Maguire								X	X	X	X	X	
	Clusia viscida Engl.								X	X	X	X	X	
Combretaceae	**Combretum cacoucia Exell ex Sandwith**	XX	X							X		X	X	X
	Combretum fruticosum (Loefl.) Stuntz	XX	X							X		X		
	Combretum glabrum DC.	XX	X							X		X		X
	Combretum laxum Jacq.	XX	X							X		X	X	X
	Combretum paraguariense (Eichler) Stace	XX	X							X				X
	Combretum pyramidatum Desv. ex Ham.	XX	X							X		X	X	X
	Combretum rohrii Exell	XX	SC							X				FG
	Combretum rotundifolium Rich.	XX	X							X		X	X	X
Connaraceae	**Cnestidium guianense (Schellenb.) Schellenb.**	XX										X	X	X
	Connarus coriaceus Schellenb.	X	X							X		X	X	X
	Connarus erianthus Benth. ex Baker	XX	X							X		X	X	X
	Connarus fasciculatus (DC.) Planch.	XX	X							X			X	X
	Connarus patrisii (DC.) Planch.	XX	X							X		X	X	X
	Connarus perrottetii (DC.) Planch.	XX	X							X		X	X	X
	Connarus punctatus Planch.	XX	X							X		X	X	X
	Pseudoconnarus macrophyllus (Poepp.) Radlk.	XX	X							X		X		

FAMILY	SPECIES	L	SC	LSW	LV	V	HEV	HEL	HET	S	T	GU	SU	FG
	Pseudoconnarus subtriplinervis (Radlk.) G.Schellenb.	XX	X							X		X	X	
	Rourea frutescens Aubl.	XX	X							X		X	X	X
	Rourea grosourdyana Baill.	XX	X							X		X		
	Rourea kappleri Lanj.	XX	X							X			X	
	Rourea ligulata Baker	XX	X							X				X
	Rourea neglecta G.Schellenb.	XX	X							X				X
	Rourea pubescens (DC.) Radlk.	XX	X							X		X	X	X
	Rourea pubescens (DC.) Radlk.	XX	X							X		X		
	Rourea surinamensis Miq.	XX	X							X		X	X	X
Convolvulaceae	Aniseia cernua Moric.					X						X	X	X
	Aniseia martinicensis (Jacq.) Choisy					X						X	X	X
	Bonamia maripoides Hallier f.	XX			X							X	X	X
	Calycobolus glaber (Kunth) House	XX										X	X	X
	Dicranostyles ampla Ducke	XXX	X							X		X		
	Dicranostyles guianensis Mennega	XXX	X							X		X	X	X
	Dicranostyles integra Ducke	XXX	X							X				X
	Dicranostyles villosus Ducke	XXX	X							X			X	X
	Evolvulus alsinoides (L.) L.					X						X	X	X
	Evolvulus cardiophyllus Schltdl.					X						X	X	X
	Evolvulus convolvuloides (Willd. ex Schult.) Stearn					X						X		
	Evolvulus filipes Mart.					X						X	X	X
	Evolvulus glomeratus Nees & C.Mart.					X						X	X	X
	Evolvulus nummularius (L.) L.					X						X	X	X
	Evolvulus sericeus Sw.					X						X	X	
	Ipomoea alba L.					X						X	X	X
	Ipomoea anisomeres B.L.Rob. & Bartlett					X						X	X	
	Ipomoea aquatica Forssk.					X						X	X	X
	Ipomoea argentea Meisn.					X						X		
	Ipomoea batatas (L.) Lam.					X						X	X	X
	Ipomoea batatoides Choisy				X	X						X	X	X
	Ipomoea bignonioides Sims					X								X
	Ipomoea cairica (L.) Sweet					X						X		
	Ipomoea carnea Jacq.	X			X							X		
	Ipomoea distans Choisy ex DC.					X								X
	Ipomoea hederifolia L.					X						X	X	
	Ipomoea horsfalliae Hook.					X							X	
	Ipomoea imperati (Vahl) Griseb.					X						X	X	X

FAMILY	SPECIES	L	SC	LSW	LV	V	HEV	HEL	HET	S	T	GU	SU	FG
	Ipomoea killipiana O'Donell					X								X
	Ipomoea leprieurii D.F.Austin					X								X
	Ipomoea mauritiana Jacq.					X						X	X	X
	Ipomoea nil (L.) Roth					X						X	X	X
	Ipomoea pes-caprae (L.) R.Br.				X	X				X		X	X	X
	Ipomoea philomega (Vell.) House	X			X							X	X	X
	Ipomoea piurensis O'Donell					X						X	X	X
	Ipomoea quamoclit L.					X						X	X	X
	Ipomoea rubens Choisy					X						X	X	X
	Ipomoea schomburgkii Choisy					X						X	X	X
	Ipomoea setifera Poir.					X						X	X	X
	Ipomoea squamosa Choisy					X						X	X	
	Ipomoea subrevoluta Choisy					X						X	X	X
	Ipomoea tiliacea (Willd.) Choisy					X						X	X	X
	Ipomoea violacea L.					X						X	X	X
	Jacquemontia agrestis (Mart. ex Choisy) Meisn.					X						X		
	Jacquemontia ciliata Sandwith					X						X	X	X
	Jacquemontia guyanensis (Aubl.) Meisn.					X						X	X	X
	Jacquemontia sphaerostigma (Cav.) Rusby					X						X		
	Jacquemontia tamnifolia (L.) Griseb.					X						X	X	X
	Lysiostyles scandens Benth.	XX										X	X	X
	Maripa densiflora Benth.	XXX										X	X	X
	Maripa glabra Choisy	XXX										X	X	X
	Maripa longifolia Sagot ex Hallier f.	XXX												X
	Maripa paniculata Barb. Rodr.	XXX										X	X	X
	Maripa reticulata Ducke	XXX										X	X	X
	Maripa scandens Aubl.	XXX										X	X	X
	Maripa violacea (Aubl.) Ooststr. ex Lanj. & Uittien	XXX										X	X	X
	Merremia aegyptia (L.) Urb.				X	X						X	X	X
	Merremia aturensis (Kunth) Hallier f.				X	X						X	X	X
	Merremia cissoides (Lam.) Hallier f.				X	X						X	X	X
	Merremia digitata (Spreng.) Hallier f.				X	X						X		
	Merremia dissecta (Jacq.) Hallier f.				X	X						X	X	X
	Merremia macrocalyx (Ruiz & Pav.) O'Donell			X	X	X						X	X	X
	Merremia quinquefolia (L.) Hallier f.			X	X	X						X	X	
	Merremia ternifoliola Pittier				X	X						X		
	Merremia umbellata (L.) Hallier f.				X	X						X	X	X

FAMILY	SPECIES	L	SC	LSW	LV	V	HEV	HEL	HET	S	T	GU	SU	FG
	Merremia wurdackii D.F.Austin & Staples				X	X						X	X	X
	Operculina hamiltonii (G. Don) D.F.Austin & Staples			X	X	X						X	X	X
	Operculina sericantha (Miq.) Ooststr.			X	X	X						X	X	X
Cucurbitaceae	Cayaponia angustiloba (Cogn.) Cogn.			X	X	X						X	X	
	Cayaponia coriacea Cogn.			X	X	X						X	X	X
	Cayaponia cruegeri (Naudin) Cogn.			X								X	X	X
	Cayaponia guianensis C.Jeffrey			X								X		
	Cayaponia jenmanii C.Jeffrey			X								X	X	X
	Cayaponia ophthalmica R.E.Schult.			X								X	X	X
	Cayaponia racemosa (Mill.) Cogn.			X								X	X	X
	Cayaponia rigida (Cogn.) Cogn.			X								X	X	X
	Cayaponia selysioides C.Jeffrey			X								X	X	X
	Cayaponia triangularis (Cogn.) Cogn.			X	X							X	X	X
	Ceratosanthes palmata (L.) Urb.			X	X							X	X	X
	Fevillea cordifolia L.					X								X
	Gurania acuminata Cogn.			X	X	X						X	X	X
	Gurania bignoniacea (Poepp. & Endl.) C.Jeffrey			X	X							X	X	X
	Gurania brevipedunculata Cogn.			X	X							X		X
	Gurania huberi Cogn.			X	X							X	X	X
	Gurania lobata (L.) Pruski			X	X							X	X	X
	Gurania nigrescens C.Jeffrey			X	X							X		
	Gurania oxyphylla C.Jeffrey			X	X							X	X	X
	Gurania reticulata Cogn.			X	X									X
	Gurania robusta Suess.			X	X								X	X
	Gurania simplicifolia (Steyerm.) C.Jeffrey			X	X							X		
	Gurania subumbellata (Miq.) Cogn.			X	X							X	X	X
	Psiguria triphylla (Miq.) C.Jeffrey			X	X							X	X	X
	Psiguria umbrosa (Kunth) C.Jeffrey			X	X									X
Cyclanthaceae	Asplundia brachyphylla Harling						X					X	X	X
	Asplundia fanshawei (Maguire) Harling						X					X	X	
	Asplundia glandulosa (Gleason) Harling						X					X	X	X
	Asplundia gleasonii Harling						X					X		
	Asplundia guianensis Harling						X					X		
	Asplundia heteranthera Harling						X						X	X
	Asplundia maguirei Harling						X					X	X	
	Evodianthus funifer (Poit.) Lindm.						X					X	X	X
	Thoracocarpus bissectus (Vell.) Harling						X					X	X	X
Dichapetalaceae	**Dichapetalum pedunculatum (DC.) Baill.**	XX										X	X	X

FAMILY	SPECIES	L	SC	LSW	LV	V	HEV	HEL	HET	S	T	GU	SU	FG
	Dichapetalum rugosum (Vahl) Prance	XX										X	X	X
	Dichapetalum schulzii Prance	XX											X	
Dilleniaceae	**Davilla alata (Vent.) Briq.**	XX										X	X	X
	Davilla kunthii A.St.-Hil.	XX										X	X	X
	Davilla lacunosa Mart.	XX												X
	Davilla nitida (Vahl) Kubitzki	XX										X	X	X
	Davilla rugosa Poir.	XX										X	X	X
	Doliocarpus amazonicus Sleumer	XXX											X	
	Doliocarpus brevipedicellatus Garcke	XXX										X	X	X
	Doliocarpus dentatus (Aubl.) Standl.	XXX										X	X	X
	Doliocarpus gracilis Kubitzki	XXX											X	X
	Doliocarpus guianensis (Aubl.) Gilg.	XXX										X	X	X
	Doliocarpus macrocarpus Mart. ex Eichler	XXX										X	X	X
	Doliocarpus major J.F.Gmel.	XXX										X	X	X
	Doliocarpus paraensis Sleumer	XXX										X	X	X
	Doliocarpus sagotianus Kubitzki	XXX												X
	Doliocarpus savannarum Sandwith	XXX										X	X	X
	Doliocarpus spraguei Cheesman	XXX										X	X	X
	Pinzona coriacea Mart. & Zucc.	XX										X	X	X
	Tetracera asperula Miq.	XX										X	X	X
	Tetracera costata Mart. ex Eichler	XX										X	X	X
	Tetracera maguirei G.A.Aymard & B.M.Boom	XX										X		
	Tetracera surinamensis Miq.	XX										X	X	X
	Tetracera tigarea DC.	XX										X	X	X
	Tetracera volubilis L.	XX										X	X	X
	Tetracera willdenowiana Steud.	XX										X	X	X
Dioscoreaceae	Dioscorea altissima Lam.				X	X						X	X	
	Dioscorea amazonum Mart. ex Griseb.				X	X						X	X	X
	Dioscorea cordifolia Laness.				X	X								X
	Dioscorea crotalariifolia Uline				X	X						X		
	Dioscorea dodecaneura Vell.				X	X						X	X	X
	Dioscorea guianensis R.Knuth				X	X						X		
	Dioscorea hastata Mill.				X	X								X
	Dioscorea megacarpa Gleason				X	X						X	X	X
	Dioscorea microura R.Knuth				X	X						X		
	Dioscorea oblonga Gleason				X	X						X	X	
	Dioscorea pilosiuscula Bertero ex Spreng.				X	X						X	X	X
	Dioscorea piperifolia Humb. & Bonpl. ex Willd.				X	X						X		

FAMILY	SPECIES	L	SC	LSW	LV	V	HEV	HEL	HET	S.	T	GU	SU	FG
	Dioscorea polygonoides Humb. & Bonpl. ex Willd.				X	X						X	X	X
	Dioscorea pomeroonensis R.Knuth				X	X						X		
	Dioscorea potarensis R.Knuth				X	X						X		
	Dioscorea syringifolia (Kunth) Kunth & R.H. Schomb. ex R.Knuth				X	X						X		X
	Dioscorea trachyandra Griseb.				X	X						X		
	Dioscorea trichanthera Gleason				X	X						X		
	Dioscorea trifida L.f.				X	X						X	X	X
	Dioscorea truncata Miq.				X	X							X	
Ericaceae	Cavendishia callista Donn.Sm.							X		X			X	X	X
	Satyria carnosiflora Lanj.							X		X			X	X	
	Satyria cerander (Dunal) A.C.Sm.							X		X				X	X
	Satyria panurensis (Benth. ex Meisn.) Benth. & Hook.f. ex Nied.							X		X			X	X	X
	Sphyrospermum cordifolium Benth.							X		X			X	X	X
	Thibaudia nutans Klotzsch ex Mansf.							X		X			X	X	
Euphorbiaceae	Acalypha scandens Benth.				X								X	X	
	Bia fendleri (Müll.Arg.) G.L.Webster					X							X		
	Bia lessertiana Baill.					X							X	X	X
	Croton pullei Lanj.	XXX								X			X	X
	Dalechampia affinis Müll.Arg.					X							X	X	X
	Dalechampia attenuistylus Armbr.					X								X	X
	Dalechampia brevicolumna Armbr.					X									X
	Dalechampia brownsbergensis G.L.Webster & Armbr.					X								X	
	Dalechampia dioscoreifolia Poepp.					X									X
	Dalechampia fragrans Armbr.					X								X	X
	Dalechampia heterobractea Armbr.					X							X	X	X
	Dalechampia parvibracteata Lanj.					X							X	X	
	Dalechampia scandens L.					X							X	X	X
	Dalechampia stipulacea Müll.Arg.					X									X
	Dalechampia tiliifolia Lam.					X							X		X
	Mabea pulcherrima Müll.Arg.	XX								X	X	X	X	X
	Mabea taquari Aubl.	XX								X	X	X	X	X
	Omphalea diandra L.	XXX										X	X	X
	Plukenetia loretensis Ule	XXX										X		
	Plukenetia polyadenia Müll.Arg.	XXX										X	X	X
	Plukenetia supraglandulosa L.J.Gillespie	XXX										X	X	X
	Plukenetia verrucosa Sm.	XXX										X	X	X
	Plukenetia volubilis L.	XXX										X	X	

FAMILY	SPECIES	L	SC	LSW	LV	V	HEV	HEL	HET	S	T	GU	SU	FG
	Tragia tabulaemontana Gillespie				X	X								X
	Tragia volubilis L.				X	X							X	X
Gesneriaceae	Columnea grisebachiana Kuntze			X								X		
	Drymonia coccinea (Aubl.) Wiehler			X								X	X	X
	Drymonia serrulata (Jacq.) Mart.			X								X	X	X
	Nematanthus savannarum (C.V.Morton) J.L.Clark			X								X		
	Paradrymonia campostyla (Leeuwenb.) Wiehler			X									X	X
	Paradrymonia maculata (Hook.f.) Wiehler			X								X		X
Gnetaceae	Gnetum camporum (Markgr.) D.W.Stev. & Zanoni	XX										X		
	Gnetum leyboldii Tul.	XX										X		
	Gnetum nodiflorum Brongn.	XX										X	X	X
	Gnetum paniculatum Spruce ex Benth.	XX										X		X
	Gnetum schwackeanum Taub. ex Schenck	XX										X		
	Gnetum urens (Aubl.) Blume	XX										X	X	X
Hernandiaceae	**Sparattanthelium aruakorum Tutin**	XX	X							X		X	X	
	Sparattanthelium guianense Sandwith	XX	X							X		X		
	Sparattanthelium tupiniquinorum Mart.	XX	X							X		X	X	X
	Sparattanthelium uncigerum (Meisn.) Kubitzki	XX	X							X		X	X	X
	Sparattanthelium wonotoboense Kosterm.	XX	X							X		X	X	X
Icacinaceae	**Casimirella ampla (Miers) R.A.Howard**	X	X							X	X			X
	Leretia cordata Vell.	XXX	X									X	X	X
	Pleurisanthes artocarpi Baill.	XXX	X									X		X
	Pleurisanthes emarginata Tiegh.	XXX	X										X	X
	Pleurisanthes flava Sandwith	XXX	X									X		
	Pleurisanthes parviflora (Ducke) R.A.Howard	XXX	X									X	X	X
Lamiaceae	Aegiphila bracteolosa Moldenke	XX	X							X	X	X		
	Aegiphila deppeana Steud.	XX	X							X	X			X
	Aegiphila elata Sw.	XX	X							X	X	X	X	
	Aegiphila integrifolia (Jacq.) B.D.Jacks.	XX	X							X	X	X	X	X
	Aegiphila laevis (Aubl.) J.F.Gmel.	XX	X							X	X	X	X	X
	Aegiphila laxiflora Benth.	XX	X							X	X	X		
	Aegiphila lhotzkiana Cham.	XX	X							X	X		X	X
	Aegiphila longifolia Turcz.	XX	X							X	X	X		
	Aegiphila macrantha Ducke	XX	X							X	X	X	X	X
	Aegiphila membranacea Turcz.	XX	X							X	X	X	X	X
	Aegiphila racemosa Vell.	XX	X							X	X	X	X	X
	Aegiphila roraimensis Moldenke	XX	X							X	X	X		

FAMILY	SPECIES	L	SC	LSW	LV	V	HEV	HEL	HET	S	T	GU	SU	FG
	Aegiphila venezuelensis Moldenke	XX	X							X	X	X		
	Aegiphila villosa (Aubl.) J.F.Gmel.	XX	X							X	X	X		X
Leguminosae	**Abrus precatorius L.**				X	X						X	X	X
	Acacia articulata Ducke	XX	X							X			X	
	Acacia tenuifolia (L.) Willd.	XX	X							X	X	X	X	X
	Bauhinia cupreonitens Ducke	XX	X							X		X	X	X
	Bauhinia glabra Jacq.	XXX	X							X		X	X	
	Bauhinia guianensis Aubl.	XXX	X							X		X	X	X
	Bauhinia kunthiana Vogel	XXX	X							X		X	X	X
	Bauhinia microstachya (Raddi) J.F.Macbr.	XXX	X							X		X		
	Bauhinia poiteauana Vogel	XXX	X							X				X
	Bauhinia scala-simiae Sandwith	XXX	X							X		X		
	Bauhinia siqueiraei Ducke	XXX	X							X		X		X
	Bauhinia smilacina (Schott) Steud.	XXX	X							X			X	X
	Bauhinia surinamensis Amshoff	XXX	X							X		X	X	X
	Clitoria arborescens R.Br.	X			X	X				X		X	X	X
	Clitoria brachycalyx Harms	X			X							X		
	Clitoria fairchildiana Howard	X			X									X
	Clitoria falcata Lam	X			X							X	X	X
	Clitoria guianensis (Aubl.) Benth.	X			X							X	X	X
	Clitoria kaieteurensis Fantz	X			X					X		X		
	Clitoria laurifolia Poir.	X			X							X	X	X
	Clitoria leptostachya Benth.	X			X					X		X		
	Clitoria pendens Fantz	X			X					X		X	X	
	Clitoria sagotii Fantz	X			X					X		X	X	X
	Clitoria ternata L.				X	X							X	X
	Dalbergia amazonica (Radlk.) Ducke	X	X							X		X		X
	Dalbergia brownei (Jacq.) Urb.	X	X											
	Dalbergia ecastaphyllum (L.) Taub.	X	X							X		X	X	X
	Dalbergia foliosa (Benth.) A.M.Carvalho	X	X							X		X	X	X
	Dalbergia frutescens (Vell.) Britton	X	X							X		X		
	Dalbergia intermedia A.M.Carvalho	X	X							X		X		
	Dalbergia inundata Benth.	X	X							X		X		
	Dalbergia monetaria L.f.	X	X							X	X	X	X	X
	Dalbergia riedelii (Benth.) Sandwith	X	X							X		X	X	X
	Dalbergia subcymosa Ducke	X	X							X			X	X
	Deguelia amazonica Killip	XXX										X	X	X
	Deguelia densiflora (Benth.) A.M.G.Azevedo ex M.Souza.	XXX										X		

FAMILY	SPECIES	L	SC	LSW	LV	V	HEV	HEL	HET	S	T	GU	SU	FG
	Deguelia negrensis (Benth.) Taub.	XXX										X		
	Deguelia nitidula (Benth.) A.M.G.Azevedo & R.A.Camargo	XXX										X	X	X
	Deguelia rariflora (Mart. ex Benth.) G.P. Lewis & Acev.-Rodr.	XXX										X		
	Deguelia scandens Aubl.	XXX										X	X	X
	Deguelia utilis (A.C.Sm.) A.M.G.Azevedo	XXX										X	X	X
	Dioclea apurensis Kunth	X			X								X	
	Dioclea coriacea Benth.	XX											X	
	Dioclea elliptica Maxwell				X	X						X	X	X
	Dioclea guianensis Benth.	X			X	X						X	X	X
	Dioclea macrantha Huber				X	X						X	X	X
	Dioclea macrocarpa Huber	XX										X	X	X
	Dioclea malacocarpa Ducke	XX										X	X	X
	Dioclea megacarpa Rolfe	XX										X		
	Dioclea reflexa Hook.f.	XX										X	X	X
	Dioclea scabra (Rich.) R.H.Maxwell	XX										X	X	X
	Dioclea violacea Benth.	XX										X	X	X
	Dioclea wilsonii Standl.	XX										X	X	X
	Dioclea virgata (Rich.) Amshoff	XX										X	X	X
	Entada polyphylla Benth.	XXX										X	X	X
	Entada polystachya (L.) DC.	XXX										X	X	X
	Lonchocarpus rufescens Benth.	XXX										X		
	Lonchocarpus chrysophyllus Kleinhoonte	XXX										X	X	X
	Lonchocarpus heptaphyllus (Poir.)DC.	XXX										X	X	X
	Lonchocarpus longifolius (Benth.) Benth.	XXX											X	
	Lonchocarpus martynii A.C.Sm.	XXX										X		
	Lonchocarpus pictus Pittier	XXX										X	X	X
	Machaerium aureiflorum Ducke	XX	X							X		X		X
	Machaerium biovulatum Micheli	XX	X							X		X		
	Machaerium ferox (Benth.) Ducke	XX	X							X		X	X	
	Machaerium floribundum Benth.	XX	X							X		X		X
	Machaerium inundatum (Benth.) Ducke	XX	X							X		X	X	X
	Machaerium isadelphum (E.Mey.) Standl.	XX	X							X		X		
	Machaerium kegelii Meisner	XX	X							X		X	X	X
	Machaerium leiophyllum (DC.) Benth.	XX	X							X		X	X	X
	Machaerium lunatum (L.f.) Ducke	XX	X							X		X	X	X
	Machaerium macrophyllum Benth.	XX	X							X		X	X	X
	Machaerium madeirense Pittier	XX	X							X		X	X	

FAMILY	SPECIES	L	SC	LSW	LV	V	HEV	HEL	HET	S	T	GU	SU	FG
	Machaerium microphyllum (E.Mey.) Standl.	XX	X							X			X	X
	Machaerium mutisii Rudd	XX	X							X	X	X	X	
	Machaerium myrianthum Benth.	XX	X							X		X	X	
	Machaerium paraense Ducke	XX	X							X			X	X
	Machaerium polyphyllum (Poir.) Benth.	XX	X							X			X	X
	Machaerium quinatum (Aubl.) Sandwith	XX	X							X	X	X	X	X
	Machaerium trifoliolatum Ducke	XX	X							X		X	X	X
	Mimosa annularis Benth.	XX	X							X		X	X	X
	Mimosa diplotricha C.Wright ex Sauvalle	XX	X							X		X	X	X
	Mimosa guilandinae (DC.) Barneby	XX	X							X				X
	Mimosa microcephala Humb. & Bonpl. ex Willd.	XX	X							X			X	X
	Mimosa myriadenia (Benth.) Benth.	XX	X							X		X	X	X
	Mimosa pigra L.	XX	X							X		X	X	X
	Mimosa quadrivalvis L.	XX	X							X		X		X
	Mucuna rostrata Benth.	XX												X
	Mucuna sloanei Fawc. & Rendle	XX										X	X	X
	Mucuna urens (L.) Medik.	XX										X	X	X
	Periandra coccinea (Schrad.) Benth.					X						X	X	X
	Piptadenia floribunda Kleinhoonte	XX											X	X
	Senna chrysocarpa (Desv.) H.S.Irwin & Barneby	XX	X							X		X	X	X
	Senna latifolia (G.Mey.) H.S.Irwin & Barneby	XX	X							X		X	X	X
	Senna quinquangulata (Rich.) H.S.Irwin & Barneby	XX	X							X		X	X	X
	Senna silvestris (Vell.) H.S.Irwin & Barneby	XX	X							X		X	X	X
Loganiaceae	Strychnos bredemeyeri (Schult.) Sprague & Sandwith	XX										X	X	X
	Strychnos cogens Benth.	XX										X	X	X
	Strychnos diaboli Sandwith	XX										X		
	Strychnos divaricans Ducke	XX											X	X
	Strychnos erichsonii M.R.Schomb. ex Progel	XXX										X	X	X
	Strychnos eugeniifolia Monach.	XX											X	X
	Strychnos glabra Sagot ex Progel	XXX										X	X	X
	Strychnos guianensis (Aubl.) Mart.	XXX										X	X	X
	Strychnos hirsuta Spruce ex Benth.		X							X		X		X
	Strychnos jobertiana Baill.	XX												X
	Strychnos medeola Sagot ex Progel	XXX											X	X
	Strychnos melinoniana Baill.	XXX										X	X	X
	Strychnos mitscherlichii M.R.Schomb.	XXX										X	X	X

FAMILY	SPECIES	L	SC	LSW	LV	V	HEV	HEL	HET	S	T	GU	SU	FG
	Strychnos oiapocensis Fróes		X							X			X	X
	Strychnos panamensis Seem.	XXX										X		X
	Strychnos panurensis Sprague & Sandwith	XXX												X
	Strychnos peckii B.L.Rob.	XXX										X	X	X
	Strychnos ramentifera Ducke	XXX											X	
	Strychnos rondeletioides Spruce ex Benth.	XXX											X	X
	Strychnos subcordata Spruce ex Benth.	XXX										X		
	Strychnos tomentosa Benth.	XXX										X	X	X
	Strychnos toxifera R.H.Schomb. ex Lindl.	XXX										X	X	X
Malpighiaceae	**Alicia macrodisca (Triana & Planch) W.R.Anderson**	XX										X	X	
	Banisteriopsis carolina W.R.Anderson	XX												X
	Banisteriopsis martiniana (A.Juss.) Cuatrec.	XX										X	X	X
	Banisteriopsis muricata (Cav.) Cuatrec.	XX										X	X	X
	Banisteriopsis nummifera (A.Juss.) B.Gates	XX												X
	Banisteriopsis pulcherrima (Sandwith) B.Gates	XX										X		
	Banisteriopsis schwannioides (Griseb.) B.Gates	XX												X
	Bronwenia cinerascens (Benth.) W.R.Anderson & C.Davis	XX										X		
	Bronwenia mathiasiae (W.R.Anderson) W.R.Anderson & C.Davis	XX										X		
	Bronwenia wurdackii (B.Gates) W.R.Anderson & C.Davis	XX												X
	Carolus sinemariensis (Aubl.) W.R.Anderson	XX										X		X
	Christianella surinamensis (Kosterm.) W.R.Anderson	XX										X	X	
	Diplopterys cristata (Griseb.) W.R.Anderson & C.Davis	XX										X	X	
	Diplopterys lucida (Rich.) W.R.Anderson & C.Davis	XX										X	X	X
	Diplopterys pauciflora (G.Mey.) Nied.	XX										X		X
	Excentradenia adenophora (Sandwith) W.R.Anderson	XX										X		
	Excentradenia propinqua (W.R.Anderson) W.R.Anderson	XX										X	X	X
	Heteropterys biglandulosa A.Juss.	XX												X
	Heteropterys cristata Benth.	XX										X		
	Heteropterys hoffmanii W.R. Anderson	XX										X		
	Heteropterys leona (Cav.) Exell	XX										X	X	X
	Heteropterys macradena (DC.) W.R.Anderson	XX										X	X	X
	Heteropterys macrostachya A.Juss.	XX										X	X	X
	Heteropterys mathewsiana A.Juss.	XX											X	
	Heteropterys nervosa A.Juss.	XX										X	X	X
	Heteropterys oligantha W.R. Anderson	XX										X	X	

FAMILY	SPECIES	L	SC	LSW	LV	V	HEV	HEL	HET	S	T	GU	SU	FG
	Heteropterys platyptera DC.	XX										X		
	Heteropterys siderosa Cuatrec.	XX										X		X
	Heteropterys subhelicina Nied.	XX										X		
	Heteropterys velutina W.R. Anderson	XX										X		
	Hiraea affinis Miq.	XX										X	X	X
	Hiraea fagifolia (DC.) A.Juss.	XX										X	X	X
	Hiraea faginea (Sw.) Nied.	XX										X	X	X
	Hiraea gracieana W.R. Anderson	XX										X	X	X
	Hiraea longipes W.R. Anderson	XX												X
	Hiraea longipilifera W.R. Anderson	XX										X		
	Hiraea morii W.R. Anderson	XX												X
	Hiraea quapara (Aubl.) Sprague	XX												X
	Jubelina riparia A.Juss.	XX												X
	Jubelina rosea (Miq.) Nied.	XX											X	X
	Lophopterys euryptera Sandwith	XX										X		
	Lophopterys splendens A.Juss.	XX												X
	Lophopterys surinamensis (Kosterm.) Sandwith	XX											X	
	Malpighiodes guianensis (W.R.Anderson) W.R.Anderson	XX										X	X	X
	Mascagnia arenicola C.E.Anderson	XX										X	X	X
	Mascagnia divaricata (Kunth) Nied.	XX											X	X
	Mascagnia schunkei W.R.Anderson	XX												X
	Mezia angelica W.R.Anderson	XX										X		X
	Mezia includens (Benth.) Cuatrec.	XX										X	X	X
	Niedenzuella acutifolia (Cav.) W.R.Anderson	XX										X	X	X
	Niedenzuella poeppigiana (A.Juss.) W.R.Anderson	XX										X		
	Stigmaphyllon bannisterioides (L.) C.E.Anderson	XX										X	X	X
	Stigmaphyllon convolvulifolium A.Juss.	XX										X	X	X
	Stigmaphyllon palmatum (Cav.) A.Juss.	XX												X
	Stigmaphyllon puberum (Rich.) A.Juss.	XX										X	X	X
	Stigmaphyllon sinuatum (DC.) A.Juss.	XX										X	X	X
	Tetrapterys calophylla A. Juss.	XX										X	X	X
	Tetrapterys crispa A.Juss.	XX										X	X	X
	Tetrapterys discolor (G.Mey.) DC.	XX										X	X	X
	Tetrapterys fimbripetala A.Juss.	XX										X	X	X
	Tetrapterys glabrifolia (Griseb.) Small	XX											X	X
	Tetrapterys maranhamensis A.Juss.	XX										X	X	

FAMILY	SPECIES	L	SC	LSW	LV	V	HEV	HEL	HET	S	T	GU	SU	FG
	Tetrapterys megalantha W.R.Anderson	XX										X		
	Tetrapterys mucronata Cav.	XX										X	X	X
	Tetrapterys oleifolia (Benth.) Griseb.	XX										X		
	Tetrapterys pusilla Steyerm.	XX										X		
	Tetrapterys rhodopteron Oliv.	XX										X		
	Tetrapterys styloptera A.Juss.	XX										X	X	X
Malvaceae	Byttneria aurantiaca Mildbr.	X			X									X
	Byttneria benensis Britton	X			X								X	X
	Byttneria catalpifolia Jacq.	X			X							X		
	Byttneria cordifolia Sagot	X			X								X	X
	Byttneria divaricata Benth.	X			X							X	X	X
	Byttneria genistella Triana & Planch.	X			X							X	X	X
	Byttneria morii L.C.Barnett & Dorr	X			X							X		X
	Byttneria scabra L.	X			X							X	X	X
	Byttneria uaupensis K.Schum.	X			X							X		
Marcgraviaceae	**Marcgravia coriacea Vahl**							X				X	X	X
	Marcgravia magnibracteata Lanj. & Heerdt							X				X	X	X
	Marcgravia maguirei de Roon							X						X
	Marcgravia pedunculosa Triana & Planch.							X				X	X	X
	Marcgravia purpurea I.W.Bailey							X				X	X	X
	Marcgravia sororopaniana Steyerm.							X				X		
	Marcgraviastrum pendulum (Lanj. & Heerdt) Bedell							X				X	X	X
	Norantea guianensis Aubl.							X				X	X	X
	Sarcopera tepuiensis (de Roon) Bedell							X				X		
	Souroubea dasystachya Gilg & Werderm.							X				X		
	Souroubea guianensis Aubl.							X				X	X	X
Melastomataceae	**Adelobotrys adscendens (Sw.) Triana**	XX						X				X	X	X
	Adelobotrys ayangannensis Wurdack	XX						X				X		
	Adelobotrys ciliata (Naudin) Triana	XX						X					X	X
	Adelobotrys monticola Gleason	XX						X					X	
	Adelobotrys permixta Wurdack	XX						X				X		
	Adelobotrys scandens (Aubl.) DC.	XX						X					X	X
	Leandra procumbens Ule	XX			X			X				X		
	Phainantha laxiflora (Triana) Gleason				X	X						X		
	Topobea parasitica Aubl.	XX						X					X	X
Menispermaceae	Abuta barbata Miers	XX										X	X	X
	Abuta bullata Moldenke	XX										X		X
	Abuta candollei Triana & Planch.	XX										X	X	X

FAMILY	SPECIES	L	SC	LSW	LV	V	HEV	HEL	HET	S	T	GU	SU	FG
	Abuta grandifolia (Mart.) Sandwith	XX										X	X	X
	Abuta imene (Mart.) Eichler	XX										X	X	X
	Abuta obovata Diels	XX										X	X	X
	Abuta rufescens Aubl.	XX										X	X	X
	Abuta sandwithiana Krukoff & Barneby	XX										X	X	X
	Abuta solimoesensis Krukoff & Barneby	XX												X
	Anomospermum chloranthum Diels	XX												X
	Anomospermum grandifolium Eichler	XX											X	
	Anomospermum steyermarkii Krukoff & Barneby	XX												X
	Borismene japurensis (Mart.) Barneby	XX											X	
	Caryomene foveolata Barneby & Krukoff	XX											X	
	Caryomene glaucescens (Moldenke) Barneby & Krukoff	XX												X
	Caryomene olivascens Barneby & Krukoff	XX												X
	Cissampelos andromorpha DC.	X			X							X	X	X
	Cissampelos fasciculata Benth.	X			X							X	X	
	Cissampelos ovalifolia DC.	X			X							X	X	
	Cissampelos pareira L.	X			X	X				X		X	X	
	Curarea candicans (Rich. ex DC.) Barneby & Krukoff	XX										X	X	X
	Disciphania moriorum Barneby	XX												X
	Disciphania unilateralis Barneby	XX												X
	Elephantomene eburnea Barneby & Krukoff	XX											X	X
	Hyperbaena domingensis (DC.) Benth.	XX										X	X	X
	Odontocarya tamoides (DC.) Miers	XX										X	X	
	Odontocarya wullschlaegelii (Eichler) Barneby	XX										X	X	X
	Orthomene prancei Barneby & Krukoff	XX												X
	Orthomene schomburgkii (Miers) Barneby & Krukoff	XX										X	X	X
	Orthomene verruculosa (Krukoff & Barneby) Barneby & Krukoff	XX												X
	Sciadotenia cayennensis Benth.	XX										X	X	X
	Sciadotenia duckei Moldenke	XX												X
	Sciadotenia eichleriana Moldenke	XX												X
	Sciadotenia sagotiana Diels	XX										X	X	X
	Telitoxicum duckei Moldenke	XX												X
	Telitoxicum inopinatum Moldenke	XX										X	X	X
	Telitoxicum krukovii Moldenke	XX										X	X	
Moraceae	Ficus albert-smithii Standl.						X	X	X	X	X	X	X	X
	Ficus amazonica (Miq.) Miq.							X	X	X	X	X	X	X

FAMILY	SPECIES	L	SC	LSW	LV	V	HEV	HEL	HET	S	T	GU	SU	FG
	Ficus americana ssp. guianensis (Desv. ex Ham.) C.C.Berg								X	X	X	X	X	X
	Ficus broadwayi Urb.								X	X	X	X	X	X
	Ficus caballina Standl.								X	X	X	X	X	X
	Ficus catappifolia Kunth & Bouché								X	X	X	X	X	X
	Ficus cremersii C.C. Berg								X	X	X			X
	Ficus donnell-smithii Standl.								X	X	X	X	X	X
	Ficus eximia Schott								X	X	X	X		
	Ficus gomelleira Kunth & Bouché								X	X	X	X	X	X
	Ficus greiffiana Dugand								X	X	X	X		X
	Ficus hebetifolia Dugand								X	X	X			X
	Ficus insipida Willd.								X	X	X	X	X	X
	Ficus krukovii Standl.								X	X	X		X	X
	Ficus leiophylla C.C. Berg								X	X	X		X	X
	Ficus malacocarpa Standl.								X	X	X	X	X	X
	Ficus maroniensis Benoist								X	X	X	X	X	X
	Ficus mathewsii (Miq.) Miq.								X	X	X	X	X	
	Ficus matiziana Dugand								X	X	X	X		X
	Ficus maxima Mill.								X	X	X	X	X	X
	Ficus nymphaeifolia Mill.								X	X	X	X	X	X
	Ficus pakkensis Standl.								X	X	X	X	X	X
	Ficus paludica Standl.								X	X	X	X	X	X
	Ficus panurensis Standl.								X	X	X	X	X	X
	Ficus paraensis (Miq.) Miq.								X	X	X	X	X	X
	Ficus pertusa L. f.								X	X	X	X	X	X
	Ficus piresiana Vázq. Avila & C.C. Berg								X	X	X	X		X
	Ficus roraimensis C.C. Berg								X	X	X	X		
	Ficus schumacheri (Liebm.) Griseb.								X	X	X	X	X	X
	Ficus subapiculata (Miq.) Miq.								X	X	X			X
	Ficus trigona L. f.								X	X	X	X	X	X
Nyctaginaceae	**Pisonia macranthocarpa (Donn.Sm.) Donn. Sm.**	XX	X							X		X		
Olacaceae	Cathedra acuminata (Benth.) Miers	XX										X		X
	Heisteria scandens Ducke	XX										X	X	X
Orchidaceae	Vanilla appendiculata Rolfe						X					X		
	Vanilla bicolor Lindl.						X					X	X	X
	Vanilla chamissonis Klotzsch						X							X
	Vanilla cristagalli Hoehne						X					X		
	Vanilla guianensis Splitg.						X					X	X	X

FAMILY	SPECIES	L	SC	LSW	LV	V	HEV	HEL	HET	S	T	GU	SU	FG
	Vanilla hartii Rolfe						X							X
	Vanilla hostmannii Rolfe						X							X
	Vanilla mexicana Mill.						X					X	X	X
	Vanilla odorata C.Presl						X							X
	Vanilla ovata Rolfe						X							X
	Vanilla palmarum (Salzm. ex Lindl.) Lindl.						X					X	X	X
	Vanilla penicillata Garay & Dunst.						X					X		
	Vanilla planifolia Jacks. ex Andrews						X					X	X	
Passifloraceae	Ancistrothyrsus tessmannii Harms	XX										X		X
	Dilkea acuminata Mast.	XX											X	
	Dilkea retusa Mast.	XX			X								X	
	Mitostemma glaziovii Mast.	XX	X		X							X		
	Mitostemma jenmanii Mast.	XX	X		X							X		
	Passiflora acuminata DC.	X		X	X							X		X
	Passiflora aimae Annonay & Feuillet			X	X	X								X
	Passiflora amicorum Wurdack			X	X	X						X		
	Passiflora amoena L.K.Escobar	X		X								X	X	X
	Passiflora angusta Feuillet & J.M.MacDougal			X	X	X						X		
	Passiflora arta Feuillet (unresolved name)			X	X	X						X		
	Passiflora ascidia Feuillet	X		X								X		
	Passiflora auriculata Kunth	X		X		X						X	X	X
	Passiflora balbis Feuillet	X		X								X		
	Passiflora candida (Poepp. & Endl.) Mast.	X		X								X	X	X
	Passiflora capparidifolia Killip			X	X	X						X		
	Passiflora cardonae Killip			X	X	X						X		
	Passiflora cerasina Annonay & Feuillet			X	X	X							X	X
	Passiflora ceratocarpa F.Silveira	X		X								X	X	
	Passiflora cirrhiflora Juss.			X	X	X						X	X	FG
	Passiflora citrifolia Salisb.	X		X										X
	Passiflora coccinea Aubl.	X		X								X	X	X
	Passiflora compar Feuillet			X	X	X						X		X
	Passiflora coriacea Juss.			X	X	X						X		
	Passiflora costata Mast.	X		X								X	X	X
	Passiflora crenata Feuillet & Cremers			X	X	X								X
	Passiflora curva Feuillet (unresolved)			X	X	X								X
	Passiflora davidii Feuillet (unresolved)			X	X	X								X
	Passiflora deficiens Mast.			X	X	X						X		
	Passiflora edulis Sims			X	X	X						X	X	X

FAMILY	SPECIES	L	SC	LSW	LV	V	HEV	HEL	HET	S	T	GU	SU	FG
	Passiflora exura Feuillet	X		X		X								X
	Passiflora fanchonae Feuillet			X	X	X						X		X
	Passiflora foetida L.			X	X	X						X	X	X
	Passiflora fuchsiiflora Hemsl.	X		X								X	X	X
	Passiflora garckei Mast.	X		X		X						X	X	X
	Passiflora glandulosa Cav.	X		X		X						X	X	X
	Passiflora gleasonii Killip			X	X	X						X		
	Passiflora kawensis Feuillet	X		X								X		X
	Passiflora laurifolia L.	X		X	X	X						X	X	X
	Passiflora leptopoda Harms	X		X								X	X	X
	Passiflora longicuspis Vanderpl. & S.E.Vanderpl.			X	X	X								X
	Passiflora longiracemosa Ducke			X	X	X						X		
	Passiflora maguirei Killip	X		X		X						X		
	Passiflora micropetala Mart. ex Mast.			X	X	X							X	
	Passiflora misera Kunth			X	X	X						X	X	X
	Passiflora nitida Kunth	XX		X	X	X						X	X	X
	Passiflora ovata DC.	X		X										X
	Passiflora pachyantha Killip			X	X	X						X	X	
	Passiflora pedata L.			X	X	X						X	X	
	Passiflora picturata Ker Gawl.			X	X	X							X	
	Passiflora plumosa Feuillet & Cremers	X		X		X								X
	Passiflora quadrangularis L.			X	X	X						X	X	X
	Passiflora quadriglandulosa Rodschied			X	X	X						X		
	Passiflora quelchii N.E.Br.	X		X		X						X		
	Passiflora retipetala Mast.			X	X	X						X	X	X
	Passiflora riparia Mart. ex Mast.			X	X	X						X		
	Passiflora rubra L.			X	X	X						X	X	X
	Passiflora rufostipulata Feuillet			X	X	X								X
	Passiflora saulensis Feuillet	X		X										X
	Passiflora sclerophylla Harms	X		X								X		
	Passiflora securiclata Mast.	X		X								X		
	Passiflora serratodigitata L.			X	X	X						X	X	X
	Passiflora spicata Mast.			X	X	X						X		
	Passiflora stipulata Aubl.			X	X	X								X
	Passiflora suberosa L.	X		X									X	
	Passiflora tecta Feuillet			X	X	X						X		
	Passiflora trialata Feuillet & J.M.MacDougal			X	X	X								X
	Passiflora tuberosa Jacq.			X	X	X						X		

FAMILY	SPECIES	L	SC	LSW	LV	V	HEV	HEL	HET	S	T	GU	SU	FG
	Passiflora variolata Poepp. & Endl.	X		X		X								X
	Passiflora vespertilio L.			X	X	X						X	X	X
Phyllanthaceae	Margaritaria nobilis L.f.				X	X						X	X	X
Phytolaccaceae	**Seguieria americana L.**	XX	X									X	X	X
	Seguieria macrophylla Benth.	XX	X									X		X
	Trichostigma octandrum (L.) H.Walter	X	X									X		
Piperaceae	Piper hispidum Sw.		X									X	X	X
	Piper hostmannianum (Miq.) C.DC.		X									X	X	X
Polygalaceae	**Barnhartia floribunda Gleason**	X										X	X	X
	Bredemeyera altissima (Poepp.) A.W.Benn.	X										X		
	Bredemeyera bracteata Klotzsch ex Hassk.	X										X		
	Bredemeyera densiflora A.W.Benn.	X										X	X	
	Bredemeyera floribunda Willd.	X										X		
	Bredemeyera lucida (Benth.) Klotzsch ex Hassk.	X										X	X	X
	Moutabea aculeata (Ruiz & Pav.) Poepp. & Endl.	XXX	X									X		X
	Moutabea guianensis Aubl.	XXX	X									X	X	X
	Moutabea longifolia Poepp. & Endl.	XXX	X										X	
	Polygala spectabilis DC.		X									X	X	X
	Securidaca cayennensis S.F.Blake	XX												X
	Securidaca coriacea Bonpl.	XX										X	X	
	Securidaca divaricata Nees & Mart.	XX										X	X	X
	Securidaca diversifolia (L.) S.F.Blake	XX										X	X	X
	Securidaca lateralis A.W. Benn.	XX										X		
	Securidaca longifolia Poepp.	XX										X		
	Securidaca macrophylla (Benth.) Wurdack	XX												X
	Securidaca marginata Benth.	XX										X		
	Securidaca paniculata Rich.	XX										X	X	X
	Securidaca pendula Bonpl.	XX										X		
	Securidaca pubescens DC.	XX										X	X	X
	Securidaca pubiflora Benth.	XX										X		
	Securidaca retusa Benth.	XX												X
	Securidaca spinifex Sandwith	XX											X	X
	Securidaca uniflora Oort	XX											X	X
	Securidaca volubilis L.	XX											X	X
Polygonaceae	Coccoloba acuminata Kunth	XX										X		
	Coccoloba ascendens Duss ex Lindau	XX										X	X	X
	Coccoloba brasiliensis Nees & Mart.	XX										X		
	Coccoloba conduplicata Maguire	XX											X	

FAMILY	SPECIES	L	SC	LSW	LV	V	HEV	HEL	HET	S	T	GU	SU	FG
	Coccoloba declinata (Vell.) Mart.	XX											X	X
	Coccoloba densifrons Mart. ex Meisn.	XX										X		
	Coccoloba excelsa Benth.	XX										X	X	X
	Coccoloba gymnorrhachis Sandwith	XX										X	X	X
	Coccoloba latifolia Poir.	XX										X	X	X
	Coccoloba lucidula Benth.	XX										X	X	X
	Coccoloba marginata Benth.	XX										X	X	X
	Coccoloba mollis Casar.	XX											X	X
	Coccoloba ovata Benth.	XX										X	X	X
	Coccoloba wurdackii R.A.Howard	XX										X		
Rhamnaceae	**Ampelozizyphus amazonicus Ducke**	XXX										X	X	X
	Gouania blanchetiana Miq.	X	X							X		X	X	X
	Gouania canescens Rich. ex Poir.	X	X							X				X
	Gouania colurnifolia Reissek	X	X							X		X		
	Gouania frangulifolia (Willd. ex Roem. & Schult.) Radlk.	X	X							X				X
	Gouania lupuloides (L.) Urb.	X	X							X				X
	Gouania polygama (Jacq.) Urb.	X	X							X		X	X	
	Gouania striata Rich.	X	X							X				X
	Gouania velutina Reissek	X	X							X		X	X	
Rubiaceae	Amaioua corymbosa Kunth	X	X							X		X	X	X
	Amaioua guianensis Aubl.	X	X							X		X	X	X
	Chiococca alba (L.) Hitchc.	X	X							X		X		X
	Chiococca multipedunculata Steyerm.	X	X							X		X		
	Chiococca nitida Benth.	X	X							X		X		X
	Diodella sarmentosa (Sw.) Bacigalupo & Cabral ex Borhidi	X	X							X		X	X	X
	Emmeorhiza umbellata (Spreng.) K.Schum.	X	X							X		X	X	X
	Guettarda aromatica Poepp.	X	X		X					X	T	X	X	X
	Hillia illustris (Vell.) K.Schum.						X					X	X	X
	Hillia parasitica Jacq.						X					X		X
	Hillia ulei K.Krause						X					X	X	
	Malanea ciliolata Steyerm.				X	X						X		
	Malanea cruzii Steyerm.				X	X						X		
	Malanea gabrielensis Müll.Arg.	X			X	X						X		
	Malanea glabra A.Rich.	X			X	X				X	T	X	X	X
	Malanea hypoleuca Steyerm.				X	X						X		X
	Malanea obovata Hochr.				X	X						X		
	Malanea sarmentosa Aubl.	X			X	X				X	T	X	X	X

FAMILY	SPECIES	L	SC	LSW	LV	V	HEV	HEL	HET	S	T	GU	SU	FG
	Malanea schomburgkii Steyerm.				X	X						X		
	Malanea sipapoensis Steyerm.				X	X						X		
	Malanea tafelbergensis Steyerm.	X			X					X	X	X	X	
	Malanea ueiensis Steyerm.				X	X						X		
	Manettia alba (Aubl.) Wernham	X			X							X	X	X
	Manettia coccinea (Aubl.) Willd.	X			X							X	X	X
	Manettia reclinata L.	X			X							X	X	X
	Notopleura guadalupensis (DC.) C.M.Taylor							X				X		X
	Randia armata (Sw.) DC.		X					X		X	X	X	X	X
	Randia asperifolia (Sandwith) Sandwith		X					X		X	X	X	X	
	Randia calycina Cham.		X					X		X	X	X		
	Randia hebecarpa Benth.		X					X		X	X	X		X
	Randia pubiflora Steyerm.		X					X		X	X			X
	Sabicea amazonensis Wernham	X	X		X			X		X			X	
	Sabicea aspera Aubl.	X	X		X			X		X		X	X	X
	Sabicea brachycalyx Steyerm.	X	X		X			X		X		X		
	Sabicea cinerea Aubl.	X	X		X			X		X			X	X
	Sabicea glabrescens Benth.	X	X		X			X		X		X		X
	Sabicea oblongifolia (Miq.) Steyerm.	X	X		X			X		X		X	X	
	Sabicea panamensis Wernham	X	X		X					X				X
	Sabicea romboutsii Bremek.	X	X		X					X			X	
	Sabicea surinamensis Bremek.	X	X		X					X		X	X	
	Sabicea velutina Benth.	X	X		X					X		X	X	X
	Sabicea villosa Willd. ex Schult.	X	X		X					X		X	X	
	Schradera nilssonii Steyerm.							X				X		
	Schradera polycephala DC.							X				X	X	X
	Schradera surinamensis Standl.							X				X	X	
	Schradera ternata Steyerm.							X				X		
	Uncaria guianensis (Aubl.) J.F.Gmel.	XXX	X							X		X	X	X
	Uncaria tomentosa (Willd. ex Schult.) DC.	XXX	X							X		X	X	X
Sapindaceae	Cardiospermum halicacabum var. microcarpum (Kunth) Blume					X						X	X	X
	Paullinia alata (Ruiz & Pav.) G.Don	X	X										X	X
	Paullinia anodonta Radlk.	X	X										X	X
	Paullinia anomophylla Radlk.	X	X											X
	Paullinia bernhardii Uittien	X	X										X	
	Paullinia bracteosa Radlk.	X	X										X	
	Paullinia caloptera Radlk.	X	X										X	X
	Paullinia cambessedesii Tria. & Planch.	X	X											X

FAMILY	SPECIES	L	SC	LSW	LV	V	HEV	HEL	HET	S	T	GU	SU	FG
	Paullinia capreolata (Aubl.) Radlk.	X	X									X	X	X
	Paullinia dasygonia Radlk.	X	X									X	X	X
	Paullinia degranvillei Acev.-Rodr.	X	X											X
	Paullinia fibulata Rich. ex Juss.	X	X											X
	Paullinia fuscescens Kunth	X	X									X	X	X
	Paullinia imberbis Radlk.	X	X											X
	Paullinia ingifolia Rich. ex Juss.	X	X									X	X	X
	Paullinia isoptera Radlk.	X	X									X		
	Paullinia latifolia Benth. ex Radlk.	X	X									X	X	
	Paullinia leiocarpa Griseb.	X	X									X		
	Paullinia lingulata Acev.-Rodr.	X	X											X
	Paullinia livescens Radlk.	X	X									X		
	Paullinia oldemanii Acev.-Rodr.	X	X											X
	Paullinia pinnata L.	X	X									X	X	X
	Paullinia plagioptera Radlk.	X	X									X	X	X
	Paullinia prevostiana Acev.-Rodr.	X	X									X		X
	Paullinia rubiginosa Cambess.	X	X											X
	Paullinia rufescens Rich. ex Juss.	X	X									X		X
	Paullinia rugosa Benth. ex Radlk.	X	X									X		X
	Paullinia sphaerocarpa Rich. ex Juss.	X	X										X	X
	Paullinia spicata Benth.	X	X										X	X
	Paullinia stellata Radlk.	X	X									X		X
	Paullinia stenopetala Sagot	X	X										X	
	Paullinia tetragona Aubl.	X	X											X
	Paullinia tricornis Radlk.	X	X									X	X	X
	Paullinia venosa Radlk.	X	X										X	X
	Paullinia verrucosa Radlk.	X	X									X	X	X
	Paullinia vespertilio Sw.	X	X											X
	Paullinia xestophylla Radlk.	X	X									X		
	Serjania adusta Radlk.	X	X									X		
	Serjania caracasana (Jacq.) Willd.	X	X									X		X
	Serjania chartacea Radlk.	X	X									X		X
	Serjania grandifolia Sagot ex Radlk.	X	X									X	X	X
	Serjania membranacea Splitg.	X	X									X	X	X
	Serjania paucidentata DC.	X	X									X	X	X
	Serjania pedicellaris Radlk.	X	X										X	X
	Serjania pyramidata Radlk.	X	X									X		X
	Thinouia myriantha Triana & Planch.	X	X									X	X	X

FAMILY	SPECIES	L	SC	LSW	LV	V	HEV	HEL	HET	S	T	GU	SU	FG
	Urvillea ulmacea Kunth	X	X									X	?	X
Schlegeliaceae	Schlegelia fuscata A.H.Gentry							X						X
	Schlegelia macrophylla Ducke							X						X
	Schlegelia paraensis Ducke							X				X	X	X
	Schlegelia roseiflora Ducke							X						X
	Schlegelia scandens (Briq. & Spruce) Sandwith							X					X	
	Schlegelia violacea Griseb.							X				X	X	X
Smilacaceae	Smilax cinnamomea Desf. ex A.DC.	X		X										X
	Smilax cordato-ovata Rich.	X		X										X
	Smilax cuspidata Duhamel	X		X										X
	Smilax domingensis Willd.	X		X								X	X	X
	Smilax guianensis Vitman	X		X								X	X	
	Smilax irrorata Mart. ex Griseb.	X		X										X
	Smilax longifolia Rich.	X		X										X
	Smilax oblongata Sw.	X		X								X	X	X
	Smilax poeppigii Kunth													X
	Smilax quinquenervia Vell.	X		X										X
	Smilax santaremensis A.DC.	X		X								X	X	X
	Smilax saulensis J.D.Mitch.	X		X										X
	Smilax schomburgkiana Kunth	X		X								X	X	X
	Smilax talbotiana A.DC.	X		X								X		
Solanaceae	Lycianthes heteroclita (Sendtn.) Bitter	X			X								X	
	Lycianthes nitida Bitter	X			X							X		
	Lycianthes pauciflora (Vahl) Bitter	X			X							X	X	X
	Markea coccinea Rich.							X				X	X	X
	Markea formicarum Dammer							X				X	X	X
	Markea longiflora Miers							X				X	X	X
	Markea sessiliflora Ducke							X				X	X	X
	Solandra longiflora Tussac							X				X	X	X
	Solandra paraensis Ducke							X						X
	Solanum coriaceum Dunal	X			X							X	X	X
	Solanum pensile Sendtn.	X			X							X	X	X
	Solanum rubiginosum Vahl	X			X							X	X	X
Trigoniaceae	Trigonia candelabra Lleras	X	X							X		X		
	Trigonia coppenamensis Stafleu	X	X							X		X		
	Trigonia hypoleuca Griseb.	X	X							X		X	X	X
	Trigonia microcarpa Sagot ex Warm.	X	X							X		X	X	X
	Trigonia nivea Cambess.	X	X							X		X		

FAMILY	SPECIES	L	SC	LSW	LV	V	HEV	HEL	HET	S	T	GU	SU	FG
	Trigonia reticulata Lleras	X	X							X		X		
	Trigonia subcymosa Benth.	X	X							X		X	X	
	Trigonia villosa Aubl.	X	X							X	X	X	X	X
Urticaceae	Coussapoa angustifolia Aubl.								X	X	X	X	X	X
	Coussapoa asperifolia Trécul								X	X	X	X	X	X
	Coussapoa ferruginea Trécul								X	X	X		X	X
	Coussapoa latifolia Aubl.								X	X	X	X	X	X
	Coussapoa leprieurii Benoist								X	X	X			X
	Coussapoa microcephala Trécul								X	X	X	X	X	X
	Coussapoa parvifolia Standl.								X	X	X	X		
	Coussapoa trinervia Spruce ex Mildbr.								X	X	X	X		
	Coussapoa viridifolia Cuatrec.								X	X	X	X		
Verbenaceae	Petrea blanchetiana Schauer	XX								X				X
	Petrea bracteata Steud.	XX								X		X	X	X
	Petrea macrostachya Benth.	XX								X		X	X	X
	Petrea sulphurea Jans.-Jac.	XX								X				X
	Petrea volubilis L.	XX								X		X	X	X
Violaceae	**Corynostylis arborea (L.) S.F.Blake**	XX										X	X	X
Vitaceae	Cissus descoingsii Lombardi			X	X	X						X		
	Cissus duarteana Cambess.			X	X	X								X
	Cissus erosa Rich.			X	X	X						X	X	X
	Cissus haematantha Miq.			X	X	X						X	X	X
	Cissus palmata Poir.			X	X	X						X	X	X
	Cissus spinosa Cambess.			X	X	X						X	X	X
	Cissus subrhomboidea (Baker) Planch.			X	X	X						X	X	
	Cissus surinamensis Desc.			X	X	X							X	X
	Cissus ulmifolia (Baker) Planch.			X	X	X							X	
	Cissus venezuelensis Steyerm.			X	X	X						X		
	Cissus verticillata (L.) Nicolson & C.E.Jarvis			X	X	X						X	X	X

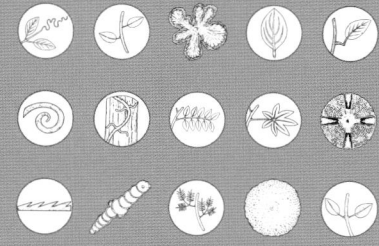

GLOSSARY

TERM	EXPLANATION
abaxial	Facing away from the stem (central axis), lower surface of a leaf (opposite of *adaxial*).
abrupt	Terminating suddenly.
achene	A dry, indehiscent, single-seeded fruit, e.g. Asteraceae.
actinomorphic flower	A flower with radial symmetry; can be cut into equal pieces like a pizza or a star-shape (also termed "regular").
aculeate	Prickly; covered with prickles.
acuminate	Apex of leaf tapering very gradually at tip; long acuminate indicates a very long, thin "drip tip".
acute	Apex of leaf pointed, with more or less straight sides.
adaxial	Facing towards the stem (central axis), upper surface of a leaf (opposite of *adaxial*).
adnate	Dissimilar parts fused or adjoining.
adventitious/aerial root	Root emerging from node on stem; may hold a climbing stem onto surface or absorb water/nutrients.
alternate leaves	Each node with one leaf along the length of a stem.
androgynophore	Stalk supporting the androecium and gynoecium in some flowers; Passifloraceae.
animal dispersal	Seeds of a plant species dispersed by birds, bats, ants, agoutis, monkeys or other animals. May also refer to secondary dispersal when one animal drops or loses seeds and they are moved by another animal. Does not include seed predation.
anther	The pollen-bearing part of a flower's stamen.
apex	The tip; the point farthest from the point of attachment.
apiculate	Leaves, petals etc. ending abruptly in a short, flexible point.
apocarpous	A flower with free (unfused) carpels as subunits of the gynoecium; e.g. Annonaceae.
appressed	Structures flattened or lying flat; commonly used here for hairs.
appressed-sericeous	Hairs lying flat and appearing "silky"; e.g. many Malpighiaceae species.
aril	Thin fleshy covering of whole or part of seed inside the fruit (inside pericarp); usually brightly colored.
aristate	Bearing an awn or bristle at the tip.
article	Section of a fruit separated from others by a constricted joint.
asclepiad	A member of the APG III delimited family Apocynaceae, subfamily Asclepiadoideae; traditionally a family of its own, the Asclepiadaceae.
asperous	Surfaces with a rough texture like a cat's tongue (synonym: scabrous); e.g. Dilleniaceae, Verbenaceae.
attenuate	Leaf apex or base gradually tapering.
auriculate	Leaf or leaflet with a pair of ear-shaped lobes at the base.
awl-shaped	Short, narrowly triangular, and sharply pointed like an awl.
axillary	Located at or in the axil of a leaf, leaf vein, or other plant part.
axil	The upper angle between a stem and petiole/leaf or primary and secondary veins.
axis	The central column or stem of a whole plant or any plant part, including a leaf, branch, inflorescence, flower, or fruit.
ballistic dispersal	Seed dispersal via an explosive mechanism in certain fruits, usually due to a release in tension upon drying.
banner	The upper and usually largest petal of a legume flower; Leguminosae-Papilionoideae.
barbellate	Minutely barbed, with tiny spear- or hook-like hairs or structures.
basal	Arising at base; e.g. basal nerve is a nerve arising at (or very close to) the base of lamina.
base	The end of a structure closest to a central axis or the ground; the part of the leaf blade

TERM	EXPLANATION
	closest to the leaf stalk (petiole or petiolule) or the point of attachment.
basilaminar	The area near the base of a leaf blade, often in reference to raised gland pairs; e.g. Euphorbiaceae, Malpighiaceae.
berry	Fleshy fruit with 1-several seeds immersed in soft flesh, seeds without a hard outer layer; e.g. Solanaceae (see *drupe*).
bifid	Deeply two-cleft or two-lobed, usually from the tip.
bilateral	Arranged on two sides, as leaves on a stem.
bilaterally symmetrical	A flower or other structure that produces a mirror image on two sides when a line is drawn through the center (synonym: zygomorphic); e.g. Orchidaceae flowers. (See *radially symmetrical* and *actinomorphic*.)
bilobed	Divided into two lobes.
bi-pinnate	Leaves compound, twice pinnately divided; e.g. Leguminosae-Mimosoideae.
bisexual	Flowers with both male and female reproductive organs.
bi-trifoliolate	Leaves compound, twice tri-foliolate divided.
blade	Broad, flat part of a leaf or leaflet, without the stalk (synonym: lamina).
bract	A general name for flaps of tissue (modified leaves) that can occur on vegetative or re-productive plant parts. Often associated with inflorescences and sometimes with bright colors mimicking flower petals or serving as nectaries; e.g. Asteraceae, Malpighiaceae, Marcgraviaceae.
bract collar	Specialized structures on the male or female reproductive cones of Gnetum species.
bracteole	A type of bract smaller than the standard bract.
branchlet	A small branch or stem that is part of the younger leafy growth of a plant and not a mature or woody stem.
branchlet climber	Climbing or clambering via horizontal or recurved branchlets.
bud	An immature shoot sometimes covered with tough scales or stipules, or an undevel-oped flower, often protected by the calyx or bracts.
bullate	Leaf surface raised in a series of domes between the veins on the upper surface.
caducous	(Stipule) short-lived, early deciduous.
calyx	Collective term for the sepals of a flower.
calyptra	A cap or covering over the reproductive structures of a flower (such a flower is termed 'calyptrate').
campanulate	Bell-shaped.
canescent	Leaf underside with gray or white color due to a layer of very short, dense and fine hairs (invisible to the naked eye).
capitate	Head-like.
capitulum	An inflorescence with a dense cluster of unstalked flowers borne on a flattened or domed receptacle, e.g. Asteraceae.
capsule	A common type of dehiscent fruit derived from two or more united carpels, splits open along several lines or pores to release seeds.
carpel	The basic unit of the gynoecium. The terms 'pistil' and 'carpel' are synonymous when the pistil or pistils in a flower are simple – comprising ovary, style and stigma. When the pistil is compound, carpels form the sub-units (syncarpous). Internally, the ovary may have walls that reveal the original number of carpels or be reduced, without walls, and appearing simple.
carpellate	A flower with only female parts (carpels); synonym: pistillate.
cartilaginous	Hard and tough but flexible.

TERM	EXPLANATION
caruncle	A protuberance or appendage near the hilum of a seed.
cauliflorous	Flowers and fruits borne on the trunk or woody stem of a plant.
channeled	Surface grooved.
chartaceous	Surface papery in texture.
ciliate	With a fringe of hairs along leaf or petal margins. Like eyelashes, rather than bearded.
cincinnus	A dense helicoid cyme; e.g. Boraginaceae (Tournefortia).
cirrus	A whip-like extension of the (pinnate-compound) leaf midrib, with 1-2 robust hooks at nodes, of the climbing palm genus, Desmoncus.
claw	A narrow, stalk-like basal portion of a petal, sepal or bract. Clawed petals have a narrow base and broad apex.
colleter glands	Tiny stipule- or finger-like glandular structures in the leaf axils and flowers of some plant families; e.g. Apocynaceae. Not to be confused with true stipules.
compound	Any plant part made up of multiple sub-units, such as leaves, inflorescences, pistils and fruits.
compound leaf	A leaf with more than one leaf blade on the axis.
concolorous	The two sides of a surface, such as a leaf, having similar, non-contrasting colors (see *discolorous*).
connate	Fusion of like parts, such as the fusion of petals or stamen filaments into a tube.
connivent	Adjoining, but not actually fused or united.
cordate	Heart-shaped, may refer to the leaf base or the entire leaf.
coriaceous	Tough, leathery.
corolla	Petals as a whole, especially when they are united to form a tube.
corolla lobe	Petal-like structures extending beyond the apex of a corolla tube; tiny to much larger than the tube itself. (see also *limb.*)
corona	A distinct ring of tissue arising from the corolla, perianth or filaments of a flower; e.g. Passifloraceae.
corymb	A raceme-like inflorescence in which the stalks of the lower flowers are longer than those of the flowers above, bringing all flowers to about the same horizontal level.
costate	A ribbed surface.
crenate	Margin round- or blunt-toothed, scalloped.
crenate-dentate	Margin variable within a leaf or between leaves, blunt- to sharp-toothed.
crenulate	Margin with very small rounded teeth.
cross-section (x-s)	The view when a stem or other structure is cut across the middle, at a 90-degree angle to the length in elongate structures.
cuneate	Leaf base wedge-shaped; becoming gradually thinner at base towards petiole.
cuspidate	Leaf apex abruptly tapering into a sharp, rigid acute point.
cylindrical	Elongate and more or less "round" in cross-section.
cyme	Branched inflorescence without a single main axis, but where each flower ends an axis and subsequent flowers develop from branches.
cystolith	Calcium carbonate crystals embedded in leaves – visible with a hand lens as white streaks on vegetation.
deciduous	(Stipule) quickly falling off, short-lived; synonym: caducous.
decurrent	A structure (petiole, leaf blade) extending downwards from the point of insertion.
decussate	Of opposite leaves, when the orientation of successive pairs are at right angles to each other.
dehiscent	Fruits that split open when ripe to release seeds; e.g. capsule, legume (opposite of *indehiscent*).

TERM	EXPLANATION
deltoid	Leaf shaped like an equilateral triangle.
dendroid	Hairs with tree-like branching (dendroid hairs); often difficult to distinguish from stellate hairs.
dentate	Leaf margin toothed; strictly the teeth should be equal-sided and pointing directly outwards (see *serrate*).
denticulate	Leaf margin small-toothed.
diadelphous	Stamens two-ranked, with two pairs of unequal length filaments.
dichasium	A cyme with two branches from each node.
dichotomous	Branching by forking into two equal branches, as if splitting perfectly into half.
didynamous	When fertile stamens occur as two pairs in a flower - one pair longer and one pair shorter. Common in lianas with showy tubular flowers such as the Bignoniaceae.
dioecious	"Two-housed"; plant species with separate male (staminate flowers) and female (carpellate flowers) individuals.
discolorous	The two sides of a surface, such as a leaf, having contrasting colors (see *concolorous*).
dissected	Deeply divided into many narrow segments.
distichous	Leaves two-ranked; regularly arranged one above the other in two opposite rows, one on each side of the stem.
distal	The part of a plant furthest from the central axis or branch; ex: the distal end of a petiole where it meets the leaf blade (opposite: *proximal*).
domatia	Tufts of hair or small pockets in vein axils thought to house insects. (sing. *domatium*)
double pulvinus	With a swelling at the base and apex of the petiole; e.g. Malvaceae.
drupe	A fleshy, indehiscent fruit with a stony endocarp surrounding a usually single seed.
elliptic	Leaf blade shape broadest at the middle and narrower at the two equal ends.
emarginate	Tip of leaf with a slight notch (inverse of *apiculate*).
endocarp	Innermost layer of fruit wall (pericarp); in a drupe, the stony layer.
endosperm	The nutritive tissue surrounding the embryo of some types of seed.
entire	Margin smooth; evenly curved or straight, without teeth or waves.
epicarp	Same as *exocarp*.
epipetalous	Attached to the petals.
epipetiolar	A stipule attached to the petiole; e.g. Malpighiaceae (Hiraea).
erose	Irregularly, shallowly toothed and/or lobed margins; appearing gnawed.
even-pinnate	A pinnately-compound leaf without a terminal leaflet.
exocarp	The outermost layer of the fruit wall, usually skin-like.
exserted	Protruding stamens, longer than corolla tube.
exudate	Watery, latex-like, clear or colored substances emitted by plants upon cutting herbaceous or woody parts.
farinose	Covered with a mealy, powdery substance.
fascicle	Cluster or bundle, especially a dense cluster of flowers arising in one place and apparently unstalked.
fenestrate	Bark with a network of raised areas and window-like openings in-between.
ferruginous	Rust- (iron) colored; i.e. reddish- or orange-brown.
filament	A stalk supporting the pollen-bearing anther of a stamen.
flat glands	Small, circular, flat glands present on the leaf blade, often marginal or sub-marginal (near the edge of the leaf).
floret	Specialized terminology for the small flower units within Asteraceae flower heads.
foetid	Having a disgusting smell, like rotting meat or excrement.

TERM	EXPLANATION
foliaceous	Leaf-like (e.g. stipules, bracts) in shape and texture.
foliate	Having leaves; leaf-like.
foliolate	Referring to the number of blades per leaf (1-foliolate, 2-foliolate, 3-foliolate, etc.)
follicle	A dehiscent, many-seeded fruit developing from one carpel, splitting along one line (legumes split along two lines).
free	Not attached to other organs.
funnelform	Gradually widening from base to apex; funnel-shaped.
geniculate	With abrupt knee-like bends and joints.
glabrous	Hairless or appearing hairless from a distance.
gland	A structure, within or on the surface of a plant, with a secretory function.
glands	Structures of many shapes and sizes on the surface of a plant that secrete substances; they commonly appear as knobs, domes, craters, or flat discs.
glandular	Covered with secretory or excretory glands.
glaucous	Surface (of leaf) with non-shiny (matte) texture like fine dust and pale blue, whitish or pinkish, due to thin waxy coating.
globose	More or less spherical, shaped like a ball. Most often a term for fruits.
gynandrous	With the stamens adnate (fused or adjoined) to the pistil.
gynoecium	All of the carpels or pistils of a flower, collectively.
gynophore	A stalk bearing the gynoecium above the level of insertion of the other floral parts.
gynostemium	A structure formed from the fusion of anthers and stigmatic region of the gynoecium, as in the Asclepiadoideae.
head	An inflorescence with unstalked flowers in a dense cluster; e.g. Asteraceae.
helicoid	Coiled like a spiral or helix, as in some one-sided cymose inflorescences in the Boraginaceae.
hexagonal	Six-sided (certain Bignoniaceae species have 6-sided stems in cross-section).
hilum	A scar on the edge of a seed where it was attached to the fruit; e.g. Leguminosae (Dioclea).
hirsute	With long, coarse, and fairly stiff hairs; hairs erect, completely covering surface.
hispid	With bristles or (very) long, stiff, bristly hairs. Hairs stiffer and slightly sparser than hirsute.
hook-climber	A plant that climbs by hooks (modified stipules, branchlets, inflor. branches); e.g. Loganiaceae (Strychnos); Rubiaceae (Uncaria).
hyaline	See *membranous*.
hypanthium	A cup-shaped basal structure in flowers with an inferior ovary – formed from the calyx, corolla, and androecium and often enclosing the gynoecium.
imbricate	Overlapping petals or other structures, like tiles on a roof.
imperfect	With either stamens or pistils, but not both; unisexual.
indehiscent	A fruit not splitting open at maturity, e.g. drupe, berry, achene (opposite: *dehiscent*).
inferior ovary	An ovary located below the point of attachment of the calyx, corolla and stamen whorls; remains of the perianth often visible at the apex of mature fruits of plant taxa with an inferior ovary.
inflorescence	Several/many flowers arranged on a single leafless axis, sometimes including many leaf-like bracts and bracteoles.
interpetiolar stipule	On opposite-leaved plants, stipules located on the stem between the two petiole bases, often large and covering the stem at the node, often short-lived and leaving a stipule scar; e.g. Rubiaceae (Uncaria, Sabicea).

TERM	EXPLANATION
intersecondary veins	Fine veins (3°) running parallel and in-between larger secondary veins.
intrapetiolar stipule	On alternate- or opposite-leaved plants, stipules located in or adjacent to the leaf axil and petiole base, often small, linear, short-lived, leaving scars.
irregular	A flower with parts of unequal shape or size; also sometimes including flowers with bilateral symmetry.
keel	The two lower united, boat-like petals of a legume flower (Leguminosae-Papilionoideae).
labium	The lower lip of a 2-lipped corolla (pl. *labia*).
lamina	The leaf blade (not including petiole).
lanate	Woolly, covered with long, curled and densely matted hairs.
lanceolate	Lance-shaped; much longer than wide, with the widest point below the middle.
latex	Paint-like exudate, white or colored, from cut herbaceous or woody plant parts; e.g. Apocynaceae, Celastraceae, Convolvulaceae.
leaflet	The leaf blade and stalk in a compound leaf.
legume	A dry, dehiscent fruit derived from a single carpel and usually opening along two lines of dehiscence, as a pea pod.
lenticel	Protuberances on woody stems and branchlets, of many shapes, colors, sizes and heights. Often oriented in horizontal or vertical lines.
lenticellate	With few to many lenticels.
lepidote	Covered in small, scurfy scales.
limb	The expanded part of a petal or tepal; the expanded part of a fused corolla, as distinct from the tube.
linear	Resembling a line; long and narrow with more or less parallel sides.
lip	One of the two projections or segments of an irregular, two-lipped corolla or calyx; a labium.
lobe	Part of a leaf, corolla or other organ, often rounded, formed by incisions.
lobed	A division of a leaf or corolla, often rounded; lobes shallow to deep.
locule	Enclosed compartments within an ovary or anther.
longitudinal section (l-s)	The view when a stem or other structure is cut along its length.
lustrous	Shiny or glossy.
Malpighiaceous hairs	T-shaped or seesaw-shaped hairs, with two long ends and a central stalk. Visible with a hand lens, commonly seen on Malpighiaceae species petioles.
margin	The edge of a leaf blade.
mealy	With the consistency of meal; powdery, dry, and crumbly.
membranous	Like a membrane. i.e. thin, soft, flexible, and more or less translucent. Leaves thinner than the 'papery' leaf texture type.
mericarp	A one-seeded portion of a many-seeded fruit that breaks at maturity into units; Leguminosae (Entada).
mesocarp	The middle layer in a fruit wall, often fleshy and edible in a drupe, sometimes missing.
midrib	The central or middle rib or vein of a leaf.
monocarp	The single-seeded, free subunits of a single flower in families with apocarpous flowers; (Annonaceae; Guatteria scandens).
monocot stem	Stems lacking true wood and with scattered vascular bundles rather than typical dicot growth (e.g. Arecaceae - Desmoncus). In dicot climbers, stems differ from both monocots and typical dicot plants.
monoecious	"One-housed"; with separate male and female flowers, but both sexes on one plant.
mucro	A short, sharp, abrupt point, usually at the tip of a leaf.

TERM	EXPLANATION
mucronate	Tip of leaf with a stiff narrow process (mucro) on end of midrib.
nectar disc	Sugar-secreting gland or glands at the base of the ovary in the form of a disc.
net-veined	In the form of a network; synonym: reticulate.
node	The site on the stem of leaf insertion, stipules, glands, and axillary buds which may grow into leaves, branches or inflorescences. The basic modular growth structure of plants consists of internodes and nodes.
non-asperous	Smooth.
nut	A hard, dry, indehiscent fruit, usually with a single seed.
nutlet	A small nut; one of the lobes or sections of the mature ovary of some members of the Boraginaceae, Verbenaceae, and Labiatae (Lamiaceae).
obcordate	Leaf shape inversely cordate, with the attachment at the narrower end; sometimes refers to any leaf with a deeply notched apex.
oblanceolate	As *lanceolate,* but broadest part is beyond the middle of the longer axis.
oblique	Having an asymmetrical base.
oblong	Leaf shape two to four times longer than broad with nearly parallel sides.
obovate	Leaf shape inversely ovate, with the attachment at the narrower end.
obtuse	Flat-ended or almost so (apex or base).
ochrea	A sheathing stipule encircling the apical bud and the node; e.g. Moraceae (Ficus); Polygonaceae (Coccoloba).
ochreole	A very small ochrea on plant subunits, subtending the inflorescence or flowers.
odd-pinnate	Pinnately compound with a terminal leaflet rather than a pair of leaflets or a tendril, so that there is an odd number of leaflets.
opposite	Each node with two, opposed leaves along the length of a stem.
orbicular	A flat shape, circular in outline.
oval	Evenly broad, symmetrical; broadly elliptic, the width over one-half of the length.
ovary	The part of the gynoecium containing ovules.
ovate	Egg-shaped in outline and attached at the broad end (applied to plane surfaces, such as a leaf).
ovoid	Egg-shaped (three-dimensional) structure. With wider portion at base; 3-dimensional object, ovate in all sections through long-axis.
ovule	Organ (inside the ovary in Angiosperms, open in Gymnosperms) that develops into the seed after fertilization.
oxidizing	Transitions in living plant characters due to exposure to oxygen upon cutting. Often diagnostic changes in sap color.
palmate	Veins or leaflets arising from a common point, like the fingers of a hand.
palmate-lobed	Simple leaf with lobes and main nerves diverging like fingers on a hand; more than three distinct lobes, whose main nerves meet at the base. If the leaf has leaflets rather than lobes, then it is palmate-compound.
panicle	A compound, much-branched inflorescence, where at least some of the branches have more than one flower.
papilionaceous	Butterfly-like, as the irregular corolla of a pea, with a banner petal, two wing petals, and two fused keel petals.
papilla	A short, rounded nipple-like bump or projection (pl. *papillae*).
papillate	Having a finely bumpy surface (see *papillae*).
pappus	Hairs, simple or feathery, or sometimes bristles or scales formed; terminology for Asteraceae flowers and achenes.

TERM	EXPLANATION
papyraceous	Papery in texture and usually color.
parallel-veined	With the main veins parallel to the leaf axis or to each other. Compare net-veined, pinnate-veined and palmate veined.
patelliform (gland)	Of a gland shaped like an open saucepan with thickened, raised edges.
pedicel	Flower stalk. The stalk of a single flower in an inflorescence.
peduncle	Inflorescence stalk; the central stalk holding all of the inflorescence branches or the stalks of inflorescence sub-units.
pellucid-punctate (gland)	Circular, flat glands on leaf blades that are translucent (like portholes) and often contain essential oils.
peltate	A structure with a stalk attached to under-surface near middle, not edge. May refer to leaf blades or to dome-shaped glands on plants.
perianth	A term including both calyx and corolla in a flower, especially used when there is little difference between petals and sepals (= tepals).
pericarp	The wall of the fruit.
persistent	Remaining attached, not falling off, even though the original function of the organ has ceased; e.g. of petals not falling after flowering; e.g. persistent calyx in Olacaceae (Heisteria).
petal	An individual segment or member of the corolla.
petiole	The leaf stalk by which the midrib or main axis (rachis) of a leaf is attached to the stem.
petiole-climber	A plant climbing via a clasping or jutting petiole.
petiolule	The stalk by which the midrib of a leaflet is attached to the axis of a compound leaf.
phloem	Tissue which conducts sugars and other nutrients and often alternates with wood within mature liana stems to form unique patterns visible to the naked eye.
phyllary	An involucral (subtending flower head) bract of the Asteraceae.
pilose	Hairy, covered with long, weak, soft, 'shaggy,' not appressed hairs.
pinna	One of the primary divisions or leaflets of a pinnate leaf (pl. *pinnae*).
pinnate	A once-compound leaf with leaflets borne along a rachis (pinnate = feather-like).
pinnate-lobed	A leaf cut deeply into lobes that diverge from the midrib.
pinnipalmate	Intermediate between pinnate and palmate, with the main veins palmate at base and becoming pinnate towards the apex.
pistil	The female reproductive organ of a flower, typically consisting of a stigma, style, and ovary. May be composed of one or more carpels.
pit	The stony endocarp of a drupe.
pod	Any dry, dehiscent fruit, especially a legume or follicle.
pollinium	A max of waxy pollen grains transported as a unit in many members of the Orchidaceae and Apocynaceae (Asclepiadoideae) (pl. *pollinia*).
poricidal	Opening by pores, as in a poppy or Melastomataceae capsule.
prophyll	Stipule-like outgrowths of the axillary bud at nodes in many Bignoniaceae lianas.
proximal	The part of a plant closest to the central axis or branch; ex: the proximal end of a petiole is where it is attached to the stem; opposite: *distal*.
puberulent	Minutely pubescent; with fine, short hairs.
puberulous	See *puberulent*.
pubescent	Hairy, but neither densely nor roughly so.
pulvinus	A swelling usually at the base of the petiole; cylindrical in Leguminosae; globose in Menispermaceae; sometimes swollen at the base and apex of the petiole and then termed a 'double pulvinus' (Malvaceae).

TERM	EXPLANATION
punctate	Surface dotted with pits, translucent, sunken glands or colored dots.
pyramidal	Tetrahedral; pyramid-shaped.
pyriform	Pear-shaped.
raceme	Unbranched, elongated inflorescence with no terminal flower and with the lowest flowers the most developed; individual flowers with pedicel.
rachis	The main axis of a pinnate leaf, 2-3 pinnate leaf or an inflorescence.
raised gland-pairs	Raised glands commonly occuring in pairs on the petiole or lamina base (basilaminar) in certain families; e.g. Euphorbiaceae, Malpighiaceae, Passifloraceae.
ranalean	A "primitive" scent of some plant families, e.g. Annonaceae, Lauraceae, Myristicaceae.
receptacle	The often dome-shaped portion of the flower stalk bearing flower parts or the peduncle-bearing flowers in Asteraceae.
regular	Radially symmetrical; said of a flower in which all parts are similar in size and arrangement on the receptacle.
reniform	Kidney-shaped.
reticulate	Arranged like strings in a net; net-veined.
revolute	Rolled downwards or backwards; of a leaf, with the margins rolled towards the abaxial (lower) surface.
rhombic	Diamond-shaped.
ribbed	With prominent ribs or veins.
root-climber	A plant that attaches to trees and climbs with small aerial roots growing from the side of the stem. This climbing mechanism is common in hemi-epiphytic climbers.
rotate	Disc-shaped; flat and circular, as a sympetalous corolla with widely spreading lobes and little or no tube.
rounded	(Leaf apex with) margins and apex forming a smooth arc.
rugose	Wrinkled.
salverform	With a slender tube and an abruptly spreading, flattened limb; trumpet-shaped.
samara	A dry, indehiscent, winged fruit.
scabrous	With a rough texture like a cat's tongue (synonym: asperous); e.g. Dilleniaceae; Verbenaceae (Petrea).
scalariform	With the appearance of ladder-like steps; e.g. Melastomataceae with secondary veins palmate (arcuate) and tertiary veins scalariform.
scandent	A shrub unable to fully support itself and climbing or clambering weakly over other plants and the ground in mostly horizontal directions.
schizocarp	A dry fruit formed from more than one carpel but breaking apart into single-seeded units (mericarps) when ripe.
scorpioid	Shaped like a scorpion's tail, as in some coiled cymes; a determinate inflorescence with a zigzag rachis.
sclereid	A thick-walled lignified plant cell, often branched in shape.
sepal	A sub-unit of the outermost whorl of a flower, the calyx, especially if green and dissimilar to petals.
septifragal	A fruit with the valves separating from the inner wall upon splitting open at maturity.
sericeous	Silky, with a covering of long, appressed, soft, flexible, straight and glossy hairs.
serrate	Saw-toothed; usually with teeth arching forwards (but cut 1/16-1/8 distance to midrib or midvein).
sessile	Attached directly, without a supporting stalk, as a leaf without a petiole.
setose	Covered with bristles.

TERM	EXPLANATION
sheath	A tubular or rolled structure often enclosing another organ; e.g. Arecaceae (Desmoncus) with base of leaf enclosing stem below junction.
silky	Silk-like in appearance or texture; sericeous.
simple leaf	A leaf with only one blade. It may be toothed, scalloped, or lobed but is never divided all the way to the leafstalk. See also *compound leaf*.
sinuate	With a strongly wavy margin.
smooth	With an even surface; not rough to the touch.
solitary	Flowers occurring singly and not borne in a cluster or group.
spadix	A spike with small flowers crowded on a thickened axis.
spathe	A large bract or pair of bracts subtending and often enclosing an inflorescence; e.g. Araceae with spathe and spadix.
spatulate	Like a spatula in shape, with a rounded blade above gradually tapering to the base.
spike	Unbranched inflorescence; individual flowers without stalks (see *raceme*).
spine	Sharp, needle-like projection from below bark, or needle-like structures leaf margins.
stamen	The male reproductive organ of a flowering plant, consisting (usually) of stalk (filament) and a pollen-bearing portion (anther).
staminode	A sterile stamen that produces no pollen and is often reduced in size and shape.
stele	The central core of the stem and root of a vascular plant, consisting of the vascular tissue (xylem and phloem) and associated supporting tissue.
stellate	Hairs star-shaped, with a cluster of hairs radiating from a point; e.g. Euphorbiaceae.
stem x-s pattern	Woody stems in liana families often have unique cross-section patterns that are diagnostic to family level in many cases. These patterns are described, categorized, illustrated and named in the illustrated glossary section of this book.
stigma	The pollen-receptive, usually sticky surface in the female part of a flower (carpel or group of fused carpels). Usually a point or small head at the summit of the style, though shape varies between species.
stigma internal	Specialized terminology for the Malpighiaceae focuses upon the shape/location of the stigma as "internal" (on the side towards the center), terminal or otherwise.
stipe	A stalk supporting a structure, as the stalk attaching the ovary to the receptacle in some flowers.
stipel	Thread-like structure in axil of leaflets on some compound leaves; e.g. Leguminosae.
stipule	Leaf-like or other small structure close to petiole at leaf node, often in opposite pairs, extending across stem in-between petioles, or as a complete sheath around stem.
strap-shaped	Elongated and flat.
striate	With many fine parallel longitudinal (lengthwise) lines.
strigillose	Minutely strigose.
strigose	Covered with stiff hairs that are slanting rather than erect, usually with a bulbous base.
style	Thin stalk or narrowed portion in female part of flower, connecting the stigma to the ovary.
sub	Not quite; almost; e.g. sub-cordate (slightly cordate).
subulate	Awl-shaped; narrow and tapering evenly from a narrow base to a fine point.
sulcate	With longitudinal grooves or furrows.
superior ovary	An ovary that is attached above the point of attachment of the other floral whorls.
suture	A line of fusion; the line of dehiscence of a fruit.
sympetalous	A corolla with the petals fused to form a tube, sometimes splitting at the apex into lobes.
syncarpous	Of or pertaining to a syncarp; with united carpels (compare *apocarpous*).

TERM	EXPLANATION
tendril	A slender, specialized stem, leaf or petiole with a twining threadlike shape, used to grasp support for climbing.
tendril-climber	A climbing plant grasping objects and lifting itself higher via tendrils.
tendril-like-shoot	A climbing plant with the apical shoots twisting and grasping onto objects in a similar way to a tendril-climber. This is in contrast to a twining climber, which climbs via a single spiraling shoot.
tepal	A segment of a perianth which is not differentiated into calyx and corolla; a sepal or petal.
terminal	Situated at the tip, apex or distal end.
ternate	Compound leaves with three leaflets attached to end of petiole. See *trifoliate* and *trifoliolate*.
tetragonal	Four-sided (certain Bignoniaceae species have 4-sided stems in cross-section).
thyrse	A compact, cylindrical, or ovate panicle with an indeterminate main axis and cymose sub-axes.
tomentellous	See *tomentulose*.
tomentose	With very dense short, matted hairs and soft texture, like felt.
tomentulose	Slightly tomentose.
translucent	Having a surface that light can shine through to a strong or slight degree.
trifoliate	Compound leaves with three leaflets attached to end of petiole (ternate). See *trifoliolate*.
trifoliolate	Compound leaves with three leaflets attached to end of petiole (ternate). See *trifoliate*.
truncate	Ending very abruptly as if cut straight across (of lamina tips as well as bases).
tubercle	A small tuber-like swelling or projection.
tuberculate	With small, smooth, blunt projections. With a warty surface. Tubercle is a small blunt outgrowth.
tubular	With the form of a tube or cylinder.
twiner	A plant with apical shoots that move in spirals to find and attach to objects.
umbel	An umbrella-shaped, racemose inflorescence, with the pedicels arising more or less from a common point.
umbelliform	An inflorescence with the general appearance, but not necessarily the structure, of a true umbel.
unarmed	Lacking spines, prickles, or thorns.
undulate	Wavy, but not so deeply or as pronounced as *sinuate*.
unisexual	With either male or female reproductive parts, but not both.
urceolate	Pitcher-like; hollow and contracted near the mouth like a pitcher or urn.
urticating	With stinging hairs.
valvate	With flower parts touching but not overlapping in bud; also for organs dehiscing with valves that do not overlap.
valve	One of the flaps/parts/sides opening on a dehiscent fruit.
vein	See *venation*. Secondary and tertiary veins, see figure.
velutinous	Velvety: densely covered with soft, short (or long), not matted erect hairs.
velvety	Like the cloth called velvet; a carpet of very dense short and erect silky hairs, forming a very soft carpet.
venation	In leaf blades, the system or pattern of the main veins (1°, 2° palmate, pinnate, parallel) and of finer veins (3°, net-veined, parallel and at right angle or slanted towards mid-vein).
verticillate	Whorled, several similar organs at the same point of an axis.
villose	See *villous*.
villous	Bearing long, soft, shaggy, but unmatted hairs.

TERM	EXPLANATION
vine	A herbaceous climber with the stem not self-supporting, but climbing or trailing on some support. Lianas are woody vines.
volcano-crater-shaped glands	Type of gland with sloping edges, low rim, and open-pit/caldera within. Common in the genus Adenocalymma (Bignoniaceae).
water-dispersed	A plant species with fruits and/or seeds dispersed by water, seeds often with hollow chambers that allow them to float.
wavy	Leaf margin with smooth, shallow curves.
whorl	A ring-like arrangement of similar parts arising from a common point or node; a verticil.
whorled	An arrangement with more than two leaves or branches at the same node.
wind-dispersed	A plant species with fruits and/or seeds dispersed by wind; fruits or seeds most often winged or very tiny.
wing	A thin, flat margin bordering or extending from a structure; one of the two lateral petals of a papilionaceous corolla.
woolly	With long, soft, entangled hairs; lanate.
zygomorphic	A flower or other structure that produces a mirror image on two sides when a line is drawn through the center (synonym: bilaterally symmetrical); e.g. Orchidaceae flowers. (Compare with actinomorphic and radially symmetrical.)
x-s	See *cross-section*.

Data sources include the author's original data and (with permission) direct or adapted entries from the 'plant characteristics' glossary of Dr. William Hawthorne at Kew Royal Botanic Gardens.

Common Name	Scientific Name(s)	Common Name	Scientific Name(s)
A		arimi arokë (Tr)	Pyrostegia venusta
abago-maka (Sa)	Smilax oblongata	ariminaimë (Tr)	Annona haematantha,
abo(ho), (white) (Ar)	Serjania paucidentata		Fridericia patellifera
abonenge tatai (Sa)	Mansoa alliacea	aripoimë (Tr)	Machaerium lunatum
abuyamibia (Ar)	Barnhartia floribunda,	aritaimë (Tr)	Paullinia dasygonia
	Mansoa alliacea	aritaritamë (Tr)	Cissampelos andromorpha
agaimargaiyik (Pat)	Senna quinquangulata	aruakabo (Ar)	Marcgravia coriacea,
agbagomaka-tatai (Ar)	Smilax schomburgkiana		Marcgravia pedunculosa
agbo-ndeku (Sa)	Tilesia baccata	asitaremu (Ca)	Desmoncus polyacanthos
aguwago-maka (Bo)	Smilax oblongata	asrikatité (Sr)	Doliocarpus sp.
agwago-maka (Au)	Smilax oblongata	atarakare (Tr)	Norantea guianensis
akapore (Tr)	Mansoa alliacea,	attakari (GU)	Peritassa laevigata
	Mansoa standleyi	aturai (Ca)	Machaerium lunatum
akalamaka (Sa)	Mimosa myriadenia	aturaimë (Tr)	Cayaponia rigida
akoacorollii (Ar)	Martinella obovata	aturia rang (Ca)	Dalbergia monetaria
akareowoi (Ca)	Connarus coriaceus,	awari imopi tëkënë (Tr)	Aegiphila laevis,
	Dalbergia ecastaphyllum,		Aegiphila racemosa,
	Dalbergia foliosa		Dichapetalum pedunculatum,
akawe pomëdë (Ca)	Maripa reticulata		Dichapetalum rugosum,
akikiwa (Ca)	Smilax oblongata		Tournefortia maculata,
akuri i'kererï (Ca)	Stigmaphyllon		Trigonia subcymosa,
	Convolvulifolium,		Trigonia villosa
	Stigmaphyllon sinuatum	awari parata (Tr)	Hiraea faginea
alakule (Ca)	Desmoncus orthacanthos	awawe ansa (Sa)	Dolichandra unguis-cati
alalalabu (Sa)	Norantea guianensis	aweja (Ca)	Ipomoea philomega
alasiku (Way)	Disciphania unilateralis	ayun tatai (Sa, Au)	Mansoa alliacea
aláta-tátai (Sa)	Hiraea faginea	aymoutabou (Ca)	Moutabea guianensis
alátu páu (Sa)	Hiraea faginea		
amaretto liaan (SD)	Bignonia nocturna		
amatobomiki (Pa)	Dioclea virgata	**B**	
ampuku-titei (Au)	Passiflora glandulosa	baakakifaia (Par)	Cissus verticillata
andïkïrï (Ca)	Curarea candicans	babunnoto (Sr)	Omphalea diandra
anuana menepuru (Ca)	Machaerium lunatum	babun-malasi (Sr)	Maripa reticulata
apotai (Ca-GU)	Strychnos tomentosa	babunnoot (SD)	Omphalea diandra
apuku-titei (Sr)	Guatteria scandens	balawito (Ww-SU)	Strychnos guianensis
araha enapiro (Tr)	Passiflora glandulosa	balauitu (Ww-GU)	Strychnos guianensis
araha enapiru (Tr)	Passiflora coccinea	bambakka (Sa)	Desmoncus spp.
arakaituran (Ca)	Connarus coriaceus	bambamakka (Sr)	Desmoncus spp.
arakure (Ca)	Desmoncus orthacanthos	bariribada (Ar)	Machaerium lunatum
arawata (Tr)	Machaerium floribundum	barudaballi (Ar)	Allamanda cathartica,
arawata kariwa (Tr)	Adenocalymma inundatum,		Amphilophium crucigerum
	Adenocalymma validum,	baskita-titei (Sr)	Bignonia prieurei,
	Bignonia aequinoctialis,		Bignonia sordida,
	Bignonia prieurei		Tournefortia bicolor
arawone simiori (Ca)	Cuspidaria inaequalis	baskita titei, witte (Sr)	Bignonia aequinoctialis
aretepe (Ca)	Cheiloclinium cognatum	baskita-titei, rode (Sr)	Tanaecium pyramidatum

Common Name	Scientific Name(s)	Common Name	Scientific Name(s)
basterd tingimoni (Sr)	Connarus punctatus	busi yamsi-titei (Sr)	Smilax oblongata
bergimanbebe (Sr)	Ampelozizyphus amazonicus	busnjamsi (Sr)	Smilax domingensis
bimititokon (Ar)	Passiflora glandulosa		
blaka-wintje (Par)	Cordia schomburgkii	**C**	
blaka-uma (Sr)	Cordia schomburgkii	cacao-titei (Sr)	Davilla nitida
blaka-uma, man (Sr)	Cordia polycephala	calajourou (FG)	Fridericia chica
bradilifi (Sr)	Coccoloba sp	cat-claw liana (En)	Jubelina rosea
bodori (Ar)	Machaerium lunatum	chewstick (Jamaica)	Gouania lupuloides
bokotaballi (Ar)	Tanaecium bilabiatum	cipó-cravo (BP)	Tynanthus polyanthus;
botermelkenkaas (SD)	Paullinia sp.	cipo d'alho (BP)	Mansoa alliacea,
bofrumaka (Sr)	Solanum rubiginosum		Mansoa standleyi
bofrusiri (Sr)	Abuta grandifolia	cipó-uíra (BP)	Guatteria scandens
bofrutiki (Sr)	Cheiloclinium cognatum	cockshun (En)	Smilax oblongata
bohoribada (Ar)	Machaerium floribundum,	cow itch (GU)	Mucuna urens
	Machaerium inundatum,	crab's eyes (GU)	Abrus precatorius
	Machaerium kegelii	crete poule (FG)	Drymonia coccinea
bokota firoberu (Ar)	Tanaecium bilabiatum	curare (GU)	Strychnos toxifera
bomia-titei (Pa)	Jubelina rosea		
bonbon gasa (Sa)	Petrea volubilis		
bos pampoen (SD)	Gurania robusta	**D**	
bos tomaat (SD)	Lycianthes pauciflora	danpuwi (Par)	Cordia schomburgkii
boskomkommer (SD)	Gurania lobata	devildoer (GU)	Aristolochia daemoninoxia,
bosolijf (SD)	Guatteria scandens		Strychnos bredemeyeri,
boyari (Ar)	Aristolochia daemoninoxia		Strychnos cogens,
brantimaka (Sr)	Machaerium inundatum,		Strychnos diaboli,
	Machaerium lunatum		Strychnos glabra,
broinati-titei (Sr)	Rourea surinamensis		Strychnos melinoniana,
brokobaka (Sr)	Mikania sp.		Strychnos mitscherlichii,
bron-titei (Sr)	Davilla kunthii		Strychnos tomentosa,
bruinhart-titei (SD)	Rourea surinamensis		Strychnos toxifera
budidiabatere (Sr)	Marcgravia pedunculosa	devildoer, black (GU)	Strychnos diaboli
buffelhout (SD)	Abuta grandifolia	devildoer, red (GU)	Strychnos erichsonii
bukuti (Pal)	Aristolochia leprieurii,	diaba-titei (Pa)	Souroubea guianensis
	Aristolochia stahelii	diabetere (Sr)	Souroubea guianensis
bunatimama, witte (Sr)	Cissus verticillata	dialopu (Sr)	Doliocarpus sp.,
bunatimama, rode (Sr)	Cissus erosa		Pinzona coriacea
búndu-tatái (Sa)	Abuta grandifolia,	dialopu-tité (Par)	Doliocarpus sp.,
	Abuta rufescens		Pinzona coriacea
busibranti (Sr)	Piptadenia floribunda	diamarpe (Ar)	Cordia schomburgkii
busi-kasterol (Ma)	Allamanda cathartica	dia-tetei (Au)	Davilla kunthii,
busi-lemki-maka (Sr)	Celtis iguanae,		Davilla nitida
	Seguieria americana	dia-titei (Sr)	Doliocarpus sp.,
businenge pogasi (Au)	Markea coccinea		Pinzona coriacea
busipesi (Pa)	Senna chrysocarpa	dia-titei (Sr)	Davilla kunthii,
busipongo (Au)	Gurania bignoniacea		Davilla nitida
busi-watramon (Sr)	Cissus spinosa	dikke bambamaka (Sr)	Desmoncus orthacanthos

Common Name	Scientific Name(s)	Common Name	Scientific Name(s)
	schomburgkii	kaikusi amararï (Ca)	Marcgravia coriacea
hikuritarafon (Ar)	Bauhinia guianensis	kakhero (Ar)	Odontadenia macrantha
hoasoropan (Ar)	Maripa scandens	kakutoru (Ar)	Mikania trinitaria
hoasoropan (Sr)	Cheiloclinium cognatum	kalapasapoã (Wpi)	Callichlamys latifolia
horotoballi (Ar)	Gurania robusta,	kalayulu (Wpi)	Fridericia candicans
	Passiflora fuchsiiflora	kamaka (Cr)	Dalbergia foliosa
horse eye (GU)	Mucuna urens	kamama (Cr)	Dalbergia foliosa
huriaballi (Ar)	Heteropterys nervosa	kamararai (Ca)	Tilesia baccata
		kamasuri (Ca)	Paullinia sp.
		kamoroballi (Ar)	Cuspidaria inaequalis,
I			Pachyptera kerere
idaballi (Ar)	Omphalea diandra	kamooraballi (GU)	Combretum laxum
ikhana (Ar)	Smilax oblongata	kamoro (Ar)	Adenocalymma inundatum,
ihipkunau (Pal)	Bignonia nocturna		Amphilophium
ilay kamwi (Pal)	Mansoa alliacea		magnoliifolium,
ineku (Tr)	Deguelia amazonica,		Bignonia prieurei,
	Duguelia scandens,		Dolichandra uncata,
	Lonchocarpus		Dolichandra unguiscati,
	chrysophyllus		Martinella obovata
inekuipë (Tr)	Clitoria arborescens	kamowa (Pat)	Desmoncus polyacanthos
inekuran (Ca)	Deguelia negrensis	kamuali (Ar)	Desmoncus spp.
ingi brumtjie (Sa)	Allamanda cathartica	kamwari (small leaf) (GU)	Desmoncus polyacanthos
ingineko (Pa)	Deguelia amazonica	kanakudiballi (Ar)	Dichapetalum pedunculatum
ituri-ishi-lokodo (Ar)	Orthomene schomburgkii	kanari-wiwiri (Sr)	Orthomene schomburgkii
		kapadula (GU)	Davilla kunthii,
			Davilla nitida,
			Doliocarpus sp.
J		kapai awara (Tr)	Petrea bracteata,
jacitara (BP)	Desmoncus spp.		Petrea volubilis
jamaraimë (Tr)	Desmoncus spp.	kapi (Tr)	Prionostemma aspera
japepuku (Ca)	Bignonia aequinoctialis	kapuwa siri (Sr)	Abuta grandifolia
jekara (Tr)	Combretum cacoucia,	karakara (Ar)	Norantea guianensis,
	Combretum rotundifolium		Souroubea guianensis
jorka-pesi (Sr)	Senna quinquangulata	karampai (Ar)	Annona haematantha
		kararawa andïkïrï (Ca)	Combretum cacoucia
		karajura (Pal)	Fridericia chica
K		karawiru (Ca)	Fridericia chica
kaboeiakoro mibikoro (Ar)	Mikania psilostachya	karia (Ar)	Stigmaphyllon sinuatum
kabuiakoro mibikoro (Ar)	Mikania psilostachya	karinama (Tr)	Diplopterys lucida
kabuduli (Ar)	Davilla kunthii,	karoshiri (Ar)	Cuervea kappleriana,
	Davilla nitida,		Hylenaea comosa
	Doliocarpus sp.	karu arib (Pal)	Combretum cacoucia
kabujakoro makuja (Ca)	Cordia schomburgkii	kasaba-titei (Sr)	Merremia macrocalyx
kadiamen (GU)	Combretum fruticosum	kashiparaballi (Ar)	Adenocalymma
kaikui amoi tëpu (Tr)	Jubelina rosea		schomburgkii
kaikui tepu (Tr)	Marcgravia pedunculosa	kashiri (Ar)	Serjania paucidentata
kaikui wenaru (Tr)	Marcgravia pedunculosa		

Common Name	Scientific Name(s)	Common Name	Scientific Name(s)
kasselerodang (Ar)	Guatteria scandens	kotiki (Sr)	Mabea taquari
katamïïmë (Tr)	Croton pullei	kotupuru (Ca)	Serjania paucidentata
kattepoot liaan (Sr)	Jubelina rosea	kréré (Ca)	Peritassa laevigata
katyusi anyarari (Ca)	Marcgravia coriacea	krintongo (Sr)	Guatteria scandens
kaw-ai (Sr)	Clitoria arborescens,	kromoko (Sr)	Machaerium inundatum
	Dioclea malacocarpa,	kuiakë wanupë (Tr)	Gnetum nodiflorum
	Dioclea scabra,	kujule-	
	Mucuna sloanei,	huhanukutpuh (Way)	Bauhinia guianensis
	Mucuna urens	kuleto (Ar)	Passiflora glandulosa
kaw-ai, pikin (Sr)	Dioclea virgata	kumarawa (Ar, Are,	
kaw-titei (Sr)	Davilla kunthii,	Mac, Pat)	Strychnos bredemeyeri,
	Doliocarpus sp.,		Strychnos cogens,
	Hiraea affinis		Strychnos diaboli,
keskesbosro (Sr)	Combretum rotundifolium		Strychnos glabra,
keskeskankan (Sr)	Amphilophium crucigerum,		Strychnos melinoniana,
	Combretum rotundifolium		Strychnos mitscherlichii
khawouieng-		kumaruballi (Ar)	Lonchocarpus chrysophyllus
gumapuiyik (Pat)	Bauhinia guianensis	kumata (Ca)	Lonchocarpus chrysophyllus
kifundu maka (Par)	Mimosa myriadenia	kumpuruni (Tr)	Aegiphila laevis,
kiintongo (Au)	Guatteria scandens		Aegiphila racemosa,
kikwe nyehnyeh (Ak)	Cissus erosa		Solandra longiflora
kiraporan (Ca)	Allamanda cathartica,	kumpuruniimë (Tr)	Drymonia coccinea,
	Odontadenia macrantha		Drymonia serrulata
kirikahu (Ar)	Guatteria scandens	kunaniimë (Sr)	Mikania sp.
kirikawa (Ar)	Guatteria scandens	kunofruktutitei (Sr)	Mansoa alliacea, Mansoa
knoflookliaan (SD)	Mansoa alliacea,		standleyi
	Mansoa standleyi	kupirisi (Sa)	Combretum laxum
knoflookwiri (Sr)	Mansoa alliacea,	kuraiwe (Tr)	Martinella iquitoensis
	Mansoa standleyi	kuraiweimë (Tr)	Adenocalymma flaviflorum,
kofiballi (Ar)	Guatteria scandens		Adenocalymma inundatum,
koko (Au)	Mabea taquari		Adenocalymma
kokomandialu (Par)	Dioclea violacea,		schomburgkii,
	Mucuna sloanei		Bignonia aequinoctialis,
kokriki (Sr)	Abrus precatorius		Callichlamys latifolia,
komo (Ca)	Odontadenia macrantha		Martinella obovata,
konikoni boontje (Sr)	Matelea denticulata		Tanaecium pyramidatum
konikoni-titei (Sr)	Prionostemma aspera	kurumï enuru (Ca)	Dioclea virgata,
konkonikasaba-titei (Sr)	Stigmaphyllon		Mucuna sloanei
	convolvulifolium,	kurumpïnpïrïkï (Tr)	Hiraea faginea,
	Stigmaphyllon sinuatum		Trigonia subcymosa,
konopo yorokorï (Ca)	Norantea guianensis,		Trigonia villosa
	Souroubea guianensis	kurumu eneru (Ca)	Clitoria arborescens
koratiwiwiri (Sr)	Stigmaphyllon sinuatum	kurumuri (Tr)	Securidaca paniculata
		kusai tatái (Sa)	Stigmaphyllon sinuatum
kôsyiton (Pal)	Machaerium paraense,	kutupa (Way)	Paullinia sp.
	Gnetum nodiflorum	kutupuran (Ca),	

Common Name	Scientific Name(s)	Common Name	Scientific Name(s)
kutupu (Ca)	Serjania paucidentata	makwariballi (Ar)	Forsteronia gracilis,
kuturi	Moutabea guianensis		Forsteronia guyanensis
kwabanaro (Ar)	Strychnos bredemeyeri,	malakopesi (Bo)	Guatteria scandens
	Strychnos cogens,	malokopesi (Pa)	Senna quinquangulata
	Strychnos diaboli,	mamakure (Tr)	Strychnos melinoniana
	Strychnos melinoniana,	mametala (Ww)	Aristolochia leprieurii
	Strychnos mitscherlichii,	mamuare (Sr)	Cheiloclinium cognatum
	Strychnos tomentosa	mananari (Ca)	Mikania psilostachya
kwabanero (Ar)	Strychnos glabra	maneko (Par)	Clitoria pendens,
kwaitaka nangra (Au)	Dolichandra unguis-cati		Connarus perrottetii,
kwata (Ar)	Dalbergia foliosa		Dalbergia monetaria
kwata-ede-sula (Sr)	Securidaca pubescens	manprasara (Sr)	Aegiphila laevis
kwatakama-titei (Sr)	Machaerium myrianthum	marakusya (Par)	Passiflora costata,
kwerimuru (Ar)	Souroubea guianensis		Passiflora glandulosa
kwipokan (Ak)	Mansoa alliacea	markusa (Sr)	Passiflora coccinea,
			Passiflora costata,
			Passiflora fuchsiiflora,
			Passiflora garckei
L			
lali, wilali (Way)	Strychnos glabra,	markusa, rode (SD)	Passiflora glandulosa
	Strychnos tomentosa	marudi oura (Ar)	Passiflora coccinea
lebikifaia (Pa)	Cissus erosa	matu kasaba (Sa)	Stigmaphyllon sinuatum
legwana titei (Sr)	Senna latifolia	matu-lémíki (Sa)	Securidaca paniculata,
lianappel (SD)	Moutabea guianensis		Securidaca pubescens
liane-ail (FG)	Mansoa alliacea	matukru (Tr)	Stigmaphyllon sinuatum
liane amére (FG)	Aristolochia leprieuri,	matukruimë (Tr)	Tetrapterys discolor
	Aristolochia stahelii	mehw-deh-tah (Tr)	Thinouia myriantha
liane gris (FG)	Petrea volubilis	meku kuware (Ca)	Omphalea diandra
liane trèfle (FG)	Aristolochia trilobata	mërii asoso (Tr)	Mabea taquari
liane panier (FG)	Bignonia aequinoctialis	merki-titei (Sr)	Rhabdadenia biflora
lickrish (GU)	Abrus precatorius	meu abesu (Pal)	Combretum rotundifolium
liba-pongo (Sa)	Cayaponia cruegeri	meudetah (Tr)	Thinouia myriantha
loango-titei (Sr)	Aristolochia staheli,	meyu akānta (Wpi)	Combretum rotundifolium
	Aristolochia surinamensis	mibi (Ar)	Bredemeyera densiflora
logoso sikada, bê (Sa)	Bauhinia guianensis	mibi tokoro (Ar)	Cissampelos andromorpha
lokonanjo (Ar)	Senna latifolia,	mibi yoro (Ar)	Dioclea macrocarpa
	Senna quinquangulata	mïkakijee (Tr)	Machaerium lunatum
losaboto kalawiru (Ca)	Passiflora glandulosa	mïrokoko ehe (Tr)	Bauhinia cupreonitens,
			Bauhinia guianensis
		misa (Way)	Gouania blanchetiana
M		molokju (Way)	Guatteria scandens
mabakubia (GU)	Coussapoa microcephala	monkey bora (GU)	Condylocarpon
mabudehi (Ar)	Banisteriopsis martiniana		intermedium
maka-titei (Sr)	Smilax oblongata,	monkey comb (GU)	Amphilophium crucigerum
	Smilax schomburgkiana	monkey ladder (GU)	Bauhinia guianensis
makui pana (Tr)	Mendoncia	monkey syrup (GU)	Maripa scandens
	hoffmannseggiana	moroko-titei (Sa/Sr)	Adenocalymma flaviflorum

Common Name	Scientific Name(s)	Common Name	Scientific Name(s)
		P	
moussi (Ca)	Fridericia candicans	pahaglaime (Way)	Serjania paucidentata
msibiu awak (Pal)	Dolichandra unguis-cati	pah-de (Tr)	Dioclea macrocarpa
murewa (Ca)	Guatteria scandens	pai tïkïtïkïkai (Tr)	Hyperbaena domingensis
		pakia tatái (Sa)	Coccoloba sp.
		pakale (Wpi)	Paullinia sp.
N		panapana (Ca)	Clitoria arborescens
nah-pe-wah (Wa)	Cissampelos andromorpha	parakatai (Ca)	Cordia schomburgkii
napëkoimë (Tr)	Cissus erosa,	paramaru (Ca)	Cissus verticillata
	Cissus verticillata	parapo (Ca)	Petrea bracteata
ndeku (Sa)	Lonchocarpus	parii (Tr)	Callichlamys latifolia
	chrysophyllus	paruisyiton (Par)	Strychnos melinoniana
ndjuka makodja (Sa)	Passiflora glandulosa	pata wanatitei (Sr)	Maripa violacea
ndulu-ndulu (Bo)	Guatteria scandens	patatatitei (Sr)	Merremia macrocalyx
nebi yoro (Ar)	Dioclea macrocarpa	patona (Ca)	Cissus verticillata
neku (Sr)	Lonchocarpus chrysophyllus,	pauisima (Ak)	Aristolochia daemoninoxia
	Serjania sp.	peigne singe rouge (FG)	Combretum rotundifolium
nere amoi (Tr)	Dolichandra uncata,	pëmenetë (Tr)	Gnetum nodiflorum
	Dolichandra unguis-cati	përaimë (Tr)	Adenocalymma validum
nere amoi (Tr)	Dolichandra uncata,	petit panacoco (FG)	Abrus precatorius
	Dolichandra unguis-cati	pikiriran (Ca)	Machaerium leiophyllum
nyamsi-maka (Sr)	Smilax oblongata,	pinyapau (Ma)	Hiraea faginea
	Smilax schomburgkiana	pipaman (Tr)	Bauhinia cupreonitens,
nyamsi-titei (Sr)	Smilax schomburgkiana		Bauhinia guianensis
		piri (Tr)	Desmoncus polyacanthos
		piri (Way)	Desmoncus polyacanthos
O		pïtïkïrïkï (Tr)	Acacia tenuifolia,
oepretete (Sr)	Bignonia aequinoctialis		Dalbergia monetaria
opu (Pa)	Tanaecium pyramidatum	pitïkitïkïkai (Tr)	Byttneria cordifolia
okó búka (Sa)	Senna chrysocarpa	piupiu (Ca)	Adenocalymma
okrai (Ar)	Dioclea reflexa		schomburgkii
old man's bark (GU)	Serjania paucidentata		
once-a-mile (GU)	Martinella obovata	pokopoko uru (Sr/Tr)	Senna chrysocarpa
oneka (Ca)	Sparattanthelium	pomme-liane (Sr)	Passiflora glandulosa
	wonotoboense	pongu (Sr)	Gurania subumbellata
ongles de chat (FG)	Dolichandra unguis-cati	popokaimaka (Sr)	Guettarda aromatica
onseballi (Ar)	Anemopaegma parkeri,	prasara-titei (Sr)	Dolichandra uncata,
	Cuspidaria inaequalis,		Dolichandra unguis-cati,
	Maripa violacea		Tanaecium bilabiatum
orari/urali (Ar, Ca, Are,	Strychnos toxifera	pugassi (Au)	Allamanda cathartica
Mac, Pat, Wap, Wna)		puspus-tere (Sr)	Combretum cacoucia
oraridan (Ca)	Strychnos melinoniana		
ou-ra-rie-yik (Pat)	Strychnos guianensis		
		R	
		rabu de arara (BP)	Combretum cacoucia
		rafrutere (Sr)	Norantea guianensis

Common Name	Scientific Name(s)	Common Name	Scientific Name(s)
rou-ah-amon (Ca, Gal)	Strychnos guianensis	sukrudyani (Sr)	Securidaca diversifolia
ruwiimë	Stizophyllum riparium	sukrutanta (Sr)	Tilesia baccata
		supple jack (GU)	Combretum laxum
		switiboontje-titei (Sr)	Dalbergia monetaria
S			
sabana (Sr)	Tetracera asperula		
sakëtaitu (Tr)	Davilla kunthii, Davilla nitida, Doliocarpus sp., Pinzona coriacea	**T**	
		taaukasi (Wpi)	Paullinia sp.
		taitetulea (Wpi)	Serjania paucidentata
sandpaper-vine (GU)	Petrea macrostachya	takusi wokuru (Ca)	Passiflora garckei
sarsaparilla (En)	Smilax oblongata	tamakaremu (Ca)	Orthomene schomburgkii
satijnblad-liaan (SD)	Bonamia maripoides	tameyu urang (Ca)	Doliocarpus sp.
schuurpapier (SD)	Petrea bracteata	tameyu'u (Ca)	Davilla kunthii, Davilla nitida
schuurpapier-liaan (SD)	Davilla kunthii, Davilla nitida		
		tamo (Tr)	Coccoloba sp.
sekema (Ca)	Combretum cacoucia, Combretum rotundifolium	tamoko enu (Tr)	Dioclea guianensis, Mucuna urens
		tamokumpë marakë (Tr)	Abuta rufescens
sekemaran (Ca)	Prestonia surinamensis	tapanapi (Tr)	Fridericia chica
sekrepatu-trapu (Sr)	Bauhinia guianensis	tapanapïmë (Tr)	Bignonia binata
sekrepatu-bita (Sr)	Tournefortia maculata	tara-titei (Sr)	Odontadenia macrantha
sepé atano (Ca)	Smilax domingensis	tariki (Ca)	Hiraea faginea
seybifinga wiwiri (Sr)	Serjania sp.	tarimakuresah (Way)	Mandevilla surinamensis
seweyuballi (Ar)	Connarus perrottetii	tawaau (Way)	Mucuna urens
sikime (Tr)	Fridericia oligantha	tawaimë (Tr)	Mandevilla surinamensis
simio (Ca)	Anemopaegma parkeri, Bignonia aequinoctialis	teteabo (Ar)	Curarea candicans
		tiapotano (Ca)	Tanaecium bilabiatum
sipatamu (Ca)	Smilax oblongata, Smilax schomburgkiana	tingi-titei (Sr)	Mansoa alliacea
		tiriki (Ca)	Hiraea faginea
sipatu (Tr)	Smilax oblongata, Smilax schomburgkiana	todofinga (Sr)	Cissus erosa
		tonolo epurlele (Ca)	Mikania trinitaria
sipokï (Ca)	Entada polystachya	topukeng (Ca)	Odontadenia macrantha, Odontadenia nitida
sipun (Tr)	Cheiloclinium cognatum, Cheiloclinium hippocrateoides, Salacia spp., Tontalea spp.		
		towa (Tr)	Gnetum urens, Gnetum nodiflorum
		towauri (Vz)	Gnetum nodiflorum, Gnetum urens
sipunuimë (Tr)	Salacia spp., Tontalea spp.		
sirito (Ca)	Dalbergia foliosa	tpsátá (Wmi)	Peritassa huanucana
sityò (Ca)	Omphalea diandra	tuhpokuimë (Tr)	Serjania sp.
skraati-titei (Sr)	Davilla nitida	tukuinetë (Tr)	Pyrostegia venusta
snekimarkusa (Sr)	Passiflora coccinea	tukukuimë (Tr)	Paullinia sp., Serjania sp.
sobolero (Ar)	Solanum rubiginosum	turtle ladder (GU)	Bauhinia guianensis
soitongo (Sa)	Cissus erosa		
sosoporo (Ca)	Passiflora auriculata		
soutmeti-udu (Sr)	Cheiloclinium cognatum		
sukerdyap (Sr)	Connarus coriaceus		

Common Name	Scientific Name(s)	Common Name	Scientific Name(s)
U		wayamu nepuru (Ca)	Bauhinia surinamensis
uralime (Wa)	Abuta grandifolia	werikë (Tr)	Prestonia annularis,
urari (Ca)	Strychnos guianensis,		Prestonia quinquangularis
	Strychnos medeola	wetibaka (Par)	Heteropterys macrostachya
urarian (Ar)	Cheiloclinium cognatum	weweimë (Tr)	Ampelozizyphus
urariballi (Ar)	Strychnos diaboli		amazonicus
uraridan (Ar)	Strychnos melinoniana	white devildoer (GU)	Strychnos melinoniana,
uquilla (Ar)	Martinella obovata		Strychnos mitscherlichii
		wilali (Way)	Strychnos guianensis
		wilali-piyu (Way)	Strychnos toxifera
V		wilali-sili (Way)	Strychnos toxifera
vijfvingerblad (SD)	Paullinia sp.	wilde markusa (SD)	Passiflora auriculata
voladora (Ven)	Desmoncus spp.	wilde napie (SD)	Cissus erosa
		wilde sopropo (SD)	Gurania bignoniacea
		wilde sukwa (SD)	Gurania bignoniacea,
W			Gurania subumbellata
wacht-een-beetje (SD)	Acacia tenuifolia,	wilkensbita (Sr)	Allamanda cathartica
	Mimosa myriadenia	wime etni kamwi (Pat)	Guatteria scandens
wajamaka finga (Sa)	Dolichandra unguis-cati	wïpore (Ca)	Mansoa alliacea
waka-weimë (Tr)	Machaerium floribundum	wirapa ijokë (Tr)	Abuta sp., Orthomene
wakagamu (Way)	Dolichandra uncata,		schomburgkii
	Dolichandra unguis-cati	wïrarï (Tr)	Strychnos guianensis,
wakapuimë (Tr)	Connarus perrottetii		Strychnos medeola
wakorokoda (Ar)	Lonchocarpus	wïrarïmë (Tr)	Strychnos guianensis
	chrysophyllus	wit bawang (Ja)	Mansoa alliacea
waktipikinso (Sr)	Acacia tenuifolia	wo'pole (Cr)	Mansoa alliacea
waktipikinso (Sr)	Smilax santaremensis	wuporėng (Cr)	Mansoa alliacea
wanëkë (Tr)	Curarea candicans,		
	Guatteria scandens,		
	Sciadotenia cayennensis,		
	Sparattanthelium	**Y**	
	uncigerum	yakki (Mac)	Strychnos bredemeyeri
warifar (Ar)	Tilesia baccata	yaoutimouta (Ca)	Bauhinia guianensis
warikë (Tr)	Omphalea diandra	yapepuku (Ca)	Adenocalymma inundatum,
wátawenú (Sa)	Bignonia nocturna		Corynostylis arborea,
watra ingi (Ca)	Hiraea faginea		Bignonia aequinoctialis,
watra-titei (Pa)	Tanaecium pyramidatum		Pachyptera kerere,
watra-titei (Sr)	Davilla kunthii,		Tanaecium pyramidatum
	Davilla nitida,	yariman (Ar)	Combretum rotundifolium
	Doliocarpus sp., Tetracera	yarimanni (Ar)	Combretum cacoucia
	asperula	yawahü merekuya (Ar)	Passiflora glandulosa
watramama kasabatiki (Sr)	Allamanda cathartica	ye no de ka (Way)	Dalbergia monetaria
wayamaka amosaitïrï (Ca)	Acacia tenuifolia	yesikushi (Mac)	Paullinia sp.
wayamaka andïkïrï (Ca)	Byttneria scabra	yoyoca (BP)	Combretum cacoucia
wayamaka erepatï (Ca)	Cordia schomburgkii	yuwana hi (Ar)	Entada polyphylla,
wayamu eresitjuru (Ca)	Tournefortia maculata		Entada polystachya,
			Mimosa myriadenia

CREDITS

illustrations and photography

PHOTOGRAPHY

[AK]	Andreas Kay [www.flickr.com/andreaskay/albums]
[AvP]	André van Proosdij
[AP]	Alex Popovkin
[BB]	Bejuco Osa Blogspot (see RA)
[BH]	Bruce Hoffman
[BH2]	Barry Hammel
[CB]	Chequita Bhikhi
[CF]	Christian Feuillet
[DC]	Dick Culbert
[DJ]	Daniel Janzen
[FRD]	Flora da Reserva Ducke Project, Brazil (Mike Hopkins)
[MP]	Mark Plotkin
[MR]	Maxime Rome (Topobea parasitica©)
[MTR]	Maria Tereza Rodrigues
[NH]	Niradj Hanoeman
[OG]	Olivier Gaubert [floredeguyane.piwigo.com]
[PT]	Pieter Teunissen
[QP]	Quinta Privada
[RA]	Reinaldo Aguilera (Costa Rica)
	[bejucososa.blogspot.com]
[RC]	Richard C. Hoyer/WINGS Birding Tours
[RR]	Rene van Raders
[RM]	Robin Morran
[RP]	Ruth Palsson
[SR]	Sofie Ruysschaert
[SI]	Smithsonian Institution, U.S. National Herbarium
[STRI]	Smithsonian Tropical Research Institute
	[biogeodb.stri.si.edu/]
[SZ]	Scott Zona
[TB]	Tom Ballinger
[TR]	Tanya Rehse
[WA]	William R. Anderson

ILLUSTRATIONS

[BA]	Bobbi Angell (many orig. published in Scott Mori. et al., 2002)
[KS]	Klei Sousa
[OM]	Omar Kasijo
[KMJ]	Kristi Michiels-Johnson [www.inkimage.com]
[NBC]	Naturalis Biodiversity Center, Netherlands
	(Wim Hekking, Hendrik Rypkema)
[UMH]	University of Michigan Herbarium (Karin Douthit)

BIBLIOGRAPHY

ABOUT THE AUTHORS

1. GENERAL DATA SOURCES

a) Books, Journals

Acevedo-Rodríguez, P. 2005. Vines and climbing plants of Puerto Rico and the Virgin Islands. Contributions from the United States National Herbarium 51, Smithsonian Institution, Washington, D.C.

_____. 1996. Flora of St. John: U.S. Virgin Islands. Memoirs of the New York Botanical Garden 78: 1-581.

Bremer, B., Bremer, K., Chase, M., Fay, M., Reveal, J., Soltis, D., & Stevens, P. 2009. An update of the Angiosperm Phylogeny Group classification for the orders and families of flowering plants: APG III. Botanical Journal of the Linnean Society 161(2): 105-121.

Feuillet, C. 2009. Checklist of the plants of the Guiana shield 1. An update to the angiosperms. Journal of the Botanical Research Institute of Texas 3(2): 799-814.

Funk, V., Hollowell, T., Berry, P.E., Kelloff, C., & Alexander, S.N. 2007. Checklist of the plants of the Guiana Shield (Venezuela: Amazonas, Bolivar, Delta Amacuro; Guyana, Surinam, French Guiana). Contributions from the United States National Herbarium 55, Smithsonian Institution, Washington, D.C.

Gentry, A. 1993. A field guide to the families and genera of woody plants of northwest South America (Colombia, Ecuador, Peru): with supplementary notes. Conservation International, Washington, D. C., USA.

Görts-van Rijn, A. R. A. & Janson-Jacobs, M. (Eds.). 1985-2011. Flora of the Guianas. Series A: Phanerogams. Koeltz Scientific Books, Koenigstein, Germany.

Isnard, S. & Silk, W. K. 2009. Moving with climbing plants from Charles Darwin's time into the 21st century. American Journal of Botany 96: 1205 -1221.

Keller, R. 2004. Identification of tropical woody plants in the absence of flowers and fruits - a field guide. 2nd Edition. Birkhauser Verlag AG, Basel, Switzerland.

Lindemann, J. C. & Molenaar, S. P. 1955. Van de bostypen in het noordelijk deel van Suriname. Dienst 'sLands Bosbeheer, Paramaribo.

Mori, S. A., Cremers, G., Gracie, C. A., Granville, J. D., Hoff, M., & Mitchell, J. D. (Eds.). 1997. Guide to the vascular plants of central French Guiana: part 1. Pteridophytes, gymnosperms, and monocotyledons. Memoirs of the New York Botanical Garden 76(1), Bronx, New York.

Mori, S. A., Cremers, G., Gracie, C. A., Granville, J. D., Heald, S., Hoff, M., & Mitchell, J. D. (Eds.). 2002. Guide to the vascular plants of central French Guiana: part 2. Dicotyledons. Memoirs of the New York Botanical Garden 76(2), Bronx, New York.

Ribeiro, J. E. L. S., Hopkins, M. J. G., Vicentini, A., Sothers, C. A. S., da Costa, M. A. S., Brito, J. M., Touza, M. A. D., Martins, L. H. P., Lohmann, L. G., Assunção, P. A. C. L., Pereira, E. C., Silva, C. F. S., Mesquita, M. R., & Procópio, L. C. (Eds.). 1999. Flora da Reserva Ducke - guia de indentificação das plantas vasculares de uma floresta de terra-firme na Amazônia Central. INPA-DFID, Manaus, Brazil.

Putz, F. E. & Mooney, H. A. (Eds.). 1991. Biology of vines. Cambridge University Press, Cambridge.

Pulle, A., Lanjouw, J., Stoffers, A. L. & Lindeman, J. C. (Eds.) 1932-1984. Flora of Suriname. J. H. de Bussy, Ltd., Amsterdam, the Netherlands.

Steyermark, J. A., Berry, P. E., Yatskjevich, K., & Holst, B. E. (Eds.). 1995-2005. Flora of the Venezuelan Guayana. Missouri Botanical Garden Press, St. Louis, USA.

van Roosmalen, M. G. M. 1985. Fruits of the Guianan Flora. University of Utrecht, National Herbarium of the Netherlands, Utrecht, the Netherlands.

b) Websites

Chicago Field Museum, Neotropical Herbarium Specimens [http://fm1.fieldmuseum.org/vrrc/].

Global Biodiversity Information Facility [www.gbif.org].

Herbier IRD de Guyane (CAY - Herbarium of French Guiana).[http://publish.plantnet-project.org/project/caypub].

The Liana Ecology Project (Stephen Schnitzer lab, U. Wisconsin-Milwaukee) [www.lianaecologyproject.com]

Milliken, W., Klitgård, B., & Baracat, A. (Eds.). Neotropikey - Interactive key and information resources for flowering plants of the Neotropics. Kew Botanic Gardens, UK. [*www.kew.org/science/tropamerica/neotropikey/families.htm*].

Mota de Oliveira, S. (Ed.). 2012 onwards. Flora of the Guianas, Series A: Phanerogams. Royal Botanic Garden, Kew, U.K. [*http://portal.cybertaxonomy.org/flora-guianas/*].

Raes, N. & R.C. Ek. 2002. Climbers of Guyana - website [*www.lianas-of-guyana.org*].

Smithsonian Institution, National Museum of Natural History, Department of Botany, Collections [*http://collections.nmnh.si.edu/search/botany/*].

Stevens, P. F. 2001 onwards. Angiosperm Phylogeny Website. Version 12, July 2012 [*http://www.mobot.org/MOBOT/research/APweb/*].

Mori, S. A., M. Tulig, J.-J. de Granville, S. Gonzalez & V. Guerin. 15 Dec 2007 onward. French Guiana e-Flora Project. The New York Botanical Garden and the Institut de Recherche pour le Développement [*http://sweetgum.nybg.org/fg/*].

Naturalis Biodiversity Center, National Herbarium of the Netherlands, Bioportal [*http://bioportal.naturalis.nl/*].

New York Botanical Garden, International Plant Center, C. V. Starr Virtual Herbarium, Vascular Plant Collections [*http://sciweb.nybg.org/Science2/hcol/allvasc/index.asp.html*].

TNRS (The Taxonomic Name Resolution Service). 2016. iPlant Collaborative. Version 4.0. [*http://tnrs.iplantcollaborative.org*]

The Plant List (2013). Version 1.1. [*www.theplantlist.org*].

Tropicos - Botanical Information System at the Missouri Botanical Garden [www.tropicos.org].

Watson, L., & Dallwitz, M. J. 1992 onwards. The families of flowering plants: Descriptions, illustrations, identification, and information retrieval. Version: 14th December 2015. [*http://biodiversity.uno.edu/delta/*].

c) Liana common names and uses

DeFilipps, R. A., Maina, S. L., & Crepin, J. 2004. Medicinal plants of the Guianas (Guyana, Suriname, French Guiana) [http://botany.si.edu/bdg/medicinal/index.html].

Fanshawe, D. B. 1949. Glossary of Arawak names in natural history. International Journal of American Linguistics, 15(1): 57-74.

Fanshawe, D. B. 1950. Forest Products of British Guiana: Part 2. Minor Forest Products. British Guiana Forestry Department, Georgetown, British Guiana.

Grenand, P., Moretti, C., Jacquemin, H., & Prévost, M. F. 2004. Pharmacopées traditionnelles en Guyane: Créoles. Wayãpi, Palikur. IRD Editions, Paris.

Hoffman, B. 2012. Exploring biocultural contexts: comparative woody plant knowledge of indigenous and Afro-American Maroon communities in Suriname, South America. In: Voeks, R. & Ashford, J. (Eds.), African Ethnobotany in the Americas, Springer, New York. Pp. 335-393.

Phillips, O. 1992. Ethnobotany and use of lianas. In: Putz, F. E. & Mooney, H. A. (Eds.), Biology of Vines. Cambridge University Press, Cambridge.

Plotkin M. J. 1986. Ethnobotany and the conservation of the tropical forest with special reference to the Indians of southern Suriname. Unpublished Ph.D. dissertation, Dept Biology, Tufts U., Medford, Mass., USA.

Schultes, R. E. and Raffauf, R. F. 1990. The healing forest: medicinal and toxic plants of northwest Amazonia. Dioscorides Press, Portland, Oregon, USA.

van Andel, T. 2000. Non-timber forest products of the north-west district of Guyana. Part II. A field guide. Ph.D. dissertation, Utrecht University, Utrecht, the Netherlands. Tropenbos-Guyana Programme, Georgetown, Guyana.

_____, & Ruysschaert, S. 2011. Medicinale en rituele planten van Suriname. K.I.T. Publishers, Amsterdam, the Netherlands.

_____, Mitchell, S., Volpato, G., Vandebroek, I., Swier, J., Ruysschaert, S., Jiménez, C. A. R., & Raes, N. 2012. In search of the perfect aphrodisiac: parallel use of bitter tonics in West Africa and the Caribbean. Journal of Ethnopharmacology 143: 840-850.

van 't Klooster, C. I. E. A., Lindeman, J. C. & Jansen-Jacobs, M. J. 2003. Index of vernacular plant names of Suriname. Blumea 15(Suppl.): 1-322.

d) Liana Biology and Conservation

Chicago Field Museum, Neotropical Herbarium Specimens [*http://fm1.fieldmuseum.org/vrrc/*].

Global Biodiversity Information Facility [*www.gbif.org*].

Herbier IRD de Guyane (CAY - Herbarium of French Guiana). [*http://publish.plantnet-project.org/project/caypub*].

The Liana Ecology Project (Stephen Schnitzer lab, U. Wisconsin-Milwaukee) [*www.lianaecologyproject.com*].

Milliken, W., Klitgård, B., & Baracat, A. (Eds.). Neotropikey - Interactive key and information resources for flowering plants of the Neotropics. Kew Botanic Gardens, UK. [*www.kew.org/science/tropamerica/neotropikey/families.htm*].

Mota de Oliveira, S. (Ed.). 2012 onwards. Flora of the Guianas, Series A: Phanerogams. Royal Botanic Garden, Kew, U.K. [*http://portal.cybertaxonomy.org/flora-guianas/*].

Raes, N. & R.C. Ek. 2002. Climbers of Guyana - website [*www.lianas-of-guyana.org*]

Smithsonian Institution, National Museum of Natural History, Department of Botany, Collections [*http://collections.nmnh.si.edu/search/botany/*].

Stevens, P. F. 2001 onwards. Angiosperm Phylogeny Website. Version 12, July 2012 [*http://www.mobot.org/MOBOT/research/APweb/*].

Mori, S. A., M. Tulig, J.-J. de Granville, S. Gonzalez & V. Guerin. 15 Dec 2007 onward. French Guiana e-Flora Project. The New York Botanical Garden and the Institut de Recherche pour le Développement [*http://sweetgum.nybg.org/fg/*].

Naturalis Biodiversity Center, National Herbarium of the Netherlands, Bioportal [*http://bioportal.naturalis.nl/*].

New York Botanical Garden, International Plant Center, C. V. Starr Virtual Herbarium, Vascular Plant Collections [*http://sciweb.nybg.org/Science2/hcol/allvasc/index.asp.html*].

TNRS (The Taxonomic Name Resolution Service). 2016. iPlant Collaborative. Version 4.0. [*http://tnrs.iplantcollaborative.org*]

The Plant List (2013). Version 1.1. [*www.theplantlist.org*].

Tropicos - Botanical Information System at the Missouri Botanical Garden [*www.tropicos.org*].

Watson, L., & Dallwitz, M. J. 1992 onwards. The families of flowering plants: Descriptions, illustrations, identification, and information retrieval. Version: 14th December 2015. [*http://biodiversity.uno.edu/delta/*].

2. LIANA ECOLOGY RESEARCH

Dewalt, S. J., Schnitzer, S. A. & Denslow, J. S. 2000. Density and diversity of lianas along a chronosequence in a central Panamanian lowland forest. Journal of Tropical Ecology 16: 1–19.

Jiménez-Castillo, M. & Lusk, C. H. 2013. Vascular performance of woody plants in a temperate rain forest: lianas suffer higher levels of freeze–thaw embolism than associated trees. Functional Ecology 27: 403–412.

Ledo, A., & Schnitzer, S. A. 2014. Disturbance and clonal reproduction determine liana distribution and maintain liana diversity in a tropical forest. Ecology 95: 2169–2178.

Letcher, S. G. & Chazdon, R. L. 2012. Life history traits of lianas during tropical forest succession. Biotropica 44: 720–727.

Lopes, W. A. L., Souza, L. A., Moscheta, I. M., Albiero, A. L. M., & Mourão, K. S. M. 2008. A comparative anatomical study of the stems of climbing plants from the forest remnants of Maringa, Brazil. Gayana Botanica 65(1): 28-38.

Phillips, O. L., Martínez, R. V., Arroyo, L. & Baker, T. R. 2002. Increasing dominance of large lianas in Amazonian forests. Nature 247.

_____, Martínez, R. V., Mendoza, A. M., Baker, T. R. & Vargas, P. N. 2005. Large lianas as hyperdynamic elements of the tropical forest canopy. Ecology 86: 1250–1258.

Putz, F. E. 1983. Liana biomass and leaf area of a "Tierra Firme" forest in the Rio Negro Basin, Venezuela. Biotropica 15: 185–189.

_____. 1984. The natural history of lianas on Barro Colorado Island, Panama. Ecology 65: 1713–1724.

_____. & Holbrook, N. M. 1986. Notes on the natural history of hemiepiphytes. Selbyana 9: 61–69.

Schnitzer, S. A., Dalling, J. W., & Carson, W. P. 2000. The impact of lianas on tree regeneration in tropical forest canopy gaps: Evidence for an alternative pathway of gap-phase regeneration. Journal of Ecology 88: 655–666.

_____, Kuzee, M. E. & Bongers, F. 2005. Disentangling above- and below-ground competition between lianas and trees in a tropical forest. Journal of Ecology 93: 1115–1125.

_____, Mangan, S. A., Dalling, J. W., Baldeck, C. A., Hubbell, S. P., Ledo, A., Muller-Landau, H., Tobin, M. F., Aguilar, S., Brassfield, D. and Hernandez, A. 2012. Liana abundance, diversity, and distribution on Barro Colorado Island, Panama. PLOS ONE, 7(12), p. e52114.

Silk, W. K. & Holbrook, N. M. 2005. The importance of frictional interactions in maintaining the stability of the twining habit. American Journal of Botany 92: 1820–1826.

Vidal, E., Johns, J., Gerwing, J. J., Barreto, P. & Uhl, C. 1997. Vine management for reduced-impact logging in eastern Amazonia. Forest Ecology Management 98: 105–114 (1997).

Wright, S. J., Calderón, O., Hernandéz, A. & Paton, S. 2004. Are lianas increasing in importance in tropical forests? A 17-year record from Panama. Ecology 85: 484–489.

Zagt, R.J., Ek, R.C. and Raes, N., 2003. Logging effects on liana diversity and abundance in Central Guyana. Tropenbos International, Utrecht, the Netherlands.

3. LIANA DATA SOURCES BY FAMILY

ACANTHACEAE (MENDONCIACEAE)

Bremekamp, C. E. B. 1938. Acanthaceae. Flora of Suriname 4(2): 166-256.

Costa, M. A. S., & de Souza, M. A. D. 1999. Acanthaceae. Flora da Reserva Ducke: 606-607.

Wasshausen, D. C. 2002. Acanthaceae. Guide to the vascular plants of central French Guiana 2: 31-39.

_____. 2006. Acanthaceae (156). Mendonciaceae (159). Flora of the Guianas A(23): 1-188.

ANNONACEAE

Fries, R. E. 1940. Annonaceae. Flora of Suriname 2(2): 341-383.

Jansen-Jacobs, M. J. 1976. Annonaceae (additions and corrections). Flora of Suriname 2(2): 658-687.

Maas, P. J. M. 2009. Neotropical Annonaceae. Neotropikey [Online: See general references section].

_____, & Maas-van de Kamer, H. 2002. Annonaceae. Guide to the vascular plants of central French Guiana 2: 53-67.

Steyermark, J. A., Maas, P. J. M., Berry, P. E., Johnson, D. M., Murray, N. A., & Heimo, R. 1995. Annonaceae. Flora of the Venezuelan Guayana 2: 413-451.

Ribeiro, J. E. L. S. 1999. Annonaceae. Flora da Reserva Ducke: 121-135.

van Roosmalen, M. G. M. 1985. Annonaceae. Fruits of the Guianan Flora: 4-21.

APOCYNACEAE

Allorge, L. 2002. Apocynaceae. Guide to the vascular plants of central French Guiana. 2: 69-84.

Endress, M. E., & Bruyns, P. V. 2000. A revised classification of the Apocynaceae s. l. The Botanical Review 66(1): 1-56.

Fallen, M. 1983. A taxonomic revision of Condylocarpon (Apocynaceae). Annals of the Missouri Botanical Garden 70: 149-169.

Goyder, D. 2009. Neotropical Apocynaceae (Asclepiadoideae). Neotropikey [Online: See general references section].

Jonker, F. P. 1937. Apocynaceae (additions and corrections). Flora of Suriname 4(1): 443-467.

Markgraf, F. 1932. Apocynaceae. Flora of Suriname 4(1): 1-65.

_____. 1940. Asclepiadaceae. Flora of Suriname 4(2): 326-352.

Morales, J. F. 1999. A synopsis of the genus Odontadenia. Series of revisions of Apocynaceae, XLV Bulletin du Jardin Botanique National de Belgique 67(1): 381-477.

Morillo, G. N. 1989. Contribución al conocimiento de las Asclepiadaceae Suramericanas, principalmente de las Guayanas. Ernstia 51: 2-15.

_____. 1995. Asclepiadaceae. Flora of the Venezuelan Guayana 2: 129-177.

_____. 2002. Asclepiadaceae. Guide to the vascular plants of central French Guiana 2: 89-93.

_____., & Carmona, J. 1995. Clave genérica para las Apocynoideae (Apocynaceae) de Venezuela y las Guayanas. Ernstia 5: 161-180.

van Roosmalen, M. G. M. 1985. Apocynaceae. Fruits of the Guianan Flora: 21-35.

Vicentini, A., & Oliveira, A. A. 1999. Apocynaceae e Asclepiadaceae. Flora da Reserva Ducke: 568-581.

Zarucchi, J. L., Morillo, G. N., Endress, M. E., Hansen, B. F., & Leeuwenberg, A. J. M. 1995. Apocynaceae. Flora of the Venezuelan Guayana 2: 471- 570.

ARECACEAE
Baker, W. J. 2009. Neotropical Arecaceae. Neotropikey [Online: See general references section].

de Granville, J.-J. 2002. Arecaceae. Guide to the vascular plants of central French Guiana 1: 190-215.

Henderson, A. 1997. Arecaceae. Flora of the Venezuelan Guayana 3: 32-122.

Küchmeister, H. D., & Hopkins, M. J. G. 1999. Arecaceae. Flora da Reserva Ducke: 653-668.

van Roosmalen, M. G. M. 1985. Palmae. Fruits of the Guianan Flora: 338-354.

Wessels Boer, J. G. 1965. Palmae. Flora of Suriname 5(1): 106-120.

ARISTOLOCHIACEAE
Barringer, K. A., & González-G, A. 1995. Aristolochiaceae. Flora of the Venezuelan Guayana: 122-129.

Edwards, W. 2009. Neotropical Aristolochiaceae. Neotropikey [Online: See general references section].

Feuillet, C., & Poncy, O. 1998. 123 Aristolochiaceae. Flora of the Guianas, Series A (10): 1-22.

_____., & Poncy, O. 2002. Aristolochiaceae. Guide to the vascular plants of central French Guiana 2: 87-88.

Ribero, J. E. L. S. 1999. Aristolochiaceae. Flora da Reserva Ducke: 188-189.

ASTERACEAE
Hind, D. J. N. 2009. Neotropical Asteraceae. Neotropikey [Online: See general references section].

Koster, J. T. 1938. Compositae. Flora of Suriname 4(2): 87-165.

Pruski, J. F. 1997. Asteraceae. Flora of the Venezuelan Guayana 3: 177-393.

_____. 2002. Asteraceae. Guide to the vascular plants of central French Guiana 2: 94-116.

Ribeiro, J. E. L. S. 1999. Asteraceae. Flora da Reserva Ducke: 648-651.

Smith, G. L. & Coile, N. C. 2007. Piptocarpha (Compositae: Vernonieae). Flora Neotropica Monographs 99: 1-94.

BIGNONIACEAE

Alcantara, S., & Lohmann, L. G.. 2010. Evolution of floral morphology and pollination system in Bignonieae (Bignoniaceae). American Journal of Botany 97(5): 782-79

De Medeiros, M. C. M. P., & Lohmann, L. G. 2015. Taxonomic Revision of Tynanthus (Bignonieae, Bignoniaceae). Phytotaxa 216(1): 1-60.

Fonseca, L. H. M., Cabral, S. M., de Fátima Agra, M., & Lohmann, L. G. 2015. Taxonomic updates in Dolichandra Cham. (Bignonieae, Bignoniaceae). PhytoKeys 46: 35-43 [doi:10.3897/phytokeys.46.8421].

Gentry, A. H. 1992. A Synopsis of Bignoniaceae Ethnobotany and Economic Botany. Annals of the Missouri Botanical Garden 79(1): 53-64

_____. 1993. Bignoniaceae. Guide to the genera of woody plants of northeastern South America. Conservation International.

_____. 1997. Bignoniaceae. In: Flora of the Venezuelan Guayana 3:403-491.

_____. 1982. Bignoniaceae. In: Flora de Venezuela. 4:1-433.

_____, & Cook K. 1984. Martinella (Bignoniaceae): a widely used eye medicine of South America. J. Ethnopharmacology 11: 337-343.

Lohmann, L. G. & Hopkins, M. J. G. 1999. Bignoniaeae. Flora da Reserva Ducke: 608-623.

_____, Brown, J. L., & Mori, S. A. 2002. Bignoniaceae. Guide to the vascular plants of central French Guiana 2: 118-139.

_____. 2006. Untangling the phylogeny of neotropical lianas (Bignonieae, Bignoniaceae). American Journal of Botany 93(2): 304 -318.

_____, & Taylor, C. M. 2014. A New Generic Classification of Tribe Bignoniea (Bignoniaceae). Annals of the Missouri Botanical Garden 99(3): 348-489.

Prance, G. T., Campbell, D. G. & Nelson, B. W. 1977. The ethnobotany of the Paumari Indians. Economic Botany 31: 129-139.

Sandwith, N. Y. 1938. Bignoniaceae. In: Flora of Suriname 4(2): 1-86.

Schultes, R. E. 1970. De plantis toxicariis e mundo novo tropicale. Commentationes VII. Bot. Mus. Leafl. Harvard Univ. 10: 345-352.

van Roosmalen, M. G. M. 1985. Bignoniaceae. Fruits of the Guianan Flora: 40-54.

BORAGINACEAE

Costa, M. A. S. 1999. Boraginaceae. Flora da Reserva Ducke: 592-595.

Feuillet, C. 2002. Boraginaceae. Guide to the vascular plants of central French Guiana 2:145-151.

Johnston, I. M. 1935. Studies in Boraginaceae X. The Boraginaceae of northeastern South America. Journal of the Arnold Arboretum 16: 1-64.

_____. 1936. Boraginaceae. Flora of Suriname 4(1): 306-333.

Stapf, M. N. S. 2009. Neotropical Boraginaceae. Neotropikey [Online: See general references section].

Uittien, H. 1937. Boraginaceae (additions and corrections). Flora of Suriname 4(1): 496-497.

van Roosmalen, M. G. M. 1985. Boraginaceae. Fruits of the Guianan Flora: 59-63.

CANNABACEAE (=ULMACEAE)

Berg, C. C. 1992. 20 Ulmaceae. Flora of the Guianas A(11): 3-9.

CELASTRACEAE (HIPPOCRATEACEAE)

Archer, R. H., & Lombardi, J. A. 2013. Neotropical Celastraceae. Neotropikey [Online: See general references section].

Brito, J. M. 1999. Celastraceae. Flora da Reserva Ducke: 471.

Coughernour, J. M., Simmons, M. P., Lombardi, J. A., Yakobson, K., & Archer, R. H. 2011. Phylogeny of Celastraceae subfamily Hippocrateoideae inferred from morphological characters and nuclear and plastid loci. Molecular Phylogenetics and Evolution 59: 320–330.

Görts-van Rijn, A. R. A. 2002. Hippocrateaceae. Guide to the vascular plants of central French Guiana 2: 347-353.

_____. & Mennega, A. M. W. 1994. 110 Hippocrateaceae. Flora of the Guianas A(16): 3-81.

Lombardi, J. A. 2012. Neotropical Hippocrateaceae. Neotropikey [Online: See general references section].

Mennega, A. M. W. & Hedin, J. P. 1999. Hippocrateaceae. Flora of the Venezuelan Guayana 5: 594-617.

Ribeiro, J. E. L. S. 1999. Hippocrateaceae. Flora da Reserva Ducke: 472-476.

Smith, A. C. 1940. The American species of Hippocrateaceae. Brittonia 3: 341-555.

van Roosmalen, M. G. M. 1985. Celastraceae. Fruits of the Guianan Flora: 74-82.

CLUSIACEAE

Gustafsson, M. H. G. 2009. Neotropical Clusiaceae. Neotropikey [Online: See general references section].

Pipoly, J. J., & Gustafsson, M. H. G. 2002. Clusiaceae. Guide to the vascular plants of central French Guiana 2: 212-223.

COMBRETACEAE

Anderson, D. M. W. 1977. The composition of the gum exudate from some Combretum species, the botanical nomenclature and systematics of the Combretaceae. Carbohydrate Research 57: 215.

Exell, A. W. & Stace, C. A. 1966. Revision of the Combretaceae. Boletim da Sociedad Broteriana, Sér. 2, 40: 5-25.

Gentry, A. H. 1993. Combretaceae. A Field Guide to the Families and Genera of Woody Plants of Northwest South America: 335-339.

Kawasaki, M. L. 2002. Combretaceae. Guide to the vascular plants of central French Guiana 2: 224-227.

Loiola, M. I. B. 2009. Neotropical Combretaceae. Neotropikey [Online: See general references section].

Ribeiro, J. E. L. S. 1999. Combretaceae. Flora da Reserva Ducke: 459-461.

Stace, C. A. 1969. The significance of the leaf epidermis in the taxonomy of the Combretaceae III. The genus Combretum in America. Brittonia 12: 130-143.

_____. 2009. 100 Combretaceae. Flora of the Guianas A(27): 40-95.

van Roosmalen, M. G. M. 1985. Combretaceae. Fruits of the Guianan Flora: 91-95.

CONNARACEAE

Brito, J. M. 1999. Connaraceae. Flora da Reserva Ducke, Brazil: 340-343.

Forero, E. 1983. Connaraceae. Flora Neotropica Monographs 36: 1-208.

_____. 2002. Connaraceae. Guide to the vascular plants of central French Guiana 2: 227-231.

_____. 2009. Neotropical Connaraceae. Neotropikey [Online: See general references section].

_____, & Görts-van Rijn, A. R. A. 1976. Connaraceae (additions and corrections). Flora of Suriname 2(2): 654-657.

_____, & Santana, E. 1998. Connaraceae. Flora of the Venezuelan Guayana 4: 365-376.

Lanjouw, J. 1940. Connaraceae. Flora of Suriname 2(2): 332-340.

van Roosmalen, M. G. M. 1985. Connaraceae. Fruits of the Guianan Flora: 95-98.

CONVOLVULACEAE

Austin, D. F. 1982. Convolvulaceae. In: Luces de Febres, Z., & Steyermark, J. A. (Eds.). Flora de Venezuela 8(3): 15-226.

_____. 2002. Convolvulaceae. Guide to the vascular plants of central French Guiana 2: 231-236.

_____. 2009. Neotropical Convolvulaceae. Neotropikey [Online: See general references section].

_____. & Cavalcante, P. B. 1982. Convolvuláceas da Amazônia. Publicaciones Avulsas Museo Paraense Emilio Goeldi 36: 1-134.

_____. & Huáman, Z. 1996. A synopsis of Ipomoea (Convolvulaceae) in the Americas. Taxon 45: 3-38.

Ribeiro, J. E. L. S. 1999. Convolvulaceae. Flora da Reserva Ducke: 588-591.

van Ooststroom, S. J. 1966. Convolvulaceae. Flora of Suriname 4(1): 66-102, 468-471.

van Roosmalen, M. G. M. 1985. Convolvulaceae. Fruits of the Guianan Flora: 98-101.

CUCURBITACEAE

Costa, M. A. S., & Martins, L. H. 1999. Flora da Reserva Ducke: 307-309.

Jeffrey, C. 1980. A revision of the Cucurbitaceae. Botanical Journal of the Linnean Society 81: 233-247.

_____. 1984. Cucurbitaceae. Flora of Suriname 5(1): 457-518.

Kearns, D. M. 1998. Cucurbitaceae. Flora of the Venezuelan Guayana 4: 431—461.

Nee, M. 2000. Cucurbitaceae. Guide to the vascular plants of central French Guiana 2: 236-246.

Taylor, N., & Zappi, D. 2009. Neotropical Cucurbitaceae. Neotropikey [Online: See general references section].

van Roosmalen, M. G. M. 1985. Cucurbitaceae. Fruits of the Guianan Flora: 100-103.

DICHAPETALACEAE

Prance, G. T. 1972. Dichapetalaceae. Flora Neotropica Monographs 10: 1-84.

_____. 1979. A new species of Dichapetalum from Suriname. Bulletin Torrey Botanical Club 106(4): 309-312.

_____. 1983. Additions to Neotropical Dichapetalaceae. Brittonia 35(1): 49-54.

_____. 2002. Dichapetalaceae. Guide to the vascular plants of central French Guiana 2: 247-250.

_____. 2009a. 113 Dichapetalaceae. Flora of the Guianas A(27).

_____. 2009b. Neotropical Dichapetalaceae. Neotropikey [Online: See general references section].

Sothers, C. A., & Brito, J. M. 1999. Dichapetalaceae. Flora da Reserva Ducke.

Stafleu, F. A. 1951. Dichapetalaceae. Flora of Suriname 3(2): 166-172.

van Roosmalen, M. G. M. 1985. Dichapetalaceae. Fruits of the Guianan Flora: 103-105.

DILLENIACEAE

Aymard, G. A. 1998. Dilleniaceae. Flora of the Venezuelan Guayana 4: 671-685.

_____, & Mori, S. A. 2002. Dillenaciaceae. Guide to the vascular plants of central French Guiana 2: 250-254.

_____. 2009. Neotropical Dilleniaceae. Neotropikey [Online: See general references section].

Jansen-Jacobs, M. J. 1986. Dilleniaceae. Flora of Suriname 3(1-2): 475-484.

Lanjouw, J., & van Heerdt, P. F. 1941. Dilleniaceae. Flora of Suriname 3(1): 386-408.

Sothers, C. A., & de Souza, M. A. D. 1999. Dilleniaceae. Flora da Reserva Ducke: 228-232.

van Roosmalen, M. G. M. 1985. Dilleniaceae. Fruits of the Guianan Flora: 105-107.

ERICACEAE
_____. 2002. Ericaceae. Guide to the vascular plants of central French Guiana 2: 261-262.

_____, & Pedraza-Peñalosa, P. 2006. Neotropical Blueberries: The Plant Family Ericaceae. New York Botanical Gardens [www.nybg.org/bsci/res/lut2 (continuously updated)].

Pedraza-Peñalosa, P. 2009. Neotropical Ericaceae. Neotropikey [Online: See general references section].

Ribeiro, J. E. L. S. 1999. Ericaceae. Flora da Reserva Ducke: 311.

EUPHORBIACEAE
Esser, H.-J. 2009. Neotropical Euphorbiaceae. Neotropikey [Online: See general references section].

Gillespie, L. J. 2002. Euphorbiaceae. Guide to the vascular plants of central French Guiana 2: 266-298.

_____. & Armbruster, W. S. 1997. A contribution to the Guianan flora: Dalechampia, Haematostemon, Omphalea, Pera, Plukenetia, and Tragia (Euphorbiaceae) with notes on subfamily Acalyphoideae. Smithsonian Contributions to Botany 86: 1-48.

Görts-van Rijn, A. R. A. 1976. Euphorbiaceae (additions and corrections). Flora of Suriname 2(2): 387- 424.

Lanjouw, J. 1932. Euphorbiaceae. Flora of Suriname 2(1): 1-101.

_____. 1939. Euphorbiaceae (additions and corrections). Flora of Suriname 2(1): 457-470.

van Roosmalen, M. G. M. 1985. Euphorbiaceae. Fruits of the Guianan Flora: 113-126.

Vicentini, A., & Cordeiro, I. 1999. Euphorbiaceae. Flora da Reserva Ducke: 484-497.

GESNERIACEAE
Chautems, A. 2009. Neotropical Gesneriaceae. Neotropikey [Online: See general references section].

Feuillet, C., & Skog, L. E. 2002. Gesneriaceae. Guide to the vascular plants of central French Guiana 2: 334-344.

Feuillet, C. & Steyermark, J. A. 1999. Gesneriaceae. Flora of the Venezuelan Guayana 5: 542-573.

Leeuwenberg, A. J. M. 1958. The Gesneriaceae of Guiana. Acta Botanica Neerlands 7: 291-444.

_____. 1984. Gesneriaceae. Flora of Suriname 5(1):592-650.

Martins, I., & Costa, M. A. S. 1999. Gesneriaceae. Flora da Reserva Ducke: 602-605.

Skog, L. E., & Feuillet, C. 2008. 155 Gesneriaceae. Flora of the Guianas A(26): 1-136.

GNETACEAE
Cavalcante, P. B. 1978. Contribuição ao conhecimento das Gnetáceas da Amazônia (Gimnospermas). Acta Amazonica 8(2): 201-215.

Markgraf, F. 1965. New discoveries of Gnetum in tropical America. Annals of the Missouri Botanical Garden 52: 379-386.

Stevenson, D. W. 1997. Gnetaceae. Guide to the vascular plants of central French Guiana 1: 162.

_____. 1999. Gnetaceae. Flora of the Venezuelan Guayana 5: 573-576.

_____, & Zanoni T. 1991. 209 Gnetaceae. Flora of the Guianas A(9): 12-18.

van Roosmalen, M. G. M. 1985. Gnetaceae. Fruits of the Guianan Flora: 131-132.

Vicentini, A. 1999. Gnetaceae. Flora da Reserva Ducke: 119-120.

HERNANDIACEAE

Kostermans, A. J. G. H. 1937. Hernandiaceae. Flora of Suriname 2(1): 338-344.

Mori, S. A., & Brown, J. L. 2002. Hernandiaceae. Guide to the vascular plants of central French Guiana 2: 344-346.

Sasaki, D. 2009. Neotropical Hernandiaceae. Neotropikey [Online: See general references section].

Sothers, C. A., & Brito, J. M. 1999. Hernandiaceae. Flora da Reserva Ducke: 180.

van Proosdij, A. S. J. 2007. Hernandiaceae. Flora of the Guianas A(24): 1-214.

van Roosmalen, M. G. M. 1985. Hernandiaceae. Fruits of the Guianan Flora: 141-143.

ICACINACEAE

Brito, J. M. 1999. Icacinaceae. Flora da Reserva Ducke: 47-48.

de Roon, A. C. 1994. 112 Icacinaceae. Flora of the Guianas A(16): 1-158.

de Roon, A. C., & Mori, S. A. 2002. Icacinaceae. Guide to the vascular plants of central French Guiana 2: 358-361.

Duno de Stefano, R. 2009. Neotropical Icacinaceae. Neotropikey [Online: See general references section].

Howard, R. A. & Duno de Stefano, R. 1999. Icacinaceae. Flora of the Venezuelan Guayana 5: 646-658.

Jansen-Jacobs, M. J. 1979. Icacinaceae. Flora of Suriname 5(1): 344-355.

van Roosmalen, M. G. M. 1985. Icacinaceae. Fruits of the Guianan Flora: 146.

LAMIACEAE (includes Aegiphila formerly of Verbenaceae)

Aymard, G. 2005. Verbenaceae. Flora of the Venezuelan Guayana. 9: 407-445.

Bramley, G., Harley, R. and Paton, A. 2009. Neotropical Lamiaceae. Neotropikey [Online: See general references section].

Costa, M.A.S. 1999. Verbenaceae. Flora da Reserva Ducke: 596-599.

França, F., & Atkins, A. 2012. Verbenaceae. Neotropikey [Online: See general references section].

Harley, R. 2002. Lamiaceae. Guide to the vascular plants of central French Guiana 2: 363-369.

Jansen-Jacobs, M.J. 1988. Verbenaceae. Flora of the Guianas A(4): 1-114.

_____. 2002. Verbenaceae. Guide to the vascular plants of central French Guiana 2: 725-728.

Moldenke, H. N. 1934. A monograph of the genus Aegiphila. Brittonia 1: 245-477.

_____. 1940. Verbenaceae. Flora of Suriname 4(2): 357-321.

van Roosmalen, M. G. M. 1985. Verbenaceae. Fruits of the Guianan Flora: 430-432.

LEGUMINOSAE (FABACEAE)

Amshoff, G. J. H. 1939. Papilionaceae. Flora of Suriname 2(2): 1-257.

Barneby, R. C. 1998. Piptadenia [Mimosaceae]. Flora of the Venezuelan Guayana 4: 665—667.

_____. 1991. Sensitivae Censitae. A description of the genus Mimosa (Mimosaceae) in the New World. Memoirs of the New York Botanical Garden 65: 1-835.

_____ , & Grimes, J. (unpublished manuscript). Mimosaceae. Flora of the Guianas, Series A.

_____, & Heald, S. V. 2002. Fabaceae. Guide to the vascular plants of central French Guiana 2: 298-318.

Camargo, R. A. and de Azevedo Tozzi, A. M. G. 2014. A synopsis of the genus Deguelia (Leguminosae, Papilionoideae, Millettieae) in Brazil. Brittonia 66 (1): 12-32.

Cowan, R. S. 1961. Leguminosae. In: Maguire, B., Wurdack, J. J., & Collaborators, Botany of the Guayana Highland part IV. Memoirs of the New York Botanical Garden 10(4): 65-87.

_____, & Lindeman, J. C. 1989. Caesalpiniaceae p.p. Flora of the Guianas A(7): 1-167.

Doyle, J. J., Chappill, J. A., Bailey, C. D. & Kajita, T. 2000. Towards a comprehensive phylogeny of legumes: evidence from rbcL sequences and non-molecular data. In: Herendeen, P. S. & Bruneau, A. (Eds.). Advances in Legume Systematics part 9: 1-20. Royal Botanic Gardens, Kew, UK.

Ducke, A. 1942. Lonchocarpus, subgenus Phacelanthus Pittier, in Brazilian Amazonia. Tropical Woods 69: 2-7.

Görts-van Rijn, A. R. A., Kramer, K. U., & Lindeman, J. C. 1976. Papilionaceae (additions and corrections). Flora of Suriname 2(2): 556-610.

Irwin, H.S., & Barneby, R. C. 1982. The American Cassiinae. A synoptical revision of Leguminosae tribe Cassieae subtribe Cassiinae in the New World. Memoirs of the New York Botanical Garden 35: 1-918.

Klitgård, B. B., & Lewis, G. P. 2010. Neotropical Leguminosae (Caesalpinioideae, Mimosoideae, Papilionoideae treated separately). Neotropikey [Online: See general references section].

Lewis, G. P., & Owen, P. E. 1989. Legumes of the Ilha de Maracá. Royal Botanic Gardens Kew, UK.

_____, & Schrire, B. D. 2005. Leguminosae or Fabaceae? In: Lewis, G. P., Schrire, B., Mackinder, B. & Lock, M. Legumes of the World, pp. 1-2. Royal Botanic Gardens, Kew, UK.

Martins, L. 1999. Caesalpinioideae. Flora da Reserva Ducke: 382-395.

Maxwell, R. H. 1969. The Genus Dioclea (Fabaceae) in the New World. Ph.D. Dissertation, Southern Illinois University, Carbondale, Illinois, USA.

Mesquita, M. R., & Hopkins, M. J. G. 1999. Papilionoideae. Flora da Reserva Ducke: 396-413.

Procópio, L. C., & Hopkins, M. J. G. 1999. Mimosoideae. Flora da Reserva Ducke: 361-381.

van Roosmalen, M. G. M. 1985. Leguminosae (Caesalpiniaceae). Fruits of the Guianan Flora: 171-172, 174-195, 197.

_____. 1985. Leguminosae (Mimosaceae). Fruits of the Guianan Flora: 228-255.

_____. 1985. Leguminosae (Papilionaceae). Fruits of the Guianan Flora: 196-229.

Wojciechowski, M.F. 2003. Reconstructing the phylogeny of legumes (Leguminosae): an early 21st century perspective. In: B.B. Klitgaard & A. Bruneau (Eds.). Advances in Legume Systematics part 10: 5-35. Higher Level Systematics. Royal Botanic Gardens, Kew.

LOGANIACEAE (STRYCHNOS)

Bisset, G. B. 2002. War and hunting poisons of the New World. Part I. Notes on the early history of curare. Journal of Pharmacology 36(1): 1-26.

Bisset, N. G. 1988. Curare-botany, chemistry, and pharmacology. Acta Amazonica 18: 255-290.

Bordenave, B. 2002. Loganiaceae. Guide to the vascular plants of central French Guiana 2: 398-405.

Cockrell, R. 1941. A comparative study of the wood structure of several South American species of Strychnos. American Journal of Botany 28: 32-41.

Fanshawe, D. B. 1954. The genus Strychnos in British Guiana. Brittonia 8: 65-68.

Krukoff, B. A. 1972. American species of Strychnos. Lloydia 35(3): 193-271.

Leeuwenberg, A. J. M., Berry, P. E., & Brant, A. E. 2001. Loganiaceae. Flora of the Venezuelan Guayana 6: 22-36.

Sandwith, N. Y. 1933. The genus Strychnos in British Guiana and Trinidad. Kew Bulletin. pp. 390-400.

Others, C. A., & Brito, J. M. 1999. Loganiaceae. Flora da Reserva Ducke: 563-565.

Uittien, H. 1937. Loganiaceae (additions and corrections). Flora of Suriname 4(1): 472-474.

van Raalte, M. H. 1932. Loganiaceae. Flora of Suriname 4(1): 103-110.

van Roosmalen, M. G. M. 1985. Loganiaceae. Fruits of the Guianan Flora: 257-260.

Zappi, D. 2009. Neotropical Loganiaceae. Neotropikey [Online: See general references section].

MALPIGHIACEAE

Anderson, C. 1997. Monograph of Stigmaphyllon (Malpighiaceae). Systematic Botany Monographs. 51: 1-313.

_____. 2001. The identity of two water-dispersed species of Heteropterys (Malpighiaceae): H. leona and H. platyptera. Contr. Univ. Michigan Herbarium 23: 35–47.

_____. 2001. Novelties in Mascagnia (Malpighiaceae). Brittonia 53: 405–415.

Anderson, W. R. 1990. The taxonomy of Jubelina (Malpighiaceae), Contr. Univ. Michigan Herbarium 17: 21-37.

_____. 1993. Notes on Neotropical Malpighiaceae - IV. Contr. Univ. Michigan Herbarium 19: 355-392.

_____. 1994. New species of Hiraea (Malpighiaceae) from the Guianas and adjacent Brazil. Brittonia 46(2): 126-133.

_____. 1997. Excentradenia, a new genus of Malpighiaceae from South America. Contr. Univ. Michigan Herbarium 21: 29-36.

_____. 2000. Malpighiaceae. In: Flora of the Venezuelan Guayana 6.

_____. 2002. Malpighiaceae. Guide to the vascular plants of central French Guiana 2: 410-427.

_____, & C. C. Davis. 2001. Monograph of Lophopterys (Malpighiaceae). Contr. Univ. Michigan Herbarium 23: 83–105.

_____, & Davis, C. C. 2007. Generic adjustments in neotropical Malpighiaceae. Contributions from the University of Michigan Herbarium 25: 137-166.

Gates, Bronwen. 1982. Banisteriopsis, Diplopterys (Malpighiaceae). Flora Neotropica Monographs 30: 1-237.

Görts-van Rijn, A. R. A. & Jansen-Jacobs. M. J. 1976. Malpighiaceae (additions and corrections). Flora of Suriname 2(2): 445-450.

Jonker, F. P. 1939. Malpighiaceae (additions and corrections). Flora of Suriname 2(1): 478-480.

Kostermans, A. J. G. H. 1936. Malpighiaceae. Flora of Suriname 2(1): 146-243.

van Roosmalen, M. G. M. 1985. Malpighiaceae. Fruits of the Guianan Flora: 260, 262-272.

Vicentini, A. 1999. Malpighiaeae. Flora da Reserva Ducke: 505-511.

Zappi, D. 2015. Neotropical Malpighiaceae. Neotropikey [Online: See general references section].

MALVACEAE-BYTTNERIACEAE

Cristóbal, C.L. 1976. Estudio taxonómico del género Byttneria Loefl. Bonplandia 4: 1-428.

Dorr, L. J. 2002. Malvaceae. Guide to the vascular plants of central French Guiana 2: 428-430.

Klitgård, B. B. 2013. Neotropical Malvaceae (Byttneriaceae). Neotropikey [Online: See general references section].

Uittien, H. 1932. Sterculiaceae. Flora of Suriname 3(1): 34-48.

_____. 1941. Sterculiaceae (additions and corrections). Flora of Suriname 3(1): 437.

Vicentini, A. 1999. Malvaceae. Flora da Reserva Ducke: 272.

MARCGRAVIACEAE

Dresler, S. 2009. Neotropical Marcgraviaceae. Neotropikey [Online: See general references section].

Heald, S. V., de Roon, A. C., & Dressler, S. 2002. Marcgraviaceae. Guide to the vascular plants of central French Guiana 2: 431-436.

Lanjouw, J., & van Heerdt, P. F. 1941. Marcgraviaceae. Flora of Suriname 3(1): 373-385.

Ribeiro, J. E. L. S. 1999. Marcgraviaceae. Flora da Reserva Ducke: 239.

van Roosmalen, M. G. M. 1985. Marcgraviaceae. Fruits of the Guianan Flora: 272-274.

MELASTOMATACEAE

Sothers, C. A., & Brito, J. M. 1999. Melastomataceae. Flora da Reserva Ducke: 438-452.

van Roosmalen, M. G. M. 1985. Melastomataceae. Fruits of the Guianan Flora: 275-284.

Woodgyer, E. M. 2009. Neotropical Melastomataceae. Neotropikey [Online: See general references section].

Wurdack, J. J., Morley, T., & Renner, S. 1993. Flora of the Guianas A(13).

_____. & Renner, S. 2002. Melastomataceae. Guide to the vascular plants of central French Guiana 2: 437-464.

MENISPERMACEAE

Barneby, R. C. 1970. Revision of Neotropical Menispermaceae tribe Tinosporeae. Memoirs of the New York Botanical Garden 20(2): 81-158.

_____. 2002. Menispermaceae. Guide to the vascular plants of central French Guiana 2: 474-483.

_____, & Krukoff, B. A. 1971. Supplementary notes in American Menispermaceae. VIII. Memoirs of the New York Botanical Garden 22: 1-89.

Carlquist, S. 1996. Wood and stem anatomy of Menispermaceae. Aliso 14: 155-170.

Diels, L. 1934. Menispermaceae. Flora of Suriname 2(1): 123-131.

Jansen-Jacobs, M. J. 1976. Menispermaceae (additions and corrections). Flora of Suriname 2(2): 430-440.

Jonker, F. P. 1939. Menispermaceae (additions and corrections). Flora of Suriname 2(1): 476-477.

Krukoff, B. A., & Barneby, R. C. 1938. Studies of American Menispermaceae, with special reference to species used in preparation of arrow-poisons. Brittonia 3: 1-74.

_____, and _____. 1970. Supplementary Notes on American Menispermaceae - VI. Memoirs of the New York Botanical Garden 20: 1-70.

Milliken, W. 2009. Neotropical Menispermaceae. Neotropikey [Online: See general references section].

Ortiz, R. (undated). América Tropical - géneros comunes de Menispermaceae. Field Museum Plant Guides. [*http://fm2.fieldmuseum.org/plantguides/guideimages. asp?ID=283*].

_____, Kellogg, A. and van der Werff, H. 2007. Molecular phylogeny of the moonseed family (Menispermaceae): implications for morphological diversification. American Journal of Botany 94:1425-1438.

Sothers, C. A., & Brito, J. M. 1999. Menispermaceae. Flora da Reserva Ducke: 190-193.

van Roosmalen, M. G. M. 1985. Menispermaceae. Fruits of the Guianan Flora: 291-295.

MORACEAE (Ficus only)

Berg. C. C. 1992. 21 Moraceae. Flora of the Guianas A(11).

NYCTAGINACEAE

Damascena, L. S., & Coelho, A. O. -P. 2009. Neotropical Nyctaginaceae. Neotropikey [Online: See general references section].

DeFilipps, R. A., & Maina, S. L. 2002. Nyctaginaceae. Guide to the vascular plants of central French Guiana 2: 551-553.

DeFilipps, R. A., & Maina, S. L. 2003. Nyctaginaceae. Flora of the Guianas A(22).

Ribeiro, J. E. L. S. 1999. Nyctaginaceae. Flora da Reserva Ducke: 220-221.

Steyermark, J. A., Geraldo, A., & Aymard, C. 2003. Nyctaginaceae. Flora of the Venezuelan Guayana 7: 101-118.

van Roosmalen, M. G. M. 1985. Nyctaginaceae. Fruits of the Guianan Flora: 332-333, 335.

OLACACEAE

Amshoff, G. J. H. 1938. Olacaceae. Flora of Suriname 1(1): 262-272.

Hiepko, P. 1993. 102 Olacaceae. Flora of the Guianas A(14): 3-35.

_____. 2002. Olacaceae. Guide to the vascular plants of central French Guiana 2: 549-561.

Lombardi, J. A. 2010. Neotropical Olacaceae. Neotropikey [Online: See general references section].

Sleumer, H. O. 1984. Olacaceae. Flora Neotropica Monographs 38: 1-159.

van Roosmalen, M. G. M. 1985. Olacaceae. Fruits of the Guianan Flora: 336-337.

Vicentini, A. 1999. Olacaceae. Flora da Reserva Ducke: 463-465.

PASSIFLORACEAE

Feuillet, C. 1986. Deux Passifloraceae nouvelles et quelques espèces rares en Guyane française. Candollea 41: 173-178.

_____. 1989. Diversity and distribution of Guianan Passifloraceae. In: Holm-Nielsen, I. B., Nielsen, I. C. & Balslev, H. (Eds.), Tropical forests, botanical dynamics, speciation and diversity. Academic Press, London.

_____. 1994. Two new species of Passiflora (Passifloraceae) from French Guiana. Novon 4(3): 236-241.

_____. 2002a. Passifloraceae. Guide to the vascular plants of central French Guiana 2: 566-570.

_____. 2002b. A new series and three new species of

Passiflora subgenus Astrophea from the Guianas. Brittonia 54(1): 18-29.

Hopkins, M. J. G., & de Souza, M. A. D. 1999. Passifloraceae. Flora da Reserva Ducke: 299-306.

Killip, E.P. 1937. Passifloraceae. Flora of Suriname 3(1): 306-327.

_____. 1938. The American species of Passifloraceae. Publications of the Field Museum of Natural History, Botanical Series 19: 1-613.

Nunes, T. 2009. Neotropical Passifloraceae. Neotropikey [Online: See general references section].

Tillett, S.S. 1988. Passionis passifloris II. Terminología. Ernstia 48: 1–40.

_____. 2003. Passifloraceae. Flora of the Venezuelan Guayana 7: 625-667.

van Roosmalen, M. G. M. 1985. Passifloraceae. Fruits of the Guianan Flora: 354-359.

PHYTOLACCACEAE

Costa, M. A. S., & Lohmann, L. G. 1999. Phytolaccaceae. Flora da Reserva Ducke: 219.

DeFilipps, R. A., & Maina, S. L. 2002. Phytolaccaceae. Guide to the vascular plants of central French Guiana 2: 571-573.

_____, & _____. 2003. 27 Phytolaccaceae. Flora of the Guianas A(22).

Rohwer, J. G. 1982. A taxonomic revision of the genera Seguieria Loefl. & Gallesia Casar. (Phytolaccaceae). Mitt. Bot. Staatssamml. München 18: 231-288.

Steinmann, V. W. 2010. Neotropical Phytolaccaceae. Neotropikey [Online: See general references section].

PIPERACEAE

Görts-van Rijn, A. R. A. 2007. 9. Piperaceae. Flora of the Guianas, Series A(24): 1-214.

POLYGALACEAE

Görts-van Rijn, A. R. A. 1974. Notes on Polygalaceae from Suriname. Acta Botanica Neerlandica 23(2): 189-191.

_____. 1976. Polygalaceae (additions and corrections). Flora of Suriname 2(2): 518-523.

Jacobs-Brouwer, A. 2002. Polygalaceae. Guide to the vascular plants of central French Guiana 2: 585-589.

Marques, M. D. C. M. 1980. Revisão das espécies do gênero Bredemeyera Willd. (Polygalaceae) do Brasil. Rodriguésia 32(54): 269-321.

Martins, L. 1999. Polygalaceae. Flora da Reserva Ducke: 516-519.

Oort, A. J. P. 1939. Polygalaceae. Flora of Suriname 2(1): 406-425.

Persson, C., & Eriksen, B. 2009. Neotropical Polygalaceae. Neotropikey [Online: See general references section].

van Roosmalen, M. G. M. 1985. Polygalaceae. Fruits of the Guianan Flora: 359-363.

POLYGONACEAE

Aymard, C. G. A., & Howard, R. A. 2004. Polygonaceae. Flora of the Venezuelan Guayana 8: 347-370.

Brandbyge, J. 2002. Polygonaceae. Guide to the vascular plants of central French Guiana 2: 590-591.

de Souza, M. A. D., & Brito, J. M. 1999. Polygonaceae. Flora da Reserva Ducke: 225-227.

Eyma, P. J. 1934. Polygonaceae. Flora of Suriname 1(1): 49-71.

Lindeman, J. C. & Görts-van Rijn, A. R. A. 1968. Polygonaceae (additions and corrections). Flora of Suriname 1(2): 303-309.

Melo, E., & França, F. 2009. Neotropical Polygonaceae. Neotropikey [Online: See general references section].

van Roosmalen, M. G. M. 1985. Polygonaceae. Fruits of the Guianan Flora: 362-365, 367.

RHAMNACEAE

Acevedo-Rodríguez, P. 2005. Rhamnaceae. Vines and climbing plants of Puerto Rico and the Virgin Islands. Contributions from the United States National Herbarium 51.

Heald, S. V. 2002. Rhamnaceae. Guide to the vascular plants of central French Guiana 2: 600-602.

Lanjouw, J. 1932. Rhamnaceae. Flora of Suriname 2(1): 102-106.

Martins, L. 1999. Rhamnaceae. Flora da Reserva Ducke: 498.

Steyermark, J. A., & Berry. P. E. 2004. Rhamnaceae. Flora of the Venezuelan Guayana 8: 473–484.

van Roosmalen, M. G. M. 1985. Rhamnaceae. Fruits of the Guianan Flora: 369, 371.

RUBIACEAE

Boom, B. M., & Delprete, P. 2002. Rubiaceae. Guide to the vascular plants of central French Guiana 2: 606-648.

Bremekamp, C .E. B. 1934. Rubiaceae. Flora of Suriname 4(1): 113-298.

_____. 1937. Rubiaceae (additions and corrections). Flora of Suriname 4(1): 475-491.

do A. Campos, M. T. V., & Brito, J. M. 1999. Rubiaceae. Flora da Reserva Ducke: 625-647.

Taylor, C. M. 1994. Revision of Hillia (Rubiaceae). Annals of the Missouri Botanical Garden 81(4): 571-609.

van Roosmalen, M. G. M. 1985. Rubiaceae. Fruits of the Guianan Flora: 371-388.

Wernham, H. F. 1914. A monograph of the genus Sabicea. British Museum, London, UK.

Zappi, D. 2009. Neotropical Rubiaceae. Neotropikey [Online: See general references section].

SAPINDACEAE

Acevedo-Rodríguez, P. 1993. Systematics of Serjania (Sapindaceae). Part 1: A revision of Serjania sect. Platycoccus. Memoirs of the New York Botanical Garden 67: 1-93.

_____. 2002. Sapindaceae. Guide to the vascular plants of central French Guiana 2: 656-668.

_____. 2009. Neotropical Sapindaceae. Neotropikey [Online: See general references section].

_____. 2012. 127 Sapindaceae. Flora of the Guianas A(29).

Costa, M. A. S. 1999. Sapindaceae. Flora da Reserva Ducke: 520-533.

Kramer, K. U. 1976. Sapindaceae (additions and corrections). Flora of Suriname 2(2): 487-511.

Uittien, H. 1937. Sapindaceae. Flora of Suriname 2(1): 345-396.

van Roosmalen, M. G. M. 1985. Sapindaceae. Fruits of the Guianan Flora: 391-403.

SMILACACEAE
Botina-Papamija, J. R. 2009. Neotropical Smilacaceae. Neotropikey [Online: See general references section].

Costa, M. A. S. 1999. Smilacaceae. Flora da Reserva Ducke: 722.

DeFilipps, R. A., & Maina, S. L. 2003. 204 Smilacaceae. Flora of the Guiana.

Gentry, A. 1993. Smilacaceae. A field guide to the families and genera of woody plants of northwest South America. pp. 200.

Mitchell, J. D. 1997. Smilacaceae. Guide to the vascular plants of central French Guiana 1: 362, 365-366.

Sipman, H. 1979. Liliaceae. Flora of Suriname 5(1): 442-456.

SOLANACEAE
Bernardello, L. M. & Hunziker, A. T. 1987. A synoptical revision of Solandra (Solanaceae). Nordic Journal of Botany 7: 639-652.

Edmonds, J. M. 1972. A synopsis of the taxonomy of Solanum sect. Solanum (Maurella) in South America. Kew Bulletin 27: 95-114.

Knapp, S. 2009. Neotropical Solanaceae. Neotropikey [Online: See general references section].

Martins, L., & Costa, M. A. S. 1999. Solanaceae. Flora da Reserva Ducke: 583-587.

Nee, M. 2002. Solanaceae. Guide to the vascular plants of central French Guiana 2: 689-699.

van Roosmalen, M. G. M. 1985. Solanaceae. Fruits of the Guianan Flora: 415-419.

TRIGONIACEAE
Berry, P. E. 2005. Trigoniaceae. Flora of the Venezuelan Guayana 9: 364-368.

Lleras, E. 1998. 124 Trigoniaceae. Flora of the Guianas A(21): 49-60.

Miguel, J. R., & Guimarães, E. F. 2011. Neotropical Trigoniaceae. Neotropikey [Online: See general references section].

Mori, S. A. 2002. Trigoniaceae. Guide to the vascular plants of central French Guiana 2: 716-717.

Stafleu, F. A. 1951. Trigoniaceae. Flora of Suriname 3(2): 173-177.

van Roosmalen, M. G. M. 1985. Trigoniaceae. Fruits of the Guianan Flora: 428.

URTICACEAE
_____. 2003. Urticaceae. Flora of the Guianas A(11).

VERBENACEAE
Aymard, G. 2005. Verbenaceae. Flora of the Venezuelan Guayana 9: 407-445.

Costa, M. A. S. 1999. Verbenaceae. Flora da Reserva Ducke: 596-599.

França, F., & Atkins, A. 2012. Neotropical Verbenaceae. Neotropikey [Online: See general references section].

Jansen-Jacobs, M. J. 1988. 148 Verbenaceae. Flora of the Guianas A(4): 1-114.

_____. 2002. Verbenaceae. Guide to the vascular plants of central French Guiana 2: 725-731.

Moldenke, H. N. 1940. Verbenaceae. Flora of Suriname 4(2): 357-321.

Rueda, R. M. 1994. Systematics and evolution of the genus Petrea. Annals of the Missouri Botanical Garden 81(4): 610-652.

van Roosmalen, M. G. M. 1985. Verbenaceae. Fruits of the Guianan Flora: 430-432.

VIOLACEAE

Hekking, W. H. A., & Mitchell, J. D. 2002. Violaceae. Guide to the vascular plants of central French Guiana 2: 732-738.

Munzinger, J. K., & Ballard, H. E. 2003. Hekkingia (Violaceae), a new arborescent Violet genus from French Guiana, with a key to genera in the family. Systematic Botany 28 (2): 345-351.

Paula-Souza, J., & Ballard Jr, H. E. 2009. Neotropical Violaceae. Neotropikey [Online: See general references section].

Sothers, C. A. 1999. Violaceae. Flora da Reserva Ducke: 295-297.

van Roosmalen, M. G. M. 1985. Violaceae. Fruits of the Guianan Flora: 433.

VITACEAE

de Souza, M. A. D. 1999. Vitaceae. Flora da Reserva Ducke: 499.

Görts-van Rijn, A. R. A. 1979. Vitaceae. Flora of Suriname 5(1): 335-343.

Heald, S. V. 2002. Vitaceae. Guide to the vascular plants of central French Guiana 2: 740-741.

Lombardi, J. A. 2000. Vitaceae: Gêneros Ampelocissus, Ampelopsis e Cissus. Flora Neotropica Monographs 80: 1-250.

_____. 2007. Systematics of Vitaceae in South America. Canadian Journal of Botany 85: 712-721.

_____. 2009. Neotropical Vitaceae. Neotropikey [Online: See general references section].

Paiva, E. A. S., Buono, R. A. & Lombardi, J. A. 2009. Food bodies in Cissus verticillata (Vitaceae): ontogenesis, structure and functional aspects. Annals of Botany 103(3): 517-524.

van Roosmalen, M. G. M. 1985. Vitaceae. Fruits of the Guianan Flora: 434-436.

ABOUT THE AUTHORS

Dr. Bruce Hoffman is a tropical field botanist and ethnobotanist from the United States with over 20 years of experience in the Guiana Shield region. He earned his Botany Ph.D. in 2009 from the University of Hawaii, based upon cross-cultural ethnobotanical dissertation research in Suriname. Since 2013, he has been employed as Research and Development Manager with the Amazon Conservation Team (ACT) - Suriname. Dr. Hoffman is active in biocultural conservation projects and is currently providing technical assistance to indigenous rangers (Amazon Conservation Rangers, ACT) on forest inventory and monitoring. Bruce has published or co-published papers on tree diversity patterns, quantitative ethnobotany and sustainable resource use. He looks forward to ongoing field tests and improvements of the Lianas of the Guianas guide.

Sofie Ruysschaert is currently working at WWF Guianas as Biodiversity Officer in Suriname. She obtained her MSc. in biology at Ghent University (Belgium) in 2002. She acquired vast field work experience in the Guianas with a multi-annual research project on non-timber forest products and biodiversity in Suriname. Since graduation, Sofie has been actively volunteering in several ethnobotanical and biodiversity related projects. She co-wrote the book, "Medicinal and Ritual Plants of Suriname". Her fascination for lianas emerged through wandering around in tropical forests with local villagers.